AF323046

GEOMETRIC CONTROL AND NONSMOOTH ANALYSIS

Series on Advances in Mathematics for Applied Sciences – Vol. 76

GEOMETRIC CONTROL AND NONSMOOTH ANALYSIS

*In Honor of the 73rd Birthday of H. Hermes
and of the 71st Birthday of R. T. Rockafellar*

Edited by

Fabio Ancona
Università di Bologna, Italy

Alberto Bressan
Penn State University, USA

Piermarco Cannarsa
Università di Roma "Tor Vergata", Italy

Francis Clarke
Université Claude Bernard Lyon I, France

Peter R Wolenski
Louisiana State University, USA

World Scientific

NEW JERSEY · LONDON · SINGAPORE · BEIJING · SHANGHAI · HONG KONG · TAIPEI · CHENNAI

Published by

World Scientific Publishing Co. Pte. Ltd.

5 Toh Tuck Link, Singapore 596224

USA office: 27 Warren Street, Suite 401-402, Hackensack, NJ 07601

UK office: 57 Shelton Street, Covent Garden, London WC2H 9HE

Library of Congress Cataloging-in-Publication Data
Geometric control and nonsmooth analysis / edited by Fabio Ancona ... [et al.].
   p. cm. -- (Series on advances in mathematics for applied sciences ; v. 76)
  Includes bibliographical references and index.
  ISBN-13: 978-981-277-606-8 (hardcover : alk. paper)
  ISBN-10: 981-277-606-0 (hardcover : alk. paper)
  1. Control theory--Research. 2. Nonsmooth optimization--Research.
  3. Systems engineering--Research. I. Ancona, Fabio, 1964–
  QA402.3.G436 2008
  515'.642--dc22

                        2008017673

British Library Cataloguing-in-Publication Data
A catalogue record for this book is available from the British Library.

Printed in Singapore by B & JO Enterprise

PREFACE

Applied mathematics and engineering abound with situations in which one wishes to influence the behavior of systems that are modeled as trajectories of differential equations. This has given rise to the large and multivalent subject known as *control*, the mathematical theory of which is thought of as having been born in the 1950's. The goal of controlling a system can be (roughly speaking) of two sorts: for optimality (as in the minimization of a cost), or for positional purposes (for example, driving the state to an equilibrium). The first of these is directly in the line of thought of the classical calculus of variations, while the second has its origins primarily in engineering systems design.

An important feature of control theory in recent decades has been the convergence of these two types of control at the level of the underlying theory. Thus the distinction between them has blurred in recent decades. A principal reason for this has been the development of tools which successfully address issues of both types, especially those stemming from geometry and from nonsmooth analysis.

The interaction between geometry and control theory rests upon the use of classical differential-geometric tools, in particular Lie algebras of vector fields, free and nilpotent Lie algebras, filtrations, symplectic geometry, exterior differential systems, infinite dimensional manifolds. Also involved are such concepts of analytic geometry and dynamical systems as subanalytic sets, stratifications, center and stable manifolds, among others.

Somewhat more recently, the use of nonsmooth analysis has revealed itself to be essential in both optimal and positional control. Nonsmooth calculus and its related circle of ideas (variational principles, penalization, differential inclusions, generalized tangency and normality...) can now be viewed as a relatively mature subject. The theory has been applied to control issues with great effect, in combination with such tools as viscosity solutions, dynamic programming, Hamilton-Jacobi inequalities, duality, discontinuous feedbacks, convexity and semiconcavity, and nonsmooth Lyapunov functions.

Geometric methods and nonsmooth analysis have together given rise to many recent developments in control theory. Among these are: new and unified formulations of necessary conditions in optimal control, the synthesis and classification of optimal or stabilizing controls, and new techniques for optimal and stabilizing feedback design, in particular sliding modes.

The conference *Geometric Control and Nonsmooth Analysis*, which took place in Rome in June 2006 at the Istituto Nazionale di Alta Matematica (INdAM), celebrated the leading contributions to control theory of two eminent researchers, Henry Hermes and R. T. Rockafellar (often known as Hank and Terry to their many friends). The topics covered a wide range of the issues mentioned above, from the calculus of variations to optimal control, from controllability to stability, from invariance to impulse systems, from approximation to feedback design. This volume, which assembles the contributions of the speakers of the conference, provides an overview which illustrates well the vitality, the range, and the importance of the subject that is control theory.

Soon after the conference, we learnt that Wijesuriya Dayawansa, whom we had just seen in Rome, had abruptly passed away. This volume is dedicated to his memory.

<table>
<tr><td>Fabio Ancona</td><td>Bologna, Italy</td></tr>
<tr><td>Alberto Bressan</td><td>Penn State, USA</td></tr>
<tr><td>Piermarco Cannarsa</td><td>Roma, Italy</td></tr>
<tr><td>Francis H. Clarke</td><td>Lyon, France</td></tr>
<tr><td>Peter W. Wolenski</td><td>Baton Rouge, USA</td></tr>
</table>

April 2008

PROFESSOR HENRY HERMES
UNIVERSITY OF COLORADO AT BOULDER

PROFESSOR R. TYRRELL ROCKAFELLAR
UNIVERSITY OF WASHINGTON

CONTENTS

MULTISCALE SINGULAR PERTURBATIONS AND HOMOGENIZATION OF OPTIMAL CONTROL PROBLEMS

O. ALVAREZ

UMR 60–85, Université de Rouen,
76821 Mont–Saint Aignan cedex, France
E–mail: olivier.alvarez@univ-rouen.fr

M. BARDI

Dipartimento di Matematica Pura ed Applicata, Università di Padova,
via Trieste 63, 35121 Padova, Italy
E–mail: bardi@math.unipd.it

C. MARCHI

Dipartimento di Matematica, Università della Calabria,
Ponte P. Bucci 30B, 87036 Rende (CS), Italy
E–mail: marchi@mat.unical.it

The paper is devoted to singular perturbation problems with a finite number of scales where both the dynamics and the costs may oscillate. Under some coercivity assumptions on the Hamiltonian, we prove that the value functions converge locally uniformly to the solution of an *effective* Cauchy problem for a limit Hamilton-Jacobi equation and that the effective operators preserve several properties of the starting ones; under some additional hypotheses, their explicit formulas are exhibited. In some special cases we also describe the effective dynamics and costs of the limiting control problem. An important application is the homogenization of Hamilton-Jacobi equations with a finite number of scales and a coercive Hamiltonian.

Keywords: Singular perturbations, viscosity solutions, Hamilton–Jacobi equations, dimension reduction, iterated homogenization, control systems in oscillating media, multiscale problems, oscillating costs.

2

1. Introduction

The controlled system

$$\dot{x}_s = f(x_s, y_s, z_s, \alpha_s), \quad x_0 = x,$$
$$\varepsilon \dot{y}_s = g(x_s, y_s, z_s, \alpha_s), \quad y_0 = y,$$
$$\varepsilon^2 \dot{z}_s = r(x_s, y_s, z_s, \alpha_s), \quad z_0 = z,$$

where α is the control and $\varepsilon > 0$ a small parameter, is a model of systems whose state variables evolve on three different time scales. Consider the cost functional

$$P^\varepsilon(t, x, y, z, \alpha) := \int_0^t l(x_s, y_s, z_s, \alpha_s)\, ds + h(x_t, y_t, z_t)$$

and the value function

$$u^\varepsilon(t, x, y, z) := \inf_\alpha P^\varepsilon(t, x, y, z, \alpha).$$

The analysis of the convergence of u^ε as $\varepsilon \to 0$ gives informations on the optimization problem after a sufficiently large time, namely, when the fast variables y and z have reached their regime behaviour.

For two time scales, i.e. $r \equiv 0$, the problem has a large mathematical and engineering literature, see the books by Kokotović et al.[24] and Bensoussan,[14] the references therein, and the more recent contributions by Artstein and Gaitsgory[10] and the authors.[3–6] For more than two time scales Gaitsgory and Nguyen[21] extended the method of limit occupational measures. In this paper we follow a method based on the Hamilton–Jacobi–Bellman equation satisfied by the value function, that is

$$\partial_t u^\varepsilon + \max_\alpha \left\{ -f \cdot D_x u^\varepsilon - g \cdot \frac{D_y u^\varepsilon}{\varepsilon} - r \cdot \frac{D_z u^\varepsilon}{\varepsilon^2} - l \right\} = 0$$

$$\text{in} \quad (0, T) \times \mathbb{R}^n \times \mathbb{R}^m \times \mathbb{R}^p.$$

It is based on the theory of viscosity solutions (for an overview, see the book Ref. 12) and was used for two-scale problems by the first and the second named authors.[3,4] They also developed it further to stochastic systems and differential games.[5] In a companion paper[6] the authors extended the method to stochastic problems with three and n scales; in those cases the value functions solve some *2nd* order degenerate parabolic equation. In the present paper, we show how to apply our method[6] to deterministic control problems and *1st* order H-J-B equations with a finite number of scales.

An important advantage of our PDE approach is that it applies naturally in the generality of *differential games*, namely, problems governed by two conflicting players, α and β, acting on the system

$$\dot{x}_s = f(x_s, y_s, z_s, \alpha_s, \beta_s),$$
$$\varepsilon \dot{y}_s = g(x_s, y_s, z_s, \alpha_s, \beta_s),$$
$$\varepsilon^2 \dot{z}_s = r(x_s, y_s, z_s, \alpha_s, \beta_s).$$

If the game is 0-sum, i.e., the second player's goal is the maximization of the cost P^ε, the (lower) value function of the game satisfies the Cauchy problem for the Isaacs PDE

$$\begin{cases} \partial_t u^\varepsilon + \min_\beta \max_\alpha \left\{ -f \cdot D_x u^\varepsilon - g \cdot \frac{D_y u^\varepsilon}{\varepsilon} - r \cdot \frac{D_z u^\varepsilon}{\varepsilon^2} - l \right\} = 0 \\ \qquad\qquad\qquad\qquad\qquad\qquad \text{in} \quad (0,T) \times \mathbb{R}^n \times \mathbb{R}^m \times \mathbb{R}^p, \\ u^\varepsilon(0, x, y, z) = h(x, y, z). \end{cases}$$

Throughout this paper, in fact, we will never assume the Hamiltonian be of the Bellman form, that is, convex in the gradient variables. Therefore, our results apply, for instance, to the robust optimal control of systems with bounded unknown disturbances. Our goal is proving that the value functions u^ε converge locally uniformly to the viscosity solution of a new Cauchy problem (called the *effective* problem)

$$\begin{cases} \partial_t u + \overline{H}(x, D_x u) = 0 & \text{in } (0,T) \times \mathbb{R}^n \\ u(0,x) = \overline{h}(x) & \text{on } \mathbb{R}^n. \end{cases} \tag{$\overline{\text{HJ}}$}$$

For the two-scale case, the effective Hamiltonian $\overline{H}$ and the effective initial condition $\overline{h}$ are obtained, respectively, as the *ergodic constant* of a stationary problem and by the time-asymptotic limit of the solution to a related Cauchy problem. For the multiscale case, this construction must be done iteratively. Moreover, owing to this procedure and to the coercivity assumption, the effective (and the *intermediate*) operators inherit several properties of the starting ones, in particular those ensuring the Comparison Principle (we refer the reader to Ref. 5 for the case of intermediate or effective operators lacking these properties). An interesting issue is to represent the effective solution u as the value function of some new control problem. In some special cases, we shall prove that the effective PDE is associated to a limit control problem whose dynamics and effective costs can be described explicitly. An example of Hamilton-Jacobi equation that fits within our theory is

$$\partial_t u^\varepsilon + F(x, y, z, D_x u^\varepsilon) + \varphi_1(x, y, z) \frac{|D_y u^\varepsilon|}{\varepsilon} + \varphi_2(x, y, z) \frac{|D_z u^\varepsilon|}{\varepsilon^2} = 0 \tag{1}$$

4

in $(0, T) \times \mathbb{R}^n \times \mathbb{R}^m \times \mathbb{R}^p$, where F is a standard Bellman-Isaacs operator, and, for some $\nu > 0$, $\varphi_1 \geq \nu$ and $\varphi_2 \geq \nu$. In this special case we can compute the effective Cauchy problem, which is

$$\begin{cases} \partial_t u + \max_{y,z} F(x, y, z, D_x u) = 0 & \text{in } (0, T) \times \mathbb{R}^n, \\ u(0, x) = \min_{y,z} h(x, y, z) & \text{on } \mathbb{R}^n. \end{cases}$$

An important byproduct of the previous theory is the homogenization of systems in highly heterogeneous media with more than two space scales. Now the system is

$$\dot{x}_s = f\left(x_s, \frac{x_s}{\varepsilon}, \frac{x_s}{\varepsilon^2}, \alpha_s, \beta_s\right)$$

and the cost

$$P^\varepsilon(t, x, \alpha, \beta) := \int_0^t l\left(x_s, \frac{x_s}{\varepsilon}, \frac{x_s}{\varepsilon^2}, \alpha_s, \beta_s\right) ds + h\left(x_t, \frac{x_t}{\varepsilon}, \frac{x_t}{\varepsilon^2}\right).$$

In this case, the value function v^ε satisfies (in the viscosity sense) the Cauchy problem

$$\begin{cases} \partial_t v^\varepsilon + \min_\beta \max_\alpha \left\{-f\left(x, \frac{x}{\varepsilon}, \frac{x}{\varepsilon^2}, \alpha, \beta\right) \cdot D_x v^\varepsilon - l(x, \frac{x}{\varepsilon}, \frac{x}{\varepsilon^2}, \alpha, \beta)\right\} = 0 \\ \qquad\qquad\qquad\qquad\qquad\qquad\qquad\qquad \text{in } (0, T) \times \mathbb{R}^n, \\ v^\varepsilon(0, x) = h\left(x, \frac{x}{\varepsilon}, \frac{x}{\varepsilon^2}\right). \end{cases}$$

Here the oscillations are in space. By setting $y = x/\varepsilon, z = x/\varepsilon^2$ this problem can be written as a singular perturbation one. Motivated by this, we will call throughout the paper x the *macroscopic* variables, y the *mesoscopic* ones, and z the *microscopic* variables. Under an assumption of coercivity of the Hamiltonian in the gradient variables we prove that the solution v^ε converge locally uniformly to the solution of an effective problem $(\overline{\text{HJ}})$, for a suitable construction of the effective operator and initial data. Our result applies to the homogenization of the *eikonal equation*

$$\partial_t v^\varepsilon + \varphi\left(x, \frac{x}{\varepsilon}, \frac{x}{\varepsilon^2}\right) |D_x v^\varepsilon| = l\left(x, \frac{x}{\varepsilon}, \frac{x}{\varepsilon^2}\right) \tag{2}$$

with $\varphi \geq \nu > 0$.

In the framework of viscosity solutions, the study of the two-scale homogenization, initiated by Lions, Papanicolaou, Varadhan[26] and improved by Evans,[18,19] has been extended to related questions: see, e.g., Capuzzo–Dolcetta and Ishii[17] for the rate of convergence, Horie and Ishii[22] and the first author[2,7] for periodic homogenization in perforated domains, Ishii,[23] Arisawa,[8] and Birindelli, Wigniolle[15] for non–periodic homogenization, Rezakhanlou and Tarver,[30] Souganidis,[31] Lions, Souganidis[27,28] for stochastic homogenization, the book Ref. 12, Artstein, Gaitsgory,[10] the first two authors,[3] and the references therein for singular perturbations in optimal control.

Let us stress that all the aforementioned papers consider only two scales and that, as far as we know, fully nonlinear problems with multiple scales have been attacked for the first time in our paper Ref. 6; in fact, iterated homogenization was addressed only in the variational setting, starting with the pioneering work of Bensoussan, J.L. Lions and Papanicolaou[13] for linear equations and, afterwards, for semilinear equations, using the Γ–convergence approach[16,29] (see also and references therein) or G–convergence techniques.[1,11,25]

The plan of the paper is as follows. The standing assumptions are listed in Section 2. Section 3 recalls the notions of ergodicity and stabilization for a Hamiltonian. Section 4 is devoted to the regular perturbations of two-scale problems because they are of independent interest and for later use. We address the multiscale singular perturbations and the multiscale homogenization respectively in Section 5 and in Section 6. Some examples arising from deterministic optimal control theory and differential games are collected in Section 7. One of them replaces the coercivity of the Hamiltonian with a non-resonance condition and allows to show that exchanging the roles of ε and ε^2 may produce a different effective PDE.

2. Standing assumptions

We consider Bellman–Isaacs Hamiltonians

$$H(x, y, p_x, p_y) := \min_{\beta \in B} \max_{\alpha \in A} L^{\alpha, \beta}(x, y, p_x, p_y),$$

for the family of linear operators

$$L^{\alpha, \beta}(x, y, p_x, p_y) := -p_x \cdot f(x, y, \alpha, \beta) - p_y \cdot g(x, y, \alpha, \beta) - l(x, y, \alpha, \beta).$$

The following assumptions will hold in Sections 3 and 4:

- The control sets A and B are compact metric spaces.
- The functions f, g and l are bounded continuous functions in $\mathbb{R}^n \times \mathbb{R}^m \times A \times B$ with values, respectively, in $\mathbb{R}^n$, $\mathbb{R}^m$ and $\mathbb{R}$.
- The drift vectors f and g are Lipschitz continuous in (x, y), uniformly in (α, β).
- The running cost l is uniformly continuous in (x, y), uniformly in (α, β).
- The initial condition h is bounded and uniformly continuous.

- The functions f, g, h and l are $\mathbb{Z}^m$-periodic in the fast variables y.
- H is coercive in p_y: there exist $\nu, C \in \mathbb{R}^+$ such that:

$$H(x, y, p_x, p_y) \geq \nu|p_y| - C(1 + |p_x|) \qquad \text{for every } x, y, p_x, p_y.$$

Let us observe that the last assumption holds provided that:

$$B_m(0, \nu) \subset \overline{\mathrm{conv}}\{g(x, y, \alpha, \beta) \mid \alpha \in A\} \qquad \forall x, y, \beta$$

where $B_m(0, \nu)$ is the ball centered in 0 with radius ν in the space $\mathbb{R}^m$.

In the deterministic control theory, this relation entails a strong form of small-time controllability of the deterministic fast subsystem, i.e. any two points can be reached from one another by the player acting on α, whatever the second player does, and within a time proportional to the distance between the points.

We introduce the *recession function* (or homogeneous part) of H in p_y by

$$H'(x, y, p_y) := \min_{\beta \in B} \max_{\alpha \in A}\{-p_y \cdot g(x, y, \alpha, \beta)\}.$$

We note that H' is positively 1-homogeneous in p_y, namely $H'(x, y, \lambda p_y) = \lambda H'(x, y, p_y)$ for $\lambda \geq 0$ and that, for every $\overline{x}, \overline{p}_x \in \mathbb{R}^n$, there is a constant C so that

$$|H(x, y, p_x, p_y) - H'(x, y, p_y)| \leq C \quad \forall\, (y, p_y) \in \mathbb{R}^m \times \mathbb{R}^m, \tag{3}$$

for every (x, p_x) in a neighborhood of $(\overline{x}, \overline{p}_x)$.

In the case of three scales, treated in Section 5.1, the Hamiltonian depends also on z and p_z. Then the linear operators $L^{\alpha, \beta}$ have the additional term $-p_z \cdot r(x, y, z, \alpha, \beta)$, whereas f, g, and l may depend on z as well. We make the same assumptions on the dependence of the data from z as from y, namely, periodicity, Lipschitz continuity of f, g, r, and uniform continuity of l. The obvious analogous assumptions are made in the general case of $j + 1$ scales studied in Section 5.2.

3. Ergodicity, stabilization and the effective problem

The aim of this Section is to recall from Ref. 5,6 the notions of ergodicity and of stabilization that are crucial in the definition of the effective problem $(\overline{\mathrm{HJ}})$. We establish some properties of the effective operators and in some cases we provide their explicit formulas.

3.1. *Ergodicity and the effective Hamiltonian*

This Subsection is devoted to recall the definition of ergodicity introduced in Ref. 4. For $(\overline{x}, \overline{p}_x)$ fixed, by the standard viscosity solution theory, the *cell δ-problem*

$$\delta w_\delta + H(\overline{x}, y, \overline{p}_x, D_y w_\delta) = 0 \quad \text{in } \mathbb{R}^m, \qquad w_\delta \text{ periodic}, \qquad \text{(CP}_\delta\text{)}$$

has a unique solution. We denote the solution by $w_\delta(y; \overline{x}, \overline{p}_x)$ so as to display its dependence on the frozen slow variables. We say that the Hamiltonian is *ergodic* in the fast variable at $(\overline{x}, \overline{p}_x)$ if

$$\delta w_\delta(y; \overline{x}, \overline{p}_x) \to \text{const} \qquad \text{as } \delta \to 0, \text{ uniformly in } y.$$

In this case, we define

$$\overline{H}(\overline{x}, \overline{p}_x) := -\lim_{\delta \to 0} \delta w_\delta(y; \overline{x}, \overline{p}_x);$$

the function H is called *effective Hamiltonian*. We say that H is ergodic if it is ergodic at every $(\overline{x}, \overline{p}_x)$. In the next Proposition we collect some properties of $\overline{H}$ and, in some special cases, also its explicit formula.

Proposition 3.1. *Under the standing assumptions there holds*

(a) the Hamiltonian H is ergodic.
(b) $\overline{H}$ is regular: there are $C \in \mathbb{R}$ and a modulus of continuity ω such that:

$$|\overline{H}(x_1, p) - \overline{H}(x_2, p)| \leq C|x_1 - x_2|(1 + |p|) + \omega(|x_1 - x_2|) \ \forall x_i, p \in \mathbb{R}^n;$$
$$|\overline{H}(x, p_1) - \overline{H}(x, p_2)| \leq C(|p_1 - p_2|) \qquad \forall x, p_i \in \mathbb{R}^n;$$

in particular, the Comparison Principle holds for the effective problem $(\overline{\text{HJ}})$.
(c) If

$$H(\overline{x}, y, \overline{p}_x, p_y) \geq H(\overline{x}, y, \overline{p}_x, 0) \qquad \forall y, p_y \in \mathbb{R}^m, \qquad (4)$$

then $\overline{H}$ has the explicit formula:

$$\overline{H}(\overline{x}, \overline{p}_x) = \max_y H(\overline{x}, y, \overline{p}_x, 0).$$

Proof. The proofs of (a) and (b) are slight adaptations of the arguments used in [4, Proposition 9], [3, Proposition 12] and [19, Lemma 2.2] so we omit them.

(c) The Comparison Principle for the cell δ-problem (CP_δ) entails: $\delta w_\delta \geq -\sup_y H(\overline{x}, y, \overline{p}_x, 0)$; as $\delta \to 0$, we infer:

$$\overline{H}(\overline{x}, \overline{p}_x) \leq \sup_y H(\overline{x}, y, \overline{p}_x, 0).$$

In order to prove the reverse inequality, we shall argue by contradiction, assuming: $\overline{H}(\overline{x}, \overline{p}_x) < H(\overline{x}, y, \overline{p}_x, 0)$ in a open set U. Therefore, the cell δ-problem reads

$$\delta w_\delta + H_0(y, D_y w_\delta) + H(\overline{x}, y, \overline{p}_x, 0) = 0$$

where $H_0(y, q) := H(\overline{x}, y, \overline{p}_x, q) - H(\overline{x}, y, \overline{p}_x, 0)$. The ergodicity of H and the relation (4) entail

$$0 \leq \overline{H}(\overline{x}, \overline{p}_x) - H(\overline{x}, y, \overline{p}_x, 0) + O(\delta) \qquad \text{in } U.$$

As $\delta \to 0$, we obtain the desired contradiction. $\qquad\qquad\square$

Remark 3.1. Let us observe that condition (4) is satisfied if the control α splits into (α_1, α_2), where α_i belongs to the compact A_i ($i = 1, 2$), the drift f and the running cost l do not depend on α_2, while the drift $g = g(x, y, \alpha_2, \beta)$ fulfills:

$$B_m(0, \nu) \subset \overline{\mathrm{conv}}\{g(x, y, \alpha_2, \beta) \mid \alpha_2 \in A_2\} \qquad \forall x, y, \beta.$$

3.2. *Stabilization and the effective initial data*

The stabilization to a constant for degenerate eqs. was introduced by the first two authors.[4] For $\overline{x}$ fixed, the *cell Cauchy problem* for the homogeneous Hamiltonian H'

$$\partial_t w + H'(\overline{x}, y, D_y w) = 0 \quad \text{in } (0, +\infty) \times \mathbb{R}^m, \qquad w(0, y) = h(\overline{x}, y) \quad \text{on } \mathbb{R}^m \tag{CP$'$}$$

has a unique bounded viscosity solution $w(t, y; \overline{x})$. Observe that by the positive homogeneity of H', the constants $\|h(\overline{x}, \cdot)\|_\infty$ and $-\|h(\overline{x}, \cdot)\|_\infty$ are respectively a super- and a subsolution. Furthermore, the Comparison Principle yields the uniform bound: $\|w(t, \cdot)\|_\infty \leq \|h(\overline{x}, \cdot)\|_\infty$ for all $t \geq 0$.

We say that the pair (H, h) is *stabilizing* (to a constant) at $\overline{x}$ if

$$w(t, y; \overline{x}) \to \mathrm{const} \quad \text{as } t \to +\infty, \text{ uniformly in } y. \tag{5}$$

In this case, we define

$$\overline{h}(\overline{x}) := \lim_{t \to +\infty} w(t, y; \overline{x}). \tag{6}$$

We say that the pair (H, h) is stabilizing if it is stabilizing at every $\overline{x} \in \mathbb{R}^n$. The function $\overline{h}$ is called the *effective initial data*.

Proposition 3.2. *Under the standing assumptions, the pair (H, h) is stabilizing. Moreover, the effective initial datum $\overline{h}$ is continuous and has the* *form:*

$$\overline{h}(\overline{x}) = \min_y h(\overline{x}, y).$$

The proof is a slight adaptation of the arguments used in [4, Proposition 10] and in [3, Theorem 8] and we shall omit it.

4. Regular perturbation of singular perturbation problems

This Section is devoted to a convergence result for the regular perturbation of a singular perturbation problem

$$\begin{cases} \partial_t u^\varepsilon + H^\varepsilon \left(x, y, D_x u^\varepsilon, \frac{D_y u^\varepsilon}{\varepsilon} \right) = 0 & \text{in } (0, T) \times \mathbb{R}^n \times \mathbb{R}^m \\ u^\varepsilon(0, x, y) = h^\varepsilon(x, y) & \text{on } \mathbb{R}^n \times \mathbb{R}^m. \end{cases} \quad \text{(HJ}^\varepsilon\text{)}$$

Regular perturbation means that $H^\varepsilon \to H$ and $h^\varepsilon \to h$ as $\varepsilon \to 0$ uniformly on all compact sets and that H, h and every H^ε, h^ε satisfy the standard assumptions of Sec. 2. For example, we have a regular perturbation when the control sets A and B are independent of ε and the functions f^ε, g^ε and l^ε converge locally uniformly to f, g and l. We suppose also that

$$|H^\varepsilon(x, y, 0, 0)| \leq C \qquad \forall (x, y), \tag{7}$$

for some constant C independent of ε small (this assumption is satisfied for instance if the running costs l^ε are equibounded). Let us note that the problem (HJ$^\varepsilon$) has a unique bounded solution (that is also periodic in y) and fulfills the Comparison Principle.[3,4]

The next result and the arguments of its proof will be used extensively in the next Sections.

Theorem 4.1. *Assume that H^ε and h^ε converge respectively to H and to h uniformly on the compact sets and that the equiboundedness condition (7) holds. Then, u^ε converges uniformly on the compact subsets of $(0, T) \times \mathbb{R}^n$ to the unique viscosity solution of $(\overline{\text{HJ}})$ where the effective Hamiltonian $\overline{H}$ and the effective initial datum $\overline{h}$ are defined respectively in Subsec. 3.1 and 3.2.*

Proof. The proof of this Theorem relies on [6, Corollary 1] (see also [4, Theorem 1]) and on the ergodicity and stabilization results stated in Proposition 3.1 and in Proposition 3.2. For the sake of completeness, let us sketch the main features of the argument. The family $\{u^\varepsilon\}$ is equibounded; indeed, the Comparison Principle gives: $\|u^\varepsilon(t,\cdot)\|_\infty \leq \sup_\varepsilon \|h^\varepsilon\|_\infty + Ct$. We can therefore define the upper semilimit $\overline{u}$ of u^ε as follows

$$\overline{u}(t,x) := \limsup_{\varepsilon \to 0,\ (t',x') \to (t,x)} \sup_y u^\varepsilon(t',x',y) \quad \text{if } t > 0,$$

$$\overline{u}(0,x) := \limsup_{(t',x') \to (0,x),\ t' > 0} \overline{u}(t',x') \quad \text{if } t = 0.$$

We define analogously the lower semilimit $\underline{u}$ by replacing limsup with liminf and sup with inf. The two-steps definition of the semilimit for $t = 0$ is needed to avoid a possible initial layer.

Let us notice that, under our hypotheses, the effective problem $(\overline{\text{HJ}})$ satisfies the Comparison Principle and admits exactly one bounded solution u. If $\overline{u}$ and $\underline{u}$ are respectively a super- and a subsolution to $(\overline{\text{HJ}})$ then the proof is accomplished. Actually, the Comparison Principle ensures: $\overline{u} \leq u \leq \underline{u}$. On the other hand the reverse inequality $\underline{u} \leq \overline{u}$ is always true. Whence, we have $\overline{u} = \underline{u} = u$ and, by standard arguments, we deduce: $u^\varepsilon \to u$ locally uniformly as $\varepsilon \to 0$.

Let us now ascertain that $\overline{u}$ is a subsolution to $(\overline{\text{HJ}})$; being similar, the other proof is omitted. We proceed by contradiction assuming that there are a point $(\bar{t}, \bar{x}) \in (0, T) \times \mathbb{R}^n$ and a smooth test function φ such that: $\overline{u}(\bar{t}, \bar{x}) = \varphi(\bar{t}, \bar{x})$, $(\bar{t}, \bar{x})$ is a strict maximum point of $\overline{u} - \varphi$ and there holds

$$\partial_t \varphi(\bar{t}, \bar{x}) + \overline{H}(\bar{x}, D_x \varphi(\bar{t}, \bar{x})) \geq 3\eta$$

for some $\eta > 0$. For every $r > 0$, we define

$$H_r^\varepsilon(y, p_y) := \min\{H^\varepsilon(x, y, D_x \varphi(t, x), p_y) \mid |t - \bar{t}| \leq r,\ |x - \bar{x}| \leq r\}.$$

We put $\overline{H} := \overline{H}(\bar{x}, \bar{p}_x)$ with $\bar{p}_x = D_x \varphi(\bar{t}, \bar{x})$ and we fix $r_0 > 0$ so that

$$|\partial_t \varphi(t, x) - \partial_t \varphi(\bar{t}, \bar{x})| \leq \eta \quad \text{as} \quad |t - \bar{t}| < r_0, |x - \bar{x}| \leq r_0.$$

Now we want to prove that, for every $r > 0$ small enough, there is a parameter $\varepsilon' > 0$ and an equibounded family of functions $\{\chi^\varepsilon \mid 0 < \varepsilon < \varepsilon'\}$ (called *approximated correctors*) so that

$$H_r^\varepsilon(y, D_y \chi^\varepsilon) \geq \overline{H} - 2\eta \quad \text{in } \mathbb{R}^m. \tag{8}$$

To this aim, taking into account the ergodicity of H, we fix a parameter $\delta > 0$ so that the solution w_δ to the cell δ-problem (CP_δ) fulfills:

$$\|\delta w_\delta + \overline{H}\|_\infty \leq \eta.$$

Since $H_r^\varepsilon(y, p_y) \to H(\bar{x}, y, \bar{p}_x, p_y)$ as $(\varepsilon, r) \to (0, 0)$ uniformly on the compact sets, the stability property entails that the solution $w_{\delta,r}^\varepsilon$ of

$$\delta w_{\delta,r}^\varepsilon + H_r^\varepsilon(y, D_y w_{\delta,r}^\varepsilon) = 0 \quad \text{in } \mathbb{R}^m, \qquad w_{\delta,r}^\varepsilon \text{ periodic},$$

converges uniformly to w_δ as $(\varepsilon, r) \to (0, 0)$. In particular, for $\varepsilon' > 0$ and $0 < r' < \min\{r_0, \bar{t}\}$, we get

$$\|\delta w^\varepsilon_{\delta, r} + \overline{H}\|_\infty \leq 2\eta \qquad \text{when} \quad 0 < \varepsilon < \varepsilon' \quad \text{and} \quad 0 < r < r'.$$

The function $\chi^\varepsilon = w^\varepsilon_{\delta, r}$ is a supersolution of (8). Moreover, by the Comparison Principle, the family $\{\chi^\varepsilon\}$ is equibounded: $\|\chi^\varepsilon\|_\infty \leq \delta^{-1} \sup\{|H^\varepsilon_r(y, 0, 0)| \mid y \in \mathbb{R}^m, \ 0 < \varepsilon < \varepsilon'\}$. Hence, our claim is proved.

We consider the perturbed test function

$$\psi^\varepsilon(t, x, y) := \varphi(t, x) + \varepsilon \chi^\varepsilon(y).$$

In the cylinder $Q_r =]\bar{t} - r, \bar{t} + r[\times B_r(\overline{x}) \times \mathbb{R}^m$, ψ^ε is a supersolution of (HJ$^\varepsilon$) (see Ref. 4 for the rigourous proof). Since $\{\psi^\varepsilon\}$ converges uniformly to φ on $\overline{Q_r}$, we obtain

$$\limsup_{\varepsilon \to 0, \ t' \to t, \ x' \to x} \ \sup_y (u^\varepsilon - \psi^\varepsilon)(t', x', y) = \overline{u}(t, x) - \varphi(t, x).$$

But $(\bar{t}, \overline{x})$ is a strict maximum point of $\overline{u} - \varphi$, so the above relaxed upper limit is negative on ∂Q_r. By compactness, one can find $\eta' > 0$ so that $u^\varepsilon - \psi^\varepsilon \leq -\eta'$ on ∂Q_r for ε small. Since ψ^ε is a supersolution in Q_r, we deduce from the Comparison Principle that $\psi^\varepsilon \geq u^\varepsilon + \eta'$ in Q_r for ε small. Taking the upper semi-limit, we get $\varphi \geq \overline{u} + \eta'$ in $(\bar{t} - r, \bar{t} + r) \times B(\overline{x}, r)$. This is impossible, for $\varphi(\bar{t}, \overline{x}) = \overline{u}(\bar{t}, \overline{x})$. Thus, we have reached the desired contradiction.

We now check that $\overline{u}$ satisfies the initial condition, that is $\overline{u} \leq \overline{h}$. Let w^ε_r be the unique solution of the following Cauchy problem

$$\begin{cases} \partial_t w^\varepsilon_r + H^{\varepsilon,'}_r(y, D_y w^\varepsilon_r) = 0 \text{ in } (0, +\infty) \times \mathbb{R}^m, \\ w^\varepsilon_r(0, y) = h^\varepsilon_r(y) \qquad \text{on } \mathbb{R}^m, \quad w^\varepsilon_r \text{ periodic in } y \end{cases}$$

where the Hamiltonian $H^{\varepsilon,'}_r$ and the initial datum h^ε_r are given by

$$H^{\varepsilon,'}_r(y, p_y) := \min\{H^{\varepsilon,'}(x, y, p_y) \mid |x - \overline{x}| \leq r\},$$

$$h^\varepsilon_r(y) := \max\{h^\varepsilon(x, y) \mid |x - \overline{x}| \leq r\}.$$

Let us claim that

$$\limsup_{r \to 0, \ \varepsilon \to 0, \ t \to \infty} \ \sup_y |w^\varepsilon_r(t, y) - \overline{h}(\overline{x})| = 0. \tag{9}$$

Fix $\eta > 0$. The stabilization ensures that the solution w to the cell Cauchy problem (CP$'$) fulfills

$$\|w(T, \cdot) - \overline{h}(\overline{x})\|_\infty \leq \eta$$

12

for some $T > 0$. Letting $(\varepsilon, r) \to (0,0)$, we have $H_r^{\varepsilon,\prime} \to H'(\overline{x}, \cdot)$ and $h_r^\varepsilon \to h(\overline{x}, \cdot)$ uniformly on the compact sets; therefore, by the stability properties of viscosity solutions, we know that $w_r^\varepsilon \to w'$ locally uniformly. Whence, there are ε' and r' so that

$$\|w_r^\varepsilon(T, \cdot) - \overline{h}(\overline{x})\|_\infty \leq 2\eta \qquad \text{for all} \quad 0 < \varepsilon < \varepsilon', \ 0 < r < r'.$$

Since $H_r^{\varepsilon,\prime}(\cdot, 0) \equiv 0$, the Comparison Principle entails that

$$\|w_r^\varepsilon(t, \cdot) - \overline{h}(\overline{x})\|_\infty \leq 2\eta \qquad \text{for all} \quad t \geq T, \ 0 < \varepsilon < \varepsilon', \ 0 < r < r'.$$

This gives (9).

Let $r > 0$, $\varepsilon' > 0$ and $T > 0$ be such that the last inequality is satified. For $Q_r^+(\overline{x}) := (0, r) \times B_r(\overline{x}) \times \mathbb{R}^m$, we fix M so that $M \geq \|u^\varepsilon\|_{\mathbf{L}^\infty(Q_r^+(\overline{x}))}$ for all $\varepsilon < \varepsilon'$ and we construct a bump function ψ_0 that is nonnegative, smooth, with $\psi_0(\overline{x}) = 0$ and $\psi_0 \geq 2M$ on $\partial B_r(\overline{x})$. Finally, we choose the constant $C > 0$ given by (3) so that

$$|H^\varepsilon(x, y, D_x\psi_0(x), p_y) - H^{\varepsilon,\prime}(x, y, p_y)| \leq C$$

for every (y, p_y), $x \in B_r(\overline{x})$, $0 < \varepsilon < \varepsilon'$. We introduce the function

$$\psi^\varepsilon(t, x, y) := w_r^\varepsilon(\varepsilon^{-1}t, y) + \psi_0(x) + Ct$$

and we observe that it is a supersolution of

$$\partial_t\psi^\varepsilon + H^\varepsilon(x, y, D_x\psi^\varepsilon, \varepsilon^{-1}D_y\psi^\varepsilon) = 0 \quad \text{in } Q_r^+(\overline{x})$$
$$\psi^\varepsilon = h^\varepsilon \quad \text{on } \{0\} \times B_r(\overline{x}) \times \mathbb{R}^m, \quad \psi^\varepsilon = M \quad \text{on } [0, r) \times \partial B_r(\overline{x}) \times \mathbb{R}^m.$$

By the Comparison Principle, we deduce that

$$u^\varepsilon(t, x, y) \leq \psi^\varepsilon(t, x, y) = w_r^\varepsilon(\varepsilon^{-1}t, y) + \psi_0(x) + Ct \quad \text{in} \quad Q_r^+(\overline{x}).$$

Taking the supremum over y and sending $\varepsilon \to 0$, we obtain the inequality

$$\overline{u}(t, x) \leq \overline{h}(\overline{x}) + 2\eta + \psi_0(x) + Ct \qquad \text{for all} \quad t > 0, \ x \in B_r(\overline{x}).$$

Sending $t \to 0^+$, $x \to \overline{x}$, we get $\overline{u}(0, \overline{x}) \leq \overline{h}(\overline{x}) + \eta$. Taking into account the arbitrariness of η, one can easily accomplish the proof. $\qquad\square$

Remark 4.1. Let us stress that the coercivity assumption has been used only for establishing the following properties: i) the starting Hamiltonian H is ergodic, ii) the pair (H, h) is stabilizing, iii) the effective Hamiltonian $\overline{H}$ is sufficiently regular to fulfill the Comparison Principle.

It is worth to recall that there exist non-coercive Hamiltonians that enjoy properties (i)–(iii) (e.g., see Sec. 7 below). It is obvious that Theorem 4.1 applies also to these Hamiltonians.

5. Singular perturbations with multiple scales

This Section is devoted to the study of singular perturbation problems having a finite number of scales. For the sake of simplicity, in the first Subsection we shall focus our attention on the three scales case, which is the simplest one, providing a detailed proof of our result. After, in the second Subsection we shall briefly give the result for a wider class of problems.

5.1. *The three scale case*

We consider the problems:

$$\begin{cases} \partial_t u^\varepsilon + H^\varepsilon \left(x, y, z, D_x u^\varepsilon, \frac{D_y u^\varepsilon}{\varepsilon}, \frac{D_z u^\varepsilon}{\varepsilon^2} \right) = 0 & \text{in } (0,T) \times \mathbb{R}^n \times \mathbb{R}^m \times \mathbb{R}^p \\ u^\varepsilon(0, x, y, z) = h^\varepsilon(x, y, z) & \text{on } \mathbb{R}^n \times \mathbb{R}^m \times \mathbb{R}^p \end{cases} \tag{10}$$

where H^ε and h^ε are 1-periodic in y and z. Each variable corresponds to a certain scale of the problem: x is the *macroscopic* (or the *slow*) variable, y is the *mesoscopic* (or the not so fast) variable and z is the *microscopic* (or the *fast*) variable.

Roughly speaking, we shall attack this problem iteratively: by virtue of the different powers of ε, one first considers *both x and y* as slow variables, freezing them and homogenizing with respect to z and after, still with x frozen, one shall homogenizes with respect to y. In other words, in a first approximation, problem (10) is a singular perturbation problem only in the variable z; under adequate assumptions of ergodicity and stabilization with respect to z, we shall achieve a *mesoscopic* effective Hamiltonian H_1 and $u^\varepsilon(t, x, y, z)$ should converge to the solution $v^\varepsilon(t, x, y)$ of the mesoscopic problem

$$\begin{cases} \partial_t v^\varepsilon + H_1 \left(x, y, D_x v^\varepsilon, \frac{D_y v^\varepsilon}{\varepsilon} \right) = 0 & \text{in } (0,T) \times \mathbb{R}^n \times \mathbb{R}^m \\ v^\varepsilon(0, x, y) = h_1(x, y) & \text{on } \mathbb{R}^n \times \mathbb{R}^m. \end{cases}$$

This problem falls within the theory of Sec. 4 (see also Ref.[4]); v^ε will converge to the solution u of the limit problem $(\overline{\text{HJ}})$ provided that H_1 is ergodic and (H_1, h_1) is stabilizing. In conclusion, we expect that $u^\varepsilon(t, x, y, z)$ will converge to $u(t, x)$ where the effective quantities are defined inductively.

For the sake of simplicity, we shall assume as before that the operator H is given by

$$H(x, y, z, p_x, p_y, p_z) := \min_{\beta \in B} \max_{\alpha \in A} L^{\alpha, \beta}(x, y, z, p_x, p_y, p_z),$$

for the family of linear operators

$$L^{\alpha,\beta}(x,y,z,p_x,p_y,p_z) := -p_x \cdot f(x,y,z,\alpha,\beta) - p_y \cdot g(x,y,z,\alpha,\beta)$$
$$- p_z \cdot r(x,y,z,\alpha,\beta) - l(x,y,z,\alpha,\beta).$$

We shall require the following assumptions:

- $H^\varepsilon \to H$ and $h^\varepsilon \to h$ as $\varepsilon \to 0$ uniformly on the compact sets. We also suppose that H, h and every H^ε, h^ε satisfy the standard assumptions of Sec. 2, i.e. they are 1-periodic in (y,z) and H^ε, H are HJI operators with the regularity in the coefficients suitably extended to the additional variable z.
- The Hamiltonians are equibounded: $|H^\varepsilon(x,y,z,0,0,0)| \le C$, for every x,y,z and ε.
- *Microscopic coercivity* The Hamiltonian H is coercive in p_z: there are $\nu, C \in \mathbb{R}^+$, such that

$$H(x,y,z,p_x,p_y,p_z) \ge \nu|p_z| - C(1 + |p_x| + |p_y|) \qquad \forall x,y,z,p_x,p_y,p_z.$$

- *Mesoscopic coercivity* For some $\nu, C \in \mathbb{R}^+$, there holds

$$H(x,y,z,p_x,p_y,0) \ge \nu|p_y| - C(1 + |p_x|) \qquad \forall x,y,z,p_x,p_y.$$

It is worth to observe that the first two assumptions ensure that problem (10) admits exactly one continuous bounded solution u^ε that is periodic in (y,z). For instance, the second assumption is guaranteed by the equiboundedness of the running costs while the last two assumptions are satisfied if there is $\nu > 0$ such that, for every (x,y,z,β), there holds

$$B_m(0,\nu) \subset \overline{\mathrm{conv}}\{g(x,y,z,\alpha,\beta) \mid \alpha \in A\}, \tag{11}$$
$$B_p(0,\nu) \subset \overline{\mathrm{conv}}\{r(x,y,z,\alpha,\beta) \mid \alpha \in A\}. \tag{12}$$

We denote by H' the recession function of H with respect to the variables (y,z):

$$H'(x,y,z,p_y,p_z) := \min_{\beta \in B} \max_{\alpha \in A} \left\{-p_y \cdot g(x,y,z,\alpha,\beta) - p_z \cdot r(x,y,z,\alpha,\beta)\right\}.$$

The function H' is positively 1-homogeneous in (p_y,p_z) and for every $(\overline{x},\overline{p}_x)$ there is a constant C such that

$$|H(x,y,z,p_x,p_y,p_z) - H'(x,y,z,p_y,p_z)| \le C \qquad \forall y, p_y \in \mathbb{R}^m,\ z, p_z \in \mathbb{R}^p \tag{13}$$

for every (x,p_x) in a neighborhood of $(\overline{x},\overline{p}_x)$. We introduce also the recession function with respect to z for x and y frozen:

$$H''(x,y,z,p_z) := H'(x,y,z,0,p_z) = \min_{\beta \in B} \max_{\alpha \in A} \left\{-p_z \cdot r(x,y,z,\alpha,\beta)\right\}.$$

We observe that H'' is positively 1-homogeneous in p_z and for every $(\overline{x}, \overline{y}, \overline{p}_x, \overline{p}_y)$ there is a constant C such that

$$|H(x, y, z, p_x, p_y, p_z) - H''(x, y, z, p_z)| \leq C \qquad \forall z, p_z \in \mathbb{R}^p \qquad (14)$$

for every (x, y, p_x, p_y) in a neighborhood of $(\overline{x}, \overline{y}, \overline{p}_x, \overline{p}_y)$.

Let us now state some results on the construction of the effective Hamiltonian $\overline{H}$ and the effective initial datum $\overline{h}$. For some special cases, explicit formulas are also provided.

Proposition 5.1. *Under the above assumptions, we have:*

(a) The Hamiltonian H is ergodic in the microscopic variable z. The effective Hamitonian $H_1 = H_1(x, y, p_x, p_y)$ is regular: there are $C \in \mathbb{R}$ and a modulus of continuity ω such that:

$$|H_1(x_1, y_1, p, q) - H_1(x_2, y_2, p, q)| \leq \omega(|x_1 - x_2| + |y_1 - y_2|)$$
$$+ C(|x_1 - x_2| + |y_1 - y_2|)(1 + |p| + |q|)$$

$$|H_1(x, y, p_1, q_1) - H_1(x, y, p_2, q_2)| \leq C(|p_1 - p_2| + |q_1 - q_2|)$$

for every $x_i, p_i \in \mathbb{R}^n$, $y_i, q_i \in \mathbb{R}^m$.

(b) H_1 is coercive in p_y; moreover, it is ergodic in y and its effective Hamiltonian $\overline{H}$ fulfills the Comparison Principle.

(c) Assume that for each $(\overline{y}, \overline{p}_y)$ there holds

$$H(\overline{x}, \overline{y}, z, \overline{p}_x, \overline{p}_y, p_z) - H(\overline{x}, \overline{y}, z, \overline{p}_x, \overline{p}_y, 0) \geq 0 \qquad \forall z, p_z \in \mathbb{R}^p, \quad (15)$$

and

$$H(\overline{x}, y, z, \overline{p}_x, p_y, 0) - H(\overline{x}, y, z, \overline{p}_x, 0, 0) \geq 0 \qquad \forall y, p_y \in \mathbb{R}^m, \quad \forall z \in \mathbb{R}^p. \qquad (16)$$

Then, $\overline{H}(\overline{x}, \overline{p}_x)$ can be written as

$$\overline{H}(\overline{x}, \overline{p}_x) = \max_{y,z} H(\overline{x}, y, z, \overline{p}_x, 0, 0).$$

(d) [6, Lemma 2] The recession function H' is ergodic in the microscopic variable z; its effective Hamiltonian is the recession function H'_1 of H_1.

Proof. *a)* By virtue of the microscopic coercivity and Proposition 3.1-(*a*), H is ergodic in z with an effective Hamiltonian H_1. The regularity of H_1 is an immediate consequence of Proposition 3.1-(*b*) (with x and p_x replaced respectively by (x, y) and (p_x, p_y)).

b) In order to prove that H_1 is coercive in p_y, we first observe that the solution w_δ to the microscopic δ-cell problem satisfies: $\delta w_\delta \leq -\inf_z H(x, y, z, p_x, p_y, 0)$. As $\delta \to 0$, we get:

$$\inf_z H(x, y, z, p_x, p_y, 0) \leq H_1(x, y, p_x, p_y)$$

and, by the mesoscopic coercivity, we deduce that H_1 is coercive in p_y. Applying Proposition 3.1, one obtains the second part of the statement.

c) Proposition 3.1-(*c*) yields: $H_1(x, y, p_x, p_y) = \max_z H(x, y, z, p_x, p_y, 0)$. For each $(\overline{x}, \overline{p}_x)$, relation (16) ensures:

$$H_1(\overline{x}, y, \overline{p}_x, p_y) - H_1(\overline{x}, y, \overline{p}_x, 0) \geq 0 \qquad \forall y, p_y \in \mathbb{R}^m.$$

Applying again Proposition 3.1-(*c*), we obtain the statement.

d) For the sake of completeness, let us recall the arguments of [6, Lemma 2]. Arguing as in (*a*), one can prove that $H_\lambda(x, y, z, p_y, p_z) := \lambda^{-1} H(x, y, z, 0, \lambda p_y, \lambda p_z)$ is ergodic in z with the effective Hamiltonian $\overline{H}_\lambda(x, y, p_y) := \lambda^{-1} H_1(x, y, 0, \lambda p_y)$. Since $H_\lambda \to H'$ as $\lambda \to +\infty$ uniformly (for x bounded), by the Comparison Principle on the cell δ-problem, one obtains that H' is ergodic in z with $H_1' := \lim_{\lambda \to +\infty} \lambda^{-1} H_1(x, y, 0, \lambda p_y)$ as effective Hamiltonian.

Let us now check that H_1' is the recession function of H_1. We first note that the positive 1-homogeneity of H' entails the one of H_1'. By estimate (13), the Comparison Principle yields:

$$|H_1(x, y, p_x, p_y) - H_1'(x, y, p_y)| \leq C \qquad \forall y, p_y \in \mathbb{R}^m$$

for every (x, p_x) in a neighborhood of $(\overline{x}, \overline{p}_x)$, namely H_1' is the recession function of H_1. $\qquad\square$

Proposition 5.2. *Under our assumptions, the pair (H'', h) is stabilizing at $(\overline{x}, \overline{y})$ to $\min_z h(\overline{x}, \overline{y}, z) =: h_1(\overline{x}, \overline{y})$. Moreover, the pair (H_1, h_1) is stabilizing at $\overline{x}$ to $\min_{y,z} h(\overline{x}, y, z) =: \overline{h}(\overline{x})$.*

The proof of this Proposition relies on the iterative application of Proposition 3.2 and we shall omit it.

Theorem 5.1. *Under the above assumptions, the solution u^ε to problem (10) converges uniformly on the compact subsets of $(0, T) \times \mathbb{R}^n$ to the unique viscosity solution of $(\overline{HJ})$ where the effective Hamiltonian $\overline{H}$ and the effective initial datum $\overline{h}$ are defined respectively in Propositions 5.1 and 5.2.*

Proof. The proof of this Theorem is based on [6, Theorem 2] and on the properties of ergodicity and stabilization established in Propositions 5.1

and 5.2. For the sake of completeness, let us just stress the crucial parts. We shall argue as in the proof of Theorem 4.1: we set

$$\overline{u}(t,x) := \limsup_{\varepsilon \to 0,\ (t',x') \to (t,x)} \sup_{y,z} u^\varepsilon(t',x',y,z) \quad \text{if } t > 0,$$

$$\overline{u}(0,x) := \limsup_{(t',x') \to (0,x),\ t' > 0} \overline{u}(t',x') \quad \text{if } t = 0,$$

and we want to show that $\overline{u}$ is a subsolution to $(\overline{\mathrm{HJ}})$. As before, it suffices to prove that, for every $r > 0$ small enough, there is a parameter $\varepsilon' > 0$ and an equibounded family of continuous correctors $\{\chi^\varepsilon \mid 0 < \varepsilon < \varepsilon'\}$ so that

$$H^\varepsilon(x,y,z,D_x\varphi(t,x),D_y\chi^\varepsilon,\frac{D_z\chi^\varepsilon}{\varepsilon}) \geq \overline{H} - 2\eta$$

in $Q_r(\overline{t},\overline{x}) := (\overline{t}-r,\overline{t}+r) \times B_r(\overline{x}) \times \mathbb{R}^m \times \mathbb{R}^p$, for every $\varepsilon < \varepsilon'$. To this aim, we consider the mesoscopic δ-cell problem

$$\delta w_\delta + H_1(\overline{x},y,\overline{p}_x,D_y w_\delta) = 0 \tag{17}$$

with $\overline{p}_x := D_x\varphi(\overline{t},\overline{x})$. For $\delta > 0$ sufficiently small, the ergodicity of H_1 (established in Proposition 5.1-(b)) ensures

$$\|\delta w_\delta + \overline{H}\|_\infty \leq \eta \tag{18}$$

where $\overline{H}$ is defined as before. For every $\varepsilon > 0$ and $r > 0$, we consider the problem

$$\delta w^\varepsilon_{\delta,r} + H^\varepsilon_r(y,z,D_y w^\varepsilon_{\delta,r},\frac{D_z w^\varepsilon_{\delta,r}}{\varepsilon}) = 0$$

with $H^\varepsilon_r(y,z,p_y,p_z) := \min_{x \in B_r(\overline{x})} H^\varepsilon(x,y,z,D_x\varphi(x),p_y,p_z)$. We note that $H^\varepsilon_r(y,z,p_y,p_z) \to H(\overline{x},y,z,\overline{p}_x,p_y,p_z)$ locally uniformly as $(\varepsilon,r) \to (0,0)$ and that the limit Hamiltonian H is ergodic in z with effective Hamiltonian H_1. Hence, applying Theorem 4.1 for the stationary eq., we obtain that $\{w^\varepsilon_{\delta,r}\}$ uniformly converge to w_δ as $(\varepsilon,r) \to (0,0)$. By (18), we deduce that there are small ε' and r' so that

$$\|\delta w^\varepsilon_{\delta,r} + \overline{H}\|_\infty \leq 2\eta \qquad \text{for all} \quad 0 < \varepsilon < \varepsilon',\ 0 < r < r'.$$

Finally, as in Theorem 4.1, it suffices to define $\chi^\varepsilon(y,z) := w^\varepsilon_{\delta,r}(y,z)$ (for $0 < r < r'$ fixed).

Now, let us check that $\overline{u}(0,\overline{x}) \leq \overline{h}(\overline{x})$. We introduce the following notations:

$$H^{\varepsilon,\prime}_r(y,z,p_y,p_z) := \min_{|x-\overline{x}|\leq r} H^{\varepsilon,\prime}(x,y,z,p_y,p_z),$$

$$h^\varepsilon_r(y,z) := \max_{|x-\overline{x}|\leq r} h^\varepsilon(x,y,z).$$

18

It can be easily checked that, as $(\varepsilon, r) \to (0,0)$, $H_r^{\varepsilon,\prime}$ and h_r^ε converge locally uniformly respectively to $H'(\overline{x}, \cdot)$ and to $h(\overline{x}, \cdot)$. Let w_r^ε be the unique solution of the Cauchy problem

$$\partial_t w_r^\varepsilon + H_r^{\varepsilon,\prime}(y, z, D_y w_r^\varepsilon, \varepsilon^{-1} D_z w_r^\varepsilon) = 0 \quad \text{in } (0, +\infty) \times \mathbb{R}^m \times \mathbb{R}^p$$
$$w_r^\varepsilon(0, y, z) = h_r^\varepsilon(y, z), \quad \text{on } \mathbb{R}^m \times \mathbb{R}^p, \quad w_r^\varepsilon \text{ periodic in } y \text{ and in } z.$$

By Proposition 3.1-(d) and Proposition 5.2, the limit Hamiltonian $H'(\overline{x}, \cdot)$ is ergodic with effective Hamiltonian H'_1 and the pair (H'', h) stabilizes with respect to the microscopic variable z. Thus, Theorem 4.1 ensures that, as $(\varepsilon, r) \to (0,0)$, the solution w_r^ε converges locally uniformly to the one of the mesoscopic cell Cauchy problem

$$\partial_t w + H'_1(\overline{x}, y, D_y w) = 0 \quad \text{in } (0, +\infty) \times \mathbb{R}^m,$$
$$w(0, y) = h_1(\overline{x}, y) \quad \text{on } \mathbb{R}^m, \quad w \text{ periodic in } y.$$

Since (H'_1, h_1) is stabilizing (still by Proposition 5.2), for every $\eta > 0$, there exists $T > 0$ such that $\|w(T, \cdot) - \overline{h}(\overline{x})\|_\infty \leq \eta$. Hence, for every $\eta > 0$ and for T sufficiently large, there exist ε' and r' so small that

$$\|w_r^\varepsilon(T, \cdot, \cdot) - \overline{h}(\overline{x})\|_\infty \leq 2\eta \quad \text{for every } \varepsilon \leq \varepsilon', r \leq r'.$$

Therefore, by the Comparison Principle, we obtain:

$$\|w_r^\varepsilon(t, \cdot, \cdot) - \overline{h}(\overline{x})\|_\infty \leq \eta \quad \text{for every } \varepsilon \leq \varepsilon', r \leq r', t \geq T$$

and we conclude as before. $\qquad\square$

Remark 5.1. This Theorem also applies to non coercive Hamiltonian that are microscopically stabilizing and ergodic with a mesoscopic Hamiltonian that fulfills the Comparison Principle, stabilizes, and is ergodic with an effective Hamiltonian that also fulfills the Comparison Principle (e.g., see Sec. 7 below). We refer the reader to the paper Ref. 6 for the case when the Comparison Principle fails either for the mesoscopic Hamiltonian or for the effective one.

5.2. *The general case*

We consider the problem having $j + 1$ scales:

$$\begin{cases} \partial_t u^\varepsilon + H^\varepsilon\left(x, y_1, \ldots, y_j, D_x u^\varepsilon, \varepsilon^{-1} D_{y_1} u^\varepsilon, \ldots, \varepsilon^{-j} D_{y_j} u^\varepsilon\right) = 0 \\ \qquad\qquad\qquad\qquad (0, T) \times \mathbb{R}^n \times \mathbb{R}^{m_1} \times \cdots \times \mathbb{R}^{m_j} \\ u^\varepsilon(0, x, y_1, \ldots, y_j) = h^\varepsilon(x, y_1, \ldots, y_j) \quad \mathbb{R}^n \times \mathbb{R}^{m_1} \times \cdots \times \mathbb{R}^{m_j}. \end{cases} \quad (19)$$

We assume the following hypotheses:

- $H^\varepsilon \to H$ and $h^\varepsilon \to h$ locally uniformly as $\varepsilon \to 0$; these functions are periodic in $(y_1, \ldots, y_j)$;
- the initial data h^ε and h are BUC;
- the H^ε are equibounded: $|H^\varepsilon(x, y_1, \ldots, y_j, 0, 0, \ldots, 0)| \leq C$;
- *iterated coercivity:* there are $\nu, C \in \mathbb{R}^+$, such that, for every $k = j, \ldots, 1$, there holds

$$H(x, y_1, \ldots, y_j, p_x, p_{y_1}, \ldots, p_{y_k}, 0, \ldots, 0) \geq \nu |p_{y_k}|$$
$$- C\left(1 + |p_x| + \sum_{i=1}^{k-1} |p_{y_i}|\right)$$

for every $x, y_1, \ldots, y_j, p_x, p_{y_1}, \ldots, p_{y_k}$;
- there exists a recession function $H^{\varepsilon,\prime} = H^{\varepsilon,\prime}(x, y_1, \ldots, y_j, p_{y_1}, \ldots, p_{y_j})$, positively 1-homogeneous in $(p_{y_1}, \ldots, p_{y_j})$, which satisfies, for some constant $C > 0$

$$\left| H^\varepsilon(x, y_1, \ldots, y_j, p_x, p_{y_1}, \ldots, p_{y_j}) - H^{\varepsilon,\prime}(x, y_1, \ldots, y_j, p_{y_1}, \ldots, p_{y_j}) \right| \leq C$$

for every $y_i, p_{y_i} \in \mathbb{R}^{m_i}$ $(i = 1, \ldots, j)$, for every (x, p_x) in a neighborhood of $(\overline{x}, \overline{p}_x)$ and for every ε.

Let us observe that the equiboundedness of the running costs ensures the third assumption; furthermore, the Hamiltonian

$$H(x, y_1, \ldots, y_j, p_x, p_{y_1}, \ldots, p_{y_j}) = \min_\beta \max_\alpha \left\{ -f \cdot p_x - \sum_{i=1}^j g_i \cdot p_{y_i} - l \right\}$$

is iteratively coercive whenever there holds

$$B_{m_i}(0, \nu) \subset \overline{\mathrm{conv}}\{g_i(x, y_1, \ldots, y_j, \alpha, \beta) \mid \alpha \in A\}$$

for every $x, y_1, \ldots, y_j, \beta$ $(i = 1, \ldots, j)$.

We write $H_j = H$ and $h_j = h$. For each $i = j, \ldots, 1$, H_i fulfills the Comparison Principle and is ergodic with respect to y_i. We denote by H_{i-1} its effective Hamiltonian. We set $\overline{H} := H_0$.

As before, one can prove that the Hamiltonian H has a recession function H' which is the uniform limit on the compact sets of $H^{\varepsilon,\prime}$ as $\varepsilon \to 0$. Moreover, as in Proposition 5.1-(d), we obtain that H_i has a recession function H_i', that every H_i' is ergodic and that its effective Hamiltonian is H_{i-1}' (for every $i = j, \ldots, 1$).

For $H_i'' := H_i'(x, y_1, \ldots, y_i, 0, \ldots, 0, p_i)$, the pair (H_i'', h_i) is stabilizing with respect to y_i at each point $(x, y_1, \ldots, y_{i-1})$ (for $i = j, \ldots, 1$). We denote by h_{i-1} its effective initial data and we put $\overline{h} = h_0$.

Theorem 5.2. *Under the above assumptions, u^ε converges uniformly on the compact subsets of $(0,T) \times \mathbb{R}^n$ to the unique viscosity solution of $(\overline{\mathrm{HJ}})$.*

We shall omit the proof of this Theorem: actually, it can be easily obtained by using iteratively the arguments followed in Theorem 5.1.

Remark 5.2. This result can be immediately extended to pde with non power-like scales:

$$\partial_t u^\varepsilon + H^\varepsilon(x, y_1, \ldots, y_j, D_x u^\varepsilon, \varepsilon_1^{-1} D_{y_1} u^\varepsilon, \ldots, \varepsilon_j^{-1} D_{y_j} u^\varepsilon) = 0,$$

with $\varepsilon_1 \to 0$ and $\varepsilon_i/\varepsilon_{i-1} \to 0$ $(i = 2, \ldots, j)$. The above eq. encompasses (19) for $\varepsilon_i = \varepsilon^i$.

6. Iterated homogenization for coercive equations

In this Section we address the study of the iterated homogenization of first order equations with multiple scales. For the sake of simplicity, we shall focus our attention on the three-scale case:

$$\begin{cases} \partial_t v^\varepsilon + F^\varepsilon\left(x, \frac{x}{\varepsilon}, \frac{x}{\varepsilon^2}, D_x v^\varepsilon\right) = 0 & \text{in } (0,T) \times \mathbb{R}^n \\ v^\varepsilon(0,x) = h^\varepsilon(x, \frac{x}{\varepsilon}, \frac{x}{\varepsilon^2}) & \text{on } \mathbb{R}^n. \end{cases} \tag{20}$$

The operator $F^\varepsilon = F^\varepsilon(x,y,z,p)$ and the function $h^\varepsilon = h^\varepsilon(x,y,z)$ are periodic in y and in z; moreover, they are respectively a regular perturbation of F and h, namely, $F^\varepsilon \to F$ and $h^\varepsilon \to h$ locally uniformly as $\varepsilon \to 0$. We assume that F is a HJI operator

$$F(x,y,z,p_x) := \min_{\beta \in B} \max_{\alpha \in A}\{-p_x \cdot f(x,y,z,\alpha,\beta) - l(x,y,z,\alpha,\beta)\}$$

where the drift f and the costs l and h fulfill the requirements stated in Sec. 2. We assume also that F is coercive with respect to p: for $\nu, C \in \mathbb{R}^+$ there holds

$$F(x,y,z,p) \geq \nu|p| - C \qquad \forall x,y,z,p; \tag{21}$$

(for instance, this condition holds if $B_n(0,\nu) \subset \overline{\mathrm{conv}}\{f(x,y,z,\alpha,\beta) \mid \alpha \in A\}$ for every (x,y,z,β)).

Let us emphasize that problem (20) encompasses the problem studied by Lions, Papanicolaou and Varadhan.[26] Actually, we extend the previous literature in two directions: we consider regular perturbations F^ε and h^ε of F and of h and, mainly, we address the three-scale problem both for the HJI eq. and for the initial condition.

Our purpose is to apply Theorem 5.1 by proving that (20) is a particular case of (10). To this aim, we introduce the shadow variables $y = x/\varepsilon$ and $z =$

x/ε^2 and consider the solution $u^\varepsilon(t, x, y, z)$ of (10) with the Hamiltonian H given by

$$H(x, y, z, p_x, p_y, p_z) = F(x, y, z, p_x + p_y + p_z).$$

The Hamiltonian H clearly satisfies the assumptions of Sec. 5. By uniqueness, one sees immediately that

$$v^\varepsilon(t, x) = u^\varepsilon(t, x, x/\varepsilon, x/\varepsilon^2).$$

By the periodicity in y, Theorem 5.1 ensures that u^ε converges uniformly on compact subsets to the unique solution of $(\overline{\text{HJ}})$. Therefore, the following result holds:

Corollary 6.1. *Under the above assumptions, v^ε converges uniformly on the compact subsets of $(0, T) \times \mathbb{R}^n$ to the unique viscosity solution of $(\overline{\text{HJ}})$.*

Remark 6.1. Arguing as in the Subsec. 5.2, one can easily extend this result to the homogenization with an arbitrary number of scales.

7. Examples

In this Section we discuss some examples arising in the optimal control theory and in deterministic games. For simplicity, we shall only address three-scale problems.

The first Subsection is devoted to a singular perturbation of a deterministic game; in some special cases, the effective problem is still a deterministic game and we shall provide the explicit formulas for the effective quantities (dynamics, pay-off, etc.). The second Subsection concerns the homogenization of a deterministic optimal control problem. In the third Subsection the coercivity of the Hamiltonian is replaced by a non-resonance condition introduced by Arisawa and Lions.[9] Here we show that the effective problem may change if the roles of ε and ε^2 are exchanged.

7.1. *Singular perturbation of a differential game*

Fix $T > 0$ and, for each $\varepsilon > 0$, consider the dynamics

$$\dot{x}_s = f(x_s, y_s, z_s, \alpha_s, \beta_s), \qquad x_0 = x,$$
$$\varepsilon \dot{y}_s = g(x_s, y_s, z_s, \alpha_s, \beta_s), \qquad y_0 = y,$$
$$\varepsilon^2 \dot{z}_s = r(x_s, y_s, z_s, \alpha_s, \beta_s), \qquad z_0 = z,$$

for $0 \leq s \leq T$. The admissible controls α_s and β_s are measurable function with value respectively in the compact sets A and B. They are governed by two different players. We consider the cost functional

$$P^\varepsilon(t, x, y, z, \alpha, \beta) := \int_0^t l(x_s, y_s, z_s, \alpha_s, \beta_s)\, ds + h(x_t, y_t, z_t).$$

The goal of the first player controlling α is to minimize P^ε, wheras the second player wishes to maximize P^ε by controlling β.

Consider the *upper* value function

$$u^\varepsilon(t, x, y, z) := \sup_{\beta \in B(t)} \inf_{\alpha \in \mathcal{A}(t)} P^\varepsilon(t, x, y, z, \alpha, \beta[\alpha]),$$

where $\mathcal{A}(t)$ denotes the set of admissible controls of the first player in the interval $[0, t]$ and $B(t)$ denotes the set of admissible strategies of the second player in the same interval (i.e., nonanticipating maps from $\mathcal{A}(t)$ into the admissible controls of the second player; see Ref. 20 for the precise definition).

Under the assumptions of Sec. 2 the upper value function is the unique viscosity solution of the HJI eq. (10) with[20]

$$H^\varepsilon = H(x, y, z, p_x, p_y, p_z) = \max_{\alpha \in A} \min_{\beta \in B} \{-p_x \cdot f - p_y \cdot g - p_z \cdot r - l\}.$$

Let us assume (11)–(12), so the Hamitonian H is microscopically and mesoscopically coercive. Suppose in addition that H has the properties (15)–(16). By Theorem 5.1, the upper value u^ε converge locally uniformly to the solution u to the effective problem

$$\begin{cases} \partial_t u + \max_{y, z, \alpha} \min_\beta \{-D_x u \cdot f(x, y, z, \alpha, \beta) - l(x, y, z, \alpha, \beta)\} = 0, \\ u(0, x) = \min_{y, z} \{h(x, y, z)\}. \end{cases}$$

Using again the theory of Evans and Souganidis,[20] one can see that u is the upper value of the following effective deterministic differential game. The effective dynamics are

$$\dot{x}_s = f(x_s, y_s, z_s, \alpha_s, \beta_s), \qquad x_0 = x,$$

where y and z are new controls, while the effective cost is

$$P(t, x, y, z, \alpha, \beta) := \int_0^t l(x_s, y_s, z_s, \alpha_s, \beta_s)\, ds + \min_{y, z} h(x_t, y, z).$$

In the effective game the first player wants to minimize P by choosing the controls y, z, α, and the second player wants to maximize P by choosing β.

We end this Subsection by giving a simple condition on the control system that implies (15)–(16). Suppose the controls of the first player are in the separated form $\alpha = (\alpha^S, \alpha^F, \alpha^V) \in A^S \times A^F \times A^V$, where α^S is the control of the slow variables x, α^F of the fast variables y, and α^V is the control of the very fast variables z. More precisely

$$f = f\left(x, y, z, \alpha^S, \beta\right), \qquad l = l\left(x, y, z, \alpha^S, \beta\right),$$
$$g = g\left(x, y, z, \alpha^F, \beta\right), \qquad r = r\left(x, y, z, \alpha^V, \beta\right).$$

In this case the conditions (11)–(12) become

$$B_m(0, \nu) \subset \overline{\mathrm{conv}}\{g(x, y, z, \alpha^F, \beta) \mid \alpha^F \in A^F\},$$
$$B_p(0, \nu) \subset \overline{\mathrm{conv}}\{r(x, y, z, \alpha^V, \beta) \mid \alpha^V \in A^V\}$$

for every x, y, z, β. Then it is easy to check that (15)–(16) hold, and the effective Hamiltonian is

$$H(x, p_x) = \max_{(y,z)\in[0.1]^2} \max_{\alpha^S \in A^S} \min_{\beta \in B} \left\{ -p_x \cdot f\left(x, y, z, \alpha^S, \beta\right) - l\left(x, y, z, \alpha^S, \beta\right) \right\}.$$

Note that in the further special case

$$g = \varphi_1(x, y, z)\alpha^F, \quad A^F = B_m(0, 1), \quad \varphi_1(x, y, z) \geq \nu > 0,$$
$$r = \varphi_2(x, y, z)\alpha^V, \quad A^V = B_p(0, 1), \quad \varphi_2(x, y, z) \geq \nu > 0,$$

the PDE in (10) becomes the model problem (1) presented in the Introduction.

7.2. *Homogenization of a deterministic optimal control problem*

For each $\varepsilon > 0$, we consider dynamics having the form

$$\dot{x}_s = f\left(x_s, \frac{x_s}{\varepsilon}, \frac{x_s}{\varepsilon^2}, \alpha_s\right), \qquad x_0 = x$$

for $0 \leq s \leq T$. The admissible controls $\alpha.$ are measurable function with value in the compact A. We denote by $\mathcal{A}(t)$ the set of admissible controls on the interval $(0, t)$. Our goal is to choose the control in order to minimize the payoff functional:

$$P^\varepsilon(t, x, \alpha) := \int_0^t l\left(x_s, \frac{x_s}{\varepsilon}, \frac{x_s}{\varepsilon^2}, \alpha_s\right) ds + h\left(x_t, \frac{x_t}{\varepsilon}, \frac{x_t}{\varepsilon^2}\right).$$

By standard theory[12] the value function

$$v^\varepsilon(t, x) := \inf_{\alpha \in \mathcal{A}(t)} P^\varepsilon(t, x, y, \alpha),$$

24

is the unique viscosity solution of the HJ eq. (20), provided that f, l and h fulfill the assumptions of Sec. 2. Moreover, let us recall that the Hamitonian F is coercive provided that

$$B_n(0,\nu) \subset \overline{\mathrm{conv}}\{f(x,y,z,\alpha) \mid \alpha \in A\} \qquad \text{for all } (x,y,z).$$

Under these assumptions, Corollary 6.1 ensures that v^ε converges locally uniformly to the solution of problem $(\overline{\mathrm{HJ}})$. Here $\overline{h}(x) = \min_{y,z} h(x,y,z)$, but $\overline{H}$ does not have an explicit representation.

Note that the eikonal equation (2) is a special case of this example. It is enough to take

$$A = B_n(0,1), \qquad f(x,y,z,\alpha) = \varphi(x,y,z)\alpha, \qquad \varphi(x,y,z) \geq \nu > 0.$$

7.3. *Multiscale singular perturbation under a nonresonance condition*

This Subsection is devoted to the case of a nonresonance condition, introduced by Arisawa and Lions[9] (see also the first two authors[5]), that ensures the ergodicity for a class of non-coercive Hamiltonian with an effective operator that fulfills the Comparison Principle and can be written explicitly.

As a byproduct, we show that the roles of ε and ε^2 can not be exchanged in general. To this aim, we first consider a three scales perturbation problem that is nonresonant in the microscopic variable z and coercive in the mesoscopic variable y, and then a problem that is coercive in z and nonresonant in y. The two effective Hamiltonians are different.

1st case: For each $\varepsilon > 0$, consider the dynamics

$$\dot{x}_s = f(x_s, y_s, z_s, \alpha_s^S), \qquad \dot{y}_s = \frac{1}{\varepsilon} g(x_s, y_s, \alpha_s^F), \qquad \dot{z}_s = \frac{1}{\varepsilon^2} r(x_s, y_s)$$

for $0 \leq s \leq T$ with initial conditions

$$x_0 = x, \qquad y_0 = y, \qquad z_0 = z.$$

The admissible control $\alpha = (\alpha^S, \alpha^F)$ splits in two control: one for the *slow* and one for the *fast* variable (the superscript recall this fact). Furthermore, α_s^S and α_s^F are measurable functions with value respectively in the compacts A^S and A^F. By the choice of α, one wants to minimize the *payoff* functional:

$$P^\varepsilon(t,x,y,z,\alpha) := \int_0^t l(x_s, y_s, z_s, \alpha_s^S)\, ds + h(x_t, y_t).$$

Then the value function u^ε solves the Cauchy problem:

$$\begin{cases} \partial_t u^\varepsilon + \max_{\alpha^S \in A^S}\{-D_x u^\varepsilon \cdot f(x,y,z,\alpha^S) - l(x,y,z,\alpha^S)\} + \\ \qquad\qquad \max_{\alpha^F \in A^F}\{-\tfrac{1}{\varepsilon}D_y u^\varepsilon \cdot g(x,y,\alpha^F)\} - \tfrac{1}{\varepsilon^2}r(x,y)\cdot D_z u^\varepsilon = 0 \\ u^\varepsilon(0,x,y,z) = h(x,y) \end{cases}$$

We require that the microscopic dynamic is nonresonant and that the mesoscopic one is coercive, namely:

$$r(x,y)\cdot k \neq 0 \ \forall k \in \mathbb{Z}^p \setminus \{0\}, \quad B_m(0,\nu) \subset \overline{\mathrm{conv}}\{g(x,y,\alpha^F)\mid \alpha^F \in A^F\}$$

for every (x,y). The Hamiltonian H is ergodic in the fast variable[5,9] and the mesoscopic Hamiltonian H_1 has the form:

$$H_1(x,y,p_x,p_y) = \int_{(0,1)^p} \max_{\alpha^S \in A^S}\{-p_x \cdot f(x,y,z,\alpha^S) - l(x,y,z,\alpha^S)\}\, dz$$
$$+ \max_{\alpha^F \in A^F}\{-p_y \cdot g(x,y,\alpha^F)\}.$$

One can easily check that H_1 fulfills the Comparison Principle, is ergodic with respect to y and satisfies (4). Therefore Theorem 5.1 ensures that u^ε converge locally uniformly to the solution of problem $(\overline{HJ})$ where the effective quantities are given by:

$$\overline{H}(x,p_x) = \max_y \int_{(0,1)^p} \max_{\alpha^S \in A^S}\{-p_x \cdot f(x,y,z,\alpha^S) - l(x,y,z,\alpha^S)\}\, dz$$
$$\overline{h}(x) = \min_y h(x,y).$$

2nd case: For each $\varepsilon > 0$, consider the dynamics

$$\dot{x}_s = f(x_s,y_s,z_s,\alpha_s), \qquad \dot{y}_s = \frac{1}{\varepsilon}g(x_s), \qquad \dot{z}_s = \frac{1}{\varepsilon^2}r(x_s,y_s,z_s,\alpha_s)$$

for $0 \leq s \leq T$ with initial conditions

$$x_0 = x, \qquad y_0 = y, \qquad z_0 = z.$$

The control α is a measurable function with value in the compact A. As before, by the choice of α, one wants to minimize the payoff:

$$P^\varepsilon(t,x,y,z,\alpha) := \int_0^t l(x_s,y_s,z_s,\alpha_s)\, ds + h(x_t,z_t).$$

The value function u^ε is the viscosity solution of the following Cauchy problem:

$$\begin{cases} \partial_t u^\varepsilon + \max_{\alpha \in A}\{-D_x u^\varepsilon \cdot f(x,y,z,\alpha) - \frac{1}{\varepsilon^2} D_z u^\varepsilon \cdot r(x,y,z,\alpha) - l(x,y,z,\alpha)\} \\ \qquad\qquad\qquad\qquad\qquad\qquad\qquad\qquad\qquad -\frac{1}{\varepsilon} D_y u^\varepsilon \cdot g(x) = 0 \\ u^\varepsilon(0,x,y,z) = h(x,z) \end{cases}$$

Assume that the dynamics are microscopically coercive and mesoscopically nonresonant:

$$g(x) \cdot k \neq 0 \quad \forall k \in \mathbb{Z}^m \setminus \{0\}, \qquad B_p(0,\nu) \subset \overline{\mathrm{conv}}\{r(x,y,z,\alpha) \mid \alpha \in A\}$$

for every (x,y,z). We require also that condition (15) is satisfied. Hence, H is microscopically ergodic and stabilizing and the mesoscopic quantities are given by:

$$H_1(x,y,p_x,p_y) = \max_{\alpha,z}\{-p_x \cdot f(x,y,z,\alpha) - l(x,y,z,\alpha)\} - p_y \cdot g(x)$$
$$h_1(x,y) = \min_z h(x,z).$$

Arguing as before, one can prove that the H_1 is ergodic, stabilizes, and fulfills the Comparison Principle. Whence, by Theorem 5.1, the value function u^ε converge locally uniformly to the solution of $(\overline{\mathrm{HJ}})$ with:

$$\overline{H}(x,p_x) = \int_{(0,1)^m} \max_{\alpha,z}\{-p_x \cdot f(x,y,z,\alpha) - l(x,y,z,\alpha)\}\,dy$$
$$\overline{h}(x) = \min_z h(x,z).$$

References

1. G.Allaire and M. Briane, *Proc. Roy. Soc. Edinburgh sect. A* **126**, 297 (1996).
2. O. Alvarez, *J. Differential Equations* **159**, 543 (1999).
3. O. Alvarez and M. Bardi, *SIAM J. Control Optim.* **40**, 1159 (2001).
4. O. Alvarez and M. Bardi, *Arch. Ration. Mech. Anal.* **170**, 17 (2003).
5. O. Alvarez and M. Bardi, Ergodicity, stabilization and singular perturbations for Bellman-Isaacs equations, tech. rep., Dipartimento di Matematica P.A., Università di Padova (Padova, Italy, 2007), available at http://cpde.iac.rm.cnr.it/ricerca.php. (January 2007).
6. O. Alvarez, M. Bardi and C. Marchi, *J. Differential Equations* **243**, 349 (2007).
7. O. Alvarez and I. Ishii, *Comm. Partial Differential Equations* **26**, 983 (2001).
8. M. Arisawa, *Adv. Math. Sci. Appl.* **11**, 465 (2001).
9. M. Arisawa and P.-L. Lions, *Comm. Partial Differential Equations* **23**, 2187 (1998).
10. Z. Artstein and V. Gaitsgory, *Appl. Math. Optim.* **41**, 425 (2000).
11. M. Avellaneda, *Comm. Pure Appl. Math.* **40**, 527 (1987).

12. M. Bardi and I. Capuzzo-Dolcetta, *Optimal Control and Viscosity Solutions of Hamilton-Jacobi-Bellman Equations* (Birkhauser, Boston, 1997).

13. A. Bensoussan, J.-L. Lions and G. Papanicolaou, *Asymptotic Analysis for periodic Structures* (North-Holland, Amsterdam, 1978).

14. A. Bensoussan, *Perturbation methods in optimal control* (Wiley/Gauthiers-Villars, Chichester, 1988).

15. I. Birindelli and J. Wigniolle, *Comm. Pure Appl. Anal.* **2**, 461 (2003).

16. A. Braides and A. Defranceschi, *Homogenizations of multiple integrals* (Clarendon Press, Oxford, 1998).

17. I. Capuzzo-Dolcetta and H. Ishii, *Indiana Univ. Math. J.* **50**, 1113 (2001).

18. L. Evans, *Proc. Roy. Soc. Edinburgh Sect. A* **111**, 359 (1989).

19. L. Evans, *Proc. Roy. Soc. Edinburgh Sect. A* **120**, 245 (1992).

20. L. Evans and P. E. Souganidis, *Indiana Univ. Math. J.* **33**, 773 (1984).

21. V. Gaitsgory and M.-T. Nguyen, *SIAM J. Control. Optim.* **41**, 954 (2002).

22. K. Horie and H. Ishii, *Indiana Univ. Math. J.* **47**, 1011 (1998).

23. H. Ishii, Almost periodic homogenization of Hamilton-Jacobi equations, in *International conference on differential equations. Proceedings of the con ference, Equadiff '99 (B. Fiedler, ed.)*, vol. *1*, (World Scientific, Singapore, 2000).

24. P. V. Kokotović, H. K. Khalil and J. O'Reilly, *Singular perturbation methods in control: analysis and design* (Academic Press, London, 1986).

25. J.-L. Lions, D. Lukkassen, L. E. Persson and P. Wall, *Chinese Ann. Math. Ser. B* **22**, 1 (2001).

26. P.-L. Lions, G. Papanicolaou and S. R. S. Varadhan, Homogeneization of Hamilton Jacobi equations, Unpublished, (1986).

27. P.-L. Lions and P. E. Souganidis, *Ann. Inst. H. Poincaré Anal. Non Linéaire* **22**, 667 (2005).

28. P.-L. Lions and P. E. Souganidis, *Comm. Partial Differential Equations* **30**, 335 (2005).

29. D. Lukkassen, *C. R. Acad. Sci. Paris Sèr. I Math.* **332**, 999 (2001).

30. F. Rezakhanlou and J. E. Tarver, *Arch. Rational Mech. Anal.* **151**, 277 (2000).

31. P. E. Souganidis, *Asymptot. Anal.* **20**, 1 (1999).

PATCHY FEEDBACKS FOR STABILIZATION AND OPTIMAL CONTROL: GENERAL THEORY AND ROBUSTNESS PROPERTIES

F. ANCONA

*Dipartimento di Matematica and C.I.R.A.M., Università di Bologna,
Piazza di Porta S. Donato 5, Bologna 40126, Italy
E-mail: ancona@ciram.unibo.it*

A. BRESSAN

*Department of Mathematics, Penn State University,
University Park, PA 16802, U.S.A.
E-mail: bressan@math.psu.edu*

This paper provides a survey of the theory of patchy feedbacks, and its applications to asymptotic stabilization and optimal control. It also contains two new results, showing the robustness of sub-optimal patchy feedbacks both in the case of (internal and external) deterministic disturbances, and of random perturbations modelled by stochastic Brownian motion.

Keywords: Optimal Feedback Control; Discontinuous Feedback; Robustness; Stochastic Disturbances.

1. Introduction

Consider a nonlinear control system on $\mathbb{R}^n$, of the form

$$\dot{x} = f(x, u) \,. \tag{1}$$

Here the upper dot denotes a derivative w.r.t. time. We assume that the control u takes values in $\mathbb{R}^m$ and that that the map $f : \mathbb{R}^n \times \mathbb{R}^m \mapsto \mathbb{R}^n$ is smooth. In connection with (1), two classical problems have been extensively studied in the control literature.

(I) Asymptotic Feedback Stabilization: Construct a feedback control $u = U(x)$ such that all trajectories of the resulting O.D.E.

$$\dot{x} \; = \; g(x) \; \doteq \; f\big(x, \, U(x)\big) \tag{2}$$

approach the origin as $t \to \infty$.

(II) Optimal Feedback Control: Given a compact set $K \subset \mathbb{R}^m$ of admissible control values, construct a feedback control $u = U(x) \in K$ such that all solutions of (2) reach a given target set $S \subset \mathbb{R}^n$ in minimum time.

In an ideal situation, these goals would be achieved by a $\mathcal{C}^1$ feedback control $x \mapsto U(x)$. In this case, for every initial state

$$x(0) = \bar{x}, \tag{3}$$

the solution of the Cauchy problem (2)-(3) has a unique classical solution, continuously depending on the initial data. However, smooth feedback controls exist only in a limited number of cases.

(I) For the problem of asymptotic stabilization, several examples of control systems are known, where every initial state can be steered toward the origin as $t \to \infty$ by an open-loop control $t \mapsto u(t)$. However, due to the presence of certain topological obstructions, no smooth (or even continuous) feedback control function $x \mapsto U(x)$ can accomplish the same task,[11,15,28,29] not even locally in a small neighborhood of the origin. In all these examples asymptotic stabilization can be achieved only by a discontinuous feedback.

(II) For nonlinear optimization problems, even in few space dimensions, it is known the optimal feedback can be discontinuous, with very complicated structure, while optimal controls can have infinitely many switchings. Moreover, a discontinuous optimal feedback may not be *robust* w.r.t. perturbations. In other words, one may find arbitrarily small functions e_1, e_2 such that the solution to the perturbed equation

$$\dot{x} = f\big(x,\, U(x + e_1(t))\big) + e_2(t) \tag{4}$$

achieves a much worse performance than the optimal one.

The use of discontinuous feedback controls $x \mapsto U(x)$ leads to the theoretical problem of how to interpret solutions for the O.D.E. (2), whose right hand side is discontinuous w.r.t. the state variable x. Different approaches can be followed:

- Consider only solutions in a strong sense. We recall that a *Carathéodory solution* of (2)-(3) is an absolutely continuous map $t \mapsto x(t)$ which satisfies (2) at a.e. time t. Equivalently,

$$x(t) = \bar{x} + \int_0^t g\big(x(s)\big)\, ds. \tag{5}$$

In general, we expect that these strong solutions will have the required asymptotic convergence properties. However, if one does not impose any regularity assumption of the feedback $U(x)$, there is no guarantee that any Carathéodory solution will exist.

- Define weaker concepts of solutions. For example, one can consider trajectories of the differential inclusion (see Refs. 6,18,19)

$$\dot{x} \in G(x) \doteq \bigcap_{\varepsilon > 0} \overline{co}\Big\{ g(y); \quad |y - x| < \varepsilon \Big\}. \tag{6}$$

It is not difficult to show that solutions in this relaxed sense always exist. However, too many solutions are now allowed. Not all of them may have the desired properties.

Of course, there is a wide variety of techniques[14,20,21] for constructing "generalized solutions" to the Cauchy problem (2)-(3). One of the paths followed in recent literature is to consider approximate solutions (or limits of approximate solutions) obtained by robust approximation methods, such as the "sample and hold" technique introduced in Ref. 13, originated from the theory of positional differential games.[22]

In the present paper we discuss an alternative approach, introduced by the authors in Ref. 1. It involves a particular class of feedback controls $U(\cdot)$, called *patchy feedbacks*. These are piecewise constant controls, whose discontinuities are sufficiently tame in order to guarantee the existence of Carathéodory solutions forward in time. At the same time, this class is sufficiently broad to solve a wide class of stabilization and optimization problems.

Section 2 contains the basic definitions of patchy vector fields and patchy feedbacks, together with a brief discussion of their basic properties. In Section 3 we show how to solve the asymptotic stabilization problem by means of a patchy feedback. Section 4 describes the construction of a patchy feedback which provides a close-to-optimal solution to a minimum time problem. In Section 5 we discuss the robustness of patchy feedbacks, showing that they still perform well also in the presence of small external perturbations, or small measurement errors in the state of the system. Finally, in Section 6 we analyze the performance of a patchy feedback in the presence of random perturbations, modelled by stochastic white noise. For the problem of reaching the origin in minimum time, we provide an estimate of the probability of entering the ε-ball B_ε around the origin within a given time T.

2. Patchy vector fields and patchy feedbacks

For a general discontinuous vector field g, the Cauchy problem for the O.D.E.

$$\dot{x} = g(x) \tag{7}$$

may not have any Carathéodory solution. Or, on the contrary, it may have infinitely many solutions, with wild behavior. We shall now describe a particular class of discontinuous vector fields g, whose corresponding trajectories are well defined and have nice properties. This is of particular interest, because a wide variety of stabilization and optimal control problems can be solved by discontinuous feedback controls (2), within this class. Throughout the paper, the closure and the boundary of a set Ω are denoted by $\overline{\Omega}$ and $\partial\Omega$, respectively.

Definition 2.1. By a *patch* we mean a pair (Ω, g) where $\Omega \subset \mathbb{R}^n$ is an open domain with smooth boundary and g is a smooth vector field defined on a neighborhood of $\overline{\Omega}$, which points strictly inward at each boundary point $x \in \partial\Omega$.

Calling $\mathbf{n}(x)$ the outer normal at the boundary point x, we thus require

$$\langle g(x),\ \mathbf{n}(x) \rangle < 0 \qquad \forall x \in \partial\Omega. \tag{8}$$

Clearly, in the case where $\Omega = \mathbb{R}^n$, no condition is required on the smooth vector field g.

Definition 2.2. We say that $g : \Omega \mapsto \mathbb{R}^n$ is a *patchy vector field* on the open domain Ω if there exists a family of patches $\{(\Omega_\alpha, g_\alpha);\ \alpha \in \mathcal{A}\}$ such that (see Fig. 1)
- $\mathcal{A}$ is a totally ordered set of indices,
- the open sets Ω_α form a locally finite covering of Ω,
- the vector field g can be written in the form

$$g(x) = g_\alpha(x) \qquad \text{if} \qquad x \in \Omega_\alpha \setminus \bigcup_{\beta > \alpha} \Omega_\beta. \tag{9}$$

By defining

$$\alpha^*(x) \doteq \max\{\alpha \in \mathcal{A}\ ;\ x \in \Omega_\alpha\}, \tag{10}$$

the identity (9) can be written in the equivalent form

$$g(x) = g_{\alpha^*(x)}(x) \qquad \forall\, x \in \Omega. \tag{11}$$

We shall occasionally adopt the longer notation $\left(\Omega,\ g,\ (\Omega_\alpha,\ g_\alpha)_{\alpha\in\mathcal{A}}\right)$ to indicate a patchy vector field, specifying both the domain and the single patches.

Remark 2.1. In general, the patches $(\Omega_\alpha,\ g_\alpha)$ are not uniquely determined by the patchy vector field g. Indeed, whenever $\alpha < \beta$, by (9) the values of g_α on the set $\Omega_\alpha \cap \Omega_\beta$ are irrelevant. In the construction of patchy vector fields, the following observation is often useful. Assume that the open sets Ω_α form a locally finite covering of Ω and that, for each $\alpha \in \mathcal{A}$, the vector field g_α satisfies (8) at every point $x \in \partial\Omega_\alpha \setminus \bigcup_{\beta>\alpha} \Omega_\beta$. Then g is still a patchy vector field. To see this, it suffices to construct vector fields $\widetilde{g}_\alpha$ which satisfy the inward pointing property (8) at every point $x \in \partial\Omega_\alpha$ and such that $\widetilde{g}_\alpha = g_\alpha$ on the closed set $\overline{\Omega}_\alpha \setminus \bigcup_{\beta>\alpha} \Omega_\beta$.

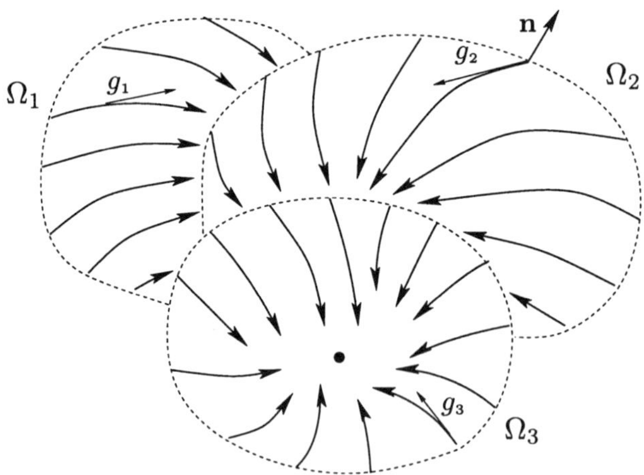

Fig. 1.   A patchy vector field.

Remark 2.2. In some situations it is useful to adopt a more general definition of patchy vector field than the one formulated above. Indeed, one can take in consideration patches $(\Omega_\alpha,\ g_\alpha)$ where the domain Ω_α has a piecewise smooth boundary (cf. Ref. 3). In this case, the inward-pointing condition (8) can be expressed requiring that

$$g(x) \in \overset{\circ}{T}_\Omega(x) \tag{12}$$

where $\overset{\circ}{T}_\Omega(x)$ denotes the interior of the (*Bouligand*) tangent cone to Ω at the point x, defined by (cfr. Ref. 14)

$$T_\Omega(x) \doteq \left\{ v \in \mathbb{R}^n \; : \; \liminf_{t\downarrow 0} \frac{d(x + tv, \; \Omega)}{t} = 0 \right\}. \tag{13}$$

Clearly, at any regular point $x \in \partial\Omega$, the interior of the tangent cone $T_\Omega(x)$ is precisely the set of all vectors $v \in \mathbb{R}^n$ that satisfy $\langle v, \; \mathbf{n}(x) \rangle < 0$ and hence (12) coincides with the inward-pointing condition (8). One can easily see that all the results concerning patchy vector fields established in Refs. 1,5 remain true within this more general formulation.

The main properties of solutions to the O.D.E. (7) when g is a patchy vector field are collected below.

Theorem 2.1 (Trajectories of a patchy vector field). *In connection with a patchy vector field $\left(\Omega, \; g, \; (\Omega_\alpha, \; g_\alpha)_{\alpha \in A}\right)$, the following holds.*

(i) *If $t \to x(t)$ is a Carathéodory solution of (7) on an open interval J, then $t \to \dot{x}(t)$ is piecewise smooth and has a finite set of jumps on any compact subinterval $J' \subset J$. The function $t \mapsto \alpha^*(x(t))$ defined at (10) is piecewise constant, left continuous and non-decreasing. Moreover there holds*

$$\dot{x}(t-) = g\left(x(t)\right) \qquad \forall \, t \in J. \tag{14}$$

(ii) *For each $\bar{x} \in \Omega$, the Cauchy problem for (7) with initial condition $x(0) = \bar{x}$ has at least one local forward Carathéodory solution and at most one backward Carathéodory solution.*

(iii) *The set of Carathéodory solutions of (7) is closed. More precisely, assume that $x_\nu : [a_\nu, b_\nu] \mapsto \Omega$ is a sequence of solutions and, as $\nu \to \infty$, there holds*

$$a_\nu \to a, \qquad b_\nu \to b, \qquad x_\nu(t) \to \hat{x}(t) \;\; \forall t \in \,]a, b[. \tag{15}$$

Then $\hat{x}(\cdot)$ is itself a Carathéodory solution of (2.1).

Sketch of the proof. We outline here the main arguments in the proof. For details see Ref. 1.

1. To prove (i), observe that on any compact interval $[a, b]$ a solution $x(\cdot)$ can intersect only finitely many domains Ω_α, say those with indices $\alpha_1 < \alpha_2 < \cdots < \alpha_m$. It is now convenient to argue by backward induction. Since Ω_{α_m} is positively invariant for the flow of g_{α_m}, the set of times $\{t \in$

$[a, b]$; $x(t) \in \Omega_{\alpha_m}\}$ must be a (possibly empty) interval of the form $]t_m, b]$. Similarly, the set $\{t \in [a, b]$; $x(t) \in \Omega_{\alpha_{m-1}}\}$ is an interval of the form $]t_{m-1}, t_m]$. After m inductive steps we conclude that

$$\dot{x}(t) = g_{\alpha_j}(x(t)) \qquad\qquad t \in]t_j, \, t_{j+1}[$$

for some times t_j with $a = t_1 \leq t_2 \leq \cdots \leq t_{m+1} = b$. All statements in (i) now follow from this fact. In particular, (14) holds because each set Ω_α is open and positively invariant for the flow of the corresponding vector field g_α.

2. To prove the local existence of a forward Carathéodory solution, consider the index

$$\bar{\alpha} \doteq \max\{\alpha \in \mathcal{A} ; \;\; \bar{x} \in \overline{\Omega}_\alpha\}.$$

Because of the transversality condition (8), the solution of the Cauchy problem

$$\dot{x} = g_{\bar{\alpha}}(x), \qquad\qquad x(0) = \bar{x} , \qquad\qquad (16)$$

remains inside $\Omega_{\bar{\alpha}}$ for all $t \geq 0$. Hence it provides also a solution of (7) on some positive interval $[0, \delta]$.

3. To show the backward uniqueness property, let $x_1(\cdot)$, $x_2(\cdot)$ be any two Carathéodory solutions to (2.1) with $x_1(0) = x_2(0) = \bar{x}$. For $i = 1, 2$, call

$$\alpha_i^*(t) \doteq \max\{\alpha \in \mathcal{A} ; \;\; x_i(t) \in \Omega_\alpha\} .$$

By (i), the maps $t \mapsto \alpha_i^*(t)$ are piecewise constant and left continuous. Hence there exists $\delta > 0$ such that

$$\alpha_1^*(t) = \alpha_2^*(t) = \bar{\alpha} \doteq \max\{\alpha \in \mathcal{A} ; \;\; \bar{x} \in \Omega_\alpha\} \qquad \forall t \in] - \delta, \, 0].$$

The uniqueness of backward solutions is now clear, because on $] - \delta, \, 0]$ both x_1 and x_2 are solutions of the same Cauchy problem with smooth coefficients

$$\dot{x} = g_{\bar{\alpha}}(x), \qquad\qquad x(0) = \bar{x} .$$

4. Finally, let $x_\nu : [a_\nu, \, b_\nu] \mapsto \mathbb{R}^n$ be a sequence of solutions satisfying (2.6). To prove that the limit $\hat{x}(\cdot)$ is also a Carathéodory solution, we observe that on any compact subinterval $J \subset]a, b[$ the functions u_ν are uniformly continuous and intersect a finite number of domains Ω_α, say with indices $\alpha_1 < \alpha_2 < \cdots < \alpha_m$. For each ν, the function

$$\alpha_\nu^*(t) \doteq \max\{\alpha \in \mathcal{A} ; \;\; x_\nu(t) \in \Omega_\alpha\}$$

is non-decreasing and left continuous, hence it can be written in the form

$$\alpha_\nu^*(t) = \alpha_j \qquad \text{if } t \in \,]t_j^\nu, \, t_{j+1}^\nu].$$

By taking a subsequence we can assume that, as $\nu \to \infty$, $t_j^\nu \to \hat{t}_j$ for all j. By a standard convergence result for smooth O.D.E's, the function $\hat{x}$ provides a solution to $\dot{x} = g_{\alpha_j}(x)$ on each open subinterval $I_j \doteq \,]\hat{t}_j, \, \hat{t}_{j+1}[$. Since the domains Ω_β are open, there holds

$$\hat{x}(t) \notin \Omega_\beta \qquad\qquad \forall \beta > \alpha_j, \ t \in I_j.$$

On the other hand, since g_{α_j} is inward pointing on the boundary $\partial\Omega_{\alpha_j}$, a limit of trajectories $\dot{x}_\nu = g_{\alpha_j}(x_\nu)$ taking values within Ω_{α_j} must remain in the interior of Ω_{α_j}. Hence $\alpha^*\big(\hat{x}(t)\big) = \alpha_j$ for all $t \in I_j$, completing the proof of (iii). $\qquad\qquad\square$

Remark 2.3. For a patchy vector field g, in general the set of Carathéodory solutions to a given Cauchy problem is not connected, not acyclic. In spite of appearances, this fact is a distinct advantage. Indeed, it avoids the topological obstructions toward the existence of stabilizing feedbacks which are otherwise encountered working with continuous O.D.E's, or differential inclusions such as (6), with upper semi-continuous, convex valued right hand side.

Example 2.1. Consider the covering of $\Omega \doteq \mathbb{R}^2$ consisting of three patches:

$$\Omega_1 \doteq \mathbb{R}^2, \quad \Omega_2 \doteq \big\{(x_1, x_2); \ x_1 < -x_2^2\big\}, \quad \Omega_3 \doteq \big\{(x_1, x_2); \ x_1 > x_2^2\big\}.$$

Moreover, consider the family of inward-pointing vector fields

$$\begin{aligned}
g_1 &: \Omega_1 \to \mathbb{R}^2, \qquad g_1(x_1, x_2) = (0, 1),\\
g_2 &: \Omega_2 \to \mathbb{R}^2, \qquad g_2(x_1, x_2) = (-1, 0),\\
g_3 &: \Omega_3 \to \mathbb{R}^2, \qquad g_3(x_1, x_2) = (1, 0).
\end{aligned}$$

Then the vector field g on $\mathbb{R}^2$ defined as

$$g(x_1, x_2) = \begin{cases} (0, 1) & \text{if} & |x_1| \le x_2^2, \\ (-1, 0) & \text{if} & x_1 < -x_2^2, \\ (1, 0) & \text{if} & x_1 > x_2^2, \end{cases}$$

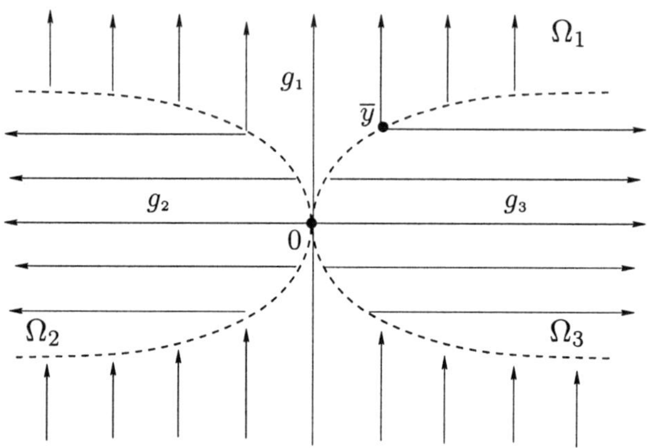

Fig. 2.   Trajectories of a patchy vector field.

is the patchy vector field associated with $(\Omega_\alpha, g_\alpha)_{\alpha=1,2,3}$ (see Fig. 2). Notice that in this case the O.D.E. (7) has exactly three forward Carathéodory solutions starting from the origin a ttime $t = 0$, namely

$$x^{(1)}(t) = (t,0), \qquad x^{(2)}(t) = (-t,0), \qquad x^{(3)}(t) = (0,t) \qquad\qquad t \geq 0.$$

The only backward Carathéodory solution is

$$x^{(1)}(t) = (0,t) \qquad\qquad t \leq 0.$$

Moreover, for every initial point of the form $\overline{y} = (\xi, \sqrt{\xi})$ with $\xi > 0$, there exist two forward Carathéodory solutions, but no solution backward in time.

On the other hand, starting from the origin, the relaxed differential inclusion (6) has infinitely many forward and backward solutions. Indeed, since

$$F(0,0) = \overline{co}\{(0,1),\,(1,0),\,(-1,0)\}\,,$$

for every $\tau < 0 < \tau'$, the function

$$x(t) = \begin{cases} (0,\,t-\tau) & \text{if} \quad t < \tau\,, \\ (0,0) & \text{if} \quad t \in [\tau,\tau']\,, \\ (t-\tau',\,0) & \text{if} \quad t > \tau'\,, \end{cases}$$

provides a solution to (6).

Because of the nice properties described in Theorem 2.1, in connection with several control problems one may seek a feedback $u = U(x)$ such that the resulting map $g(x) = f(x, U(x))$ is a patchy vector field. This leads to the following:

Definition 2.3. Let $(\Omega, g, (\Omega_\alpha, g_\alpha)_{\alpha \in \mathcal{A}})$ be a patchy vector field. Assume that there exists control values $k_\alpha \in K$ such that, for each $\alpha \in \mathcal{A}$

$$g_\alpha(x) \doteq f(x, k_\alpha) \qquad \forall x \in \Omega_\alpha \setminus \bigcup_{\beta > \alpha} \Omega_\beta. \qquad (17)$$

Then the piecewise constant map

$$U(x) \doteq k_\alpha \qquad \text{if} \qquad x \in \Omega_\alpha \setminus \bigcup_{\beta > \alpha} \Omega_\beta \qquad (18)$$

is called a *patchy feedback* control on Ω.

Defining $\alpha^*(x)$ as in (10), the feedback control can be written in the equivalent form

$$U(x) = k_{\alpha^*(x)} \qquad x \in \bigcup_\alpha \Omega_\alpha. \qquad (19)$$

3. Stabilizing feedback controls

In this section we describe the use of patchy feedbacks to solve a problem of asymptotic stabilization. Throughout the following we assume that the control set $K \subset \mathbb{R}^m$ is compact. We first recall a basic definition.[1,13,26]

Definition 3.1. The control system

$$\dot{x} = f(x, u) \qquad u(t) \in K \qquad (20)$$

is said to be *globally asymptotically controllable* to the origin if the following holds.

1 - Attractivity. For each $y \in \mathbb{R}^n$ there exists an open-loop control $t \mapsto u^y(t)$ such that the corresponding solution of

$$\dot{x}(t) = f(x(t),\, u^y(t)), \qquad x(0) = y \qquad (21)$$

either tends to the origin as $t \to \infty$ or reaches the origin in finite time.

2 - Lyapunov stability. For each $\varepsilon > 0$ there exists $\delta > 0$ such that the following holds. For every $y \in \mathbb{R}^n$ with $|y| < \delta$ there is an admissible control u^y as in **1**, steering the system from y to the origin, such that the corresponding trajectory of (21) satisfies $|x(t)| < \varepsilon$ for all $t \geq 0$.

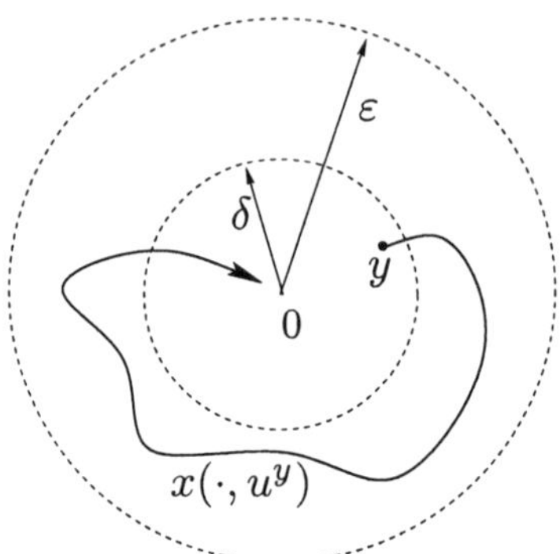

Fig. 3.   Lyapunov stability.

A major result established in Ref. 13 is that if a control system is globally asymptotically controllable then there exists a feedback $u = U(x)$ for which all trajectories converge asymptotically to the origin. Since this stabilizing feedback may well be discontinuous, rather than strong solutions in Ref. 13 the authors consider approximate solutions obtained by a "sample and hold" technique.

As proved in Ref. 1, the asymptotic stabilization can also be achieved by a patchy feedback. In this case, one can work with solutions defined in the classical Carathéodory sense.

Theorem 3.1 (Stabilization by a patchy feedback). *Assume that the control system* (20) *is globally asymptotically controllable to the origin. Then there exists a patchy feedback control* $U : \mathbb{R}^n \setminus \{0\} \mapsto K$ *such that every Carathéodory solution of* (2) *either tends asymptotically to the origin, or reaches the origin in finite time.*

Sketch of the proof. The main part of the proof of Theorem 3.1 consists in establishing the semi-global practical stabilization of system (20), i.e. in showing that, given two closed balls $B' \subset B$ centered at the origin, there exists a patchy feedback that steers every point $\bar{x} \in B$ inside B' within finite time. The basic idea of the construction is rather simple (see Fig. 4-5). Consider an open-loop control $t \mapsto u^y(t)$ steering a point $y \in B$ to the interior of B'. By continuity, we can construct a "flow tube" Γ^y around the trajectory $x(\cdot, u^y)$, steering all points of a neighborhood of y into B'. Finally, we can patch together a finite number of these tubes Γ^{y_i} covering the entire ball B. The basic steps of this construction are sketched below. Further details can be found in Ref. 1.

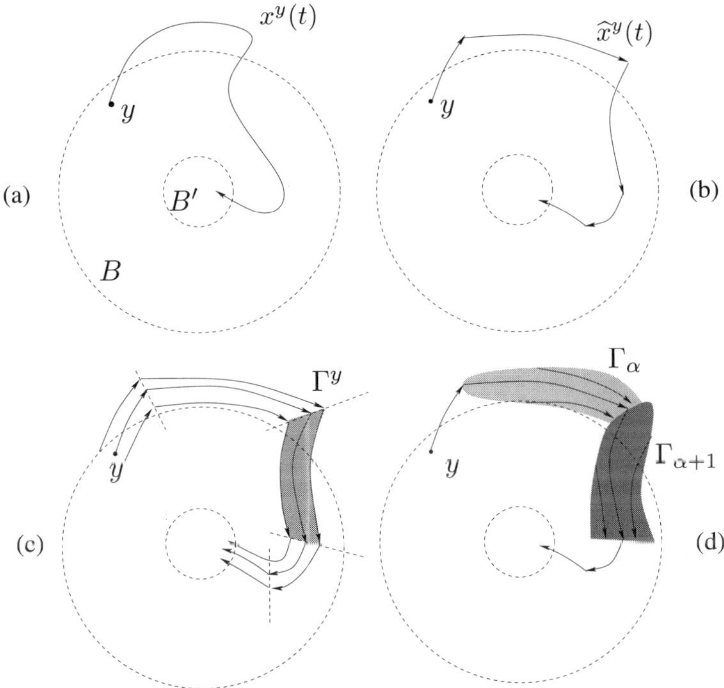

Fig. 4.   Constructing a flow tube around a given trajectory.

1. By assumption, for each point $y \in B$, there exists an open-loop control $t \mapsto u^y(t)$ that steers the system from y to a point y' in the interior of B' in finite time τ^y (Fig. 4-a). By a density and continuity argument, we can replace u^y with a piecewise constant open-loop control $\widehat{u}^y$ (Fig. 4-b), say

$$\widehat{u}^y(t) = k_\alpha \in K \qquad \text{if } t \in \,]t_{\alpha-1},\, t_\alpha], \tag{22}$$

for some finite partition $0 = t_0 < t_1 < \cdots < t_m = \tau^y$. Notice that it is not restrictive to assume that the corresponding trajectory $t \mapsto \widehat{x}^y(t) \doteq x(t, \widehat{u}^y)$ has no self-intersections.

2. We can now define a piecewise constant feedback control $u = U(x)$, taking the constant values $k_{\alpha_1}, \ldots, k_{\alpha_m}$ on a narrow tube Γ around the trajectory

$$\Big\{ \big(t, \widehat{x}^y(t)\big) ;\;\; t \in [0, \widehat{\tau}^y] \Big\}, \tag{23}$$

so that all trajectories starting inside Γ eventually reach the interior of B' (Fig. 4-c).

3. By slightly bending the outer surface of each section of the tube Γ, we can arrange so that the vector fields $g_\alpha(x) \doteq f(x, k_\alpha)$ point strictly inward along the portion of boundary $\partial \Gamma_\alpha \setminus \Gamma_{\alpha+1}$. Recalling Remark 2.1, we thus obtain a patchy vector field (Fig. 4-d) defined on a small neighborhood of the trajectory (23), which steers all points of a neighborhood of y into the interior of B'.

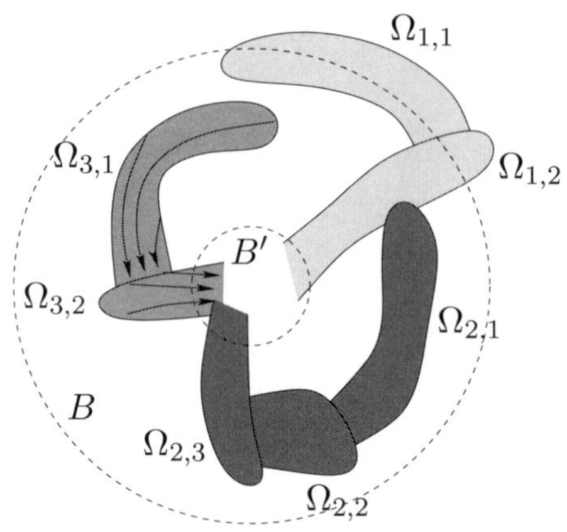

Fig. 5.   Patching together several flow tubes.

4. The above construction can be repeated for every initial point y in the compact set B. We now select finitely many points $y_1, \ldots, y_N \in B$ and patchy vector fields, $\left(\Omega_i,\, g_i,\, (\Omega_{i,\alpha},\, g_{i,\alpha})_{\alpha \in \mathcal{A}_i}\right)$ defined on the corresponding tube-like sets Γ^{y_i}, with the following properties. The union of all domains Γ^{y_i} covers B, and every trajectory of every vector field g_i on Γ^{y_i} eventually reaches the interior of B'.

5. Finally, we define the patchy feedback obtained by the superposition of the $g_{i,\alpha}$, arranged in lexicographic order:

$$g(x) = g_{i,\alpha}(x) \qquad \text{if } x \in \Gamma^{y_i}_\alpha \setminus \bigcup_{(j,\beta) \succ (i,\alpha)} \Gamma^{y_j}_\beta . \qquad (24)$$

This achieves a patchy feedback control (Fig. 5), defined on a neighborhood of $B \setminus B'$, which steers each point of B into the interior of B' in finite time.

6. For every integer ν, call B^ν the closed ball centered at the origin with radius $2^{-\nu}$. By the previous steps, for every ν there exists a patchy feedback control U_ν steering each point in B_ν inside $B_{\nu+1}$, say

$$U_\nu(x) = k_{\nu,\alpha} \qquad \text{if} \qquad x \in \Omega_{\nu,\alpha} \setminus \bigcup_{\beta > \alpha} \Omega_{\nu,\beta}. \tag{25}$$

The property of Lyapunov stability guarantees that the family of all open sets $\left\{ \Omega_{\nu,\alpha} ; \quad \nu \in \mathbb{Z}, \quad \alpha = 1, \ldots, N_\nu \right\}$ forms a locally finite covering of $\mathbb{R}^n \setminus \{0\}$. We now define the patchy feedback control

$$U_\nu(x) = k_{\nu,\alpha} \qquad \text{if} \qquad x \in \Omega_{\nu,\alpha} \setminus \bigcup_{(\mu,\beta) \succ (\nu,\alpha)} \Omega_{\mu,\beta}, \tag{26}$$

where the set of indices (ν, α) is again ordered lexicographically. By construction, the patchy feedback (26) steers each point $x \in B^\nu$ into the interior of the smaller ball $B^{\nu+1}$ within finite time. Hence, every trajectory either tends to the origin as $t \to \infty$ or reaches the origin in finite time. $\qquad \square$

4. Nearly optimal patchy feedbacks

Consider again the nonlinear control system

$$\dot{x} = f(x, u) \qquad u(t) \in K, \tag{27}$$

where f is a smooth map with sublinear growth

$$\left| f(x, u) \right| \leq c \left(1 + |x| \right) \qquad \text{for all } u \in K, \tag{28}$$

and the set of admissible control values $K \subset \mathbb{R}^m$ is bounded. For sake of definiteness, in this section we discuss a specific optimization problem, namely, how to steer the system to the origin in minimum time, where it is not restrictive to assume that K is compact and that the sets of velocities $\{f(x, u) ; u \in K\}$ is convex. Indeed, allowing the set of controls to range in the closure of K does not affect the minimum time function. Moreover, if the sets of velocites are not convex, we can replace the original system (27) by a chattering one (see Ref. 7), so that the resulting time optimization problem yields exactly the same value function.

In general, a feedback control $u = U(x)$ that accomplishes this task can have a very complicated structure. Already for systems in two space dimensions, an accurate description of all generic singularities of a time optimal synthesis involves the classification of eighteen topological equivalence classes of singular points.[8,24] In higher dimensions, an even larger number of different singularities arises, and the optimal synthesis can exhibit pathological behavior such as the the famous "Fuller phenomenon", where every optimal control has an infinite number of switchings. In alternative, one may construct a more regular feedback, which still achieves a close-to-optimal performance. As shown in Refs. 5,10, this task can also be accomplished by patchy feedbacks.

For each initial state $y \in \mathbb{R}^n$, call $T(y)$ the minimum time needed to steer the system from y to the origin. As usual, B_ε denotes the closed ball centered at the origin with radius ε, while $\overset{\circ}{B}_\varepsilon$ denotes its interior. The following result was established in Ref. 5:

Theorem 4.1 (Nearly time optimal patchy feedback). *Under the above assumptions, let $\varepsilon > 0$, $\tau > 0$ be given. Then there exists a suboptimal patchy feedback $U(x) = k_{\alpha^*(x)}$, as in (19), defined on the sub-level set*

$$\mathcal{R}_\tau \doteq \left\{ y \in \mathbb{R}^n ; \ T(y) \le \tau \right\}, \tag{29}$$

such that the following holds.

For every $y \in \mathcal{R}_\tau$, $|y| > \varepsilon$, any solution of

$$\dot{x} = f\big(x, U(x)\big) \qquad\qquad x(0) = y, \tag{30}$$

reaches a point in the ball B_ε within time $T(y) + \varepsilon$.

Remark 4.1. For each $y \in \mathcal{R}_\tau$, one can consider a time-optimal trajectory steering y to the origin. As in the proof of Theorem 3.1, we could then cover the compact set $\mathcal{R}_\tau \setminus B_\varepsilon$ with finitely many tubes $\Gamma_1, \ldots, \Gamma_N$ and, for each $\alpha = 1, \ldots, N$, construct a patchy feedback $U_\alpha : \Gamma_\alpha \mapsto K$ steering every point $y \in \Gamma_\alpha$ into the ball B_ε within time $T(y) + \varepsilon$. Unfortunately, when we patch together all these feedbacks U_α (say, using a lexicographic order), there is no guarantee that the resulting feedback

$$u(x) = U_{\alpha^*(x)} \qquad\qquad \alpha^*(x) \doteq \max\left\{\alpha ; \ x \in \Gamma_\alpha\right\}$$

is nearly-optimal (Fig. 6). Indeed, call $T_\alpha(y)$ the time taken by the control U_α to steer the point $y \in \Omega_\alpha$ inside B_ε. Let $t \mapsto x(t)$ be a trajectory of the composite feedback

$$\dot{x} = f\big(x, U_{\alpha^*(x)}(x)\big), \qquad\qquad x(0) = y,$$

that reaches the ball B_ε within time τ. Assume $\alpha^*(t) = \alpha$ for $t \in]t_{\alpha-1}, t_\alpha]$. The near-optimality of each feedback U_α implies $T(x) \leq T_\alpha(x) \leq T(x) + \varepsilon$ for every $x \in \Gamma_\alpha$. Moreover

$$T_\alpha(x(t_{\alpha-1})) - T_\alpha(x(t_\alpha)) = (t_\alpha - t_{\alpha-1}).$$

Unfortunately, from the above inequalities one can only deduce

$$T(x(t_{\alpha-1})) - T(x(t_\alpha)) \geq (t_\alpha - t_{\alpha-1}) - \varepsilon.$$

and hence $\tau \leq T(y) + N\varepsilon$. This is a useless information, because we have no control on the number N of tubes. Indeed, we expect $N \to \infty$ as $\varepsilon \to 0$.

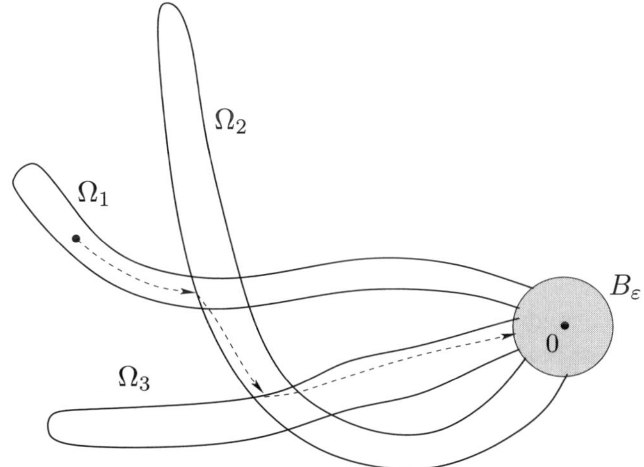

Fig. 6.   Patching together optimal trajectories does not yield a near-optimal feedback.

Sketch of the proof. As a starting point, instead of open-loop optimal controls, it is convenient to use here the value function $x \mapsto T(y)$. We describe here the main steps in the proof of Theorem 4.1. For all details we refer to Ref. 5.

1. We begin by relaxing the optimization problem. For $\varepsilon > 0$, define the function

$$\varphi_\varepsilon(x) \doteq \begin{cases} \dfrac{|x|^2}{\varepsilon^2 - |x|^2} & \text{if} \quad |x| < \varepsilon, \\ \infty & \text{if} \quad |x| \geq \varepsilon, \end{cases} \tag{31}$$

and consider the following ε-approximate minimization problem

$$\inf_{t,\,u(\cdot)} \left\{ t + \varphi_\varepsilon\big(x(t)\big) \right\}, \tag{32}$$

subject to

$$x(0) = y\,, \qquad \dot{x} = f(x,u) \qquad u(t) \in K\,. \tag{33}$$

For every initial point $y \in \mathbb{R}^n$, we denote this infimum by $V(y)$, and refer to $y \mapsto V(y) \in [0, \infty]$ as the *value function* for (32). Observe that

$$V(y) \le T(y) \qquad \forall\, y \in \mathbb{R}^n. \tag{34}$$

Thus, for a fixed time $\tau > 0$, the set of points that can be steered to the origin within time τ is contained in the sub-level set

$$\Lambda_\tau \doteq \left\{ y \in \mathbb{R}^n \,;\ V(y) \le \tau \right\}.$$

Hence, to establish Theorem 4.1 it will be sufficient to construct a sub-optimal patchy feedback on the set

$$\Lambda_{\tau,\varepsilon} \doteq \left\{ y \in \mathbb{R}^n \,;\ V(y) \le \tau \,;\ |y| \ge \varepsilon \right\}. \tag{35}$$

The assumptions on the dynamics (27) imply that the sub-level set Λ_τ (and hence $\Lambda_{\tau,\varepsilon}$) is compact. Moreover, by the regularity of the functions f and φ_ε, it follows that the restriction of the value function V to Λ_τ is Lipschitz continuous and locally semi-concave (see Ref. 5, Lemma 1).

2. It is convenient to further approximate the value function $V(\cdot)$ with a semi-concave, piecewise quadratic function of the form

$$\widetilde{T}(x) \;=\; \min_{j=1,\ldots,N} \left\{ a_j + \mathbf{b}_j \cdot x \right\} + M \cdot |x|^2, \tag{36}$$

with the following properties:

(i) For every $x \in \Lambda_{\tau,\varepsilon}$, there holds

$$T(x) \le \widetilde{T}(x) \le V(x) + \varepsilon_0\,, \tag{37}$$

for some positive constant $\varepsilon_0 < \varepsilon$.

(ii) For every $x \in \Lambda_{\tau,\varepsilon}$ where $\widetilde{T}$ is differentiable, one has

$$\min_{\omega \in K}\ \nabla \widetilde{T}(x) \cdot f(x,\omega) \;\le\; -1 + \frac{\varepsilon_0}{3}\,. \tag{38}$$

By the special form of the function $\widetilde{T}$, it follows that

- The set of points where $\widetilde{T}$ is not differentiable is contained in a finite union of hyperplanes, namely

$$H_{ij} \doteq \left\{ a_i + \mathbf{b}_i \cdot x = a_j + \mathbf{b}_j \cdot x \right\}.$$

- Each level set $\Sigma_h \doteq \{x \; ; \; \widetilde{T}(x) = h\}$ is contained in the union of finitely many spherical surfaces, say $S_{i_1}, \dots S_{i_h}$.

Moreover, observe that by (37) one has

$$\Lambda_{\tau_0,\varepsilon} \subset \left\{ x \; ; \; \widetilde{T}(x) \leq \tau_0; \quad |x| \geq \varepsilon \right\}, \qquad \tau_0 \doteq \tau + \varepsilon_0 . \tag{39}$$

The construction of the patchy feedback is now performed by induction, progressively working on larger sub-level sets of $\widetilde{T}(\cdot)$. Namely, assume to have already constructed a patchy feedback $U = U(x)$ on the sub-level set

$$\widetilde{\Lambda}_h \doteq \left\{ x \; ; \; \widetilde{T}(x) \leq h \right\}, \qquad h < \tau_0 , \tag{40}$$

so that

$$\nabla \widetilde{T}(x) \cdot f(x, U(x)) \; \leq -1 + \varepsilon_0 \qquad \forall x \in \widetilde{\Lambda}_h \setminus B_\varepsilon . \tag{41}$$

We wish to define a patchy feedback with the same property around the level set Σ_h.

3. If the surface $\{x \; ; \; \widetilde{T}(x) = h\}$ is smooth, i.e. if

$$\widetilde{T}(x) = a_j + \mathbf{b}_j \cdot x + M \cdot |x|^2 \qquad \forall \, x \in \Sigma_h ,$$

for some j, the level set Σ_h is exactly a sphere S_j. Then, because of (38),

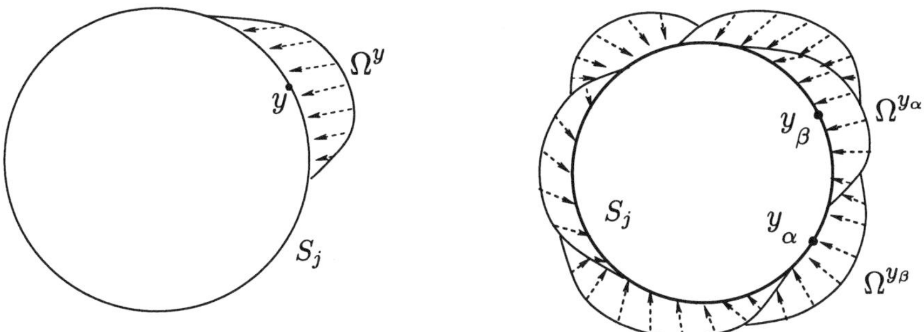

Fig. 7.   Construction of the sub-optimal patchy feedback around a sphere.

for each point $y \in S_j$ we can find a lens-shaped patch Ω^y containing y in its interior, and a constant control $\omega^y \in K$ such that

$$\nabla \widetilde{T}(y) \cdot f(y, \omega^y) \leq -1 + \varepsilon_0 . \tag{42}$$

By continuity, possibly choosing a smaller domain Ω^y, we deduce that (see Fig. 7, left)

$$\nabla \widetilde{T}(x) \cdot f(x, \omega^y) \leq -1 + \frac{2\varepsilon_0}{3} \qquad \forall x \in \Omega^y . \tag{43}$$

Moreover, we construct the outer surface of the patch Ω^y so that the vector field $x \mapsto f(x, \omega^y)$ points strictly inward on the upper boundary $\partial \Omega^y \setminus S_j$.

By compactness of the sphere S_j, we can cover it with finitely many patches $\Omega^{y_1}, \ldots, \Omega^{y_N}$ constructed as above. On the union of these patches we define the feedback

$$U(x) = \omega^{y_\alpha} \qquad \text{if } x \in \Omega^{y_\alpha} \setminus \bigcup_{\beta > \alpha} \Omega^{y_\beta} , \quad \widetilde{T}(x) > h ,$$

while we assign an higher index order to the patchy already constructed on the set $\widetilde{\Lambda}_h$. Recalling Remark 2.1, this provides a patchy feedback defined on a larger sub-level set $\widetilde{\Lambda}_{h+\delta}$, $\delta > 0$, enjoying the property (41).

4. If the boundary Σ_h is not smooth, the above construction requires more care. Indeed, consider a point y where $\widetilde{T}$ is not differentiable. For sake of illustration, assume that y lies on the intersection of exactly two spheres, say

$$\widetilde{T}(y) \;=\; a_1 + \mathbf{b}_1 \cdot y + M \cdot |y|^2 \;=\; a_2 + \mathbf{b}_2 \cdot y + M \cdot |y|^2,$$

while

$$a_j + \mathbf{b}_j \cdot y + M \cdot |y|^2 > \widetilde{T}(y) \qquad j \geq 3 .$$

Consider the half-spaces

$$\Gamma_1 \doteq \left\{ x ; \; a_1 + \mathbf{b}_1 \cdot x < a_2 + \mathbf{b}_2 \cdot x \right\} ,$$

$$\Gamma_2 \doteq \left\{ x ; \; a_2 + \mathbf{b}_2 \cdot x < a_1 + \mathbf{b}_1 \cdot x \right\} ,$$

and the dividing hyperplane

$$H \doteq \left\{ x ; \; (a_1 - a_2) + (\mathbf{b}_1 - \mathbf{b}_2) \cdot x = 0 \right\} .$$

For $i = 1, 2$ we can again choose a lens-shaped patch Ω_i and a constant control $\omega_i \in K$ such that, as in (43), there holds

$$(\mathbf{b}_i + 2M x) \cdot f(x, \omega_i) \leq -1 + \frac{2\varepsilon_0}{3} \qquad \forall x \in \Omega_i . \tag{44}$$

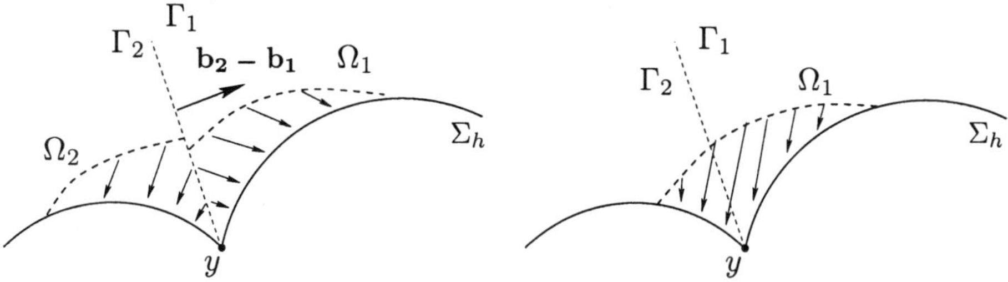

Fig. 8.   Construction of the feedback near a corner point. Two different cases.

Notice that by construction one has $(\mathbf{b}_i + 2M\,x) = \nabla\widetilde{T}(x)$ in $\Omega_i \cap \Gamma_i$. In order to combine these two patches and construct a feedback in a whole neighborhood of the point y, we need to consider two different cases.

CASE 1: The vector $f(y,\omega_1)$ points toward Γ_1, while $f(y,\omega_2)$ points toward Γ_2. More precisely,

$$f(y,\omega_1) \cdot (\mathbf{b}_2 - \mathbf{b}_1) > 0\,, \qquad\qquad f(y,\omega_2) \cdot (\mathbf{b}_2 - \mathbf{b}_1) < 0\,.$$

In this case, we can choose the hyperplane H as boundary between the two patches. Setting

$$U(x) = \begin{cases} \omega_1 & if \quad x \in \Omega_1 \cap \Gamma_1\,, \\[2mm] \omega_2 & if \quad x \in \Omega_2 \cap \overline{\Gamma}_2\,, \end{cases}$$

because of (44), and by Remark 2.1, we thus obtain a patchy feedback, defined on a small neighborhood of the point y (see Fig. 8, left), that satisfies the estimate (41).

CASE 2: One of the two vectors points toward the opposite half-space. To fix the ideas, assume

$$f(y,\omega_1) \cdot (\mathbf{b}_2 - \mathbf{b}_1) \leq 0\,.$$

By possibly choosing a smaller lens-shaped neighborhood Ω_1 so that,

$$f(x,\omega_1) \cdot (\mathbf{b}_2 - \mathbf{b}_1) \leq \frac{\varepsilon_0}{3} \qquad \forall x \in \Omega_1\,,$$

48

and relying on (44), one achieves

$$\nabla \widetilde{T}(x) \cdot f(x,\omega_1) = (\mathbf{b}_1 + 2M\,x) \cdot f(x,\omega_1)$$

$$\leq -1 + \frac{2\varepsilon_0}{3} \qquad \forall x \in \Omega_1 \cap \Gamma_1 \,,$$

$$\nabla \widetilde{T}(x) \cdot f(x,\omega_1) = (\mathbf{b}_1 + 2M\,x) \cdot f(x,\omega_1) + (\mathbf{b}_2 - \mathbf{b}_1) \cdot f(x,\omega_1)$$

$$\leq -1 + \varepsilon_0 \qquad \forall x \in \Omega_1 \cap \overline{\Gamma}_2 \,.$$

$$(45)$$

In other words, the same control ω_1 used on $\Omega_1 \cap \Gamma_1$ now works well also in the other region Γ_2. In this case, we can cover a neighborhood of the point y with the single patch Ω_1 (see Fig. 8, right) in which the estimate (41) is verified.

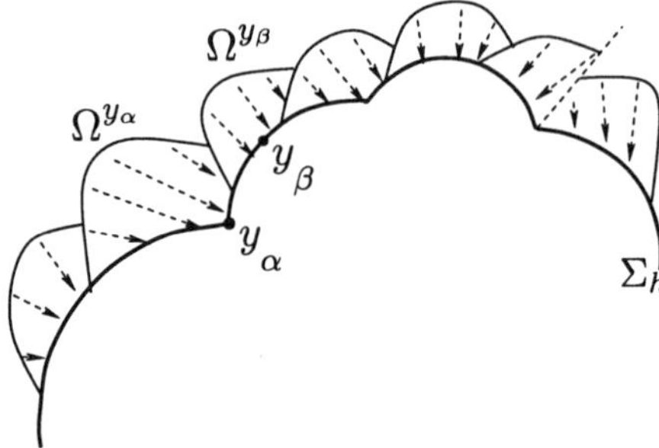

Fig. 9.   Construction of the sub-optimal patchy feedback around a surface with corners.

The remainder of the construction is performed as before. We can cover a neighborhhood of each point $y \in \Sigma_h$ by one or more patches. By compactness, the entire level set S_h can be covered with finitely many patches (Fig. 9). This yields a patchy feedback defined on a larger sub-level set $\widetilde{\Lambda}_{h+\delta}$ satisfying the decrease property (41).

5. A key step in the analysis is to show that, at each inductive step, we can construct a patchy feedback, enjoying the property (41), on a sub-level set $\widetilde{\Lambda}_{h+\delta}$ with some $\delta > 0$ uniformly positive. This ensures that the inductive procedure terminates after a finite number of steps. One can accomplish this task by providing an accurate estimate of the size of lens-shaped domains constructed around a finite collection of spheres (associated to a family of inward-pointing vector fields) and cutted along the hyperplanes passing through their intersections (cf. Ref. 5, Lemmas 3-4).

If now $t \mapsto x(t)$ is any trajectory of the closed-loop system determined by our feedback control passing through points of $\widetilde{\Lambda}_{\tau_0}$, thanks to (41) we deduce that

$$\frac{d}{dt}\, \widetilde{T}\big(x(t)\big) \leq -1 + \varepsilon_0\,, \tag{46}$$

as long as $|x(t)| > \varepsilon$. Therefore, if $x(\cdot)$ starts at a point y and reaches the ball B_ε, say at a point x_ε, in a time t_ε, one has

$$\widetilde{T}(y) - \widetilde{T}(x_\varepsilon) \geq (1 - \varepsilon_0)\, t_\varepsilon\,. \tag{47}$$

Hence, (47) together with (37), (39), implies that every initial point $y \in \Lambda_{\tau_0,\varepsilon}$ is steered into the ball B_ε within time

$$\frac{\widetilde{T}(y)}{1 - \varepsilon_0} \leq \frac{V(y) + \varepsilon_0}{1 - \varepsilon_0}\,. \tag{48}$$

Since one can choose $\varepsilon_0 < \varepsilon$ so that the right hand-side of (48) is less than $V(y) + \varepsilon$ for all $y \in \widetilde{\Lambda}_{\tau_0}$, and because of (34), this proves the theorem. $\qquad\square$

5. Robustness

For practical applications, one has to take into account the presence of several perturbations, which may degrade the performance of the feedback control. For example:

(i) The model equation, described by the function f in (1), may not be precisely known.

(ii) The evolution of the system may be affected by (possibly random) external perturbations.

(iii) While implementing the feedback, the state of the system may not be accurately measured.

As a result, instead of the planned dynamics

$$\dot{x} = g(x) \doteq f\big(x, U(x)\big)\,, \tag{49}$$

the system will actually evolve according to

$$\dot{x} = f\big(x, U(x + \varepsilon_1(t))\big) + \varepsilon_2(t)\,, \tag{50}$$

for some small perturbations $\varepsilon_1, \varepsilon_2$. Here the "inner perturbation" ε_1 accounts for measurement errors, while the "outer perturbation" ε_2 models external disturbances.

In the above framework, it is important to design a feedback control which still accomplishes the desired task in the presence of (sufficiently small) perturbations. Since we are dealing with a discontinuous O.D.E., one cannot expect the full robustness of the feedback $U(x)$ with respect to errors in the measurement of the state vectors because of possible chattering behaviour that may arise at discontinuity points.[16,21,25] For trajectories constructed by the "sample and hold" technique in Ref. 13, a "relative robustness" with respect to measurement errors was shown in Refs. 12,27, meaning that the magnitude of the error must not exceed the maximum step size of the underlying discretization of the sampling solution taken in consideration (in this case, one is not concerned with the robustness property of the limit solution as the step size vanishes, but only of the approximate sampling solution with a fixed step size).

Instead, for patchy feedbacks, we have established in Ref. 2 a robustness property of the Carathéodory trajectories with respect to measurement errors which have sufficiently small total variation (to avoid possible chattering phenomena around discontinuities), while the external disturbances were assumed to be small in $\mathbf{L}^1$ norm. We describe here the main results in this direction.[2-4] Together with the O.D.E. determined by a patchy feedback

$$\dot{x} \;=\; g(x) \;=\; f(x, k_{\alpha^*(x)}), \tag{51}$$

we thus consider the perturbed equation

$$\dot{y} = g\big(y + e_1(t)\big) + e_2(t). \tag{52}$$

We recall that the total variation norm of a BV function $\phi : [0, T] \mapsto \mathbb{R}^n$ is defined as

$$\|\phi\|_{BV} \doteq \mathrm{Tot.Var.}\{\phi\} + \|\phi\|_{\mathbf{L}^1}.$$

The Cauchy problem for (52), with initial point $\overline{y}$ in the interior of the domain of g, admits a forward Carathéodory solution $y(t) = \overline{y} + \int_0^t \big[g\big(y(s) + e_1(s)\big) + e_2(s)\big]ds$, $t \geq 0$, whenever $\|e_1\|_{BV}, \|e_2\|_{\mathbf{L}^1}$ are sufficiently small.[2] Notice that, since the Cauchy problem for (51) does not have forward uniqueness and continuous dependence, one clearly cannot expect that a single solution of (51) be stable under small perturbations. What we extablished in Ref. 2 is a different stability property, involving not a single trajectory but the whole solution set: if the perturbations e_1, e_2 are small in the BV and $\mathbf{L}^1$ norm, respectively, then every solution of (52) is close to some solution of (51). This is essentially an upper semicontinuity property of the solution set. Namely, the main robustness result for the flow of a patchy vector field can be stated as follows.[2–4]

Theorem 5.1 (Flow stability of a patchy vector field). *In connection with a patchy vector field* $g : \Omega \mapsto \mathbb{R}^n$, *given any compact subset* $C_0 \subset \Omega$, *and any* $T, \varepsilon > 0$, *there exists* $\delta > 0$ *such that the following holds. If* $y : [0, T] \mapsto \Omega$ *is a solution of the perturbed system* (52), *with* $y(0) \in C_0$, *and*

$$\|e_1\|_{BV} \leq \delta, \qquad \|e_2\|_{\mathbf{L}^1} \leq \delta, \qquad (53)$$

then there exists a solution $x : [0, T] \mapsto \Omega$ *of the unperturbed equation* (51) *with*

$$|x(t) - y(t)| < \varepsilon \qquad \forall\, t \in [0, T]. \qquad (54)$$

Proof. An outline of the main arguments of the proof can be found in Ref. 4. For details we refer to Ref. 2. The meaning of the theorem is illustrated in Fig. 10. $\qquad\square$

Remark 5.1. As shown in Fig. 10, the solution $z(\cdot)$ to the initial value problem

$$\dot{z} = g(z), \qquad z(0) = \bar{z} = y(0), \qquad (55)$$

may be unique, but very different from a solution $y(\cdot)$ of the perturbed system (52) with the same initial data. In order to find a trajectory $x(\cdot)$ of the original system (51) which remains always close to $y(\cdot)$, one may need to start from a different initial point.

Remark 5.2. If we consider a solution $y(\cdot)$ of the perturbed system (52) on a time interval $[0, T_y[$, with $T_y \leq T$, $y(0) \in C_0$, and e_1, e_2 satisfying (53), by the same arguments of Theorem 5.1 one can find a solution $x(\cdot)$ of the unperturbed equation (51) on a interval $[0, T_x[$ so that (54) holds on $[0, \min\{T_x, T_y\}[$. If we assume that the boundary of the domain Ω is (piecewise) smooth, we can choose $x(\cdot)$ so that either $T_x = T_y$, or $|T_x - T_y| < \varepsilon$ and $\lim_{t \to T_x-} x(t) \in \partial\Omega$.

Relying on Theorem 5.1, one can show that a stabilizing patchy feedback as the one provided by Theorem 3.1 still performs well in the presence of small (internal and esternal) disturbances. Namely, we obtain the following robustness property formulated within the context of a semi-global practical stabilization problem.[2,4]

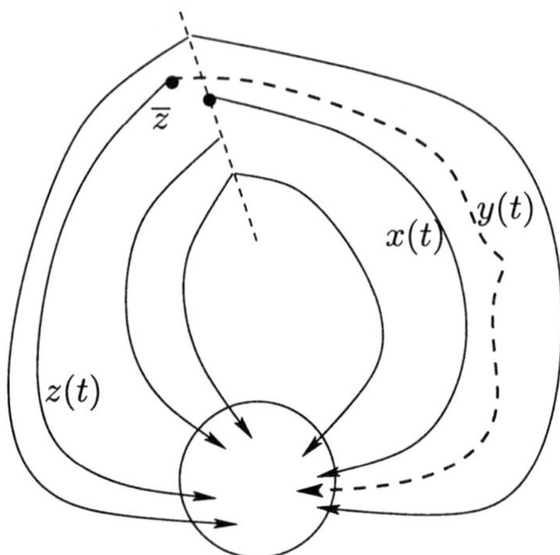

Fig. 10.   The flow stability property stated in Theorem 5.1.

Theorem 5.2 (Robustness of a stabilizing patchy feedback).   *Let*
$u = U(x)$ *be a patchy feedback for the control system* (1), *defined on an
open set* $\Omega \subseteq \mathbb{R}^n$ *containing the anular region* $\{x \in \mathbb{R}^n;\; r \le |x| \le r'\}$, *and
assume that every solution of* (49) *starting inside a compact set* $C_0 \subset \Omega$
reaches the closed ball B_r *within time* T. *Then, given* $\varepsilon > 0$, *there exists*
$\delta > 0$ *such that, if the perturbations* $\varepsilon_1, \varepsilon_2 : [0, T] \mapsto \mathbb{R}^n$ *satisfy*

$$\left\|\varepsilon_1\right\|_{BV} \le \delta, \qquad\qquad \left\|\varepsilon_2\right\|_{\mathbf{L}^1} \le \delta \tag{56}$$

then any solution $x^\varepsilon(\cdot)$ *of* (50) *with initial data* $x^\varepsilon(0) \in C_0$ *enters the ball*
$B_{r+\varepsilon}$ *within time* T.

Proof. To analyze the performance of a patchy feedback $U(x)$ in the
presence of (internal and external) disturbances, for every given solution
$t \mapsto x^\varepsilon(t)$, $t \in [0, T]$, of the perturbed system (50), writing

$$f\big(x,\, U(x + \varepsilon_1(t))\big) + \varepsilon_2(t) = g\big(x + e_1(t)\big) + e_2(t), \tag{57}$$

where

$$e_1(t) \doteq \varepsilon_1(t)\,,$$

$$e_2(t) \doteq \varepsilon_2(t) + \Big[f\big(x^\varepsilon(t),\, U(x^\varepsilon(t) + \varepsilon_1(t))\big) \tag{58}$$

$$- f\big(x^\varepsilon(t) + \varepsilon_1(t),\, U(x^\varepsilon(t) + \varepsilon_1(t))\big)\Big],$$

we find that $x^\varepsilon(\cdot)$ results to be a solution of the perturbed patchy vector field (52), with e_1, e_2 as in (58). Observe that, by the regularity of f

and since all trajectories of (50) starting from C_0 and defined on $[0, T]$ are uniformly bounded, one can choose δ sufficiently small so that, if $\varepsilon_1, \varepsilon_2$ satisfy (56) then e_1, e_2 satisfy the bounds (53) stated in Theorem 5.1. Hence, the conclusion of Theorem 5.1 guarantees the existence of a solution $t \mapsto x(t)$, $t \in [0, T]$, of the unperturbed system (49) such that

$$\left| x^\varepsilon(t) - x(t) \right| \leq \varepsilon \qquad \forall\, t \in [0, T]. \tag{59}$$

Since, by assumption, $x(\tau) \in B_r$ for some $\tau \in [0, T]$, we conclude that $x^\varepsilon(\tau) \in B_{r+\varepsilon}$, as claimed. $\qquad\square$

With the same arguments used in the proof of Theorem 5.2, one can apply Theorem 5.1 to derive the robustness of a sub-optimal patchy feedback. In fact, consider the patchy feedback $U(x)$ constructed in Ref. 10 for the control system (27), whose trajectories are all nearly optimal solutions for the general optimization problem

$$\min_{T, u(\cdot)} \left\{ \psi(x(T)) + \int_0^T L(x(s), u(s))\, ds \right\}. \tag{60}$$

The following result shows that all the trajectories of the perturbed system (50) remain nearly optimal with the same order of accuracy as the ones of the unperturbed system (49), provided that the disturbances are sufficiently small. As in Section 4 we assume here that f is smooth and satisfy the sublinear growth condition (28), and that the set of admissible control values K is compact, while the cost functional ψ, L are smooth, and the running cost is strictly positive $L(x, u) \geq \bar{\ell} > 0$. For every initial point $y \in \mathbb{R}^n$, we let $V(y)$ denote the value function for the optimization problem (60), where the minimization is taken over all $T \geq 0$ and over all solutions $x(\cdot)$ of (27) starting from y and corresponding to a measurable control u taking values in K. A solution $t \mapsto x(t)$ defined on a maximal interval $[0, \tau[$, is extended by continuity setting $x(\tau) \doteq \lim_{t \to \tau-} x(t)\ (\in \partial\Omega)$.

Theorem 5.3 (Robustness of a sub-optimal patchy feedback).
Under the above assumptions, let $\varepsilon > 0$ and a compact set C_0 be given. Then, there exist a patchy feedback $U(x)$ defined on the domain

$$\Omega \doteq \left\{ x \in \mathbb{R}^n;\ V(x) < \psi(x) - \varepsilon \right\}, \tag{61}$$

and a constant $\delta > 0$ such that, if the perturbations $\varepsilon_1,\, \varepsilon_2 : [0, \infty[\mapsto \mathbb{R}^n$ satisfy (56), then any solution $x^\varepsilon(\cdot)$ of (50), with initial data $x^\varepsilon(0) \in C_0$, *enters the terminal set* $S \doteq \mathbb{R}^n \setminus \Omega$ *within a finite time* τ_ε, *for which there holds*

$$\psi\big(x^\varepsilon(\tau_\varepsilon)\big) + \int_t^{\tau_\varepsilon} L\big(x^\varepsilon(s), U(x^\varepsilon(s))\big)\, ds \leq V\big(x^\varepsilon(t)\big) + \varepsilon \qquad \forall\, t \in [0, \tau_\varepsilon]. \tag{62}$$

Proof. We start by observing as in Ref. 10 that we can replace $f(x, u)$ with $h(x, u) \doteq L^{-1}(x, u) f(x, u)$, so that the optimization problem (60) takes the equivalent form

$$\min_{\tau, u(\cdot)} \big\{ \tau + \psi(x(\tau)) \big\}, \tag{63}$$

with dynamics

$$\dot{x} = h(x, u) \qquad\qquad x(0) = y, \tag{64}$$

where h satisfies a sublinear growth condition as f because of the uniform lower bound on L. Hence, without loss of generality we shall assume throughout the following that the running cost is $L(x, u) \equiv 1$.

1. Let still V denote the value function for the minimization problem (63), with dynamics (27). Observe that V is locally Lipschitz continuous (cfr. Ref. 10), and by possibly enlarging the set C_0 assume that

$$C_0 = \{x \in \mathbb{R}^n;\ V(x) \leq \kappa\} \subset B_r, \tag{65}$$

for some $\kappa, r > 0$. Moreover, we can assume that

$$V(x) \geq \kappa + 2\varepsilon_0 \qquad \forall\, x \in \mathbb{R}^n \setminus B_r, \tag{66}$$

for some positive constant $\varepsilon_0 < \varepsilon$. By the same arguments in the proof of Theorem 4.1 (cfr. Ref. 10) one then constructs a Lipschitz continuous, piecewise smooth approximation $\widetilde{V}$ of V, and a patchy feedback $U(x)$ on the domain

$$\widetilde{\Omega} \doteq \{x \in B_r;\ \widetilde{V}(x) < \psi(x)\}, \tag{67}$$

with the following properties:

(i) For every $x \in \widetilde{\Omega}$ there holds

$$V(x) \leq \widetilde{V}(x) \leq V(x) + \varepsilon_0. \tag{68}$$

(ii) For every $x \in \widetilde{\Omega}$ one has

$$\zeta \cdot f(x, U(x)) \leq -1 + \varepsilon_0 \qquad \forall\, \zeta \in \partial \widetilde{V}(x), \tag{69}$$

where $\partial \widetilde{V}(x)$ denotes the generalized gradient of $\widetilde{V}$ at x (see Ref. 14).

Notice that, differentiating $\widetilde{V}$ along any solution $t \mapsto x(t)$, $t \geq 0$, of (49) as

$$\frac{d}{dt}\widetilde{V}(x(t)) = \zeta(t) \cdot f\big(x(t), U(x(t))\big) \qquad \zeta(t) \in \partial \widetilde{V}(x(t)), \tag{70}$$

because of (ii) we find

$$\widetilde{V}(x(t'')) - \widetilde{V}(x(t')) = \int_{t'}^{t''} \zeta(s) \cdot f\big(x(s), U(x(s))\big)\ ds$$
$$\leq (t'' - t')(-1 + \varepsilon_0) \qquad \forall\, 0 \leq t' < t''. \tag{71}$$

Hence, relying on (65), (68), (71), we deduce that every trajectory $x(\cdot)$ of the unperturbed system (49) starting from a point $x(0) \in C_0 \cap \widetilde{\Omega}$ reaches in finite time the set

$$\widetilde{S} \doteq \{x \in \mathbb{R}^n;\ \widetilde{V}(x) \geq \psi(x)\}, \tag{72}$$

since otherwise it would cross the boundary of B_r at some time $\tau > 0$ and we would have

$$\widetilde{V}(x(\tau)) < \widetilde{V}(x(0)) < V(x(0)) + \varepsilon_0 \leq \kappa + \varepsilon_0$$

which is in contrast with (66). From (71) and (65), (68) it follows that the minimum time needed by any trajectory of (49), starting from C_0, to reach the set $\widetilde{S}$ is certainly less than $\overline{T} \doteq \frac{\kappa + \varepsilon_0}{1 - \varepsilon_0}$.

2. Consider now a solution $t \mapsto x^\varepsilon(t)$, $t \geq 0$, of the perturbed system (50), with $x^\varepsilon(0) \in C_0$, which by (57) can also be seen as a solution of the perturbed patchy vector field (52), with e_1, e_2 defined in (58). Choose the bound δ in (56) on the perturbations $\varepsilon_1, \varepsilon_2$ sufficiently small so that the quantities e_1, e_2 are, in turn, small enough to guarantee, by Theorem 5.1, that $x^\varepsilon(\cdot)$ lies in an ε_0-neighborhood of a trajectory of the unperturbed patchy vector field g in (51). Let $[0, \tau_\varepsilon[$ be the maximal interval of definition of x^ε, so that one has $x^\varepsilon(\tau_\varepsilon) \in B(\partial \widetilde{S}, \varepsilon_0)$, which implies

$$\widetilde{V}(x^\varepsilon(\tau_\varepsilon)) \geq \psi(x^\varepsilon(\tau_\varepsilon)) - \mathrm{Lip}(\widetilde{V}) \cdot \varepsilon_0, \tag{73}$$

letting $\mathrm{Lip}(\widetilde{V})$ denote a Lipschitz constant for $\widetilde{V}$ over the ball B_r. Then, recaling Remark 5.2, we deduce that there is a solution $x : [0, \tau[\mapsto \widetilde{\Omega}$ of the unperturbed system (49), such that

$$x(\tau) \in \partial \widetilde{S}, \qquad |\tau - \tau_\varepsilon| < \varepsilon_0,$$
$$|x^\varepsilon(t) - x(t)| \leq \varepsilon_0 \qquad \forall\, t \in \big[0, \min\{\tau, \tau_\varepsilon\}\big]. \tag{74}$$

In particular, (74) yields

$$|x^\varepsilon(\tau_\varepsilon) - x(\tau)| \leq (M + 1) \cdot \varepsilon_0, \tag{75}$$

56

with $M \doteq \|g\|_{\mathbf{L}^\infty(B_r)}$. To fix the ideas, assume that $\tau_\varepsilon \leq \tau$. Then, for any given time $t \in [0, \tau_\varepsilon]$, relying on (68), (71), (73)-(75), we find

$$\psi\big(x^\varepsilon(\tau_\varepsilon)\big) + (\tau_\varepsilon - t) \leq \widetilde{V}\big(x^\varepsilon(\tau_\varepsilon)\big) + \mathrm{Lip}(\widetilde{V}) \cdot \varepsilon_0 + (\tau_\varepsilon - t)$$

$$\leq \widetilde{V}\big(x(\tau)\big) + \big(1 + \mathrm{Lip}(\widetilde{V})(M+1)\big) \cdot \varepsilon_0 + (\tau - t)$$

$$\leq \widetilde{V}\big(x(t)\big) + \big((\tau - t) + 1 + \mathrm{Lip}(\widetilde{V})(M+1)\big) \cdot \varepsilon_0$$

$$\leq V\big(x^\varepsilon(t)\big) + 2\big(\overline{T} + 1 + \mathrm{Lip}(\widetilde{V})(M+1)\big) \cdot \varepsilon_0 \,,$$

$$\tag{76}$$

which establishes (62) (when $L \equiv 1$), provided that ε_0 be sufficiently small so that the second term on the right-hand side of the last inequality in (76) is $< \varepsilon$. This completes the proof of the theorem observing that, by (65), (68), comparing the definitions (61), (67), (72), we derive $C_0 \setminus \widetilde{\Omega} \subset \widetilde{S} \subset S$, $\Omega \subset \widetilde{\Omega}$. $\qquad\square$

6. Stochastic perturbations

Aim of this section is to analyze the performance of a patchy feedback in the presence of random perturbations. The basic setting is as follows. We consider the problem discussed in Section 4 of reaching the origin in minimum time, for the control system (27). We assume that the map $f : \mathbb{R}^n \times K \mapsto \mathbb{R}^n$ is smooth and satisfies the growth conditions (28), while $K \subset \mathbb{R}^m$ is compact.

We denote by $T(y)$ the minimum time needed to steer the system from y to the origin, and consider the sub-level set

$$\mathcal{R}_{T^*} \doteq \big\{ y \in \mathbb{R}^n ; \ T(y) \leq T^* \big\}, \tag{77}$$

which can be assumed to be compact as observed in Section 4. We know by Theorem 4.1 that we can construct a near time-optimal feedback $u = U(x)$ on the set $\mathcal{R}_{T^*}$, so that every initial point $y \in \mathcal{R}_{T^*}$, $|y| > \varepsilon$, is steered by the resulting O.D.E. (30) into the closed ball B_ε within time $T(y) + \varepsilon$.

We wish to investigate now the effect of a stochastic disturbance on the trajectories of the closed-loop system (30). Namely, we shall consider a stochastic differential equation, obtained by adding a random perturbation to the equation (27), say

$$dX = f(X, u)\, dt + A(X)\, dB \,, \tag{78}$$

where $B = (B_1, \ldots, B_n)$ is an n-dimensional Brownian motion, while $A = A(x)$ is an $n \times n$ matrix valued function, locally Lipschitz continuous, and satisfying the sublinear growth restriction $|A(x)| \le c(1 + |X|)$. Here and in the following $|A| \doteq (\mathrm{Tr.}\,\{AA'\})^{1/2}$ denotes the Euclidean norm of a matrix A, where A' is the transpose of A and $\mathrm{Tr.}\{A\}$ is its trace.

Under the above assumptions, for a given initial condition

$$X(0) = y, \tag{79}$$

and for any measurable control input $u(t)$ with values in K, the Cauchy problem (78)-(79) admits a unique (stochastic process) solution $t \mapsto X(t)$ defined for all $t \ge 0$ on a certain complete probability space $(\Theta, \mathcal{F}, \mathcal{P})$, with the filtration $\{\mathcal{F}_t\}$ generated by the Brownian motion B (see Ref. 23).

In connection with the patchy feedback $u = U(x)$, for every given time $\tau > T(y) + \varepsilon$, we wish to estimate the probability that a solution $t \mapsto X(t)$ of the corresponding closed-loop system

$$dX = f(X, U(X))\,dt + A(X)\,dB, \tag{80}$$

starting from y, reaches the ball B_ε within time τ. The next result shows, in the same spirit of Ref. 17, that we can provide an estimate of the distance from 1 of such a probability in terms of the supremum $\sup_{x} |A(x)|^2$ of the infinitesimal covariance function of the driving noise $A\,dB$. For any fixed $T > 0$, we set $\|A\|_T \doteq \sup_{x \in \mathcal{R}_T} |A(x)|^2$.

Theorem 6.1. (Stochastic stability of a nearly time optimal patchy feedback). *Under the above assumptions, let $T^* > 0$ be given, and fix $T' > T^*$. Then, there exist a patchy feedback $U(x)$ defined on the sublevel-set $\mathcal{R}_{T^*}$, and constants $c_1, c_2, \eta > 0$, so that, if $\|A\|_{T'} < \eta$ the following holds.*

(i) *For every $y \in \mathcal{R}_{T^*}$, $|y| > \varepsilon > 0$, any solution $X_y(\cdot)$ of (79)-(80) satisfies*

$$\mathrm{Prob.}\Big\{\big|X_y(\widehat{t}_y)\big| \le \varepsilon\Big\} \ge 1 - c_1 \cdot \|A\|_{T'}^{1/3}, \tag{81}$$

for some time $\widehat{t}_y \le T(y) + c_2\|A\|_{T'}^{1/3}$.

(ii) *There exists a continuous, strictly increasing function $\chi : [0, \infty[\mapsto [0, \infty[$ vanishing at zero, such that*

$$\lim_{\substack{|y| \to \varepsilon+ \\ y \in \mathcal{R}_{T^*}}} \mathrm{Prob.}\Big\{\inf_t |X_y(t)| \le \varepsilon\Big\} \ge 1 - \left(\frac{c_1 \cdot |A(0)|}{\varepsilon} + \chi(\varepsilon)\right). \tag{82}$$

Proof.

1. Fix some constant $T' > T^*$. By the proof of Theorem 4.1 we can construct a semiconcave, piecewise quadratic Lyapunov function $\widetilde{T}$ as in (36), and a patchy feedback $U(x)$ on the compact sub-level set

$$\widetilde{\Lambda}_{T'} \doteq \left\{ x \in \mathbb{R}^n \, ; \ \ \widetilde{T}(x) \leq T' \right\}, \tag{83}$$

with the following property. For every point $x \in \widetilde{\Lambda}_{T'}$, with $|x| > \varepsilon$, and where $\widetilde{T}$ is differentiable, there holds

$$T(x) \leq \widetilde{T}(x) \leq T(x) + \varepsilon_0 \,, \tag{84}$$

$$\nabla \widetilde{T}(x) \cdot f\big(x, U(x)\big) \leq -1 + \varepsilon_0 \,, \tag{85}$$

for some positive constant $\varepsilon_0 \ll \min\{\varepsilon, T' - T^*\}$. Observe that by (84) one has $\mathcal{R}_{T^*} \subset \widetilde{\Lambda}_{T_0}$, for $T_0 \doteq T^* + \varepsilon_0$, and $\widetilde{\Lambda}_{T'} \subset \mathcal{R}_{T'}$. Hence, to establish the theorem it will be sufficient to prove that statetments (i)-(ii) hold when y varies in $\widetilde{\Lambda}_{T_0}$, and $\|A\|_{T'}$ is redefined as $\|A\|_{T'} \doteq \sup_{x \in \widetilde{\Lambda}_{T'}} |A(x)|^2$.

2. Next, for a given initial point $y \in \widetilde{\Lambda}_{T_0}$, $|y| > \varepsilon$, consider a solution $t \mapsto X(t)$ of the stochastic differential equation (80), with initial condition (79). Observe that, since the feedback is not defined on the whole space $\mathbb{R}^n$, while we have to take into account also the possibility that the trajectoty $X(t)$ touches the boundary $\partial \widetilde{\Lambda}_{T'}$, it will be appropriate to introduce the stopping time

$$\tau \doteq \min \left\{ t \geq 0 \, ; \ \ |X(t)| = \varepsilon \ \ \text{or} \ \ \widetilde{T}(X(t)) = T' \right\}. \tag{86}$$

Our main goal is to provide an estimate of the probability that $X(\cdot)$ reaches the ball B_ε within any given time $t > 0$, i.e. of

$$\text{Prob.}\left\{ \tau \leq t, \ \ |X(\tau)| = \varepsilon \right\}. \tag{87}$$

To this end, consider the scalar random variable $Y(t) = \widetilde{T}(X(t))$, $t \leq \tau$. According to (36), at every point where $\widetilde{T}$ is smooth, the $n \times n$ Hessian matrix of second order partial derivatives of $\widetilde{T}$ is $2M\,I$, where I denotes the $n \times n$ identity matrix. By Îto's formula (see Ref. 23, p. 48), as long as $t < \tau$ and $X(t)$ remains on an open region where $\widetilde{T}$ is smooth, we have

$$dY(t) = \nabla \widetilde{T}(X(t)) \cdot f(X(t), U(X(t))) \, dt + \nabla \widetilde{T}(X(t)) \cdot A\big(X(t)\big) \, dB(t)$$

$$+ M \cdot \text{Tr.}\Big(A'(X(t)) \, A(X(t)) \Big) \, dt \,. \tag{88}$$

Indeed, since the set of points where $\widetilde{T}$ is not differentiable is contained in a finite union of $n-2$ dimensional manifolds of $\mathbb{R}^n$, the probability that $\widetilde{T}$ be differentiable at $X(t)$, $t < \tau$, is one, and hence we may assume that $Y(t)$ is a solution of (88) for all $t < \tau$. We can extend the random process $Y(t)$ for all $t \geq \tau$ by letting $Y(t)$, $t \geq 0$, be the solution of

$$
\begin{aligned}
dY(t) = {}& \nabla\widetilde{T}\big(X(t \wedge \tau)\big) \cdot f\big(X(t \wedge \tau), U(X(t \wedge \tau))\big)\, dt \\
&+ \nabla\widetilde{T}\big(X(t \wedge \tau)\big) \cdot A\big(X(t \wedge \tau)\big)\, dB(t) \\
&+ M \cdot \mathrm{Tr}.\Big(A'(X(t \wedge \tau))\, A(X(t \wedge \tau))\Big)\, dt ,
\end{aligned}
\tag{89}
$$

where $t \wedge \tau \doteq \min\{t, \tau\}$. Notice that this extension is well defined, again because the probability that $\widetilde{T}$ is not differentiable at $X(\tau)$ is zero.

3. Observe now that, from the definition (86) it follows

$$
\mathrm{Prob.}\Big\{\tau \leq t,\ |X(\tau)| = \varepsilon\Big\} \geq \mathrm{Prob.}\Big\{Y(t) \leq 0\Big\} - \mathrm{Prob.}\Big\{\tau \leq t,\ Y(\tau) = T'\Big\}.
\tag{90}
$$

Towards an estimate of the first term on the right-hand side of (90), to simplify the computations it is convenient to introduce a further random variable Z, such that

$$
\begin{aligned}
Z(t) \doteq {}& Y(t) - \Big(1 - \varepsilon_0 - M \cdot \|A\|_{T'}\Big) t \\
&- \int_0^t \Big\{ \nabla\widetilde{T}\big(X(s \wedge \tau)\big) \cdot f\big(X(s \wedge \tau), U(X(s \wedge \tau))\big) \\
&\qquad + M \cdot \mathrm{Tr}.\Big(A'(X(s \wedge \tau))\, A(X(s \wedge \tau))\Big)\Big\}\, ds ,
\end{aligned}
\tag{91}
$$

where

$$
\kappa \doteq \sup\Big\{ |\nabla\widetilde{T}(x)| ;\quad x \in \tilde{\Lambda}_{T'}\Big\}.
\tag{92}
$$

By the above definitions, thanks to (85) we deduce $Z(t) \geq Y(t)$, so that

$$
\mathrm{Prob.}\Big\{Y(t) \leq 0\Big\} \geq \rho(t) \doteq \mathrm{Prob.}\Big\{Z(t) \leq 0\Big\} \qquad \forall\, t \geq 0 .
\tag{93}
$$

Moreover, from (89) it follows that Z provides a solution to the linear stochastic differential equation

$$
dZ(t) = \big(-1 + \varepsilon_0 + M \cdot \|A\|_{T'}\big)\, dt + \beta(t)\, dB(t) ,
\tag{94}
$$

with

$$
\beta(t) \doteq \nabla\widetilde{T}\big(X(t \wedge \tau)\big) \cdot A\big(X(t \wedge \tau)\big) .
\tag{95}
$$

60

Therefore, the expectation and the variance of $Z(t)$

$$\alpha(t) \doteq E\big[Z(t)\big]\,, \qquad\qquad \gamma(t) \doteq \text{Var.}\{Z(t)\}\,, \qquad (96)$$

satisfy the differential equations

$$\dot{\alpha}(t) = \big(-1 + \varepsilon_0 + M \cdot \|A\|_{T'}\big)^{\textbf{.}}, \qquad \dot{\gamma}(t) = (\beta(t))^2 \leq \kappa^2 \cdot \|A\|_{T'}\,, \quad (97)$$

with initial conditions

$$\alpha(0) = \widetilde{T}(y)\,, \qquad \gamma(0) = 0\,. \qquad (98)$$

Relying on (97)-(98), we derive

$$\alpha(t) = \widetilde{T}(y) - \big(1 - \varepsilon_0 - M \cdot \|A\|_{T'}\big)\,t\,, \qquad \gamma(t) \leq \big(\kappa^2 \cdot \|A\|_{T'}\big)\,t\,. \quad (99)$$

Hence, observing that when $t > \widetilde{T}(y)/(1 - \varepsilon_0 - M \cdot \|A\|_{T'})$ one has

$$\text{Prob.}\Big\{Z(t) > 0, \ \ \big|Z(t) - \alpha(t)\big| \geq \big|\alpha(t)\big|\Big\} = 1\,,$$

from (99) we deduce

$$\big[\widetilde{T}(y) - (1 - \varepsilon_0 - M \cdot \|A\|_{T'})\,t\big]^2 \cdot \big(1 - \rho(t)\big) \leq \big(\kappa^2 \cdot \|A\|_{T'}\big)\,t\,, \quad (100)$$

which, in turn, yields

$$\rho(t) \geq 1 - \frac{\big(\kappa^2 \cdot \|A\|_{T'}\big)\,t}{\big[(1 - \varepsilon_0 - M \cdot \|A\|_{T'})\,t - \widetilde{T}(y)\big]^2}\,, \qquad (101)$$

for all $t > \widetilde{T}(y)/(1 - \varepsilon_0 - M \cdot \|A\|_{T'})$.

Concerning the probability that the trajectory hits the outer boundary

$$p \doteq \text{Prob.}\Big\{Y(\tau) = T'\Big\}\,, \qquad (102)$$

observe that, by (99) one has

$$\big(T' - \alpha(\tau)\big)^2\,p \leq \big(\kappa^2 \cdot \|A\|_{T'}\big)\tau\,. \qquad (103)$$

Thus, assuming

$$\varepsilon_0 < \frac{1}{4}\,, \qquad \|A\|_{T'} < \frac{1}{4M}\,, \qquad (104)$$

since $y \in \widetilde{\Lambda}_{T_0}$ from (99), (103) we derive

$$p \leq \max_{t \geq 0} \frac{\big(\kappa^2 \cdot \|A\|_{T'}\big)t}{\big(T' - \alpha(t)\big)^2} \leq \max_{t \geq 0} \frac{\big(\kappa^2 \cdot \|A\|_{T'}\big)t}{\big(T' - \widetilde{T}(y) + t/2\big)^2}$$

$$= \frac{\kappa^2 \cdot \|A\|_{T'}}{2\big(T' - \widetilde{T}(y)\big)} \leq \frac{\kappa^2}{2(T' - T_0)} \cdot \|A\|_{T'}\,. \qquad (105)$$

Hence, (90), (93), (101), (105) together yield

$$\text{Prob.}\Big\{\tau \le t, \ \ |X(\tau)| = \varepsilon\Big\} \ge 1 - \frac{\left(\kappa^2 \cdot \|A\|_{T'}\right) t}{\left[(1 - \varepsilon_0 - M \cdot \|A\|_{T'})\, t - \widetilde{T}(y)\right]^2}$$

$$- \frac{\kappa^2}{2(T' - T_0)} \cdot \|A\|_{T'}\,,$$

(106)

for all $t > \widetilde{T}(y)/(1 - \varepsilon_0 - M \cdot \|A\|_{T'})$. Since we may repeat the same reasoning for an arbitrary small $\varepsilon_0 < \varepsilon$, we thus obtain from (106) the estimate

$$\text{Prob.}\Big\{\tau \le t, \ \ |X(\tau)| = \varepsilon\Big\} \ge 1 - \frac{\left(\kappa^2 \cdot \|A\|_{T'}\right) t}{\left[(1 - M \cdot \|A\|_{T'})\, t - T(y)\right]^2}$$

$$- \frac{\kappa^2 \cdot \|A\|_{T'}}{2(T' - T^*)} \qquad \forall\, t > \frac{T(y)}{(1 - M \cdot \|A\|_{T'})}$$

(107)

Then, taking

$$\widehat{t} \doteq \frac{T(y) + \|A\|_{T'}^{1/3}}{(1 - M \cdot \|A\|_{T'})}\,,$$

and recalling (104), we deduce from (107) that

$$\text{Prob.}\Big\{\tau \le \widehat{t}, \ |X(\tau)| = \varepsilon\Big\} \ge 1 - \kappa^2 \left(\frac{4(T^* + 4M^{-1/3})}{3} + \frac{1}{2(T' - T^*)}\right) \cdot \|A\|_{T'}^{1/3},$$

(108)

which proves the statement (i) of the theorem, since

$$\widehat{t} \le \left(1 + 2M \cdot \|A\|_{T'}\right)\left(T(y) + \|A\|_{T'}^{1/3}\right)$$

(109)

$$\le T(y) + \left(2MT' + 3/2\right) \cdot \|A\|_{T'}^{1/3}\,.$$

4. Concerning (ii), let B_R be a ball containing $\widetilde{\Lambda}_{T'}$. One can easily see that the approximate minimum time function $\widetilde{T}$ defined as in (36), (with a possible slight modificaton) beside (84)-(85) enjoys the further property:

$$\frac{1}{c_3}|x|^q \le \widetilde{T}(x) \le c_3|x| + \varepsilon_0 \qquad \forall\, |x| \le R\,,$$

(110)

for some constants $c_3 > 0$, $1 < q < 2$. Next, observe that setting

$$\mu_\varepsilon \doteq \max\Big\{\widetilde{T}(x)\,; \ \ |x| \le 2\varepsilon\Big\}\,,$$

(111)

62

by (110) we have

$$\mu_\varepsilon' \doteq \mu_\varepsilon + \varepsilon < 2(c_3 + 1)\varepsilon\,,$$

$$B_{2\varepsilon} \subset \widetilde{\Lambda}_{\mu_\varepsilon} \subset \widetilde{\Lambda}_{\mu_\varepsilon'} \subset B_{c_4\,\varepsilon^{1/q}}\,, \tag{112}$$

with $c_4 \doteq c_3(2(c_3+1))^{1/q}$. Hence, letting $\mathrm{Lip}(A)$ denote a Lipschitz constant for A over the ball B_R, we find

$$\left|A(x)\right| \le \left|A(0)\right| + \mathrm{Lip}(A) \cdot c_4\,\varepsilon^{1/q} \qquad \forall\, x \in \widetilde{\Lambda}_{\mu_\varepsilon'}\,, \tag{113}$$

which yields

$$\|A\|_{\mu_\varepsilon'}^{1/3} \le \left|A(0)\right|^{2/3} + \chi_1(\varepsilon)\,, \tag{114}$$

where

$$\chi_1(\varepsilon) \doteq \left(\left|A(0)\right| + \mathrm{Lip}(A) \cdot c_4\,\varepsilon^{1/q}\right)^{2/3} - \left|A(0)\right|^{2/3}\,, \qquad \varepsilon \ge 0\,,$$

defines a continuous, strictly increasing function vanishing at zero. Finally, applying the estimates (107)-(108), with $\mu_\varepsilon, \mu_\varepsilon'$ in place of T^*, T', respectively, and relying on (112), (114) (and on $\|A\|_{\mu_\varepsilon'} < 1$), we derive

$$\lim_{\substack{|y|\to\varepsilon+ \\ y\in\mathcal{R}_{T^*}}} \mathrm{Prob.}\left\{\inf_t \left|X_y(t)\right| \le \varepsilon\right\} = \lim_{\substack{|y|\to\varepsilon+ \\ y\in\widetilde{\Lambda}_{\mu_\varepsilon'}}} \mathrm{Prob.}\left\{\inf_t \left|X_y(t)\right| \le \varepsilon\right\}$$

$$\ge 1 - c_1' \cdot \|A\|_{\mu_\varepsilon'}^{1/3} - c_1'' \cdot \frac{\|A\|_{\mu_\varepsilon'}}{\mu_\varepsilon' - \mu_\varepsilon}$$

$$\ge 1 - c_1' \cdot \left(\left|A(0)\right|^{2/3} + \chi_1(\varepsilon)\right) - c_1'' \cdot \frac{\|A\|_{\mu_\varepsilon'}}{\varepsilon}$$

$$\ge 1 - \left(c_1' \cdot \left|A(0)\right|^{2/3} + c_1''' \cdot \frac{\left|A(0)\right|}{\varepsilon}\right) - \chi_2(\varepsilon)$$

$$\ge 1 - c_1'^v \cdot \frac{\left|A(0)\right|}{\varepsilon} - \chi_2(\varepsilon)\,, \tag{115}$$

for suitable constants $c_1', c_1'', c_1''', c_1'^v > 0$, and with

$$\chi_2(\varepsilon) \doteq c_1' \cdot \chi_1(\varepsilon) + c_1''' \cdot \varepsilon^{2/q-1}\,. \tag{116}$$

This establishes (82) thus completing the proof of the theorem. $\qquad\square$

References

1. F. Ancona and A. Bressan, *ESAIM - Control, Optimiz. Calc. Var.* **4**, 445 (1999).
2. F. Ancona and A. Bressan, *SIAM J. Control Optim.* **41**, 1455 (2002).
3. F. Ancona and A. Bressan, *ESAIM - Control, Optimiz. Calc. Var.* **10**, 168 (2004).
4. F. Ancona and A. Bressan, Stabilization by patchy feedbacks and robustness properties, in *Optimal control, stabilization and nonsmooth analysis*, eds. M.S. de Queiroz, M. Malisoff and P. Wolenski, Lecture Notes in Control and Inform. Sci., Vol. 301 (Springer, Berlin, 2004), pp. 185–199.
5. F. Ancona and A. Bressan, *Ann. Inst. H. Poincaré Anal. Non Linéaire* **24**, 279 (2007).
6. J.-P. Aubin and A. Cellina, *Differential inclusions*. Set-valued maps and viability theory. Grundl. der Mathemat. Wissensch., Vol. 264 (Springer-Verlag, Berlin, 1984).
7. L. D. Berkovitz, *SIAM J. Control Optim.* **27**, 991 (1989).
8. A. Bressan and B. Piccoli, *SIAM J. Control Optim.* **36**, 12 (1998).
9. A. Bressan and B. Piccoli, *Introduction to the mathematical theory of control*, AIMS Series on Applied Mathematics, Vol. 2, first edn. (AIMS, Springfield MO, 2007).
10. A. Bressan and F. Priuli, Nearly optimal patchy feedbacks for minimization problems with free terminal time, *Discrete Contin. Dyn. Syst.* (to appear), available at http://cpde.iac.rm. cnr.it/ricerca.php (July 2007).
11. R. W. Brockett, *Asymptotic stability and feedback stabilization*, in *Differential geometric control theory (Houghton, Mich., 1982)*, eds. R. W. Brockett, R. S. Millman and H. J. Sussmann, Progr. Math., Vol. 27 (Birkhäuser Boston, Boston, MA, 1983), pp. 181–191.
12. F. H. Clarke, Y. S. Ledyaev, L. Rifford and R. J. Stern, *SIAM J. Control Optim.* **39**, 25 (2000).
13. F. H. Clarke, Y. S. Ledyaev, E. D. Sontag and A. I. Subbotin, *IEEE Trans. Automat. Control* **42**, 1394 (1997).
14. F. H. Clarke, Y. S. Ledyaev, R. J. Stern and P. R. Wolenski, *Nonsmooth analysis and control theory*, Graduate Texts in Mathematics, Vol. 178 (Springer-Verlag, New York, 1998).
15. J.-M. Coron, *Systems Control Lett.* **14**, 227 (1990).
16. J.-M. Coron and L. Rosier, *J. Math. Systems Estim. Control* **4**, 67 (1994).
17. H. Deng, M. Krstić and R. J. Williams, *IEEE Trans. Automat. Control* **46**, 1237 (2001).
18. A. Filippov, *Transl., Ser. 2, Am. Math. Soc.* **42**, 199 (1964).
19. A. Filippov, *Differential equations with discontinuous righthand sides*, Mathematics and its Applications (Soviet Series), Vol. 18 (Kluwer Academic Publishers Group, Dordrecht, 1988). Translated from the Russian.
20. O. Hájek, *J. Differential Equations* **32**, 149 (1979).

21. H. Hermes, *Discontinuous vector fields and feedback control*, in *Differential Equations and Dynamical Systems (Proc. Internat. Sympos., Mayaguez, P. R., 1965)*, eds. J. K. Hale and J. P. LaSalle (Academic Press, New York, 1967), pp. 155-165.

22. N. N. Krasovskii and A. I. Subbotin, *Positional Differential Games* [in Russian] (Izdat. Nauka, Moscow, 1974). Revised English translation: Game-Theoretical Control Problems (1988). Springer-Verlag, New York.

23. B. Øksendal, *Stochastic differential equations.* An introduction with applications. Universitext, sixth edn. (Springer-Verlag, Berlin, 2003).

24. B. Piccoli, *SIAM J. Control Optim.* **34**, 1914 (1996).

25. E. P. Ryan, *SIAM J. Control Optim.* **32**, 1597 (1994).

26. E. D. Sontag, *Mathematical control theory.* Deterministic finite-dimensional systems. Texts in Applied Mathematics, Vol. 6, second edition (Springer-Verlag, New York, 1998).

27. E. D. Sontag, Stability and stabilization: discontinuities and the effect of disturbances, in *Nonlinear analysis, differential equations and control (Montreal, QC, 1998)*, eds. F. Clarke and R. Stern (Kluwer Acad. Publ., Dordrecht, 1999), pp. 551-598.

28. E. D. Sontag and H. J. Sussmann, *Remarks on continuous feedback*, in *Proc. IEEE Conf. Decision and Control*, (Albuquerque, NM, 1980): 916-921.

29. H. J. Sussmann, *J. Differential Equations* **31**, 31 (1979).

SENSITIVITY OF CONTROL SYSTEMS WITH RESPECT TO MEASURE-VALUED COEFFICIENTS

Z. ARTSTEIN *

Department of Mathematics
The Weizmann Institute of Science
Rehovot 76100, Israel
E-mail: zvi.artstein@weizmann.ac.il

Measure-valued coefficients arise when limits of rapidly varying parameters of control systems are examined. The rapidly varying parameter is considered then a perturbation of a measure-valued parameter. We provide quantitative estimates for the sensitivity of the optimal value and near optimal solutions, with respect to such perturbations. As a particular case, which is of interest for its own sake, we establish quantitative estimates for the sensitivity of ordinary control systems with respect to relaxed controls.

1. Introduction

This paper examines control systems of the form

$$\text{minimize} \quad \int_a^b Q(x, t, u, \rho(t))dt \tag{1}$$

$$\text{subject to} \quad \frac{dx}{dt} = f(x, t, u, \rho(t)), \quad x(a) = x_0.$$

The state variable is $x \in \mathbb{R}^n$, the n-dimensional Euclidean space. The control variable u belongs to a set U in $\mathbb{R}^m$; the admissible controls are measurable functions $u = u(\cdot)$ from $[a, b]$ into U. The function $\rho(\cdot)$ is a prescribed measurable map from $[a, b]$ into a complete separable metric space Γ; the interpretation is that $\rho(t) : [a, b] \to \Gamma$ is an external parameter of the system. We are interested in the dependence of the value and the optimal solutions of the system on this external parameter. Of particular interest to us in this paper is to get, under appropriate conditions, quantitative estimates in the

*Incumbent of the Hettie H. Heineman Professorial Chair in Mathematics. Research supported by the Israel Science Foundation.

form of Lipschitz estimates, on this dependence. Furthermore, the case we concentrate on is where the external parameter $\rho(\cdot)$ may change abruptly and frequently, for instance, when it oscillates very rapidly.

A natural approach to address sensitivity issues is to identify a nominal system, for instance, to identify a parameter $\rho_0(\cdot)$ and examine the influence on the value and on the optimal solutions when a perturbation $\rho(\cdot)$ of $\rho_0(\cdot)$ occurs. This approach does not fit the case of rapidly oscillating parameters since a sequence of rapidly oscillating functions cannot be represented as a perturbation which tends to zero from a fixed function. Indeed, a sequence of more and more rapidly oscillating functions does not converge strongly, while a weak limit does not capture the oscillatory nature which may, in turn, play a role in the optimization. An appropriate nominal system, however, is one with the parameters being measure-valued, namely, Young measures. A theory of such nominal systems was offered in Artstein,[2] and was applied in various frameworks, including the case where the fast parameter is ignited by a singularly perturbed control system. See also Artstein[3,4] and Artstein and Gaitsgory[5,6] and references therein. We briefly recall the main ingredients of the theory in Section 3. Young measures play a similar role in other theories; a non exhaustive sample of references is Balder,[7] Ball,[8] Pedregal[10–12] Tartar,[14] Valadier,[15] Warga[16] and Young;[17] the latter two references concentrate on relaxed controls and generalized curves. The available theory in the applications of Young measures is primarily qualitative, due to the need to work in the highly nonsmooth space of probability measures (see though Pedregal[11] for a variational theory). The present paper is devoted to the derivation of the quantitative estimates of the sensitivity problem.

The paper is organized as follows. In the next section we display some technical assumptions on the data; these are used throughout the paper. In Section 3 the chattering parameters model is recalled. The sensitivity estimates are established with respect to the Prohorov metric which is displayed in Section 4. In Section 5 we establish the sensitivity in the particular case of relaxed controls without external parameters. A matching lemma which is of interest for its own sake, is displayed in Section 6; it serves as an important tool in verifying the general sensitivity result in Section 7. The closing section offers few comments and concrete examples illustrating the main results.

2. Standing hypotheses

We display here the technical assumptions employed throughout the paper. In order not to blur the main contribution of the paper, no attempt is done to display the most general assumptions possible; see, though, the closing section for some remarks in this respect.

Assumption 2.1.

(i) *The set of controls U in $\mathbb{R}^m$ and the space Γ in which the parameter function $\rho(\cdot)$ takes values are compact; the space Γ is endowed with a prescribed metric $d(\cdot, \cdot)$.*

(ii) *The functions $Q(x, t, u, \gamma)$ and $f(x, t, u, \gamma)$ are Lipschitz in the four variables x, t, u and γ, with Lipschitz constant, say, κ.*

Under the Standing Hypotheses, standard considerations guarantee that for any fixed control function $u(\cdot)$ and any parameter function $\rho(\cdot)$, the differential equation in (1) has a unique solution defined on the entire interval $[a, b]$. Furthermore, there is a common bound on all these trajectories. Consequently, the functions $f(x, t, u, \gamma)$ and $Q(x, t, u, \gamma)$ are uniformly bounded on all these trajectories. In particular, the cost is well defined.

Notation 2.2.

(iii) *We denote by β the common bound, guaranteed by Assumption 2.1, on the values $|f(x, t, u, \gamma)|$ when evaluated on the trajectories generated by feasible parameter and control functions.*

3. The chattering parameters model

We briefly review in this section the relevant ingredients of the chattering parameters model. We start with the definition of Young measures (newcomers to the Young measures paradigm may wish to look at Balder,[7] Valadier;[15] the examples in the closing section may also help in clarifying the structure).

Let M be a complete metric space, with a metric which we again denote by $d(\cdot, \cdot)$. A probability measure on M is a σ-additive set function, say $\mu(\cdot)$, assigning to each Borel subset B of M a value $\mu(B) \in [0, 1]$ with $\mu(M) = 1$. The family of all probability measures on M is denoted $\mathcal{P}(M)$. We later consider $\mathcal{P}(M)$ as a metric space.

We consider mappings $\mu(t) : [a, b] \to \mathcal{P}(M)$. We write $\mu(t, B)$ for the value which the measure $\mu(t)$ assigns to the set B. A Young measure into M (pertinent to the framework in (1)) is a mapping $\mu(t) : [a, b] \to \mathcal{P}(M)$

which is measurable in the sense that for each fixed Borel set B in M the real-valued map $\mu(\cdot, B) : [a, b] \to [0, 1]$ is measurable. The space of such Young measures is denoted by $\mathcal{Y} = \mathcal{Y}([a, b], M)$. We consider a convergence structure on $\mathcal{Y}$ given by: The sequence $\mu_j(\cdot)$ converges as $j \to \infty$, in the sense of Young measures, to $\mu_0(\cdot)$ if (here m is the independent variable in M)

$$\int_a^b \int_M h(t, m)\mu_j(t, dm) \, dt \to \int_a^b \int_M h(t, m)\mu_0(t, dm) \, dt \qquad (2)$$

whenever $h(t, m)$ is a bounded and continuous real-valued function defined on $[a, b] \times M$. (Notice that in accordance with the previous notation we write $\mu(t, dm)$ for the integration with respect to the variable m with respect to the measure $\mu(t)$ for a fixed t.)

The Young measure $\mu(t) : [a, b] \to \mathcal{P}(M)$ is be denoted either by $\mu(\cdot)$ or in bold face, namely, $\boldsymbol{\mu}$.

We consider in this paper three spaces of Young measures: into the parameter space Γ, into the control set U, and into $\Gamma \times U$.

The space of Young measures, namely, the space of probability measure-valued maps with the aforementioned convergence, arises in a variety of applications. A sample of applications can be found in the following papers and monographs, and references therein: Balder,[7] Ball,[8] Pedregal[10–12] Tartar,[14] Valadier,[15] Warga,[16] Young.[17] In particular, it was established that when M is compact, the space $\mathcal{Y}([a, b], M)$ with the convergence in the sense of Young measures is induced by a metric which makes it a compact space (in the sequel we concentrate on a specific metric). The space of point-valued functions $m(t) : [a, b] \to M$, is a subspace of $\mathcal{Y}([a, b], M)$, where $m(\cdot)$ is interpreted as a Young measure whose values are Dirac measures, namely $m(t)$ is interpreted as a probability measure supported on the singleton $\{m(t)\}$. The subspace of Dirac-valued measures is dense in $\mathcal{Y}([a, b], M)$ (this property relies on the Lebesgue measure, which is the base measure on $[a, b]$, not having atoms).

The following definition introduces Young measures parameters into the systems (1); it was introduced in Artstein.[2]

Definition 3.1. The parameter in (1) may be a chattering parameter, namely a Young measure $\mu(\cdot)$ into Γ; the optimization problem takes then the form

$$\text{minimize} \qquad \int_a^b \int_\Gamma Q(x,t,u,\rho)\mu(t,d\rho)\,dt \tag{3}$$
$$\text{subject to} \qquad \frac{dx}{dt} = \int_\Gamma f(x,t,u,\rho)\mu(t,d\rho), \quad x(a) = x_0,$$

where the optimization is done over all admissible controls (namely, measurable function into U) of the form $u(t,\gamma) : [a,b] \times \Gamma \to U$.

It is clear that (1) is a particular case of (3), namely, when the point-valued parameter $\rho(\cdot)$ is interpreted as a Young measure. In fact, the motivation for introducing the more complicated model (3) is that the elements with a chattering parameter form appropriate limits of sequences of problems with ordinary parameters, thus providing an appropriate closure of the space.

As in the standard control problem, the optimal control problem (3) may not have an optimal solution even if the control set U is compact. Relaxed controls, as introduced by J. Warga (see Ref. 16) may yield a solution. The induced structure is as follows.

Definition 3.2. A relaxed control for the system (3) is a mapping, say $v = v(t,\gamma)$, which assigns to each pair (t,γ) a probability measure on U, and as such it is measurable. The resulting cost function and differential equation are then

$$\int_a^b \int_\Gamma \int_U Q(x,t,u,\gamma)v(t,\gamma,du)\,\mu(t,d\gamma)\,dt \tag{4}$$
$$\frac{dx}{dt} = \int_\Gamma \int_U f(x,t,u,\gamma)v(t,\gamma,du)\,\mu(t,d\gamma), \quad x(a) = x_0.$$

We refer to the cost given via the integration of $Q(x,t,u,\gamma)$ in (4) and the solution of the associated differential equation as the *cost* and, respectively, the *trajectory* generated by the parameter $\rho(\cdot)$ and the control $v(\cdot,\cdot)$

Relaxed controls were developed by Warga, see Ref. 16, and were used in the chattering model in Artstein.[2] Notice that a relaxed control is a Young measure into U. A key observation is as follows.

Definition 3.3 (and observation). Let $\rho(\cdot)$ be a chattering parameter of (3) and let $v(\cdot,\cdot)$ be a relaxed control. The *effective input* induced by $\rho(\cdot)$ and $v(\cdot,\cdot)$ on (3) is the Young measure which assigns to each $t \in [a,b]$

the probability measure, say $p(\cdot)$, on $\Gamma \times U$ defined by

$$p(t)(C) = \int_\Gamma v(t, \gamma, C_\gamma) \rho(t, d\gamma) \tag{5}$$

(where C_γ is the γ-section of C in U, recall that $v(t, r, C_\gamma)$ is the value the measure $v(t, \gamma)$ assigns to C_γ). What counts in regard to solving the control equations and computing the cost when a relaxed control is applied to a chattering parameter is the effective input. Indeed, the system (4) takes then the form

$$\int_a^b \int_{\Gamma \times U} Q(x, t, u, \gamma) p(t, du \times d\gamma) \, dt$$
$$\frac{dx}{dt} = \int_{\Gamma \times U} f(x, t, u, \gamma) p(t, du \times d\gamma), \quad x(a) = x_0, \tag{6}$$

where $p(\cdot)$ is the effective input.

Proposition 3.4. *Let $\rho(\cdot)$ be a chattering parameter in (3) and let $v(\cdot, \cdot)$ be a relaxed control. There exists a sequence of ordinary controls $u_i(t, \gamma)$: $[a, b] \times \Gamma \to U$ such that the effective inputs induced by $u_i(\cdot, \cdot)$ converge, in the sense of Young measures, to the effective input induced by $v(\cdot, \cdot)$. Consequently, the trajectories $x_i(\cdot)$ and the costs c_i, generated by applying $u_i(\cdot, \cdot)$ to (4), converge uniformly on $[a, b]$ and in, respectively, $\mathbb{R}$, to the trajectory and cost generated by applying $v(\cdot, \cdot)$ to (4).*

Proof. We consider the Young measure $\rho = \rho(\cdot)$ as a base measure on the space $[a, b] \times \Gamma$. It is an atom-less measure. It follows from the basic properties of Young measures that the Young measure $v(\cdot, \cdot) : [a, b] \times \Gamma \to \mathcal{P}(U)$ from the base space $[a, b] \times \Gamma$ into U can be approximated in the sense of Young measures by a sequence of ordinary functions $u_i(\cdot, \cdot) : [a, b] \times \Gamma \to U$. It follows directly from the definition (2) of convergence according to Young measures and the definition (5) of the effective input, that the induced effective inputs also converge in the sense of Young measures. The effective inputs can now be viewed as relaxed controls of the control system (without parameters); thus the convergence of the trajectories and the costs follows from the classical work of J. Warga, see Ref. 16. $\qquad\square$

4. The Prohorov metric

We introduce here the metric on the space of Young measures, with respect to which the quantitative estimates are established. It is based on the Prohorov distance between two probability measures, see e.g., Billingsley,[9] hence we use the same term. We first recall the Prohorov metric in general spaces (although the definition in Ref. 9 refers to probability measures, it is clear that it is applicable to all finite measures). In what follows M is a metric space with a metric $d(\cdot, \cdot)$; the η-neighborhood of the set B in M is denoted by $B^{(\eta)}$.

Definition 4.1. Let M be a complete metric space. The Prohorov distance between two finite measures μ and ν on M is the infimum of all $\eta > 0$ satisfying $\nu(B) \leq \mu(B^{(\eta)}) + \eta$ and $\mu(B) \leq \nu(B^{(\eta)}) + \eta$ for every Borel set B of M. We denote the Prohorov distance by $\mathrm{Proh}(\mu, \nu)$.

The Prohorov distance reflects the weak convergence of measures. It is known that with the Prohorov distance the space of all finite measures on M is a complete metric space. These properties are established in Ref. 9 for the space $\mathcal{P}(M)$ of probability measures, but the arguments go through for all finite measures.

We now identify the space of Young measures $\mathcal{Y}([a, b], M)$ as a space of measures to which the Prohorov distance applies. As a preliminary step, with any $\mu(\cdot) \in \mathcal{Y}([a, b], M)$ an equivalent measure $\boldsymbol{\mu}$ on $[a, b] \times M$ is associated, as follows. The measure of a Borel subset C of $[a, b] \times M$ is given by

$$\boldsymbol{\mu}(C) = \int_a^b \mu(t, C_t)dt, \tag{7}$$

where C_t is the t-section of C. Notice that then $\boldsymbol{\mu}(T \times M) = \lambda(T)$ for every subset T of the interval $[a, b]$ (and where λ is the Lebesgue measure). It is easy to see that this property characterizes the space $\mathcal{Y}([a, b], M)$. In particular, when M is compact, the space $\mathcal{Y}([a, b], M)$ is a compact subset of the space of measures on $[a, b] \times M$.

Definition 4.2. The Prohorov distance on $\mathcal{Y}([a, b], M)$ is defined to be the Prohorov distance on the associated measures on $[a, b] \times M$, where the metric on $[a, b] \times M$ is the $\mathbf{L}^1$-metric, namely, the distance between (t_1, m_1) and (t_2, m_2) is $|t_1 - t_2| + d(m_1, m_2)$. We use $\mathrm{Proh}(\boldsymbol{\mu}, \boldsymbol{\nu})$, or $\mathrm{Proh}(\mu(\cdot), \nu(\cdot))$, to denote the Prohorov distance between $\boldsymbol{\mu}$ and $\boldsymbol{\nu}$.

In the sequel the Prohorov distance on the three spaces of Young measures, namely, into Γ, into U and into $\Gamma \times U$, will be invoked.

5. Sensitivity for relaxed controls

This section establishes the promised sensitivity estimate in a particular case, namely, the classical situation of control systems with relaxed controls as follows.

$$
\begin{aligned}
\text{minimize} \quad & \int_a^b Q(x,t,u)dt \\
\text{subject to} \quad & \frac{dx}{dt} = f(x,t,u), \quad x(a) = x_0 .
\end{aligned}
\tag{8}
$$

Such estimates seem not to be available in the literature. The general case of systems with chattering parameters, treated in the next section, will in fact be reduced to the setting in (8).

Recall that a relaxed control for (8) is a Young measure $v(t) : [a,b] \to \mathcal{P}(U)$, denoted either by $\mathbf{v}$ or by $v(\cdot)$. We write then $f(x,t,v(t))$ for $\int_U f(x,t,u)v(t,du)$.

Theorem 5.1. *Under Assumption 2.1 there exists a constant η, such that given any two relaxed controls $\mathbf{v}_1$ and $\mathbf{v}_2$, the resulting trajectories and costs in (8), say $x_1(\cdot)$ and $x_2(\cdot)$, and c_1 and c_2, satisfy $|x_1(t) - x_2(t)| \leq \eta \operatorname{Proh}(\mathbf{v}_1,\mathbf{v}_2)$, and $|c_1 - c_2| \leq \eta \operatorname{Proh}(\mathbf{v}_1,\mathbf{v}_2)$. The constant η depends only on the length $b - a$ of the interval, the Lipschitz constant κ and the bound β guaranteed by the assumption.*

Proof. On the space $\mathcal{C}([a,b],\mathbb{R}^n)$ of continuous functions from $[a,b]$ to $\mathbb{R}^n$ consider the distance induced by the norm

$$
\|x(\cdot)\|_\kappa = \max_{t \in [a,b]} |x(t)| e^{-\kappa(t-a)},
\tag{9}
$$

where κ is the Lipschitz constant given in Assumption 2.1. The trajectories $x_1(\cdot)$ and $x_2(\cdot)$ are fixed points of, respectively, the operators from $\mathcal{C}([a,b],\mathbb{R}^n)$ to itself given by

$$
T_i(x(\cdot))(t) = x_0 + \int_a^t f(x(s),s,v_i(s))ds
\tag{10}
$$

for $i = 1,2$. Each of these two operators is a contraction on $\mathcal{C}([a,b],\mathbb{R}^n)$ with respect to the norm (9); this is a classical result, see Reid [13, p. 56]. Let the contraction factor be denoted by α, namely, $\alpha < 1$ and

$$
\|T_i(x(\cdot)) - T_i(y(\cdot))\|_\kappa \leq \alpha \|x(\cdot) - y(\cdot)\|_\kappa.
\tag{11}
$$

It follows from Reid [13, p. 56] that α depends only on $b - a$ and κ.

First we give an estimate for the distance $\|x_1(\cdot) - x_2(\cdot)\|_\kappa$. Since $x_i(\cdot)$ is a fixed point of the corresponding operator it follows that

$$\|x_1(\cdot) - x_2(\cdot)\|_\kappa \leq \|T_1(x_1(\cdot)) - T_2(x_1(\cdot))\|_\kappa + \|T_2(x_1(\cdot)) - T_2(x_2(\cdot))\|_\kappa. \tag{12}$$

The contraction property (11) implies that last term in (12) is less than or equal to $\alpha\|x_1(\cdot) - x_2(\cdot)\|_\kappa$. Hence, by shifting this estimate to the left of the equation, we get that $(1 - \alpha)\|x_1(\cdot) - x_2(\cdot)\|_\kappa$ is bounded by the first term in the right hand side of (12). Since $|x_1(t) - x_2(t)|$ is bounded by $e^{\kappa(b-a)}\|x_1(\cdot) - x_2(\cdot)\|_\kappa$, and since the norm in (9) is less than or equal to the sup norm, it follows that

$$|x_1(t) - x_2(t)| \leq e^{\kappa(b-a)}(1 - \alpha)^{-1}\|T_1(x_1(\cdot)) - T_2(x_1(\cdot))\| \tag{13}$$

where the norm in (13) is the sup norm on continuous functions.

The claimed estimate for the trajectories would follow from (13) once we verify that there exists a constant η_0 (depending only on $b - a$, κ and β) such that

$$|T_1(x_1(\cdot))(\tau) - T_2(x_1(\cdot))(\tau)| \leq \eta_0 \operatorname{Proh}(\mathbf{v}_1, \mathbf{v}_2) \tag{14}$$

for all $\tau \in [a, b]$. Verifying (14) is what we do next.

Since β is a bound on $|f(x, t, u)|$ it follows that any trajectory generated by applying a relaxed control to (8) is Lipschitz in the time variable with Lipschitz constant β. We show that (14) holds when $x_1(\cdot)$ can in fact be replaced by any β-Lipschitz map, hence in the sequel we drop the subscript from $x_1(\cdot)$.

Recall that, in general, when $h(\xi) : M \to [0, \infty)$ is a bounded nonnegative measurable function on a metric space M, say bounded by 2β, and when μ is a probability measure on M then

$$\int_M h(\xi)\mu(d\xi) = \int_0^{2\beta} \mu(\{\xi : h(\xi) \geq \lambda\}) \, d\lambda, \tag{15}$$

where $d\lambda$ indicates integration with respect to the Lebesgue measure; see, e.g., Billingsley.[9] We apply this representation when $M = [a, \tau] \times U$ and when $h(t, u) = f_j(x(t), t, u) + \beta$ where the subscript denotes the j-th coordinate (the addition of β makes it nonnegative; it is clearly bounded by 2β). We deduce therefore that the difference between the j-th coordinates of $T_1(x(\cdot))(\tau)$ and $T_2(x(\cdot))(\tau)$ (appearing in (14)) is

$$\int_0^{2\beta} (\mathbf{v}_1(\{(t, u) : h(t, u) \geq \lambda\}) - \mathbf{v}_2(\{(t, u) : h(t, u) \geq \lambda\}))d\lambda, \tag{16}$$

74

where, as mentioned, $h(t, u) = f_j(x(t), t, u) + \beta$. The Lipschitz constant guaranteed in Assumption 2.1 and the β-Lipschitz property of $x(\cdot)$ imply that $h(t, u)$ is Lipschitz, with Lipschitz constant $\kappa\beta$.

The inequalities in Definition 4.1 which define the Prohorov distance imply that

$$\mathbf{v}_1(\{(t, u) : h(t, u) \geq \lambda\}) \leq \mathbf{v}_2(\{(t, u) : h(t, u) \geq \lambda\}^{(\delta)}) + \delta \qquad (17)$$

with $\delta = \mathrm{Proh}(\mathbf{v}_1, \mathbf{v}_2)$, and the similar inequality where $\mathbf{v}_1$ and $\mathbf{v}_2$ are swapped. The $\kappa\beta$-Lipschitz property of $h(t, u)$ together with (17) imply the inequality

$$\mathbf{v}_1(\{(t, u) : h(t, u) \geq \lambda\}) \leq \mathbf{v}_2(\{(t, u) : h(t, u) + \kappa\beta\delta \geq \lambda\} + \delta \qquad (18)$$

and the similar inequality with $\mathbf{v}_1$ and $\mathbf{v}_2$ swapped.

Replacing the first term of the integrand in (16) by the right hand side of (18), changing then the variable of integration (from λ to $\lambda - \kappa\beta\delta$), also taking care of the extra interval of integration (of length $\kappa\beta\delta$) and the error term δ from (17) integrated over the 2β interval, implies that (16) is less than or equal to $(2 + \kappa)\beta\delta$. Applying the same argument with $\mathbf{v}_1$ and $\mathbf{v}_2$ swapped implies that the absolute value of (16) is bounded by $(2 + \kappa)\beta\delta$. The argument can now be repeated for the n-coordinates $j = 1, \ldots, n$. Since $\delta = \mathrm{Proh}(\mathbf{v}_1, \mathbf{v}_2)$ the existence of η_0 in the estimate (14) and its dependence on $b - a$, κ and β only is established; hence the claimed estimate on the distance between the trajectories is verified.

We turn now to the claimed estimate on the distance between the costs c_1 and c_2. We need to show that

$$\left| \int_a^b (Q(x_1(t), t, v_1(t)) - Q(x_2(t), t, v_2(t))) dt \right| \leq \eta \mathrm{Proh}(\mathbf{v}_1, \mathbf{v}_2) \qquad (19)$$

for some constant η which depends only on $b - a$, κ and β. When $x_2(\cdot)$ in (19) is replaced by $x_1(\cdot)$, the same estimate follows from the argument we used to verify (14); namely, we get

$$\left| \int_a^b (Q(x_1(t), t, v_1(t)) - Q(x_1(t), t, v_2(t))) dt \right| \leq \eta_1 \mathrm{Proh}(\mathbf{v}_1, \mathbf{v}_2) \qquad (20)$$

for an appropriate η_1. When $v_1(\cdot)$ in (19) is replaced by $v_2(\cdot)$ the same estimate follows from the estimate on the distance between the trajectories and since $Q(x, t, u)$ is κ-Lipschitz; namely, we have

$$\left| \int_a^b (Q(x_1(t), t, v_2(t)) - Q(x_2(t), t, v_2(t))) dt \right| \leq \eta_2 \mathrm{Proh}(\mathbf{v}_1, \mathbf{v}_2) \qquad (21)$$

for an appropriate η_2. The latter two estimates combined with the triangle inequality verify that (19) holds with $\eta = \eta_1 + \eta_2$. This completes the proof of the theorem. $\qquad\square$

6. A matching result

We display here a matching result, related to the Monge-Kantorowitz transport problem; it assures, roughly, the existence of a relaxed transport which is of the order of the Prohorov distance. The result will serve in the derivation of the estimate established in the next section.

Recall the variational distance, which we denote by $\|\mu_1 - \mu_2\|$, between two probability measures μ_1 and μ_2 on a metric space M; it is the maximum of $|\mu_1(B) - \mu_2(B)| + |\mu_1(M\backslash B) - \mu_2(M\backslash B)|$ over the Borel subsets B of M. Recall that $\mathcal{P}(M)$ denotes the space of probability measures on the space M. If μ is a measure on M and $p(\cdot) : M \to \mathcal{P}(M)$ is a Young measure then (in full analogy with the point-valued case) $p(\mu)$ is the induced measure, namely, $p(\mu)(B) = \int_M p(s)(B)\mu(ds)$.

Definition 6.1. Let M be a complete separable metric space with metric $d(\cdot, \cdot)$. Let μ and ν be two probability measures on M. A Young measure $p(s) : M \to \mathcal{P}(M)$ is a relaxed η-matching from μ to ν if for every s the measure $p(s)$ is supported on the η-neighborhood of s and if $\|\nu - p(\mu)\| \le 2\eta$.

Remark 6.2. When $p(s) : M \to \mathcal{P}(M)$ is a relaxed η-matching from μ to ν, then the equivalent measure (see Section 4) $\mathbf{p}$ on $M \times M$ is supported on $\{(s_1, s_2) : d(s_1, s_2) \le \eta\}$. Furthermore, the disintegration of $\mathbf{p}$ with respect to the second coordinate is a relaxed η-matching from ν to μ. We may say then that $\mathbf{p}$ is a relaxed η-matching between ν and μ.

Proposition 6.3. *Let M be a compact metric space with metric $d(\cdot, \cdot)$. Let μ and ν be two probability measures on M such that $\mathrm{Proh}(\mu, \nu) \le \eta$. Then there exists a relaxed η-matching between μ and ν.*

Proof. We rely on a result established in Artstein.[1] We add to M an isolated point, say s_0, and let both μ and ν assign to it the weight η. (Now both μ and ν are not probability measures, but this will not change the argument). With each $s \in S$ we associate the set $F(s) = \{s\}^{(\eta)} \cup \{s_0\}$, namely, $F(s)$ is the union of the closed η-neighborhood of s and the singleton s_0. Then $F(s)$ is compact. Denote by $F(B)$ the union of $\{F(s) : s \in B\}$. It is easy to see that for every Borel set B in $M \cup \{s_0\}$ the inequality

$$\mu(B) \leq \nu(F(B)) \tag{22}$$

holds. Indeed, since $\mathrm{Proh}(\mu, \nu) \leq \eta$ it follows that $\mu(B) \leq \nu(B^{(\eta)}) + \eta$; then (22) follows since $\nu(\{s_0\}) = \eta$. By [1, Theorem 3.1], a mapping $p(\cdot) : M \cup \{s_0\} \to \mathcal{P}(M \cup \{s_0\})$ exists such that the equality $\mu = p(\nu)$ holds on $M \cup \{s_0\}$ We modify now $p(\cdot)$ as follows. Whenever $p(s)$ assigns a positive weight to $\{s_0\}$ we shift this weight to $\{s\}$. Now $p(\cdot)$ is from M to $\mathcal{P}(M)$. Since both measures assign to $\{s_0\}$ a weight η, it is clear that the variational distance between $p(\mu)$ and ν is less than or equal to 2η; this completes the proof. $\qquad\square$

Observation 6.4. It is clear that the previous result holds also when the two measures, rather than being probability measures, satisfy $\mu(M) = \nu(M)$.

7. Sensitivity for chattering parameters

Here is the quantitative sensitivity estimate for the system (1) with measure-valued coefficients, namely, the system (3).

Theorem 7.1. *Under Assumption 2.1 there exists a constant θ such that: Let ρ_1 be a chattering parameter for (4) and let $\mathbf{v}_1$ be a relaxed control that when applied with ρ_1 it results in the trajectory $x_1(\cdot)$ and the cost c_1. Let ρ_2 be another chattering parameter. Then there exists a relaxed control $\mathbf{v}_2$, that when applied with ρ_2 it induces a trajectory $x_2(\cdot)$ and a cost c_2, such that $|x_1(t) - x_2(t)| \leq \theta\mathrm{Proh}(\rho_1, \rho_2)$, and $|c_1 - c_2| \leq \theta\mathrm{Proh}(\rho_1, \rho_2)$. The constant θ depends only on the length $b - a$ of the interval, the Lipschitz constant κ and the bound β guaranteed by the assumption.*

Proof. Consider the space $M = [a, b] \times \Gamma$ with the $\mathbf{L}^1$-metric on the product, namely, the distance between (t_1, γ_1) and (t_2, γ_2) is $|t_1 - t_2| + d(\gamma_1, \gamma_2)$. Consider the Young measures ρ_1 and ρ_2 as measures on M. Denote their Prohorov distance by η. In view of Proposition 6.3 (see Observation 6.4) there is a relaxed η-matching $p_2(\cdot) : M \to \mathcal{P}(M)$ from ρ_2 to ρ_1. (We put the subscript in $p_2(\cdot)$ in order to emphasize that it is a matching from ρ_2 to ρ_1). We use this matching in the definition of $\mathbf{v}_2$.

For the simplicity of the argument we first work out the case where the control $\mathbf{v}_1$ is an ordinary control, i.e., the values $v_1(s)$ are points in U (recall that $s \in M$ is a pair $(t, \gamma) \in [a, b] \times \Gamma$). We define $v_2(s)$ to be the probability measure on U which is the image by the function $v_1(\cdot)$ of the probability measure $p_2(s)$ on M; namely, $v_2(s)(C) = p_2(s)(\{\sigma : v_1(\sigma) \in C\})$. We claim that then the effective inputs to (4) (see Definition 3.3) induced by $(\boldsymbol{\rho}_1, \mathbf{v}_1)$ and $(\boldsymbol{\rho}_2, \mathbf{v}_2)$ are at most 2η-apart in the Prohorov distance.

To this end we first compare the effective inputs of $(\boldsymbol{\rho}_2, \mathbf{v}_2)$ and $(p_2(\boldsymbol{\rho}_2), \mathbf{v}_1)$, namely, when $\mathbf{v}_1$ is applied to the chattering parameter which is the image of $\boldsymbol{\rho}_2$ by $p_2(\cdot)$. Denote the two effective inputs by $\mathbf{e}_2$ and $\mathbf{e}_1'$ (each of these is a measure on $[a, b] \times \Gamma \times U$). Let D be a Borel subset of $[a, b] \times \Gamma \times U$. The η neighborhood of D includes the set, say D', given by $\{(t', r', u) : d((t', r'), (t, r)) \leq \eta \text{ for some } (t, r, u) \in D\}$. Since $p_2(s)$ is supported on an η-neighborhood of s it is clear that $\mathbf{e}_2(D) \leq \mathbf{e}_1'(D')$. In particular,

$$\mathbf{e}_2(D) \leq \mathbf{e}_1'(D^{(\eta)}). \tag{23}$$

The estimate can be reversed by applying the disintegration of $\mathbf{p}_2$ with respect to the second coordinate (see Remark 6.2); since the latter is also supported on the same set, it follows also that for each Borel subset D the inequality

$$\mathbf{e}_1'(D) \leq \mathbf{e}_2(D^{(\eta)}) \tag{24}$$

holds as well. The two displayed inequalities form a stronger version of the Prohorov distance, between, however, $\boldsymbol{\rho}_2$ and $p_2(\boldsymbol{\rho}_2)$ as chattering parameters. Now we take advantage of the fact that the variational distance between $p_2(\boldsymbol{\rho}_2)$ and $\boldsymbol{\rho}_1$ is less than or equal to 2η. Hence on each set D the values of the effective input, say $\mathbf{e}_1$, of $(\boldsymbol{\rho}_1, \mathbf{v}_1)$, and the effective input $\mathbf{e}_1'$ may differ by at most 2η. This results in the inequalities

$$\mathbf{e}_2(D) \leq \mathbf{e}_1(D^{(\eta)}) + 2\eta \tag{25}$$

and

$$\mathbf{e}_1(D) \leq \mathbf{e}_2(D^{(\eta)}) + 2\eta \tag{26}$$

(compare with (23) and (24)); these establish the claimed Prohorov distance. Once the Prohorov distance between the effective inputs is less than or equal to 2η, Theorem 5.1 together with Definition 3.3 imply the existence of the claimed estimate θ. This completes the proof in the special case of a point-valued control.

To establish the existence of θ in the general case where $v_1(s)$ may be a relaxed control we may proceed in several ways. One possibility is to approximate the relaxed control by an ordinary one (see Proposition 3.4) and apply the previous considerations to the approximation. (The resulting effective inputs may then be of distance slightly greater than 2η; a limit argument would produce a relaxed control with 2η being the distance among the effective inputs.) Another possibility is to extend the previous construction to relaxed controls. For instance, use the same construction when $v_1(s)$ is interpreted as a function to the metric space $\mathcal{P}(U)$. Then $v_2(s)$ is a Young measure into $\mathcal{P}(U)$, namely, a probability distribution on measures. The desired relaxed control should then assign to each set C the expectation of $\mu(C)$ according to this probability distribution. One way or another, the proof is complete. $\qquad\square$

Corollary 7.2. *The optimal value of problem* (3) *is continuous with respect to the chattering parameter ρ when the latter is endowed with the Prohorov metric; in fact, the optimal value is Lipschitz with respect to the Prohorov metric.*

8. Remarks and examples

In this section we provide some comments on the conditions in Assumption 2.1, and examples demonstrating the main results.

Remark 8.1. The Lipschitz dependence on the time variable demanded in Assumption 2.1 cannot be replaced by, say, continuity (this in contrast to many developments in ordinary differential equations). The reason is that the Prohorov metric on Young measures is linear with respect to shifts in time. As an example consider the scalar control system $\frac{dx}{dt} = t^{\frac{1}{2}}u$, $x(0) = x_0$. Consider the control $u_\tau(\cdot)$ defined by $u_\tau(t) = 0$ if $t \le \tau$ and $u_\tau(t) = 1$ if $t > \tau$. Then the Prohorov distance between $u_\tau(\cdot)$ and $u_\sigma(\cdot)$ (say $\tau > \sigma$) is $\tau - \sigma$, while the distance between the respected solutions is $\int_\sigma^\tau t^{\frac{1}{2}}dt$. Clearly, the Lipschitz dependence on the Prohorov metric near $t = 0$ is violated.

Remark 8.2. The requirement that $Q(x, t, u, \gamma)$ be Lipschitz can be relaxed to being locally Lipschitz. Indeed, as noted in Section 2 boundedness of the relevant trajectories is guaranteed by $f(x, t, u, \gamma)$ being Lipschitz. The Lipschitz condition on $f(x, t, u, \gamma)$ cannot be replaced by its local analog since then the solutions may not stay bounded.

We illustrate now the main result concerning relaxed controls.

Example 8.3. Consider the scalar system

$$\text{minimize} \qquad \int_0^1 (x(t)^2 + |u(t) - 1|^2)dt \tag{27}$$

$$\text{subject to} \qquad \frac{dx}{dt} = f(x,t)u, \quad x(0) = 0,$$

with $f(x,t)$ satisfying Assumption 2.1 (we employ here the observation in Remark 8.2). A direct inspection reveals that in general an optimal control of (27) must be relaxed; indeed, the Young measure whose value at each t is the probability measure which assigns equal probabilities to $\{1\}$ and $\{-1\}$, is optimal. A classical approximation to this relaxed control is the ordinary control, say $u_h(\cdot)$, which alternates between the values 1 and -1 on intervals of length h. The Prohorov distance between the optimal relaxed control and $u_h(\cdot)$ is h. Now, Theorem 5.1 assures that the error in the cost and in the trajectories, resulting when the approximation $u_h(\cdot)$ is used, is of order h. (This can be inspected directly in the classical version of the example, namely when $f(x,t) = 1$.)

Next we illustrate the developments concerning the chattering parameters. In particular, the discussion serves as a motivation to the structure presented in Section 3.

Example 8.4. Let $[a,b] = [0,1]$ and let the parameter function depend on a positive parameter ε and given by

$$\rho_\varepsilon(t) = g(t)p(\varepsilon^{-1}t), \tag{28}$$

where $g(t)$ is a prescribed continuous positive function, and $p(t) : [0, \infty) \to \mathbb{R}$ is periodic. In particular, the parameter function oscillates, having amplitude varying with time. We are interested in the case where ε is small (and its exact quantity may not even be known). Then, a nominal parameter function, say $\rho_0(t)$, of which $\rho_\varepsilon(t)$ is a relevant perturbation, does not exist. The Young measure limit captures the limit distribution of the values of the parameter function. Denote by μ the probability measure which is the distribution of the values of $p(\cdot)$ over one period. The Young measure limit, as $\varepsilon \to 0$, of the functions in (28) is the map $\nu(t)$ with values being probability measures on the real line $\mathbb{R}$, given by $\nu(t)(B) = \mu(g(t)^{-1}B)$ (here $rB = \{rx : x \in B\}$). As a particular example consider

$$\rho_\varepsilon(t) = \sin(\varepsilon^{-1}t) \tag{29}$$

Then the limit distribution is

$$\nu(t)(dr) = \frac{dr}{g(t)\pi(g(t) - r^2)^{\frac{1}{2}}}. \tag{30}$$

Consider now an optimization problem based on the example, say, the scalar system

$$
\begin{aligned}
\text{minimize} \quad & \int_0^1 (x^2 + u^2)\,dt \\
\text{subject to} \quad & \frac{dx}{dt} = x + \sin(\varepsilon^{-1}t)u, \quad x(0) = 1,
\end{aligned}
\tag{31}
$$

with u in a bounded interval. For a fixed ε the optimization problem is time-varying, yet can be solved using standard techniques. However, as $\varepsilon \to 0$ the computations become difficult to execute due to the oscillations of the parameter function. The chattering parameters limit is given by

$$
\begin{aligned}
\text{minimize} \quad & \int_0^1 \left(x^2 + \int_{[-1,1]} u^2(t,r)\nu(dr)\right) dt \\
\text{subject to} \quad & \frac{dx}{dt} = x + \int_{[-1,1]} u(t,r)\nu(dr), \quad x(0) = 1
\end{aligned}
\tag{32}
$$

where $\nu(dr) = \frac{dr}{\pi(1-r^2)^{\frac{1}{2}}}$ (we examine a simple version of (28), namely, with $g(t) = 1$). Notice that the chattering limit in this example is time-invariant; (the solution $u = u(t,r)$ may, however, depend on time).

The theory presented in Ref. 2 guarantees that the value of (32) is the limit, as $\varepsilon \to 0$, of the values of (31). It is easy to see that the Prohorov distance between the limit Young measure ν and the rapidly oscillating parameter is $\varepsilon 2\pi$. Hence, Theorem 7.1 implies that the error caused by first solving (32) and then obtaining an appropriate approximation, rather than solving (31), is of order ε.

References

1. Z. Artstein, *Israel J. Math.* **46**, 313 (1983).
2. Z. Artstein, *Forum Math.* **5**, 369 (1993).
3. Z. Artstein, The chattering limit of singularly perturbed optimal control problems, in *Proc. 39 IEEE Conf. Decision and Control (Sydney, Australia, 2000)*: 564-569.
4. Z. Artstein, On the value function of singularly perturbed optimal control systems, in *Proc. 43 IEEE Conf. Decision and Control (Paradise Island, Bahamas, 2004)*: 432-437.
5. Z. Artstein and V. Gaitsgory, *SIAM J. Control Optim.* **35**, 1487 (1997).
6. Z. Artstein and V. Gaitsgory, *Appl. Math. Optim.* **41**, 425 (2000).
7. E.J. Balder, *Rend. Istit. Mat. Univ. Trieste* **31**, suppl. 1, 1 (2000).

8. J.M. Ball, *A version of the fundamental theorem for Young measures*, in *Partial Differential Equations Models of Phase Transitions, (Nice, France, 1988)*, eds. M. Rascle, D. Serre and M. Slemrod, Lecture Notes in Physics, Vol. 344 (Springer, Berlin, 1989): 207-215.

9. P. Billingsley, *Convergence of Probability Measures*, Wiley Series in Probability and Statistics: Probability and Statistics, second edition (Wiley, New York, 1999). A Wiley-Interscience Publication.

10. P. Pedregal, *Parametrized Measures and Variational Principles*, Progress in Nonlinear Differential Equations and their Applications, 30 (Birkhäuser Verlag, Basel, 1997).

11. P. Pedregal, *SIAM J. Control Optim.* **36**, 797 (1998).

12. P. Pedregal, *Bull. Amer. Math. Soc.* **36**, 27 (1999).

13. W.T. Reid, *Ordinary Differential Equations* (Wiley, New York, 1971).

14. L. Tartar, *Compensated compactness and applications to partial differential equations*, in *Nonlinear Analysis and Mechanics: Heriot-Watt Symposium, Vol. IV*, eds. R.J. Knops, Research Notes in Mathematics, Vol. 39 (Pitman, Boston, 1979): 136-212.

15. M. Valadier, *Rend. Istit. Mat. Univ. Trieste* **26**, suppl., 349 (1994).

16. J. Warga, *Optimal Control of Differential and Functional Equations* (Academic Press, New York, 1972).

17. L.C. Young, *Lectures on the Calculus of Variations and Optimal Control Theory*, Foreword by Wendell H. Fleming (Saunders Co., Philadelphia, 1969).

SYSTEMS WITH CONTINUOUS TIME AND DISCRETE TIME COMPONENTS

A. BACCIOTTI

Dipartimento di Matematica, Politecnico di Torino
Corso Duca degli Abruzzi 24, 10129 Torino, Italy
E-mail: andrea.bacciotti@polito.it

We introduce a class of difference/differential equations which is sufficiently
large to include systems of interest for applications, but at the same time
sufficiently easy to handle. In this framework, we give in particular a detailed
and rather complete study of the asymptotic behavior of pairs of oscillators.
We finally introduce appropriate notions of stability and extend Liapunov first
and second theorem.

Keywords: Hybrid systems; Stability; Liapunov functions.

1. Introduction

Motivated by a wide range of industrial and technological applications,
there has been a rapid growth of interest in the recent engineering literature
about hybrid systems. The term "hybrid" often is applied informally to
denote systems which combine subsystems of different nature, and whose
time evolution is characterized by discontinuities in the state (impulses)
and/or in the velocity (switches). A large variety of systems with very
complex behavior fall in this class. For this reason, it is hard to figure
an axiomatic definition of hybrid system. Recently, there have been some
interesting attempts, see Refs. 2,8,10,13,15,17. However, we remark that all
these definitions turn out to be extremely formal and abstract (this is the
obvious price to be payed if we want to include more and more general
classes of systems), and hence difficult to handle.

In this paper we consider a class of systems, large enough to include,
as particular cases, finite dimensional continuous time systems, discrete
time systems, open loop switched systems, feedback systems with quantized
control, sample data systems, certain types of delayed differential equations
and certain types of hybrid systems with timed automata. They will be

called *systems with continuous time and discrete time components.*

Systems with continuous time and discrete time components admit a simple mathematical representation and hence, although their generality is limited (for instance, they cannot account for impulse effects), they have the advantage of being rather concrete. On the other hand, we notice that certain aspects of systems with continuous time and discrete time components are not covered by the definitions given in Refs. 10 (where switches can be interpreted as state discontinuities but not as changes of the dynamical rules), and in Refs. 2,8,13,15 (where the state space of the discrete time component is finite).

As in Refs. 10,13,17, we address the stability problem. Far from being surprising, our results are natural generalizations of Liapunov first and second theorems. Nevertheless, their proofs are not completely obvious and cannot be deduced from the existing literature. In particular, we note that our results, compared with Ref. 17, are more precise and require less conservative assumptions, in spite of a less general setting.

We now shortly explain the organization of the paper. Section 2 contains the definition of system with continuous time and discrete time components, comments on its generality and further comparisons with analogous definitions available in the literature. In Section 3 we discuss in detail an example: we will see that in spite of a very simple structure, the asymptotic behavior of a system with continuous time and discrete time components may be very complex. The notions of stability and asymptotic stability are stated in Section 4. Section 5 is devoted to the extension of Liapunov first theorem: we give a proof of it and some other remarks. Two slight different extensions of Liapunov second theorem are finally presented in Section 6.

2. Description of the model

We are interested in objects defined by the following set of data:

- an integer $n \geq 1$;
- a locally compact metric space Q
- a continuous map $f(x, q) : \mathbb{R}^n \times Q \to \mathbb{R}^n$;
- a continuous map $g(x, q) : \mathbb{R}^n \times Q \to Q$;
- a sequence $\{d_k\}$, such that $d_k > 0$ for each $k = 0, 1, 2, \ldots$ and $\lim_{k \to +\infty} \sum_{i=0}^{k} d_i = +\infty$.

Throughout this paper, such objects are called *systems with continuous time and discrete time components* (in short, CTDTC-systems), and conventionally represented by writing

$$\begin{cases} \dot{x} = f(x, q_k) \\ q_{k+1} = g(x, q_k) \ . \end{cases} \tag{1}$$

For any given $\bar{t} \in \mathbb{R}$, we set $\tau_0 = \bar{t}$, $\tau_1 = \tau_0 + d_0$, $\tau_2 = \tau_1 + d_1, \ldots, \tau_{k+1} = \tau_k + d_k, \ldots$. By a *solution of (1) corresponding to the initial condition* $(\bar{t}, \bar{x}, \bar{q}) \in \mathbb{R} \times \mathbb{R}^n \times Q$, we mean any pair $(\varphi(t), \{u_k\})$ such that:

- $\varphi(t) : [\bar{t}, +\infty) \to \mathbb{R}^n$ is a curve with $\varphi(\bar{t}) = \bar{x}$, which is assumed to be continuous at every $t \geq \bar{t}$;
- $\{u_k\}$ is a sequence in Q, with $u_0 = \bar{q}$;
- for each $k = 0, 1, 2, \ldots$ and each $t \in (\tau_k, \tau_{k+1})$, $\varphi(t)$ is differentiable and

$$\dot{\varphi}(t) = f(\varphi(t), u_k) \ ;$$

- for each $k = 0, 1, 2, \ldots$

$$u_{k+1} = g(\varphi(\tau_{k+1}), u_k) \ .$$

The idea underlying this notion of solution can be intuitively described in this way. Starting from the point $\bar{x}$, the continuous time component evolves according to the differential equation

$$\dot{x} = f(x, \bar{q}) = f(x, u_0)$$

on the interval $[\tau_0, \tau_1]$, while the discrete time component remains unchanged. At the instant τ_1, the discrete time component is updated, according to

$$u_1 = g(\varphi(\tau_1), u_0) \ .$$

Then, on the subsequent interval $[\tau_1, \tau_2]$ the continuous time component evolves according to the new equation

$$\dot{x} = f(x, u_1)$$

and so on. To clear up the notation, it is convenient to introduce the map

$$h(t) = k \quad \text{for} \ \ t \in [\tau_k, \tau_{k+1}) \ , \qquad (k = 0, 1, 2, \ldots)$$

so that a solution can be written as a curve

$$t \mapsto (\varphi(t), u_{h(t)}) : [\bar{t}, +\infty) \to \mathbb{R}^n \times Q .$$

Note that such a curve is not continuous, in general. Note also that in order to guarantee existence of solutions which are actually defined for each $t \geq \bar{t}$, the continuity of f is not sufficient in general, due to the possible finite escape time phenomenon. To prevent it, we assume that the vector field $f(\cdot, q)$ is complete for each $q \in Q$ (sufficient conditions for completeness are well known and can be found on the more popular handbooks about ordinary differential equations). In what follows, uniqueness of solutions plays no role at all.

Remark 2.1. The following remarks illustrate the generality of CTDTC-systems.

(a) If Q reduces to a singleton, then (1) reduces to a finite dimensional, time-invariant system of ordinary differential equations.

(b) If $n = 0$ and $Q = \mathbb{R}^m$, then (1) reduces to a finite dimensional, discrete time dynamical system.

(c) If $d_k = 1$ for each $k = 0, 1, 2, \ldots$, $Q = \mathbb{R}^n$, $g(x, q) = x$ and $\bar{t} = 0$, then (1) reduces to

$$\dot{x}(t) = f(x(t), x([t])) \tag{2}$$

where $[t]$ denotes the greatest integer less than or equal to t. Note that (2) is a retarded differential equation of the type considered in Ref. 7.

(d) If $g(x, q) = g(x) : \mathbb{R}^n \to Q$, then (1) describes a system with a feedback connection, where the actuator is a digital device which is able to change its value only at the prescribed instants $\tau_0, \tau_1, \ldots$. This situation is similar to what happens in the so-called *quantized control problems*, where one initially starts with a feedback $g(x)$ which can vary continuously and free of constraints, but then the levels of quantization must be found in such a way to preserve the achievement of the control goal.

(e) If $d > 0$ is a fixed time size and $d_k = d$ for each $k = 0, 1, 2, \ldots$, a CTDTC-system can be thought of as sample-data systems.[9]

(f) If Q is a finite set (endowed with the discrete metric) and $g(x, q) = g(q) : Q \to Q$, then the sequence $\{u_k\}$ is independent of the evolution of x and can be computed in advance. Then (1) reduces to a switched system of the type considered in Refs. 6,16.

Remark 2.2. One feature of our definition of CTDTC-system is that changes in the continuous time dynamics can occur only at the prescribed instants $\tau_0, \tau_1, \ldots$. This feature is shared by similar notions available in the literature (see for instance Refs. 2,8,10,13). We point out that if Q is finite, our definition of CTDTC-system can be viewed as a special case of the definition of hybrid system studied in Refs. 8,13, the differences being that here the discrete state transitions are uniquely determined by a function, rather than by a relation (or by a set valued map as in Ref. 10), and the *reset map* is the identity (which implies in particular that systems with impulsive effects are not comprised in (1)). On the other hand, the generalization to sets Q which are not necessarily finite sets, enables us to include a wider range of applications, as indicated by the examples above.

Remark 2.3. Functions f and g do not depend explicitly on time. Nevertheless, because of the constraint on the updating times, the translation of a solution $(\varphi(t + T), u_{h(t+T)})$ in general is no more a solution. In other words, the semigroup property does not hold. Accordingly, we should not expect that a CTDTC-system behaves as a time-invariant one.

3. Oscillatory systems: an example

In spite of its simplicity, a CTDTC-system may exhibit very complex and unexpected behaviors. In this section we discuss with some details the system

$$\begin{cases} \dot{x} = -q_k y \\ \dot{y} = \dfrac{x}{q_k} \\ q_k = \dfrac{1}{q_{k-1}} \end{cases} \tag{3}$$

where $n = 2$, $Q = \{\omega, \frac{1}{\omega}\}$, and $\omega > 1$ is a given real number. Equivalently, we can look at (3) as a switched system formed by the pair of harmonic oscillators

$$\begin{cases} \dot{x} = -\omega y \\ \dot{y} = \dfrac{x}{\omega} \end{cases} \tag{4}$$

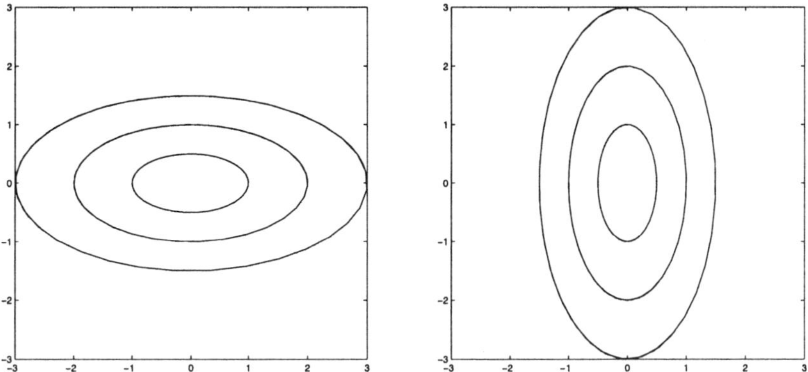

Fig. 1. *Some trajectories of system (4) on the left and some trajectories of system (5) on the right ($\omega = 2$)*

and

$$\begin{cases} \dot{x} = -\dfrac{y}{\omega} \\ \dot{y} = \omega x \, . \end{cases} \tag{5}$$

Some trajectories of these systems are plotted in Figure 1. Note that both (4) and (5) are stable at the origin. For simplicity, we limit ourselves to the case where $\bar{t} = 0$, $d_k = T$ (i.e., $\tau_k = kT$), T being a fixed real number, $T \in (0, 2\pi]$. Moreover, without loss of generality we agree that $\bar{q} = \omega$. It is clear and well known[6] that if we choose $\tau_k = k\pi/2$, then the behavior of the solutions depends on the initial condition $(\bar{x}, \bar{y}, \bar{q})$. For instance, if $(\bar{x}, \bar{y}, \bar{q}) = (1, 0, \omega)$ then it is natural to guess that $(x(t), y(t))$ converges to the origin for $t \to +\infty$, while if $(\bar{x}, \bar{y}, \bar{q}) = (0, 1, \omega)$ then $(x(t), y(t))$ becomes larger and larger as $t \to +\infty$. In fact, examples similar to the present one are often invoked in order to show that a switched system may exhibit features which are not recognizable in the singular subsystems.

Here, our purpose is to analyze how the behavior of the system actually depends on the choices of ω and T. We show in particular that for "many" values of ω the system is actually stable*, and that for the remaining values of ω the behavior of the solution corresponding to a fixed initial condition is extremely sensitive to the choice of T: in particular, we will see that the

---

*Stability of systems with continuous time and discrete time components will be formally defined later; for the moment, the term "stable" is used in the obvious heuristic meaning.

occurrence of trajectories convergent to the origin is extremely rare, and practically impossible to simulate in machine experiments.

We start by computing, for $t = T$, the fundamental matrix of system (4)

$$\Phi_{(4)}(T) = \begin{pmatrix} \cos T & -\omega \sin T \\ \dfrac{\sin T}{\omega} & \cos T \end{pmatrix}$$

and the fundamental matrix of system (5)

$$\Phi_{(5)}(T) = \begin{pmatrix} \cos T & -\dfrac{\sin T}{\omega} \\ \omega \sin T & \cos T \end{pmatrix} .$$

The idea is to look at (3) as a discrete time system of $\mathbb{R}^2$, whose state is updated at the instants $0, 2T, 4T, 6T, \ldots$. More precisely, we study the system

$$\begin{pmatrix} x_k \\ y_k \end{pmatrix} = \Phi(T) \begin{pmatrix} x_{k-1} \\ y_{k-1} \end{pmatrix}, \qquad \begin{pmatrix} x_0 \\ y_0 \end{pmatrix} = \begin{pmatrix} \bar{x} \\ \bar{y} \end{pmatrix} \tag{6}$$

where

$$\Phi(T) = \Phi_{(5)}(T)\Phi_{(4)}(T) = \begin{pmatrix} \cos^2 T - \dfrac{\sin^2 T}{\omega^2} & -\dfrac{1+\omega^2}{\omega} \sin T \cos T \\ \dfrac{1+\omega^2}{\omega} \sin T \cos T & \cos^2 T - \omega^2 \sin^2 T \end{pmatrix} .$$

It is clear, and not difficult to prove, that the stability properties of (3) can be deduced from those of (6) (see [4, Ch. 8]). To compute the eigenvalues of $\Phi(T)$, we must solve the equation

$$p_{\Phi(T)}(\lambda) = \lambda^2 - \left[2\cos^2 T - \frac{1+\omega^2}{\omega^2} \sin^2 T \right] \lambda + \frac{(1+\omega^2)^2}{\omega^2} \sin^2 T \cos^2 T$$

$$+ \left(\cos^2 T - \frac{\sin^2 T}{\omega^2} \right) \left(\cos^2 T - \omega^2 \sin^2 T \right)$$

whose discriminant is

$$\Delta(T) = \frac{(1+\omega^2)^2}{\omega^2}\sin^2 T\left[\frac{(1+\omega^2)^2}{\omega^2}\sin^2 T - 4\right].$$

Note that $\Delta(T) = 0$ only in the following two cases:

(C_1) $\sin^2 T = 0$, that is $T = \pi$ or $T = 2\pi$.

(C_2) $0 < \sin^2 T = \dfrac{4\omega^2}{(1+\omega^2)^2} < 1$, which gives rise to exactly 4 distinct solutions in $(0, 2\pi)$.

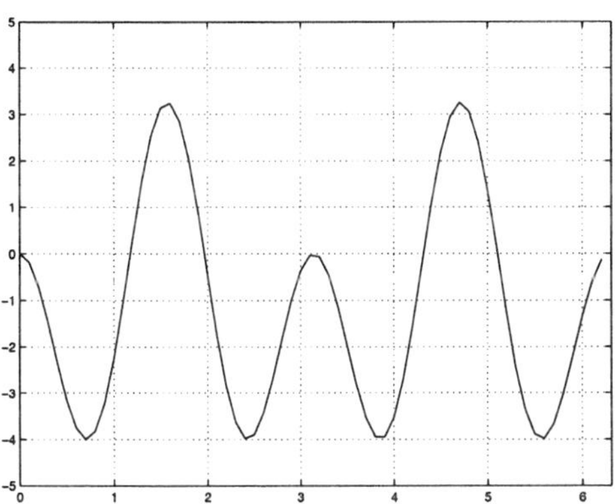

Fig. 2. *Graph of $\Delta(T)$*

The graph of $\Delta(T)$ is plotted in Figure 2 for $\omega = 1.5$. Let us examine first the case $\Delta < 0$. It is not difficult to check that in this case $\Phi(T)$ has a pair of conjugate (distinct) eigenvalues, lying exactly on the boundary of the unit circle of the complex plane. Hence, system (3) is stable.

Now, we pass to consider the case $\Delta(T) > 0$. Here, we have real eigenvalues. It is not difficult to see that one of them is always less than -1, while the other is inside the interval $(-1, 1)$. Hence in this case the system is unstable. More precisely, system (6) has a saddle point at the origin: the stable manifold coincides with the x-axis and the unstable manifold with the y-axis. If we assign an initial condition on the x-axis, we therefore expect that the trajectory converge toward the origin. Surprisingly, this prediction seems to be contradicted by numerical experiments: see Figure 3, where $T = \pi/2$. What actually happens is that, π being an irrational number, round off errors are inevitable in machine computations; as a consequence, it is impossible to keep a simulated trajectory inside the stable manifold when k becomes larger and larger. Note that $\Delta(\pi/2) > 0$ for every $\omega > 1$, and that the measure of $\{T \in (0, 2\pi) : \Delta(T) > 0\}$ goes to zero as $\omega \to 1^+$.

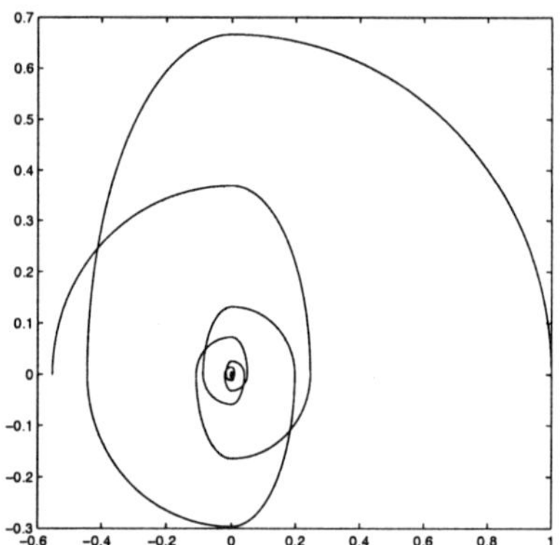

Fig. 3.　*Trajectory with $\omega = 1.5$, $T = \pi/2$ starting from $(1, 0)$*

Finally, when $\Delta(T) = 0$ the eigenvalues of $\Phi(T)$ coincide: they are both equal to 1 in case (C_1), and equal to -1 in case (C_2). Moreover, the eigenvalue is simple in case (C_1), so that the system is stable, but not in case (C_2), so that the system is not stable.

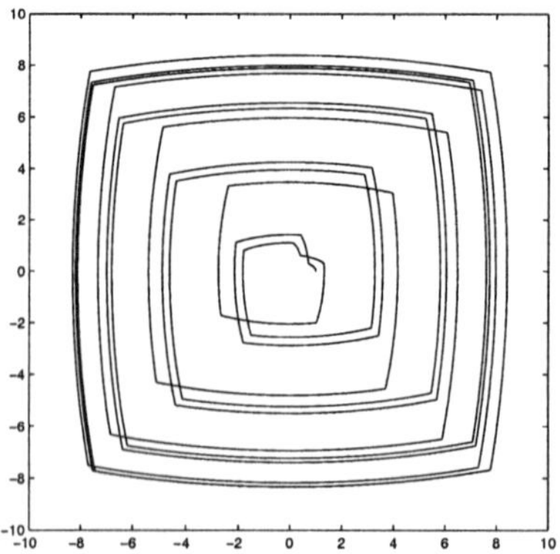

Fig. 4.　*Trajectory with $\omega = 2.4$, $T = \pi/4$ starting from $(1, 0)$*

Of course, we can look at the problem from an other point of view; for instance we can fix T and take ω as a parameter. Let us considered for instance the choice $T = \pi/4$. Our investigation reveals that with this choice, the system is stable only if $\omega < \sqrt{3 + \sqrt{8}} = 2.4142\ldots$ (see Figures 4 and 5).

Remark 3.1. We know that for certain values of T there are trajectories of system (4) which diverge. Because of periodicity, the same happens of course if we replace T by $T + 2k\pi$ (for each k). This shows that for simple stability we do not have an analogue of Lemma 2 in Ref. 14.

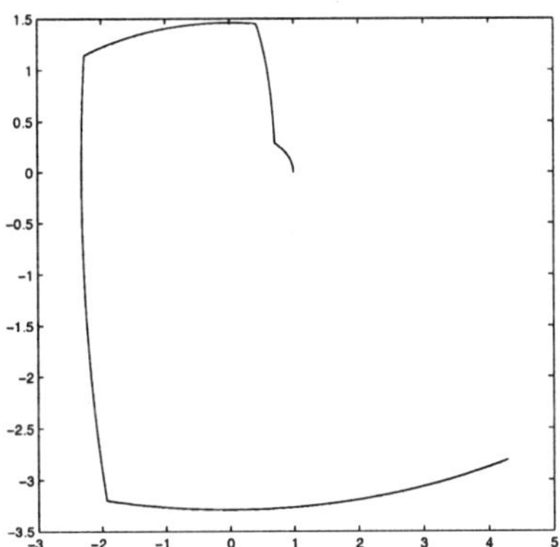

Fig. 5.   *Trajectory with $\omega = 2.5$, $T = \pi/4$ starting from $(1, 0)$*

4. Stability notions

Motivated by the example of the previous section, we give some definitions of stability which seem to be appropriate for CTDTC-systems.

Let M be a compact subset of Q. Let us denote $\mathbf{d}_M(q) = \min_{p \in M} \mathbf{d}(p, q)$, where $\mathbf{d}$ is the distance function of Q. The euclidian norm of $x \in \mathbb{R}^n$ is denoted by $|x|_{\mathbb{R}^n}$. Moreover, we set $E = \mathbb{R}^n \times Q$. For $(x, q) \in E$, let us denote $N(x, q) = \max\{|x|_{\mathbb{R}^n}, \mathbf{d}_M(q)\}$. We also denote $B_{\mathbb{R}^n}(r) = \{x : |x|_{\mathbb{R}^n} < r\}$, $B_Q(r) = \{q : \mathbf{d}_M(q) < r\}$ and $B_E(r) = \{(x, q) : N(x, q) < r\}$, where $r > 0$. The subscripts $\mathbb{R}^n$, Q and E will be dropped out, when there is no risk of ambiguity. From now on, we assume

(A_1) $f(0, p) = 0$ for each $p \in M$
(A_2) $g(0, M) \subseteq M$

that is, the origin of $\mathbb{R}^n$ is an equilibrium point for the continuous time component of the system for each $p \in M$, and M is a positively invariant set for the discrete time component, when $x = 0$.

Definition 4.1. A CTDTC-system is *uniformly stable* with respect to $\{0\} \times M$ if for each $\varepsilon > 0$ there exists $\eta > 0$ such that

$$N(\bar{x}, \bar{q}) < \eta \implies N(\varphi(t), u_{h(t)}) < \varepsilon$$

for each $\bar{t} \in \mathbb{R}$, each $t \geq \bar{t}$ and each solution $(\varphi(t), u_{h(t)})$ corresponding to the initial condition $(\bar{t}, \bar{x}, \bar{q})$.

A CTDTC-system is *uniformly-uniformly stable* with respect to $\{0\} \times M$ if it is uniformly stable for each choice of the sequence $\{d_k\}$.

Definition 4.2. A CTDTC-system is *(locally) uniformly-uniformly asymptotically stable* with respect to $\{0\} \times M$ if it is uniformly-uniformly stable and, in addition, there exists $\delta_0 > 0$ such that for each $\bar{t} \in \mathbb{R}$

$$N(\bar{x}, \bar{q}) < \delta_0 \implies \lim_{t \to +\infty} N(\varphi(t), u_{h(t)}) = 0$$

for each solution $(\varphi(t), u_{h(t)})$ corresponding to the initial condition $(\bar{t}, \bar{x}, \bar{q})$.

We emphasize that the definitions above depend on the choice of the origin, as a special steady state of $\mathbb{R}^n$, and of the set $M \subseteq Q$: in what follows, we omit to mention them explicitly for the sake of simplicity, since no ambiguity is possible. We remark also that Definitions 4.1 and 4.2 could be referred to a more general set $M_0 \times M$, where M_0 is a compact subsets of $\mathbb{R}^n$ not reduced to the origin. Again, the origin has been chosen to simplify the exposition. Instead, as far as the discrete dynamics are concerned, an analogous simplification is not convenient. The reason is that in some applications Q might be a finite set with no distinguished elements. If in addition Q is endowed with the discrete topology, then by (A_2), M plays no role at all in checking stability: in particular, when M is a singleton, the problem becomes trivial. Note that if we are interested in stability of the continuous time component alone, we can take $M = Q$ and look at the discrete time component as a stabilizing device.

5. A sufficient condition for stability

The following result is the natural extension of Liapunov First Theorem to nonlinear CTDTC-systems. With respect to the well known classical case, the interplay between the continuous time dynamics and the discrete time one requires some more care in the proof.

Theorem 5.1. *Let the CTDTC-system (1) be given. Assume that there exists $r > 0$ and a continuous map $V : B_E(r) \to \mathbb{R}$ such that:*

(i) $V(x,q)$ is positive definite at $\{0\} \times M$, that is $V(x,q) \geq 0$ and $V(x,q) = 0$ for each $(x,q) \in B_E(r)$ implies $(x,q) \in \{0\} \times M$;

(ii) for each $q \in B_Q(r)$, the map $x \mapsto V(x,q)$ is of class C^1 on $B_{\mathbb{R}^n}(r)$;

(iii) $\nabla_x V(x,q) f(x,q) \leq 0$, for each $(x,q) \in B_E(r)$;

(iv) $V(x, g(x,q)) \leq V(x,q)$, for each $(x,q) \in B_E(r)$.

Then, the system is uniformly-uniformly stable.

Proof. Let $0 < R < r$ so that V is defined and continuous on $\overline{B_E(R)}$. Let $m_R = \inf_{N(x,q)=R} V(x,q)$. It is clear that the set $\{(x,q) : N(x,q) = R\}$ is compact. Hence, m_R is actually a minimum and, by (i), $m_R > 0$. The set

$$\Omega = \{(x,q) : V(x,q) < m_R\}$$

is open, and $\{0\} \times M \subset \Omega$. Let Ω_0 be the connected component (i.e., the largest connected subset) of Ω which contains $\{0\} \times M$. Of course, $\Omega_0 \subset B_E(R)$. Let us consider the continuous map $\tilde{g}(x,q) = (x, g(x,q)) : E \to E$.

We claim that $\tilde{g}(\Omega_0) \subset \Omega_0$. Indeed, from $(x,q) \in \Omega_0$ and (iv) it follows

$$V(x, g(x,q)) \leq V(x,q) < m_R$$

which in turn implies $\tilde{g}(x,q) \in \Omega$. Moreover, if $p \in M$, then by virtue of (A_2) we have $\tilde{g}(0,p) = (0, g(0,p)) \in \{0\} \times M$; on the other hand, $\tilde{g}(0,p) \in \tilde{g}(\Omega_0)$, since $\{0\} \times M \subset \Omega_0$. This means that

$$\tilde{g}(\Omega_0) \cap \{0\} \times M \neq \emptyset .$$

The claim is proven, since the continuous image $\tilde{g}(\Omega_0)$ of the connected set Ω_0 is connected.

Pick now $\varepsilon > 0$ such that $B_E(\varepsilon) \subset \Omega_0 \subset B_E(R)$. Let

$$m_\varepsilon = \inf_{\varepsilon \leq N(x,q) \leq R} V(x,q) .$$

Again, we have that m_ε is a minimum and $m_\varepsilon > 0$. Let $\delta > 0$ such that

$$N(x,q) < \delta \implies V(x,q) < m_\varepsilon .$$

Of course, $\delta < \varepsilon$. Let $\bar{t} \in \mathbb{R}$, $(\bar{x}, \bar{q}) \in B_E(\delta)$, and let $(\varphi(t), u_{h(t)})$ be any solution of (1) such that $(\varphi(\tau_0), u_0) = (\bar{x}, \bar{q})$. We want to prove that

94

$(\varphi(t), u_{h(t)}) \in B_E(\varepsilon)$ for each $t \geq \tau_0$. To this purpose, using the mathematical induction principle, we show that the statement

$$(\varphi(t), u_{h(t)}) \in B_E(\varepsilon) \quad \text{for each} \quad t \in [\tau_k, \tau_{k+1})$$

is true for each $k = 0, 1, 2, \ldots$. We proceed according to the following pattern.

First step $(k = 0)$. Using the fact that $V(\bar{x}, \bar{q}) < m_\varepsilon$, we prove that

$$(\varphi(t), u_0) \in B_E(\varepsilon) \quad \text{for each} \quad t \in [\tau_0, \tau_1] \tag{7}$$

and, in addition,

$$(\varphi(\tau_1), u_1) \in B_E(\varepsilon) \quad \text{and } V(\varphi(\tau_1), u_1) < m_\varepsilon . \tag{8}$$

Inductive step $(k > 0)$. Assuming that $(\varphi(\tau_k), u_k) \in B_E(\varepsilon)$ and $V(\varphi(\tau_k), u_k) < m_\varepsilon$, we prove that

$$(\varphi(t), u_k) \in B_E(\varepsilon) \tag{9}$$

for each $t \in [\tau_k, \tau_{k+1}]$,

$$(\varphi(\tau_{k+1}), u_{k+1}) \in B_E(\varepsilon) \quad \text{and} \quad V(\varphi(\tau_{k+1}), u_{k+1}) < m_\varepsilon . \tag{10}$$

Proof of the first step. Let $T \in (\tau_0, \tau_1]$ be such that $(\varphi(t), u_0) \in B_E(\varepsilon)$ for each $t \in [\tau_0, T]$. Because of (iii), we obviously have

$$V(\varphi(t), u_0) \leq V(\bar{x}, \bar{u}) < m_\varepsilon \tag{11}$$

for each $t \in [\tau_0, T]$. Since $\varphi(t)$ is continuous, using (11) and arguing by contradiction, it is immediate to check the validity of (7) (note in particular that since $d_M(u_0) < \varepsilon$, $|\varphi(T_0)| = \varepsilon$ for some T_0 implies $N(\varphi(T_0), u_0) = \varepsilon$). In fact, we have $V(\varphi(\tau_1), u_0) < m_\varepsilon$.

From $(\varphi(\tau_1), u_0) \in B_E(\varepsilon) \subset \Omega_0$ it follows $(\varphi(\tau_1), u_1) \in \Omega_0$. Next, by (iv)

$$V(\varphi(\tau_1), u_1) \leq V(\varphi(\tau_1), u_0) < m_\varepsilon .$$

This in turn implies that $(\varphi(\tau_1), u_1)$ cannot belong to $\Omega_0 \setminus B_E(\varepsilon)$. The validity of (8) is so proven.

Proof of the inductive step. Taking into account the inductive assumption, the proof that $(\varphi(t), u_k)$ remains in $B_E(\varepsilon)$ for $t \in [\tau_k, \tau_{k+1}]$ can be carried out as in the case $k = 0$. In particular, we can conclude that

$$(\varphi(\tau_{k+1}), u_k) \in B_E(\varepsilon) \tag{12}$$

and

$$V(\varphi(\tau_{k+1}), u_k) < m_\varepsilon . \tag{13}$$

From (12) it follows that $(\varphi(\tau_{k+1}), u_{k+1}) \in \Omega_0$, and from (13) and (iv) it follows $V(\varphi(\tau_{k+1}), u_{k+1}) < m_\varepsilon$. Hence, $(\varphi(\tau_{k+1}), u_{k+1})$ cannot belong to $\Omega_0 \setminus B_E(\varepsilon)$.

The statement is proven, taking into account that no special role is played in the proof by the sequence $\{d_k\}$. $\qquad\qquad\square$

Remark 5.1. Conditions (iii), (iv) imply that for each solution $(\varphi(t), u_{h(t)})$, the map $\gamma(t) = V(\varphi(t), u_{h(t)})$ is nonincreasing, at least as far as $(\varphi(t), u_{h(t)})$ remains in $B_E(r)$. This fact will be used later, in the proofs of Theorems 6.1 and 6.2. However, it does not allow us to obtain a simpler proof of Theorem 5.1 (on the contrary of what happens for the classical Liapunov first theorem) since in general $(\varphi(t), u_{h(t)})$ is not continuous.

Remark 5.2. Basically, stability is a local notion. It is transformed in a global one when combined with Lagrange stability.[1] A corresponding global version of Theorem 5.1 can be stated under the additional assumption

(v) for each $q \in Q$, the function $x \mapsto V(x, q)$ is defined for each $x \in \mathbb{R}^n$ and radially unbounded.

It is worthwhile to mention that assumption (v) is actually made in Ref. 17 to obtain a merely local result: it also enables the authors to bypass the more involved aspects of the proof (due to the co-existence of discrete and continuous dynamics) and all become simpler.

As in the classical Liapunov theory, the assumption about the differentiability of $x \mapsto V(x, u)$ can be weakened; in fact, it is sufficient to ask that it is lower semicontinuous; accordingly, an appropriate notion of generalized derivative must be used in (ii) (see Ref. 3).

In a topological context, even the monotonicity condition (ii) can be relaxed.[17] We finally point out that, by allowing time-varying Liapunov functions, a converse of Theorem 5.1 has been obtained in Ref. 17.

Remark 5.3. We can interpret $V(x, q)$ as a family of Liapunov functions indexed by $q \in Q$, and condition (iv) as a kind of compatibility condition for multiple Liapunov functions.[5,6,11,12] To this respect, we point out that the condition imposed in Refs. 5,6,11,12 ore formally more general but they require an explicit knowledge of the solutions. Our condition is more conservative, but easier to apply in practice.

Remark 5.4. For the case where Q is finite, conditions (iii) and (iv) of Theorem 5.1 are the same as the conditions 1) and 2) of Theorem IV.1 of Ref. 13. To this respect, we remark that in Ref. 13 the authors are mainly interested in attractivity of certain compact sets (generalization of LaSalle invariance principle); in a sense, our Theorem 5.1 complements Theorem IV.1 of Ref. 13, showing that the selected compact set is not only attractive, but also stable.

Example 5.1. We can apply Theorem 5.1 in order to prove stability for the following system:

$$\begin{cases} \dot{x} = -(1 + q_k^2)y \\ \dot{y} = x \\ q_k = \dfrac{q_{k-1}}{2} \end{cases} \tag{14}$$

where $n = 2$, $Q = \mathbb{R}$ and $M = \{0\}$. The continuous time component of this system can be viewed as a family of harmonic oscillators (however, we notice that the eigenvalues here are different from those of example treated in Section 3). Define $V(x, y, q) = x^2 + (1 + q^2)y^2 + q^2$. It is not difficult to check directly then the assumptions of Theorem 5.1 are satisfied (a trajectory is shown in Figure 6).

6. Sufficient conditions for asymptotic stability

In order to obtain sufficient conditions for asymptotic stability, we need strengthened forms of conditions (ii), (iii) and (iv). Recall that a function $\alpha : [0, r_0) \to [0, +\infty)$ (where r_0 is some positive real number possibly dependent on α) is of class $\mathcal{K}$ if it is continuous, strictly increasing and

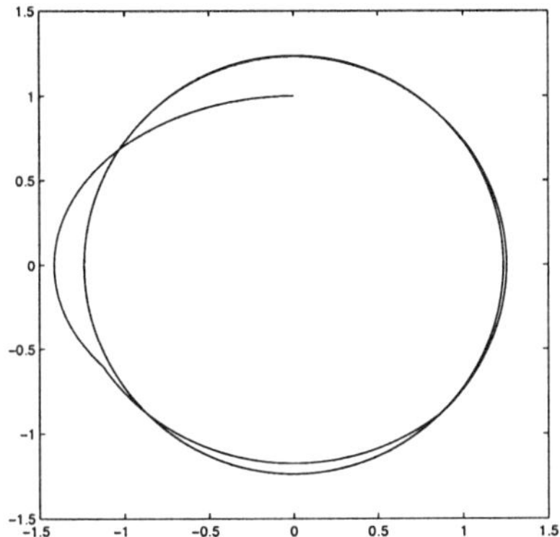

Fig. 6.   *Trajectory of system (14) with $d_k \equiv \pi/2$ starting from $(0, 1, 1)$*

such that $\alpha(0) = 0$. We are able to prove two theorems under alternative assumptions.

Theorem 6.1. *Let a CTDTC-system (1) be given, and assume that (i), (ii), (iv) hold. Assume further that:*

(ii') for each $q \in B_Q(r)$, the map $x \mapsto V(x,q)$ is of class C^1 on $B_{\mathbb{R}^n}(r)$ moreover, the function $W(x,q) = \nabla_x V(x,q) f(x,q) = $ is continuous on $B_E(r)$;

(iii') $W(x,q)$ is negative definite i.e., $W(x,q) \le 0$ for each $(x,q) \in B_E(r)$ and $W(x,q) = 0$ if and only if $(x,q) \in \{0\} \times M$.

Then, the CTDTC-system is uniformly-uniformly asymptotically stable.

Proof. According to Theorem 5.1, the system is uniformly-uniformly stable; thus, for any positive fixed number $r_0 < r$ we can find $\delta_0 > 0$ such that for each triple $(\bar{t}, \bar{x}, \bar{q})$ and each solution $(\varphi(t), u_{h(t)})$ such that $\varphi(\bar{t}) = \bar{x}$, $u_0 = \bar{q}$, one has

$$N(\bar{x}, \bar{q}) < \delta_0 \implies N(\varphi(t), u_{h(t)}) < r_0$$

for each $t \ge \bar{t}$. Let let $\gamma(t) = V(\varphi(t), u_{h(t)})$. Because of (iii), (iv), $\gamma(t)$ is nonincreasing on $[\bar{t}, +\infty)$, so that

$$\lim_{t \to +\infty} \gamma(t) = L \ge 0$$

98

and $\gamma(t) \geq L$ for each $t \geq \bar{t}$. Assume by contradiction that $L > 0$. Using (i), we can find $\mu \in (0, \delta_0)$ such that $V(x, q) < L$ for $N(x, q) < \mu$. Of course, we must have

$$\mu \leq N(\varphi(t), u_{h(t)}) \leq r_0$$

for each $t \geq \bar{t}$. Let now $c = \sup_{\mu \leq N(x,q) \leq r_0} W(x, q)$. By (v), c is actually a maximum and $c < 0$. Now,

$$\gamma(t) = \gamma(\tau_0) + \sum_{k=0}^{h(t)-1} \int_{\tau_k}^{\tau_{k+1}} W(\varphi(s), u_k)\, ds$$
$$+ \int_{\tau_{h(t)}}^{t} W(\varphi(s), u_{h(t)})\, ds < \gamma(\bar{t}) + ct \ .$$

This would imply $\lim_{t \to +\infty} \gamma(t) = -\infty$, which is impossible since $\gamma(t) \geq L > 0$. $\qquad\square$

Theorem 6.2. *Let a CTDTC-system (1) be given, and assume that (i), (ii), (iv) hold. Assume further that:*

(iv') there exists a map $\rho \in \mathcal{K}$ such that for each $(x, q) \in B_E(r)$

$$V(x, g(x, q)) \leq V(x, q) - \rho(N(x, q)) \ .$$

Then, the CTDTC-system is uniformly-uniformly asymptotically stable.

Proof. Let $\gamma(t)$ and $L \geq 0$ be as in the first part of the proof of Theorem 6.1. Conditions (iii), (iv) imply that

$$\gamma(\tau_i) = V(\varphi(\tau_i), u_i) \geq V(\varphi(\tau_{i+1}), u_i) \geq V(\varphi(\tau_{i+1}), u_{i+1}) = \gamma(\tau_{i+1}) \quad (15)$$

for each $i = 0, 1, 2, \ldots$. This yields $\lim_{i \to +\infty} V(\varphi(\tau_{i+1}), u_i) = \lim_{i \to +\infty} V(\varphi(\tau_{i+1}), u_{i+1}) = L$. Now, using (iv') we obtain

$$0 = L - L = \lim_{i \to +\infty} [V(\varphi(\tau_{i+1}), u_{i+1}) - V(\varphi(\tau_{i+1}), u_i)]$$
$$\leq - \lim_{i \to +\infty} \rho(N(\varphi(\tau_{i+1}), u_i)) \leq 0 \ .$$

It follows $\lim_{i \to +\infty} \rho(N(\varphi(\tau_{i+1}), u_i)) = 0$ and hence $\lim_{i \to +\infty} N(\varphi(\tau_{i+1}), u_i) = 0$. Using again (15) and the continuity of V, we conclude that also $\lim_{i\to+\infty} V(\varphi(\tau_{i+1}), u_{i+1}) = \lim_{i\to+\infty} \gamma(\tau_{i+1}) = L = 0$. Recalling that $\gamma(t)$ is nonincreasing, we finally infer $\lim_{t\to+\infty} \gamma(t) = 0$. It is now easy to conclude the proof. $\qquad\square$

References

1. V. Andriano, A. Bacciotti and G. Beccari, *Nonlinear Anal.* **28**, 1167 (1997).
2. Z. Artstein, *Examples of stabilization with hybrid feedback*, in *Hybrid Systems III: Verification and Control*, eds. R. Alur, T.A. Henzinger and E.D. Sontag, Lecture Notes in Computer Science, Vol. 1066 (Springer Verlag, Berlin, 1996): 173–185.
3. A. Bacciotti and L. Rosier, *Liapunov Functions and Stability in Control Theory*, Communications and Control Engineering Series, second edition (Springer Verlag, London, 2005).
4. A. Bacciotti, *Analisi della stabilità*, Quaderni dell'Unione Matematica Italiana, N. 49 (Pitagora Editrice, Bologna, 2006). In Italian.
5. A. Bacciotti and L. Mazzi, *Nonlinear Anal.* **67**, 2167 (2007).
6. M. Branicky, *IEEE Trans. Automat. Control* **43**, 475 (1998).
7. K. Cooke and J. Wiener, *A survey of differential equations with piecewise continuous arguments*, in *Delay Differential Equations and Dynamical Systems*, eds. S. Busemberg and M. Martelli, Lecture Notes in Mathematics, Vol. 1475 (Springer Verlag, Berlin, 1991): 1–15.
8. E. De Santis, M. Di Benedetto and L. Berardi, *IEEE Trans. Automat. Control* **49**, 184 (2004).
9. B. Francis and T. Georgiou, *IEEE Trans. Automat. Control* **33**, 820 (1988).
10. R. Goebel, J. Hespanha, A. Teel, C. Cai and R. Sanfelice, Hybrid systems: Generalized solutions and robust stability, in *Proceedings IFAC-NOLCOS (Stuttgart, 2004)*: 1–12.
11. D. Liberzon, *Switching in Systems and Control*, Systems & Control: Foundations & Applications (Birkhäuser, Boston, 2003).
12. D. Liberzon and A. Morse, *IEEE Control Syst. Mag.* **19**, 59 (1999).
13. J. Lygeros, K. Johansson, S. Simić, J. Zhang and S.S. Sastry, *IEEE Trans. Automat. Control* **48**, 2 (2003).
14. A. Morse, *IEEE Trans. Automat. Control* **41**, 1413 (1996).
15. H. Schumacher and A. van der Schaft, *An Introduction to Hybrid Dynamical Systems*, Lecture Notes in Control and Inform. Sci., Vol. 251 (Springer Verlag, 2000).
16. H. Ye, A. Michel and L. Hou, Stability analysis of switched systems, in *Proceedings of the 35th Conference on Decision and Control (Kobe, Japan, 1996)*: 1208–1212.
17. H. Ye, A. Michel and L. Hou, *IEEE Trans. Automat. Control* **43**, 461 (1998).

A REVIEW ON STABILITY OF SWITCHED SYSTEMS FOR ARBITRARY SWITCHINGS

U. BOSCAIN*

SISSA-ISAS, via Beirut 2-4, 34014 Trieste, Italy
and
Le2i, CNRS, Université de Bourgogne, B.P. 47870, 21078 Dijon Cedex, France,
E-mail: boscain@sissa.it

In this paper we review some recent results on stability of multilinear switched systems, under arbitrary switchings. An open problems is stated

Keywords: Switched systems, Asymptotic stability, Lyapunov functions

1. Introduction

By a switched system, we mean a family of continuous–time dynamical systems and a rule that determines at each time which dynamical system is responsible of the time evolution. More precisely, let $\{f_u : u \in U\}$ (where U is a subset of $\mathbb{R}^m$) be a finite or infinite set of sufficiently regular vector fields on a manifold M, and consider the family of dynamical systems,

$$\dot{x} = f_u(x), \quad x \in M. \tag{1}$$

The rule is given by assigning the so-called switching function, i.e., a function $u(.) : [0, \infty[\to U \subset \mathbb{R}^m$. Here, we consider the situation in which the switching function is not known a priori and represents some phenomenon (e.g., a disturbance) that is not possible to control. Therefore, the dynamics defined in (1) also fits into the framework of uncertain systems (cf. for instance Ref. 10). In the sequel, we use the notations $u \in U$ to label a fixed individual system and $u(.)$ to indicate the switching function. These kind of systems are sometimes called "n-modal systems", "dynamical polysystems", "input systems". The term "switched system" is often reserved to situations in which the switching function $u(.)$ is piecewise continuous

---

*This research has been financed by "Région Bourgogne", via a project FABER.

or the set U is finite. For the purpose of this paper, we only require $u(.)$ to be a measurable function. When all the vector fields f_u, $u \in U$ are linear, the switched system is called multilinear. For a discussion of various issues related to switched systems, we refer the reader to Refs. 6,12,14,16.

A typical problem for switched systems goes as follows. Assume that, for every fixed $u \in U$, the dynamical system $\dot{x} = f_u(x)$ satisfies a given property (P). Then one can investigate conditions under which property (P) still holds for $\dot{x} = f_{u(t)}(x)$, where $u(.)$ is an arbitrary switching function. In this paper, we focus on multilinear systems and (P) is the asymptotic stability property.

The structure of the paper is the following. In Section 2 we give the definitions of stability we need, we state the stability problem and we recall some sufficient conditions for stability in dimension n due to Agrachev, Hespanha, Liberzon and Morse.[1,7,15] In Section 3 we discuss the problem of existence of Lyapunov functions in certain functional classes (in particular in the class of polynomials). This problem was first studied by Molchanov and Pyatnitskii.[19–21] More recently new results were obtained by Dayawansa, Martin, Blanchini, Miani,[5,6,12] and in Ref. 18 in collaboration with Chitour and Mason. In Section 4, we discuss the necessary and sufficient conditions for asymptotic stability for bidimensional single input systems (bilinear systems) that were found in Ref. 7 (see also Ref. 18) and, in collaboration with Balde, in Ref. 4. In Section 5 we state an open problem.

2. General properties of multilinear systems

By a multilinear switched systems (that more often are simply called switched linear systems) we mean a system of the form,

$$\dot{x}(t) = A_{u(t)}x(t), \quad x \in \mathbb{R}^n, \quad \{A_u\}_{u \in U} \subset \mathbb{R}^{n \times n}, \tag{2}$$

where $U \subset \mathbb{R}^m$ is a compact set, $u(.) : [0, \infty[\to U$ is a (measurable) switching function, and the map $u \mapsto A_u$ is continuous (so that $\{A_u\}_{u \in U}$ is a compact set of matrices). For these systems, the problem of asymptotic stability of the origin, uniformly with respect to switching functions was investigated, in Refsw. 1,4,7,12,15.

A particular interesting class is the one of *bilinear* (or single-input) systems,

$$\dot{x}(t) = u(t)Ax(t) + (1 - u(t))Bx(t), \quad x \in \mathbb{R}^n, \quad A, B \in \mathbb{R}^{n \times n}. \tag{3}$$

Here the set U is equivalently $[0, 1]$ or $\{0, 1\}$ (see Remark 2.2 below).

Let us state the notions of stability that are used in the following.

Definition 2.1. For $\delta > 0$ let B_δ be the unit ball of radius δ, centered in the origin. Denote by $\mathcal{U}$ the set of measurable functions defined on $[0, \infty[$ and taking values on the compact set U. Given $x_0 \in \mathbb{R}^n$, we denote by $\gamma_{x_0, u(.)}(.)$ the trajectory of (2) based in x_0 and corresponding to the control $u(.)$. The *accessible set from* x_0, denoted by $\mathcal{A}(x_0)$, is

$$\mathcal{A}(x_0) = \cup_{u(.) \in \mathcal{U}} \mathrm{Supp}(\gamma_{x_0, u(.)}(.)) \,.$$

We say that the system (2) is

- **unbounded** if there exist $x_0 \in \mathbb{R}^n$ and $u(.) \in \mathcal{U}$ such that $\gamma_{x_0, u(.)}(t)$ goes to infinity as $t \to \infty$;
- **uniformly stable** if, for every $\varepsilon > 0$, there exists $\delta > 0$ such that $\mathcal{A}(x_0) \subset B_\varepsilon$ for every $x_0 \in B_\delta$;
- **globally uniformly asymptotically stable (GUAS**, for short) if it is uniformly stable and globally uniformly attractive, i.e., for every $\delta_1, \delta_2 > 0$, there exists $T > 0$ such that $\gamma_{x_0, u(.)}(T) \in B_{\delta_1}$ for every $u(.) \in \mathcal{U}$ and every $x_0 \in B_{\delta_2}$;

In this paper we focus on the following problem.

Problem 1 For the system (2) (resp. for the system (3)), find under which conditions on the compact set $\{A_u\}_{u \in U}$ (resp. on the pair (A, B)) the system is **GUAS**.

Next we always assume the following hypothesis,

(H0) for the system (2) (resp. for the system (3)) all the matrices of the compact set $\{A_u\}_{u \in U}$ (resp. A, and B) have eigenvalues with strictly negative real part (Hurwitz in the following),

otherwise Problem 1 is trivial.

Remark 2.1. Under our hypotheses (multilinearity and compactness) there are many notions of stability equivalent to the ones of Definition 2.1. More precisely since the system is multilinear, local and global notions of stability are equivalent. Moreover, since $\{A_u\}_{u \in U}$ is compact, all notions of stability are automatically uniform with respect to switching functions (see for instance Ref. 3). Finally, thanks to the multilinearity, the GUAS property is equivalent to the more often quoted property of GUES (global exponential stability, uniform with respect to switching), see for example Ref. 2 and references therein.

Remark 2.2. Whether systems of type (3) are GUAS or not is independent on the specific choice $U = [0,1]$ or $U = \{0,1\}$. In fact, this is a particular instance of a more general result stating that the stability properties of systems (2) depend only on the convex hull of the set $\{A_u\}_{u \in U}$; see for instance Ref. 18.

Example One can find many examples of systems such that each element of the family $\{A_u\}_{u \in U}$ is GUAS, while the switched system is not. Consider for instance a bidimensional system of type (3) where,

$$A = \begin{pmatrix} -.03 & -2 \\ 1/2 & -.03 \end{pmatrix}, \quad B = \begin{pmatrix} -.03 & -1/2 \\ 2 & -.03 \end{pmatrix}. \tag{4}$$

Notice that both A and B are Hurwitz. However it is easy to build trajectories of the switched systems that are unbounded, as shown in Figure 1.

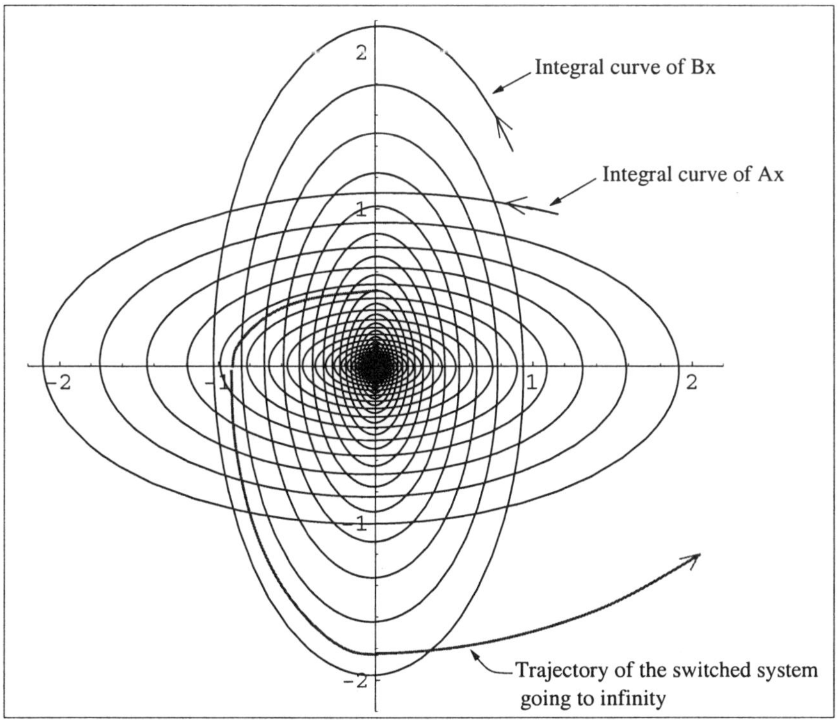

Fig. 1. For a system of type (3), where A and B are given by formula (4), this picture shows an integral curve of Ax, an integral curve of Bx, and a trajectory of the switched system, that is unbounded.

Let us recall some results about stability of systems of type (2), subject to **(H0)**. In Refs. 1,15 it is shown that the structure of the Lie algebra generated by the matrices A_u,

$$\mathbf{g} = \{A_u \; : \; u \in U\}_{L.A.},$$

is crucial for the stability of the system (2). For instance one can easily prove that if all the matrices of the family $\{A_u\}_{u \in U}$ commute, then the switched system is GUAS. The main result of Ref. 15 is the following.

Theorem 2.1. *(Hespanha, Morse, Liberzon) If g is a solvable Lie algebra, then the switched system (2) is GUAS.*

In Ref. 1 a generalization was given. Let $\mathbf{g} = \mathbf{r} \ltimes \mathbf{s}$ be the Levi decomposition of $\mathbf{g}$ in its radical (i.e., the maximal solvable ideal of $\mathbf{g}$) and a semi–simple sub–algebra, where the symbol $\ltimes$ indicates the semidirect sum.

Theorem 2.2. *(Agrachev, Liberzon) If s is a compact Lie algebra then the switched system (2) is GUAS.*

Theorem 2.2 contains Theorem 2.1 as a special case. Anyway the converse of Theorem 2.2 is not true in general: if $\mathbf{s}$ is non compact, the system can be stable or unstable. This case was also investigated. In particular, if $\mathbf{g}$ has dimension at most 4 as Lie algebra, Agrachev and Liberzon were able to reduce the problem of the asymptotic stability of the system (2) to the problem of the asymptotic stability of an auxiliary bidimensional system. We refer the reader to Ref. 1 for details. For this reason the bidimensional problem assumes particularly interest and, for the single input case, was solved in Ref. 7 (see also Ref. 18) and, in collaboration with Balde, in Ref. 4.

Before stating the stability conditions found in Refs. 4,7,18, let us recall the advantages and disadvantage of the most used method to check stability, i.e. the method of common Lyapunov functions.

3. Common Lyapunov functions

For a system (2), it is well known that the GUAS property is a consequence of the existence of a common Lyapunov function.

Definition 3.1. A common Lyapunov function (LF, for short) $V : \mathbb{R}^n \longrightarrow \mathbb{R}^+$, for a switched system (S) of the type (2), is a continuous function such that V is positive definite (i.e. $V(x) > 0$, $\forall x \neq 0$, $V(0) = 0$) and V is strictly decreasing along nonconstant trajectories of (S).

Vice-versa, it is known that, given a GUAS system of the type (2) subject to **(H0)**, it is always possible to build a C^∞ common Lyapunov function (see for instance Refs. 12,19–21 and the bibliographical note in Refs. 14).

Clearly is much more natural to use LFs to prove that a given system is GUAS (one has just to find one LF), than to prove that a system is unstable (i.e. proving that a LF does not exist). Indeed this is the reason why, usually, is much more easy to find (nontrivial) sufficient conditions for GUAS than necessary conditions. (Notice that all stability results given in the previous section in terms of the Lie algebra **g**, are, in fact, sufficient conditions for GUAS.)

Indeed the concept of LF is useful for practical purposes when one can prove that, for a certain class of systems, if a LF exists, then it is possible to find one of a certain type and possibly as simple as possible (e.g. polynomial with a bound on the degree, piecewise quadratic etc.). Typically one would like to work with a class of functions identified by a finite number of parameters. Once such a class of functions is identified, then in order to verify GUAS, one could use numerical algorithms to check (by varying the parameters) whether a LF exists (in which case the system is GUAS) or not (meaning that the system is not GUAS).

The idea of identifying a class of function where to look for a LF (i.e. sufficient to check GUAS), was first formalized by Blanchini and Miani in Ref. 6. They called such a class a "universal class of Lyapunov functions."

For instance, a remarkable result for a given class C of systems of type (2) (for instance the class of single input system (3), in dimension n) could be the following.

Claim: there exists a positive integer m (depending on n) such that, whenever a system of C admits a LF, then it admits one that is polynomial of degree less than or equal to m. In other words, the class of polynomials of degree at most m is sufficient to check GUAS (i.e. the class of polynomials of degree at most m is universal, in the language of Blanchini and Miani).

The problem of proving if this claim is true or false attracted some attention from the community.

For single input bidimensional systems of type (3), with $n = 2$, Shorten and K. Narendra provided in Ref. 23 a necessary and sufficient condition on the pair (A, B) for the existence of a *quadratic* LF, but it is known (see for instance Ref. 11,12,22,24) that there exist GUAS bilinear bidimensional systems not admitting a quadratic LF. See for instance the paper by Dayawansa and Martin[12] for a nice example.

In Ref. 12, for systems of type (3), Dayawansa and Martin assumed that the Claim above is true and posed the problem of finding the minimum m. More precisely, the problem posed by Dayawansa and Martin is the following:

Problem 2 (Dayawansa and Martin): Let Ξ be the set of systems of type (3) in dimension n satisfying **(H0)**. Define $\Xi_{GUAS} \subset \Xi$ as the set of GUAS systems. Find the minimal integer m such that every system of Ξ_{GUAS} admits a polynomial LF of degree less or equal than m.

Remark 3.1. In the problem posed by Dayawansa and Martin, it is implicitly assumed that a GUAS system always admits a polynomial common Lyapunov function. This fact was first proved by Molchanov and Pyatnitskii in Refs. 19,20, for systems of type (2), under the assumption that the set $\{A_u\}_{u \in U}$ is of the form $\{(a_{ij})_{i,j=1,\ldots n} : a_{ij}^- \leq a_{ij} \leq a_{ij}^+\}$. In Ref. 21 Molchanov and Pyatnitskii state the result, with no further details, under the more general hypothesis $\{A_u\}_{u \in U}$ just compact. In the case in which the convex hull of $\{A_u\}_{u \in U}$ is finitely generated, the existence of a polynomial common Lyapunov function for GUAS systems was proved by Blanchini and Miani,[5,6] in the context of uncertain systems. In Ref. 18, in collaboration with Chitour and Mason, a simple proof for a set $\{A_u\}_{u \in U}$ satisfying the weaker hypothesis that its convex hull is compact, (without necessarily requiring U to be compact) is provided.

The core of the paper[18] consists of showing that the Problem of Dayawansa and Martin does not have a solution, i.e. the minimum degree of a polynomial LF cannot be uniformly bounded over the set of all GUAS systems of the form (3). More precisely, we have the following:

Theorem 3.1. *If (A, B) is a pair of $n \times n$ real matrices giving rise to a system of Ξ_{GUAS}, let $m(A, B)$ be the minimum value of the degree of any polynomial LF associated to that system. Then $m(A, B)$ cannot be bounded uniformly over Ξ_{GUAS}.*

In other words the set of polynomials of fixed degree is not a universal class of Lyapunov functions. Finding which is the right functional class where to look for LFs is indeed a very difficult task. Sometimes, it is even easier to prove directly that a system is GUAS or unstable. Indeed this is the case for bidimensional single-input switched systems.

4. Two-dimensional bilinear systems

In Ref. 7 (see also Ref. 18) and Ref. 4, we studied conditions on A and B for the following property to be true:

($\mathcal{P}$) The switched system given by

$$\dot{x}(t) = u(t)Ax(t) + (1 - u(t))Bx(t),$$
$$\text{where } x \in \mathbb{R}^2, \quad A, B \in \mathbb{R}^{2\times 2}, \quad u(.) : [0, \infty[\to \{0, 1\}, \qquad (5)$$

is GUES at the origin.

The idea (coming from optimal control, see for instance Ref. 9) is that many information on the stability of (5) are contained in the set $\mathcal{Z}$ where the two vector fields Ax and Bx are linearly dependent. This set is the set of zeros of the function $Q(x) := \det(Ax, Bx)$. Since Q is a quadratic form, we have the following cases (depicted in Figure 2):

A. $\mathcal{Z} = \{0\}$ (i.e., Q is positive or negative definite). In this case the vector fields preserve always the same orientation and the system is GUAS. This fact can be proved in several way (for instance building a common quadratic Lyapunov function) and it is true in much more generality (even for nonlinear systems, see the paper Ref. 8, in collaboration with Charlot and Sigalotti).

B. $\mathcal{Z}$ is the union of two noncoinciding straight lines passing through the origin (i.e., Q is sign indefinite). Take a point $x \in \mathcal{Z} \setminus \{0\}$. We say that $\mathcal{Z}$ is *direct* (respectively, *inverse*) if Ax and Bx have the same (respectively, opposite) versus. One can prove that this definition is independent of the choice of x on $\mathcal{Z}$. Then we have the two subcases:

  B1. $\mathcal{Z}$ is inverse. In this case one can prove that there exists $u_0 \in]0, 1[$ such that the matrix $u_0 Ax + (1 - u_0)Bx$ has an eigenvalue with positive real part. In this case the system is unbounded since it is possible to build a trajectory of the convexified system going to infinity with constant control. (This type of instability is called *static instability*.)

  B2. $\mathcal{Z}$ is direct. In this case one can reduce the problem of the stability of (5) to the problem of the stability of a single trajectory called *worst-trajectory*. Fixed $x_0 \in \mathbb{R}^2 \setminus \{0\}$, the worst-trajectory γ_{x_0} is

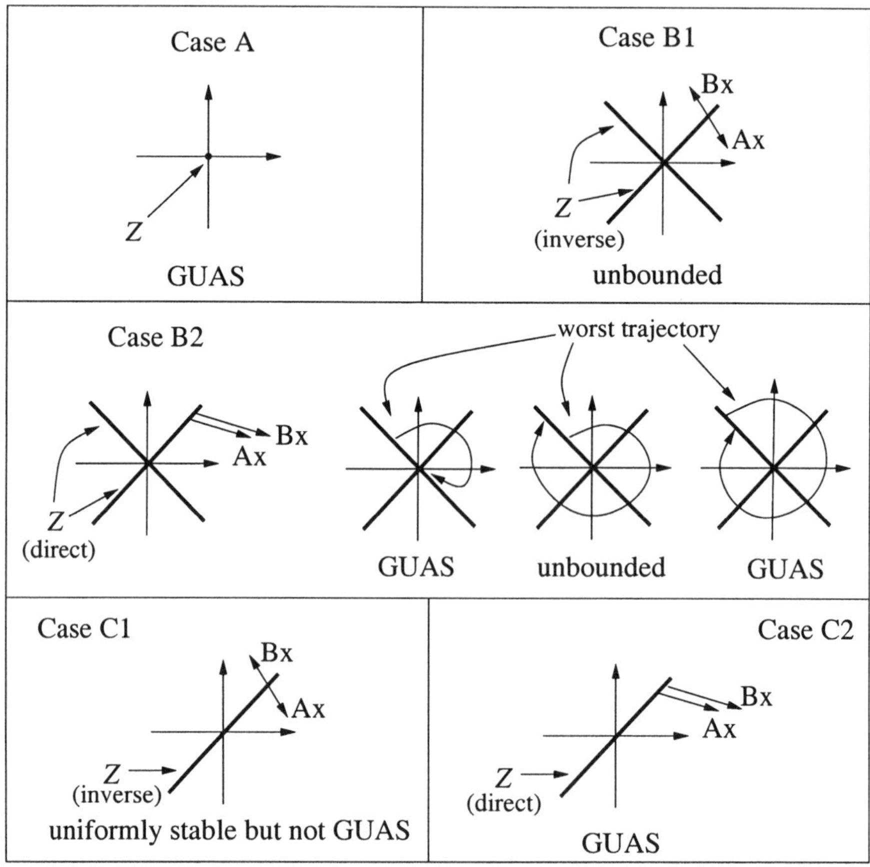

Fig. 2.

the trajectory of (5), based at x_0, and having the following property. At each time t, $\dot{\gamma}_{x_0}(t)$ forms the smallest angle (in absolute value) with the (exiting) radial direction (see Figure 3). Clearly the worst-trajectory switches among the two vector fields on the set $\mathcal{Z}$. If it does not rotate around the origin (i.e., if it crosses the set $\mathcal{Z}$ a finite number of times) then the system is GUAS. On the other side, if it rotates around the origin, the system is GUAS if and only if after one turn the distance from the origin is decreased. (see Figure 2, Case B2). If after one turn the distance from the origin is increased then the system is unbounded (in this case, since there are no trajectories of the convexified system going to infinity with constant control, we call this instability *dynamic instability*). If γ_{x_0} is periodic then the system is uniformly stable, but not GUAS.

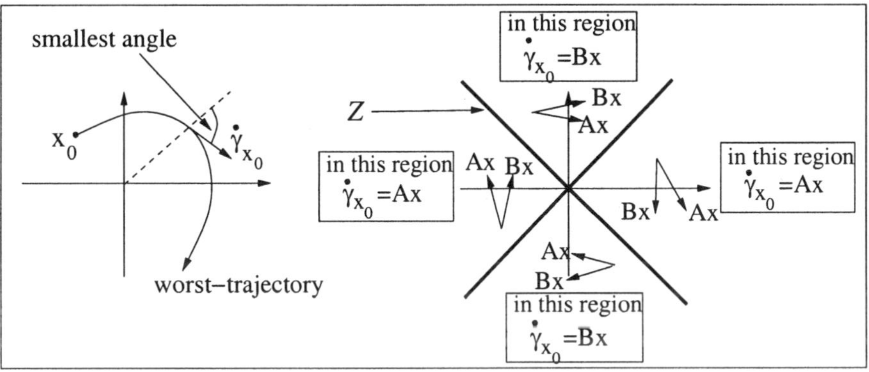

Fig. 3.

C. In the degenerate case in which the two straight lines of $\mathcal{Z}$ coincide (i.e., when Q is sign semi-definite), one see that the system is GUAS (resp. uniformly stable, but not GUAS) if and only if $\mathcal{Z}$ is direct (resp. inverse). We call these cases respectively **C2** and **C1**.

A consequence of these ideas is that the stability properties of the system (5), depend only on the shape of the integral curves of Ax and Bx and not on the way in which they are parameterized. More precisely we have:

Lemma 4.1. *If the switched system* $\dot{x} = u(t)Ax + (1 - u(t))Bx$, $u(.) :$ $[0,\infty[\to \{0,1\}$, *has one of the stability properties given in Definition 2.1, then the same stability property holds for the system* $\dot{x} = u(t)(A/\alpha_A)x + (1 - u(t))(B/\alpha_B)x$, *for every* $\alpha_A, \alpha_B > 0$.

The main point to get stability conditions is to translate the ideas above in terms of coordinate invariant parameters. To this purpose one have first to find good normal forms for the two matrices A and B, in which these parameters appear explicitly.

We treat separately the case in which the two matrices are diagonalizable (called diagonalizable case) and the case in which one or both are not (called nondiagonalizable case).

110

4.1. *The diagonalizable case*

In the case in which both A and B are diagonalizable, we assume

H1: Let λ_1, λ_2 (resp., λ_3, λ_4) be the eigenvalues of A (resp., B). Then $\text{Re}(\lambda_1)$, $\text{Re}(\lambda_2)$, $\text{Re}(\lambda_3)$, $\text{Re}(\lambda_4) < 0$.

H2: $[A, B] \neq 0$ (that implies that neither A nor B is proportional to the identity).

H3: A and B are diagonalizable in $\mathbb{C}$. (Notice that if **(H2)** and **(H3)** hold, then $\lambda_1 \neq \lambda_2$, $\lambda_3 \neq \lambda_4$.)

H4: Let $\mathbf{V}_1, \mathbf{V}_2 \in \mathbb{C}P^1$ (resp., $\mathbf{V}_3, \mathbf{V}_4 \in \mathbb{C}P^1$) be the eigenvectors of A (resp., B). Then $\mathbf{V}_i \neq \mathbf{V}_j$ for $i \in \{1, 2\}$, $j \in \{3, 4\}$. (Notice that, from **(H2)** and **(H3)**, the V_i are uniquely defined, $\mathbf{V}_1 \neq \mathbf{V}_2$ and $\mathbf{V}_3 \neq \mathbf{V}_4$, and **(H4)** can be violated only when both A and B have real eigenvalues.)

Condition **(H1)** is just the condition that A and B are Hurwitz (cf. condition **(H0)** in Section 2). Condition **(H2)** is required otherwise the system is GUAS as a consequence of Theorem 2.1. The case in which **(H1)** and **(H2)** hold but **(H3)** does not is treated in the next section. The case in which **(H1)**, **(H2)** and **(H3)** hold but **(H4)** does not can be treated with arguments similar to those of Ref. 7, and it possible to conclude that $(\mathcal{P})$ is true.

Theorem 4.1 below, gives necessary and sufficient conditions for the stability of the system (5) in terms of three (coordinates invariant) parameters given in Definition 4.1 below. The first two parameters, ρ_A and ρ_B, depend on the eigenvalues of A and B, respectively, and the third parameter $\mathcal{K}$ depends on $Tr(AB)$, which is a Killing-type pseudoscalar product in the space of 2×2 matrices. As explained in Ref. 7, the parameter $\mathcal{K}$ contains the interrelation between the two systems $\dot{x} = Ax$ and $\dot{x} = Bx$, and it has a precise geometric meaning. It is in $1 : 1$ correspondence with the cross ratio of the four points in the projective line $\mathbb{C}P^1$ that corresponds to the four eigenvectors of A and B. For more details, see Ref. 7.

Definition 4.1. Let A and B be two 2×2 real matrices and suppose that **(H1)**, **(H2)**, **(H3)**, and **(H4)** hold. Moreover, choose the labels $(_1)$ and $(_2)$ (resp., $(_3)$ and $(_4)$) so that $|\lambda_2| > |\lambda_1|$ (resp., $|\lambda_4| > |\lambda_3|$) if they are real or $\text{Im}(\lambda_2) < 0$ (resp., $\text{Im}(\lambda_4) < 0$) if they are complex. Define

$$\rho_A := -i\frac{\lambda_1 + \lambda_2}{\lambda_1 - \lambda_2};$$

$$\rho_B := -i\frac{\lambda_3 + \lambda_4}{\lambda_3 - \lambda_4};$$

$$\mathcal{K} := 2\frac{Tr(AB) - \frac{1}{2}Tr(A)Tr(B)}{(\lambda_1 - \lambda_2)(\lambda_3 - \lambda_4)}.$$

Moreover, define the following function of $\rho_A, \rho_B, \mathcal{K}$:

$$\mathcal{D} := \mathcal{K}^2 + 2\rho_A\rho_B\mathcal{K} - (1 + \rho_A^2 + \rho_B^2). \tag{6}$$

Notice that ρ_A is a positive real number if and only if A has nonreal eigenvalues and $\rho_A \in i\mathbb{R}$, $\rho_A/i > 1$ if and only if A has real eigenvalues. The same holds for B. Moreover, $\mathcal{D} \in \mathbb{R}$.

4.1.1. *Normal forms in the diagonalizable case*

Under hypotheses **(H1)** to **(H4)**, using a suitable 3-parameter changes of coordinates, it is always possible to put the matrices A and B, up the their norm, in the normal forms given in the following Proposition, where $\rho_A, \rho_B, \mathcal{K}$ appear explicitly. (See Ref. 7 and Ref. 18 for the proof.) We will call, respectively, **(CC)** the case where both matrices have nonreal eigenvalues, **(RR)** the case where both matrices have real eigenvalues, and **(RC)** the case where one matrix has real eigenvalues and the other nonreal eigenvalues.

Proposition 4.1. *Let A, B be two 2×2 real matrices satisfying conditions **(H1)**, **(H2)**, **(H3)**, and **(H4)** . In the case in which one of the two matrices has real and the other nonreal eigenvalues (i.e., the **(RC)** case), assume that A is the one having real eigenvalues. Then there exists a 3-parameter change of coordinates and two constant $\alpha_A, \alpha_B > 0$ such that the matrices A/α_A and B/α_B (still denoted below by A and B) are in the following normal forms.*

Case in which A and B have both nonreal eigenvalues ((CC) case):

$$A = \begin{pmatrix} -\rho_A & -1/E \\ E & -\rho_A \end{pmatrix}, \qquad B = \begin{pmatrix} -\rho_B & -1 \\ 1 & -\rho_B \end{pmatrix},$$

where $\rho_A, \rho_B > 0$, $|E| > 1$. In this case, $\mathcal{K} = \frac{1}{2}(E + \frac{1}{E})$. Moreover, the eigenvalues of A and B are, respectively, $-\rho_A \pm i$ and $-\rho_B \pm i$.

Case in which A has real and B nonreal eigenvalues ((RC) case):

$$A = \begin{pmatrix} -\rho_A/i + 1 & 0 \\ 0 & -\rho_A/i - 1 \end{pmatrix}, \qquad B = \begin{pmatrix} -\rho_B - \mathcal{K}/i & -\sqrt{1 - \mathcal{K}^2} \\ \sqrt{1 - \mathcal{K}^2} & -\rho_B + \mathcal{K}/i \end{pmatrix},$$

where $\rho_B > 0$, $\rho_A/i > 1$, $\mathcal{K} \in i\mathbb{R}$. In this case, the eigenvalues of A and B are, respectively, $-\rho_A/i \pm 1$ and $-\rho_B \pm i$.

Case in which A and B have both real eigenvalues ((RR) case):

$$A = \begin{pmatrix} -\rho_A/i + 1 & 0 \\ 0 & -\rho_A/i - 1 \end{pmatrix}, \qquad B = \begin{pmatrix} \mathcal{K} - \rho_B/i & 1 - \mathcal{K} \\ 1 + \mathcal{K} & -\mathcal{K} - \rho_B/i \end{pmatrix},$$

where $\rho_A/i, \rho_B/i > 1$ and $\mathcal{K} \in \mathbb{R} \setminus \{\pm 1\}$. In this case, the eigenvalues of A and B are, respectively, $-\rho_A/i \pm 1$ and $-\rho_B/i \pm 1$.

Using these normal forms, following the ideas presented at the beginning of this section, one gets the following stability conditions (see Ref. 7,18 for the proof).

4.1.2. *Stability conditions in the diagonalizable case*

Theorem 4.1. *Let A and B be two real matrices such that (H1), (H2), (H3), and (H4), given in section 4, hold and define $\rho_A, \rho_B, \mathcal{K}, \mathcal{D}$ as in Definition 4.1. We have the following stability conditions.*

Case (CC) *If A and B have both complex eigenvalues, then*

    Case (CC.1). *if $\mathcal{D} < 0$, then $(\mathcal{P})$ is true;*
    Case (CC.2). *if $\mathcal{D} > 0$, then*

        Case (CC.2.1). *if $\mathcal{K} < -1$, then $(\mathcal{P})$ is false;*
        Case (CC.2.2). *if $\mathcal{K} > 1$, then $(\mathcal{P})$ is true if and only if it holds the following condition:*

$$\rho_{CC} := \exp\left[-\rho_A \arctan\left(\frac{-\rho_A \mathcal{K} + \rho_B}{\sqrt{\mathcal{D}}} \right) \right.$$

$$\left. -\rho_B \arctan\left(\frac{\rho_A - \rho_B \mathcal{K}}{\sqrt{\mathcal{D}}} \right) - \frac{\pi}{2}(\rho_A + \rho_B) \right]$$

$$\times \sqrt{ \frac{(\rho_A \rho_B + \mathcal{K}) + \sqrt{\mathcal{D}}}{(\rho_A \rho_B + \mathcal{K}) - \sqrt{\mathcal{D}}} } < 1. \tag{7}$$

Case (CC.3). *If $\mathcal{D} = 0$, then $(\mathcal{P})$ holds true or false whether $\mathcal{K} > 1$ or $\mathcal{K} < -1$.*

Case (RC). *If A and B have one of them complex and the other real eigenvalues, define $\chi := \rho_A \mathcal{K} - \rho_B$, where ρ_A and ρ_B are chosen in such a way $\rho_A \in i\mathbb{R}$, $\rho_B \in \mathbb{R}$. Then*

Case (RC.1). *if $\mathcal{D} > 0$, then $(\mathcal{P})$ is true;*

Case (RC.2). *if $\mathcal{D} < 0$, then $\chi \neq 0$ and we have:*

> **Case (RC.2.1).** *if $\chi > 0$, then $(\mathcal{P})$ is false. Moreover, in this case $\mathcal{K}/i < 0$;*
>
> **Case (RC.2.2).** *if $\chi < 0$, then*
>
> > **Case (RC2.2.A).** *if $\mathcal{K}/i \leq 0$, then $(\mathcal{P})$ is true;*
> >
> > **Case (RC2.2.B).** *if $\mathcal{K}/i > 0$, then $(\mathcal{P})$ is true if and only if it holds the following condition:*

$$\rho_{RC} :- \left(\frac{m^+}{m^-}\right)^{-\frac{1}{2}(\rho_A/i-1)} e^{-\rho_B \bar{t}} \tag{8}$$

$$\times \left(\sqrt{1-\mathcal{K}^2}\, m^- \sin\bar{t} - \left(\cos\bar{t} - \frac{\mathcal{K}}{i}\sin\bar{t}\right)\right) < 1,$$

> *where*

$$m^\pm := \frac{-\chi \pm \sqrt{-\mathcal{D}}}{(-\rho_A/i - 1)\mathcal{K}/i}, \qquad \bar{t} = \arccos\frac{-\rho_A/i + \rho_B\mathcal{K}/i}{\sqrt{(1-\mathcal{K}^2)(1+\rho_B^2)}}.$$

Case (RC.3). *If $\mathcal{D} = 0$, then $(\mathcal{P})$ holds true whether $\chi < 0$ or $\chi > 0$.*

Case (RR). *If A and B have both real eigenvalues, then*

Case (RR.1). *if $\mathcal{D} < 0$, then $(\mathcal{P})$ is true; moreover we have $|\mathcal{K}| > 1$;*

Case (RR.2). *if $\mathcal{D} > 0$, then $\mathcal{K} \neq -\rho_A\rho_B$ (notice that $-\rho_A\rho_B > 1$) and*

> **Case (RR.2.1).** *if $\mathcal{K} > -\rho_A\rho_B$, then $(\mathcal{P})$ is false;*
>
> **Case (RR.2.2).** *if $\mathcal{K} < -\rho_A\rho_B$, then*
>
> > **Case (RR.2.2.A).** *if $\mathcal{K} > -1$, then $(\mathcal{P})$ is true;*
> >
> > **Case (RR.2.2.B).** *if $\mathcal{K} < -1$, then $(\mathcal{P})$ is true if and only if the following condition holds:*

$$\rho_{RR} := -f^{sym}(\rho_A,\rho_B,\mathcal{K})f^{asym}(\rho_A,\rho_B,\mathcal{K}) \tag{9}$$

$$\times f^{asym}(\rho_B,\rho_A,\mathcal{K}) < 1,$$

114

where

$$f^{sym}(\rho_A, \rho_B, \mathcal{K}) := \frac{1 + \rho_A/i + \rho_B/i + \mathcal{K} - \sqrt{\mathcal{D}}}{1 + \rho_A/i + \rho_B/i + \mathcal{K} + \sqrt{\mathcal{D}}};$$

$$f^{aym}(\rho_A, \rho_B, \mathcal{K}) := \left(\frac{\rho_B/i - \mathcal{K}\rho_A/i - \sqrt{\mathcal{D}}}{\rho_B/i - \mathcal{K}\rho_A/i + \sqrt{\mathcal{D}}}\right)^{\frac{1}{2}(\rho_A/i-1)}.$$

Case (RR.3). *If $\mathcal{D} = 0$, then $(\mathcal{P})$ holds true or false whether $\mathcal{K} < -\rho_A\rho_B$ or $\mathcal{K} > -\rho_A\rho_B$.*

Finally, if $(\mathcal{P})$ is not true, then in case CC.2.2 with $\rho_{CC} = 1$, case (RC.2.2.B), with $\rho_{RC} = 1$, case (RR.2.2.B), with $\rho_{RR} = 1$, case (CC.3) with $\mathcal{K} < -1$, case (RC.3) with $\chi > 0$ and case (RR.3) with $\mathcal{K} > -\rho_A\rho_B$, the origin is just stable. In the other cases, the system is unstable.

Remark 4.1. Notice that cases **(CC.2.1)**, **(RC.2.1)** and **(RR.2.1)** correspond to $\mathcal{Z}$ inverse, while cases **(CC.2.2)**, **(RC.2.2)** and **(RR.2.2)** correspond to $\mathcal{Z}$ direct.

4.2. *The nondiagonalizable case*

In the case in which one or both the matrices are nondiagonalizable, we assume

(H5) A and B are two 2×2 real Hurwitz matrices. Moreover A is nondiagonalizable and $[A, B] \neq 0$.

In this case new difficulties arises. The first is due to the fact that eigenvectors of A and B are at most 3 noncoinciding points on $\mathbb{C}P^1$. As a consequence the cross ratio is not anymore the right parameter describing the interrelation among the systems. It is either not defined or completely fixed. For this reason new coordinate-invariant parameters should be identified and new normal forms for A and B should be constructed. These coordinate invariant parameters are the three real parameters defined in Definition 4.2 below. One (η) is, up to time reparametrization, the (only) eigenvalue of A, the second (ρ) depends on the eigenvalues of B and the third (k) plays the role of the cross ratio of the diagonalizable case. For $x \in \mathbb{R}$ define

$$sign(x) = \begin{cases} +1 \text{ if } x > 0 \\ 0 \text{ if } x = 0 \\ -1 \text{ if } x < 0. \end{cases}$$

Definition 4.2. Assume **(H5)** and let δ be the discriminant of the equation $det(B - \lambda \mathrm{Id}) = 0$. Define the following invariant parameters:

$$\eta = \begin{cases} \dfrac{Tr(A)}{\sqrt{|\delta|}} & \text{if } \delta \neq 0 \\ \dfrac{Tr(A)}{2} & \text{if } \delta = 0, \end{cases} \qquad \rho = \begin{cases} \dfrac{Tr(B)}{\sqrt{|\delta|}} & \text{if } \delta \neq 0 \\ \dfrac{Tr(B)}{2} & \text{if } \delta = 0, \end{cases}$$

$$k = \begin{cases} \dfrac{4}{|\delta|} \left(Tr(AB) - \dfrac{1}{2} Tr(A) Tr(B) \right) & \text{if } \delta \neq 0 \\ Tr(AB) - \dfrac{1}{2} Tr(A) Tr(B) & \text{if } \delta = 0. \end{cases}$$

Remark 4.2. Notice that $\delta = (\lambda_1 - \lambda_2)^2 \in \mathbb{R}$, where λ_1 and λ_2 are the eigenvalues of B. Notice moreover that B has non real eigenvalues if and only if $\delta < 0$. Finally observe that $\eta, \rho < 0$ and $k \in \mathbb{R}$.

Definition 4.3. In the following, under the assumption **(H5)**, we call regular case (**R**-case for short), the case in which $k \neq 0$ and singular case (**S**-for short), the case in which $k = 0$.

4.2.1. *Normal forms in the nondiagonalizable case*

We have the following (see Ref. 4 for the proof):

Lemma 4.2. *(**R-case**) Assume **(H5)** and $k \neq 0$. Then it is always possible to find a linear change of coordinates and a constant $\tau > 0$ such that A/τ and B/τ (that we still call A and B) have the following form:*

$$A = \begin{pmatrix} \eta & 1 \\ 0 & \eta \end{pmatrix}, \tag{10}$$

$$B = \begin{pmatrix} \rho & sign(\delta)/k \\ k & \rho \end{pmatrix}. \tag{11}$$

Moreover in this case $[A, B] \neq 0$ is automatically satisfied.

Lemma 4.3. *(**S-case**) Assume **(H5)** and $k = 0$. Then $\delta > 0$ and it is always possible to find a linear change of coordinates and a constant $\tau > 0$ such that A/τ and B/τ (that we still call A and B) have the following form,*

$$A = \begin{pmatrix} \eta & 1 \\ 0 & \eta \end{pmatrix}, \quad B = \begin{pmatrix} \rho - 1 & 0 \\ 0 & \rho + 1 \end{pmatrix}, \quad \text{called } \boldsymbol{S_1}\text{-case}, \tag{12}$$

or the form,

$$A = \begin{pmatrix} \eta & 1 \\ 0 & \eta \end{pmatrix}, \quad B = \begin{pmatrix} \rho + 1 & 0 \\ 0 & \rho - 1 \end{pmatrix}, \quad \text{called } \boldsymbol{S_{-1}}\text{-case}. \tag{13}$$

4.2.2. *Stability conditions in the nondiagonalizable case*

First we need to define some functions of the invariants η, ρ, k. Set $\Delta = k^2 - 4\eta\rho k + sign(\delta)4\eta^2$. By direct computation one gets that if $k = 2\eta\rho$ then $\Delta = -4 \det A \det B < 0$. It follows,

Lemma 4.4. *Assume **(H5)**. Then $\Delta \geq 0$ implies $k \neq 2\eta\rho$.*

Moreover, when $\Delta > 0$ and $k < 0$, define

$$
\mathcal{R} = \begin{cases} \left|\left(\dfrac{-k+\sqrt{\Delta}}{-k-\sqrt{\Delta}}\right)\dfrac{2k\rho^2 - sign(\delta)(k+2\eta\rho+\sqrt{\Delta})}{2k(\rho-sign(\delta)\frac{\eta}{k})\sqrt{\rho^2-sign(\delta)}}\right| \exp\left(\dfrac{\sqrt{\Delta}}{k} + \rho\theta_{sign(\delta)}\right), \\ \quad \text{if} \ \ \rho - sign(\delta)\frac{\eta}{k} \neq 0, \\[1em] \dfrac{-2\eta}{\sqrt{k^2+\eta^2}} \exp\left(\dfrac{\sqrt{\Delta}}{k} + \rho\theta_{-1}\right), \\ \quad \text{if} \ \ \rho - sign(\delta)\frac{\eta}{k} = 0 \ \ \text{(which implies} \ \ sign(\delta) = -1\text{)}, \end{cases}
\tag{14}
$$

where

$$
\theta_{-1} = \begin{cases} \arctan \dfrac{\sqrt{\Delta}}{k(\rho+\frac{\eta}{k})+\eta} & \text{if} \ \ k(\rho+\tfrac{\eta}{k})+\eta \neq 0 \\ \pi/2 & \text{if} \ \ k(\rho+\tfrac{\eta}{k})+\eta = 0, \end{cases}
$$

$$
\theta_1 = \text{arctanh}\dfrac{\sqrt{\Delta}}{k(\rho - \frac{\eta}{k}) - \eta},
$$

$$
\theta_0 = \dfrac{\sqrt{\Delta}}{k\rho}.
$$

Notice that when $k < 0$, then $k(\rho - \frac{\eta}{k}) - \eta > 0$ and $k\rho > 0$. Hence θ_1 and θ_0 are well defined.

The following Theorem states the stability conditions in the case in which A is nondiagonalizable. For the proof see Ref. 4. The letters **A.**, **B.**, and **C.** refer to the cases described at the beginning of this section and in Figure 2. Recall Lemma 4.4.

Theorem 4.2. *Assume **(H5)**. We have the following stability conditions for the system (5).*

A. *If $\Delta < 0$, then the system is GUAS.*
B. *If $\Delta > 0$, then:*

> **B1.** *if $k > 2\eta\rho$, then the system is unbounded,*
> **B2.** *if $k < 2\eta\rho$, then*

- *in the regular case ($k \neq 0$), the system is GUAS, uniformly stable (but not GUAS) or unbounded respectively if*

$$\mathcal{R} < 1, \mathcal{R} = 1, \mathcal{R} > 1.$$

- *In the singular case ($k = 0$), the system is GUAS.*

C. *If $\Delta = 0$, then:*

C1 *If $k > 2\eta\rho$, then the system is uniformly stable (but not GUAS),*
C2 *if $k < 2\eta\rho$, then the system is GUAS.*

5. An open problem

Many problems connected to the ideas presented in this paper as still open. In the following we present a problem that, in our opinion, is of great interest.

Open Problem Find necessary and sufficient conditions for the stability of a bilinear switched system in dimension 3. This problem seems to be quite difficult, and the techniques presented in this paper cannot be applyed. (The reason is that these techniques use implicitly the Jordan separation lemma.) Even if studying all possible cases is probably too complicated, one would like to see if the set of all pairs of 3×3 matrices giving rise to a a GUAS system can be defined with a finite number of inequalities involving analytic functions, exponentials and logarithm. This was the case for bilinear planar systems, and it would be already very interesting to see if the same holds in dimension 3.
Another formulation of this problem is to extend to bilinear systems in dimension 3 the following corollary of the results given in the previous section.

Corollary 5.1. *The set of all pairs of 2×2 real matrices giving rise to a GUAS system is a set in the log-exp category.*

For the precise definition of the log-exp category, see for instance Refs. 13,17.

Remark 5.1. Notice that the topological boundary of the set described by Corollary 5.1 is related to the problem of finding the right functional class where to look for LFs. Another interesting problem is to clarify this relation. This problem is already interesting in dimension 2.

Acknowledgments

I would like to thank Andrei A. Agrachev. Indeed the open problem above comes essentially from him.

References

1. A. A. Agrachev and D. Liberzon, *SIAM J. Control Optim.* **40**, 253 (2001).
2. D. Angeli, A note on stability of arbitrarily switched homogeneous systems, *Systems Control Lett.* (to appear).
3. D. Angeli, B. Ingalls, E. D. Sontag, and Y. Wang, *J. Dynam. Control Systems* **10**, 391 (2004).
4. M Balde, U. Boscain, *Commun. Pure Appl. Anal.* **7**, 1 (2008).
5. F. Blanchini, *Automatica J. IFAC* **31**, 451 (1995).
6. F. Blanchini and S. Miani, *IEEE Transactions on Automatic Control* **44**, 641 (1999).
7. U. Boscain, *SIAM J. Control Optim.* **41**, 89 (2002).
8. U. Boscain, G. Charlot, M. Sigalotti, *Discrete Contin. Dyn. Syst.* **15**, 415 (2006).
9. U. Boscain, B. Piccoli, *Optimal Synthesis for Control Systems on 2-D Manifolds*, SMAI, Vol. 43, (Springer-Verlag, Berlin, 2004).
10. Z. Bubnicki, *Analysis and Decision Making in Uncertain Systems*, Communications and Control Engineering Series (Springer-Verlag, London, 2004).
11. G. Chesi, A. Garulli, A. Tesi, A. Vicino, *Automatica J. IFAC* **39**, 1027 (2003).
12. W. P. Dayawansa and C. F. Martin, *IEEE Trans. Automat. Control* **44**, 751 (1999).
13. L. van den Dries, A. Macintyre, D. Marker, *Ann. of Math. (2)* **140**, 183 (1994).
14. D. Liberzon, *Switching in systems and control*, Systems & Control: Foundations & Applications (Birkhäuser, Boston, 2003).
15. D. Liberzon, J. P. Hespanha, and A. S. Morse, *Systems Control Lett.* **37**, 117 (1999).
16. D. Liberzon and A. S. Morse, *IEEE Control Syst. Magaz.*, **19** 59 (1999).
17. J.M. Lion, J. P. Rolin, *Ann. Inst. Fourier* **47**, 859 (1997).
18. P. Mason, U. Boscain, and Y. Chitour, *SIAM J. Control Optim.* **45**, 226 (2006).
19. A. P. Molchanov, E. S. Pyatnitskii, *Avtomat. i Telemekh.* **3**, 63 (1986) [in Russian]. English translation: *Automat. Remote Control* **47**, 344 (1986).
20. A. P. Molchanov, E. S. Pyatnitskii, *Avtomat. i Telemekh.* **4**, 5 (1986) [in Russian]. English translation: *Automat. Remote Control* **47**, 443 (1986).
21. A. P. Molchanov, E. S. Pyatnitskii, *Systems Control Lett.* **13**, 59 (1989).
22. A. Olas, On robustness of systems with structured uncertainties, in *Mechanics and control (Proc. 4th Workshop on Control Mechanics, Los Angeles, CA, 1991)*, eds. J. M. Skowronski, H. Flashner and R. S. Guttalu, Lecture Notes in Control and Inform. Sci. Vol. 170 (Springer, Berlin, 1992): 156-169.

23. R. Shorten and K. Narendra, Necessary and sufficient conditions for the existence of a common quadratic Lyapunov function for two stable second order linear time-invariant systems, in *Proceeding 1999 American Control Conf. (San Diego, CA, 1999)*: 1410-1414.
24. A. L. Zelentsovsky, *IEEE Trans. Automat. Control* **39**, 135 (1994).

REGULARITY PROPERTIES OF ATTAINABLE SETS UNDER STATE CONSTRAINTS

P. CANNARSA[†] and M. CASTELPIETRA[‡]

Dipartimento di Matematica, Università di Roma Tor Vergata,
00133 Roma, Italia
[†] *E-mail: cannarsa@mat.uniroma2.it*
[‡] *E-mail: castelpi@mat.uniroma2.it*
http://www.mat.uniroma2.it

P. CARDALIAGUET

Laboratoire de Mathématiques, Université de Bretagne Occidentale
29285 Brest, France
E-mail: Pierre.Cardaliaguet@univ-brest.fr
http://maths2.univ-brest.fr

The Maximum principle in control theory provides necessary optimality conditions for a given trajectory in terms of the co-state, which is the solution of a suitable adjoint system. For constrained problems the adjoint system contains a measure supported at the boundary of the constraint set. In this paper we give a representation formula for such a measure for smooth constraint sets and nice Hamiltonians. As an application, we obtain a perimeter estimate for constrained attainable sets.

Keywords: Maximum principle; Interior sphere property; Perimeter.

1. Introduction

The Maximum principle is a fundamental result in optimal control theory. Not only does it provide necessary conditions for candidate solutions of an optimal control problem, but it can also be used to obtain regularity results for optimal trajectories.

To fix ideas, consider a control system of the form

$$\begin{cases} x'(t) = f(t, x(t), u(t)), & u(t) \in U \quad \text{a.e. } t \geq 0 \\ x(0) = x_0, \end{cases} \tag{1}$$

where $u(\cdot)$ is a control, and $x(\cdot)$ denotes the corresponding trajectory.

Consider an open set $\Omega \subseteq \mathbb{R}^n$. For a point $x_0 \in \Omega$ we say that a control $u(\cdot)$ is *admissible* in x_0 on $[0, T]$ if

$$x(t) \in \overline{\Omega} \qquad \forall t \in [0, T],$$

where $x(\cdot)$ is the corresponding (then, *admissible*) trajectory. $\overline{\Omega}$ is called the *constraint set*.

Denote by $\mathcal{A}(t)$ the *attainable set* in time t from a closed set $\mathcal{K} \subseteq \Omega$. We call *optimal* an admissible trajectory $x(\cdot)$ on the interval $[0, T]$ if $T = \inf\{t \geq 0 : x(T) \in \mathcal{A}(t)\}$, that is, if the trajectory $x(\cdot)$ minimizes the time to reach the point $x(T)$.

Every optimal trajectory satisfies the Maximum Principle: this result was obtained by several authors in the smooth case,[8,10,12] and, as a natural evolution, also in the nonsmooth case.[2,7,9,11,13]

In constrained problems there are singularity effects for the trajectories that touch the boundary of Ω. These effects are due to an additional term (containing a measure) that appears in the Maximum Principle. More precisely, introducing the Hamiltonian

$$H(t, x, p) = \sup_{u \in U} \langle f(t, x, u), p \rangle,$$

we have that, if $x(\cdot)$ is an optimal trajectory, then there exist an arc $p(\cdot)$ and a Radon measure μ such that for a.e. $t \geq 0$

$$(h'(t), -p'(t), x'(t)) \in \partial H(t, x(t), p(t) + \psi(t)), \qquad (2)$$

where

$$\psi(t) = \int_{[0,t)} \nu_\Omega(x(s)) \mu(ds),$$
$$h(t) = H(t, x(t), p(t) + \psi(t)).$$

Here, $\nu_\Omega(x)$ belongs to the normal cone to Ω at x, and the measure μ is supported only by the set of times t for which $x(t) \in \partial\Omega$ (see Ref. 7).

If we have a "nice" Hamiltonian (in Sec. 2 we clarify the meaning of nice), then the generalized gradient of H turns to be the triplet $(D_t H, D_x H, D_p H)$, and inclusion (2) becomes an equality.

The main result of this paper, developed in the next section, gives a representation of the additional term ψ in the Maximum Principle. That is, we show that, if $\partial\Omega$ is smooth, then $x'(t) = D_p H(t, x(t), p(t))$ and

$$-p'(t) = D_x H(t, x(t), p(t)) - \lambda(t) \nu_\Omega(x(t)) \mathbf{1}_{\partial\Omega}(x(t)),$$

where λ is a measurable function, depending on $\partial\Omega$ and H, that we explicitly compute.

In the last section of this paper, we apply the above formulation of the Maximum Principle to extend a result by Cardaliaguet and Marchi[6] for perimeters of attainable sets.

2. Maximum principle under state constraints

Let $\Omega \subseteq \mathbb{R}^n$ be an open set with uniformly $\mathcal{C}^2$ boundary. We define the *signed distance* from $\partial\Omega$ by

$$d(x) = d_\Omega(x) - d_{\Omega^c}(x) \qquad x \in \mathbb{R}^n \,,$$

where we have denoted by $d_S(x) = \inf\{|x-y| : y \in S\}$ the distance function from a set $S \subseteq \mathbb{R}^n$. The boundary of Ω is of class $\mathcal{C}^2$ if and only if there exists some $\eta > 0$ such that

$$d(\cdot) \in \mathcal{C}_b^2 \text{ on } \partial\Omega + B_\eta = \{y \in B(x,\eta) : x \in \partial\Omega\}, \tag{3}$$

where $\mathcal{C}_b^2$ is the set of functions of class $\mathcal{C}^2$ with bounded derivatives of first and second order.

We consider the following controlled system, subject to state constraints,

$$\begin{cases} x'(t) = f(t, x(t), u(t)), & \text{a.e. } t \geq 0 \\ x(t) \in \overline{\Omega}, & t \geq 0 \\ x(0) = x_0, \end{cases} \tag{4}$$

where $U \subseteq \mathbb{R}^m$ is a compact set, and $u : \mathbb{R}_+ \to U$ is measurable function, in short an admissible control. We assume that $f : \mathbb{R}_+ \times \mathbb{R}^n \times U$ is a continuous function such that, for some positive constants L and k,

$$\begin{cases} |f(t,x,u) - f(t,y,u)| \leq L|x-y| & \forall x,y \in \mathbb{R}^n, \forall(t,u) \in \mathbb{R}_+ \times U, \\ |f(t,x,u)| \leq k(1+|x|) & \forall(t,x,u) \in \mathbb{R}_+ \times \mathbb{R}^n \times U. \end{cases} \tag{5}$$

Let $\mathcal{K} \subseteq \Omega$ be a given closed set. The set of *admissible trajectories from* $\mathcal{K}$ (at time t) is

$$\mathcal{T}_{\mathrm{ad}}(\mathcal{K},t) = \{\text{trajectories } x(\cdot) \text{ that solve (4) in } [0,t], \text{ such that } x_0 \in \mathcal{K}\}.$$

The *attainable set from* $\mathcal{K}$ at time t is defined by

$$\mathcal{A}(\mathcal{K},t) = \{x(t) : x(\cdot) \in \mathcal{T}_{\mathrm{ad}}(\mathcal{K},t)\}.$$

We introduce the *minimum time function*

$$\tau_\mathcal{K}(x) = \inf\{t : x \in \mathcal{A}(\mathcal{K},t)\},$$

that is the time needed to reach a point x, starting from $\mathcal{K}$.

For state constrained systems the definition of "extremal solutions" differs, somewhat, from the unconstrained case. In fact, it is not sufficient to request that trajectories be on the boundary of $\mathcal{A}(\mathcal{K}, t)$, because a trajectory can stay on the boundary of Ω without being "optimal".

Example 2.1. Take $\Omega = \{x \in \mathbb{R}^2 : \langle x, \mathbf{e}_2 \rangle < 1\}$, where $\{\mathbf{e}_1, \mathbf{e}_2\}$ is an orthonormal basis of $\mathbb{R}^2$. Let $\mathcal{K} = \{0\}$, and $f(t, x, u) = u$ with $U = \overline{B}$, and $T = 3$.

Consider the trajectories $x_1(\cdot)$ and $x_2(\cdot)$ associated, respectively, to the controls

$$u_1(t) = \mathbf{e}_1, \qquad t \in [0, 3]$$

and

$$u_2(t) = \begin{cases} \mathbf{e}_2, & t \in [0, 1] \\ \frac{1}{2}\mathbf{e}_1, & t \in (1, 2] \\ 0, & t \in (2, 3]. \end{cases}$$

Then both x_1 and x_2 are in $\partial \mathcal{A}(\mathcal{K}, t)$ for every $t \in [0, 3]$, but "morally" only x_1 is really extremal. Indeed, the point $y = x_2(2) = x_2(3)$ is reached in time $\sqrt{3}/\sqrt{2}$ by the trajectory $x_3(\cdot)$ associated to the control

$$u_3(t) = \frac{1}{2}\mathbf{e}_1 + \frac{\sqrt{3}}{2}\mathbf{e}_2.$$

On the other hand, the point $z = x_1(3)$ cannot be reached in any time $t < 3$.

Definition 2.1. A solution $x(\cdot)$ of control system (4) is an *optimal trajectory* (or *extremal solution*) on $[0, T]$ if $\tau_{\mathcal{K}}(x(t)) = t$ for all $t \in [0, T]$. We denote the *set of optimal trajectories* on $[0, T]$ by $\mathcal{X}(T)$.

Note that the relation $\tau_{\mathcal{K}}(x(T)) = T$ suffices to guarantee that, for all t in $[0, T]$, we also have $\tau_{\mathcal{K}}(x(t)) = t$.

Now, we can focus our attention on the Maximum Principle. Define the Hamiltonian function as

$$H(t, x, p) = \sup_{u \in U} \langle f(t, x, u), p \rangle \qquad \forall (t, x, p) \in \mathbb{R}_+ \times \mathbb{R}^n \times \mathbb{R}^n.$$

Proposition 2.1. *If $x(\cdot)$ is an optimal trajectory on $[0, T]$, then there exist an arc $p(\cdot)$, with $|p(0)| > 0$, and a Radon measure μ such that for a.e. $t \in [0, T]$*

$$(h'(t), -p'(t), x'(t)) \in \partial H(t, x(t), p(t) + \psi(t)), \tag{6}$$

where $\psi(t) = \int_{[0,t)} Dd(x(s))\mu(ds)$ and $h(t) = H(t, x(t), p(t) + \psi(t))$. Here, measure μ is supported only by the set of times t for which $x(t) \in \partial\Omega$.

For a proof, see Ref. 7. In general, we don't know if ψ (hence h) is continuous. To handle this boundary term, we need some regularity of the Hamiltonian. We will assume that

$$\begin{cases} H \text{ is of class } \mathcal{C}^2 \text{ on } \mathbb{R}_+ \times \mathbb{R}^n \times (\mathbb{R}^n \backslash \{0\}), \\ M|p| \geq H(t,x,p) \geq \alpha|p|, \\ \Gamma I_n \geq D^2_{pp} H^2(t,x,p) \geq \gamma I_n \end{cases} \tag{7}$$

for some constants $\alpha, \gamma, \Gamma, M > 0$, and for all $(t,x,p) \in \mathbb{R}_+ \times \mathbb{R}^n \times (\mathbb{R}^n \backslash \{0\})$.

Remark 2.1. Note that the last assumption of (7) is made for H^2, and not for H. In fact, such an assumption, if imposed on H, would be too restrictive (see example 2.2).

Moreover, we will impose the following growth conditions for the derivatives of H: for all $(t,x,p) \in \mathbb{R}_+ \times \mathbb{R}^n \times (\mathbb{R}^n \backslash \{0\})$,

$$\begin{cases} |D_t H(t,x,p)| \leq M|p| \\ |D_x H(t,x,p)| \leq M|p| \\ |D_p H(t,x,p)| \leq M \\ |D^2_{pt} H(t,x,p)| \leq M, \\ \|D^2_{px} H(t,x,p)\| \leq M. \end{cases} \tag{8}$$

It is not restrictive to consider the same constant M of (7).

Remark 2.2. We can give sufficient conditions for f to satisfy some of the assumptions in (8). If we assume that, for all $(t,x,u) \in \mathbb{R}_+ \times \mathbb{R}^n \times U$, we have $|f(t,x,u)| \leq M$ and $|D_t f(t,x,u)| \leq M$ and $\|D_x f(t,x,u)\| \leq M$, then the first three bounds in (8) are satisfied.

Moreover, if $\{f(t,x,U)\}_{(t,x)\in\mathbb{R}_+\times\mathbb{R}^n}$ is a family of uniformly convex sets of class $\mathcal{C}^2$, then $H \in \mathcal{C}^2(\mathbb{R}_+ \times \mathbb{R}^n \times (\mathbb{R}^n \backslash \{0\}))$, see, e.g., Ref. 5.

Under the above regularity assumptions for the Hamiltonian, the Maximum Principle takes the more precise form described below.

Theorem 2.1. *Assume that (3), (5), (7) and (8) hold true. Let $x(\cdot)$ be an optimal trajectory on the time interval $[0,T]$. Then there exists a Lipschitz continuous arc $p : [0,T] \to \mathbb{R}^n$, with $|p(t)| > 0$, and a bounded measurable function $\lambda(t) \geq 0$ such that*

$$\begin{cases} x'(t) = D_p H(t,x(t),p(t)) & \text{for all } t \in [0,T], \\ -p'(t) = D_x H(t,x(t),p(t)) - \lambda(t) Dd(x(t)) & \text{for a.e. } t \in [0,T]. \end{cases}$$

Moreover, λ is the function given below:

$$\lambda = \frac{1}{\langle D^2_{pp}H(t,x,p)Dd(x), Dd(x)\rangle} \left[\langle D^2_{pp}H(t,x,p)D_xH(t,x,p), Dd(x)\rangle \right.$$
$$- \langle D^2 d(x)D_pH(t,x,p), D_pH(t,x,p)\rangle$$
$$- \langle D^2_{xp}H(t,x,p)D_pH(t,x,p), Dd(x)\rangle$$
$$\left. - \langle D^2_{pt}H(t,x,p), Dd(x)\rangle \right] \mathbf{1}_{\partial\Omega}(x),$$

where the term $\langle D^2_{pp}H(t,x,p)Dd(x), Dd(x)\rangle$ is bounded from below by a positive constant on the set $\{t \in [0,T] : x(t) \in \partial\Omega\}$.

Proof. The idea of the proof is to approximate system (4) by a penalized control system without state constraints. Then we will apply the nonsmooth Maximum Principle to such a system. Finally, we will retrieve useful information for the original system (4).

Let $\varepsilon < \eta/2$ be a positive fixed constant (where η is defined in (3)), let

$$f_\varepsilon(t,x,u) := f(t,x,u)\left(1 - \frac{1}{\varepsilon}d_\Omega(x)\right)_+,$$

and consider the unconstrained system

$$\begin{cases} x'(t) = f_\varepsilon(t,x(t),u(t)), & \text{a.e. } t \geq 0 \\ x(0) = x_0. \end{cases} \tag{9}$$

The associated perturbed Hamiltonian and the set of optimal trajectories are denoted as follows

$$H_\varepsilon(t,x,p) = \max_{u\in U}\langle f_\varepsilon(t,x,u), p\rangle \qquad \forall(t,x,p) \in \mathbb{R}_+ \times \mathbb{R}^n \times \mathbb{R}^n,$$

$$\mathcal{X}_\varepsilon(T) = \{\text{optimal trajectories of system (9) on } [0,T]\}.$$

We note that f_ε is Lipschitz continuous, but nonsmooth on the boundary of Ω, and that

$$H_\varepsilon(t,x,p) = H(t,x,p)\left(1 - \frac{1}{\varepsilon}d_\Omega(x)\right)_+.$$

We shall prove that, as soon as ε is small enough, any trajectory of $\mathcal{X}(T)$ is actually a trajectory of $\mathcal{X}_\varepsilon(T)$, i.e. $\mathcal{X}(T) \subseteq \mathcal{X}_\varepsilon(T)$. This is not an obvious fact. In fact, it is clear that any constrained solution of (4) is still a solution of system (9), but an optimal trajectory $x(\cdot) \in \mathcal{X}(T)$ may fail to be optimal for (9). Indeed, system (9) can have more trajectories that arrive at the point $x(T) \in \overline{\Omega}$.

We split the proof in 3 steps. The first and second step are devoted to show that $\mathcal{X}(T) \subseteq \mathcal{X}_\varepsilon(T)$ (so that we can use the nonsmooth Maximum Principle). In the third step we use the nonsmooth Maximum Principle to recover the conclusion in the constrained case.

Step 1 Let us check that, for a suitable choice of ε, any optimal trajectory of system (9) that stays in $\overline{\Omega}$ at time $T > 0$, remains in $\overline{\Omega}$ for all $t < T$. Equivalently, we want to prove that

$$\mathcal{X}_\varepsilon^{\overline{\Omega}}(T) := \{x(\cdot) \in \mathcal{X}_\varepsilon(T) : \ x(T) \in \overline{\Omega}\} \subseteq \mathcal{T}_{\mathrm{ad}}(\mathcal{K}, T).$$

Take $x_\varepsilon(\cdot) \in \mathcal{X}_\varepsilon^{\overline{\Omega}}(T)$. By the definition of f_ε,

$$d_\Omega(x_\varepsilon(t)) < \varepsilon < \frac{\eta}{2} \qquad \forall t \in [0, T]. \tag{10}$$

Since $x_\varepsilon(\cdot)$ is optimal for an unconstrained problem, we can use the Maximum Principle of proposition 2.1 for nonsmooth hamiltonians without the boundary term $\psi(t)$. So, we find that there is some adjoint map $p_\varepsilon : [0, T] \to \mathbb{R}^n$ such that, for a.e. $t \in [0, T]$,

$$\begin{cases} x_\varepsilon'(t) = D_p H(t, x_\varepsilon(t), p_\varepsilon(t))(1 - \dfrac{1}{\varepsilon} d_\Omega(x_\varepsilon(t)))_+ \\ -p_\varepsilon'(t) \in \partial_x H_\varepsilon(t, x_\varepsilon(t), p_\varepsilon(t)). \end{cases} \tag{11}$$

We observe that, for each t such that $x_\varepsilon(t) \notin \partial\Omega$, the generalized gradient $\partial_x H_\varepsilon(t, x_\varepsilon(t), p_\varepsilon(t))$ reduces to a singleton and

$$-p_\varepsilon'(t) = D_x H(t, x_\varepsilon(t), p_\varepsilon(t)) \left(1 - \frac{1}{\varepsilon} d_\Omega(x_\varepsilon(t)) \right)_+ \\ - \frac{\lambda_\varepsilon(t) H(t, x_\varepsilon(t), p_\varepsilon(t))}{\varepsilon} Dd(x_\varepsilon(t)), \tag{12}$$

where

$$\begin{cases} \lambda_\varepsilon(t) = 0 & \text{if } x_\varepsilon(t) \in \Omega, \\ \lambda_\varepsilon(t) = 1 & \text{if } x_\varepsilon(t) \notin \overline{\Omega}. \end{cases} \tag{13}$$

Note that $|p_\varepsilon(t)| > 0$ for all $t \in [0, T]$. In fact we can normalize p_ε so that $|p_\varepsilon(0)| = 1$. Moreover, setting $h_\varepsilon(t) = H_\varepsilon(t, x_\varepsilon(t), p_\varepsilon(t))$, we have that

$$h_\varepsilon'(t) = D_t H_\varepsilon(t, x_\varepsilon(t), p_\varepsilon(t)) \geq -M|p_\varepsilon(t)| \left(1 - \frac{d_\Omega(x_\varepsilon(t))}{\varepsilon} \right)_+,$$

thanks to (6) and (8). We recall that, in view of (7), $H_\varepsilon(t, x, p) \geq \alpha|p|(1 - d_\Omega(x)/\varepsilon)_+$. Applying Gronwall's lemma we get $h_\varepsilon(t) \geq e^{-(M/\alpha)t} h_\varepsilon(0)$, and, since $H(t, x_\varepsilon(t), p_\varepsilon(t)) \geq H_\varepsilon(t, x_\varepsilon(t), p_\varepsilon(t))$,

$$H(t, x_\varepsilon(t), p_\varepsilon(t)) \geq e^{-\frac{M}{\alpha}t} H(0, x_\varepsilon(0), p_\varepsilon(0)) \geq e^{-\frac{M}{\alpha}t} \alpha > 0. \tag{14}$$

In particular, this implies that $|p_\varepsilon(t)| > 0$ on $[0, T]$.

Now, in order to conclude the proof of step 1, suppose, by contradiction, that there exists an interval (a, b) on which $x_\varepsilon(t) \notin \overline{\Omega}$, and $x_\varepsilon(a), x_\varepsilon(b) \in \partial\Omega$. Then, using right and left derivative of $t \mapsto d(x_\varepsilon(t))$ in, respectively, a and b, we obtain

$$\begin{cases} \frac{d^+}{dt^+} d(x_\varepsilon(t))|_{t=a} = \langle D_p H(a, x_\varepsilon(a), p_\varepsilon(a)), Dd(x_\varepsilon(a)) \rangle \geq 0 \text{ and} \\ \frac{d^-}{dt^-} d(x_\varepsilon(t))|_{t=b} = \langle D_p H(b, x_\varepsilon(b), p_\varepsilon(b)), Dd(x_\varepsilon(b)) \rangle \leq 0. \end{cases}$$

Since H is nonnegative, these inequalities can also be rewritten as

$$\begin{cases} \langle D_p H^2(a, x_\varepsilon(a), p_\varepsilon(a)), Dd(x_\varepsilon(a)) \rangle \geq 0 \text{ and} \\ \langle D_p H^2(b, x_\varepsilon(b), p_\varepsilon(b)), Dd(x_\varepsilon(b)) \rangle \leq 0. \end{cases}$$

Observe that there exists a constant C (not depending on ε) such that, for all $t \in [a, b]$,

$$\|D^2 d(x_\varepsilon(t))\| \leq C, \tag{15}$$

since, thanks to (10), $x_\varepsilon(t)$ is in a set in which $d(\cdot)$ is of class $\mathcal{C}^2_b$. Hence

$$0 \geq \int_a^b \frac{d}{dt} \langle D_p H^2(t, x_\varepsilon(t), p_\varepsilon(t)), Dd(x_\varepsilon(t)) \rangle \, dt$$

$$= \int_a^b \langle D^2_{pt} H^2(t, x_\varepsilon, p_\varepsilon) + D^2_{xp} H^2(t, x_\varepsilon, p_\varepsilon) x'_\varepsilon + D^2_{pp} H^2(t, x_\varepsilon, p_\varepsilon) p'_\varepsilon, Dd(x_\varepsilon) \rangle \, dt$$

$$+ \int_a^b \langle D_p H^2(t, x_\varepsilon, p_\varepsilon), D^2 d(x_\varepsilon) x'_\varepsilon \rangle \, dt$$

$$\geq - \int_a^b \|D^2_{xp} H^2(t, x_\varepsilon, p_\varepsilon)\| \, |f(t, x_\varepsilon, u)| \, dt$$

$$- \int_a^b \left(\|D^2_{pp} H^2(t, x_\varepsilon, p_\varepsilon)\| \, |D_x H(t, x_\varepsilon, p_\varepsilon)| + |D^2_{pt} H^2(t, x_\varepsilon, p_\varepsilon)| \right) dt$$

$$- \int_a^b |D_p H^2(t, x_\varepsilon, p_\varepsilon)| \, \|D^2 d(x_\varepsilon)\| \, |f(t, x_\varepsilon, u)| \, dt$$

$$+ \int_a^b \frac{H(t, x_\varepsilon, p_\varepsilon)}{\varepsilon} \langle D^2_{pp} H^2(t, x_\varepsilon, p_\varepsilon) Dd(x_\varepsilon), Dd(x_\varepsilon) \rangle \, dt$$

$$\geq \int_a^b \left(\frac{H(t, x_\varepsilon, p_\varepsilon)}{\varepsilon} \gamma - M'(1 + |p_\varepsilon|) \right) dt \,,$$

where γ is defined by (7) and $M' = 2M^3(2 + C) + 4M^2 + \Gamma M$.

Finally, we get a contradiction setting

$$\varepsilon := \frac{\alpha \gamma e^{-\frac{M}{\alpha} T}}{4M'}. \tag{16}$$

Indeed $H(t, x_\varepsilon, p_\varepsilon)\gamma/\varepsilon - M'(1 + |p_\varepsilon|) > 0$ if, for $|p_\varepsilon| \leq 1$, we choose $\varepsilon < \alpha\gamma e^{-(M/\alpha)T}/(2M')$, and, for $|p_\varepsilon| > 1$, we choose $\varepsilon < \alpha\gamma/(2M')$.

Step 2 We want to show that an extremal solution of the constrained problem (4) is also an extremal solution of (9), i.e., that $\mathcal{X}(T) \subseteq \mathcal{X}_\varepsilon(T)$.

Let $x(\cdot) \in \mathcal{X}(T)$, and let T_ε the minimum time to reach the point $x(T)$ for the perturbed problem (9). Since any solution of the constrained problem (4) is also a solution of the unconstrained problem (because $f_{\varepsilon|\Omega} = f_{|\Omega}$), we have that $T_\varepsilon \leq T$.

Now, let $x_\varepsilon(\cdot)$ be a trajectory of the perturbed problem (9) such that $x_\varepsilon(T_\varepsilon) = x(T)$. Then $x_\varepsilon(\cdot)$ is in $\mathcal{X}_\varepsilon^{\overline{\Omega}}(T_\varepsilon)$, and, by step 1, we know that it is also a solution of the original problem (4), as it ends in $\overline{\Omega}$. Then $T_\varepsilon \geq T$, and we have that $T_\varepsilon = T$ and $x(\cdot) \in \mathcal{X}_\varepsilon(T)$.

Step 3 Finally, let $x(\cdot) \in \mathcal{X}(T)$ and $p(\cdot)$ be an associated adjoint map such that $(x(\cdot), p(\cdot))$ satisfies (11). We want to find an explicit expression for $p'(t)$. For this we define the set of times t for which the trajectory $x(\cdot)$ stays on the boundary, that is

$$E_x = \{t \in [0, T] : x(t) \in \partial\Omega\}.$$

Observe that, by standard properties of the generalized gradient,

$$\partial_x H_\varepsilon(t, x, p) \subseteq \partial_x H(t, x, p)\left(1 - \frac{d_\Omega(x)}{\varepsilon}\right)_+ + H(t, x, p)\partial_x\left(1 - \frac{d_\Omega(x)}{\varepsilon}\right)_+$$

for all $x \in \partial\Omega$ (see Ref. 7). Moreover, for all $x \in \partial\Omega$,

$$\partial_x H(t, x, p) = D_x H(t, x, p), \qquad \partial_x\left(1 - \frac{d_\Omega(x)}{\varepsilon}\right) = -Dd(x)[0, \frac{1}{\varepsilon}].$$

Recalling (11) and (12), we have that there is a measurable function $\lambda_\varepsilon : [0, T] \to [0, 1]$ such that, for a.e. $t \in [0, T]$,

$$-p'(t) = D_x H(t, x(t), p(t)) - \frac{\lambda_\varepsilon(t)H(t, x(t), p(t))}{\varepsilon}Dd(x(t)), \qquad (17)$$

where $\lambda_\varepsilon(t)$ is given by (13) for $t \notin E_x$. Now define the function

$$\varphi(t) := d(x(t)) \qquad t \in [0, T].$$

Since Ω is of class C^2 and $x'(t)$ exists for a.e. t, we have that φ is differentiable for a.e. $t \in E_x$. Moreover, for a.e. $t \in E_x$,

$$\langle Dd(x(t)), D_p H(t, x(t), p(t))\rangle = \langle Dd(x(t)), x'(t)\rangle = \varphi'(t) = 0. \qquad (18)$$

In fact, let $t \in E_x$. If $\varphi'(t) > 0$ then, by continuity, there exists some $\eta > 0$ such that $\varphi(s) > 0$ for all $s \in (t, t + \eta)$, in contrast to $x(t) \in \overline{\Omega}$

for all $t \in [0, T]$. Similarly, if $\varphi'(t) < 0$ we will have that $\varphi(s) > 0$ for all $s \in (t - \eta, t)$.

In E_x, we can differentiate (18) with respect to t. We obtain

$$0 = \langle D^2 d(x(t))x'(t), D_p H(t, x(t), p(t)) \rangle + \langle Dd(x(t)), D^2_{pt} H(t, x(t), p(t)) \rangle$$
$$+ \langle Dd(x(t)), D^2_{xp} H(t, x(t), p(t))x'(t) \rangle$$
$$+ \langle Dd(x(t)), D^2_{pp} H(t, x(t), p(t))p'(t) \rangle.$$

We focus our attention on the last term of this equation, seeing that

$$\langle\, Dd(x(t)), D^2_{pp} H(t, x(t), p(t))p'(t) \rangle$$
$$= \langle Dd(x(t)), D^2_{pp} H(t, x(t), p(t)) \Big[H(t, x(t), p(t)) \frac{\lambda_\varepsilon(t)}{\varepsilon} Dd(x(t)) - D_x H(t, x(t), p(t)) \Big] \rangle$$
$$= \frac{\lambda_\varepsilon(t)}{\varepsilon} H(t, x(t), p(t)) \langle Dd(x(t)), D^2_{pp} H(t, x(t), p(t)) Dd(x(t)) \rangle$$
$$- \langle Dd(x(t)), D^2_{pp} H(t, x(t), p(t)) D_x H(t, x(t), p(t)) \rangle.$$

We claim that, for $|p(t)| > 0$, $\langle Dd(x(t)), D^2_{pp} H(t, x(t), p(t)) Dd(x(t)) \rangle > 0$. Indeed, for a unit vector $\zeta \in \mathbb{R}^n$, we have that

$$H(t, x, p) \langle D^2_{pp} H(t, x, p)\zeta, \zeta \rangle = \frac{\langle D^2_{pp} H^2(t, x, p)\zeta, \zeta \rangle}{2} - \langle D_p H(t, x, p), \zeta \rangle^2$$
$$\geq \frac{\gamma}{2} - \langle D_p H(t, x, p), \zeta \rangle^2$$

thanks to (7). Then,

$$\langle \frac{D_p H(t, x, p)}{|D_p H(t, x, p)|}, \zeta \rangle^2 \leq \frac{\gamma}{4M^2} \quad \Longrightarrow \quad H(t, x, p)\langle D^2_{pp} H(t, x, p)\zeta, \zeta \rangle \geq \frac{\gamma}{4}.$$

As a consequence of (18), there exists a constant $\delta > 0$ such that

$$\langle Dd(x(t)), D^2_{pp} H(t, x(t), p(t)) Dd(x(t)) \rangle > \frac{\delta}{|p(t)|} \qquad \text{a.e. } t \in E_x.$$

Finally, recalling the expression of p' in (17), we find a representation of λ by setting

$$\lambda(t) := \frac{\lambda_\varepsilon(t)}{\varepsilon} H(t, x(t), p(t)).$$

Moreover, since

$$e^{-M(1+1/\varepsilon)T} \leq |p(t)| \leq e^{M(1+1/\varepsilon)T},$$

(where ε is defined in (16)) we have that λ is bounded and $p(\cdot)$ is Lipschitz continuous. Observing that the right-hand side of the equality

$x' = D_pH(t, x, p)$ is continuous, we conclude that this equality holds for all t in $[0, T]$. $\qquad\qquad\square$

Example 2.2. We discuss a class of control systems that satisfy the assumptions of this section for the maximum principle. Consider the control system

$$\begin{cases} x'(t) = f(t, x(t))u(t), & u(t) \in \overline{B} \\ x(0) = x_0 \in \mathcal{K}, \end{cases}$$

where $f : \mathbb{R}_+ \times \mathbb{R}^n \to M_n(\mathbb{R})$ is of class $\mathcal{C}^2$, and the matrix $f(t, x)$ is bounded and invertible, with bounded inverse.

Moreover, the Hamiltonian is

$$H(t, x, p) = \max_{u \in \overline{B}_1} \langle f(t, x)u, p \rangle = \max_{u \in \overline{B}_1} \langle u, f^*(t, x)p \rangle = |f^*(t, x)p|.$$

So, assumptions (7) are satisfied, since f and f^{-1} are bounded, and

$$D_{pp}^2 H^2(t, x, p) = 2f(t, x)f^*(t, x).$$

Since we can consider (t, x) in a compact set, then $D_t f(t, x)$ and $D_x f(t, x)$ are bounded, and, thanks to remark 2.2, $D_t H(t, x, p)$ and $D_x H(t, x, p)$ satisfy assumptions (8).

Moreover,

$$D_p H(t, x, p) = \frac{f(t, x)f^*(t, x)p}{|f^*(t, x)p|}. \tag{19}$$

Then we have

$$|D_{pt}^2 H(t, x, p)| \leq 2\|D_t f(t, x)\| + \|f(t, x)\| \, \|f^{-1}(t, x)\| \, \|D_t f(t, x)\|,$$

and

$$\|D_{px}^2 H(t, x, p)\| \leq 2\|D_x f(t, x)\| + \|f(t, x)\| \, \|f^{-1}(t, x)\| \, \|D_x f(t, x)\|.$$

Hence all the assumptions in (8) are satisfied.

Finally note that, for any vector $\zeta \in \mathbb{R}^n$,

$$\langle D_{pp}^2 H(t, x, p)\zeta, \zeta \rangle = \frac{1}{|f^*(t, x)p|} \left(|f^*(t, x)\zeta|^2 - \langle \frac{f^*(t, x)p}{|f^*(t, x)p|}, f^*(t, x)\zeta \rangle^2 \right),$$

so that it would be impossible to satisfy $\Gamma I_n \geq D_{pp}^2 H(t, x, p) \geq \gamma I_n$, even for this simple control system.

3. Perimeter estimates for the attainable set

In this section we will study the special case of control systems in $\mathbb{R}^2$ of the form

$$\begin{cases} x'(t) = f(t, x(t))u(t), & \text{a.e. } t \geq 0 \\ x(t) \in \overline{\Omega}, & t \geq 0 \\ x(0) = x_0 \in \mathcal{K}, \end{cases} \tag{20}$$

where $u : \mathbb{R}_+ \to \overline{B}$ is a measurable function, and $f : \mathbb{R}_+ \times \mathbb{R}^2 \to M_2(\mathbb{R})$ is such that

- f is of class $\mathcal{C}^2$ in $\mathbb{R}_+ \times \overline{\Omega}$
- $f(t, x)$ is invertible for any $(t, x) \in \mathbb{R}_+ \times \overline{\Omega}$
- $f(t, x)$ and $f^{-1}(t, x)$ are bounded by $M > 0$ for any $(t, x) \in \mathbb{R}_+ \times \overline{\Omega}$.

The Hamiltonian for this dynamic is

$$H(t, x, p) = \max_{u \in \overline{B}} \langle f(t, x)u, p \rangle = |f^*(t, x)p|.$$

The aim of this section is to estimate the perimeter of the attainable set

$$\mathcal{A}(t) := \mathcal{A}(\mathcal{K}, t).$$

Cardaliaguet and Marchi[6] proved such an estimate for $f(t, x) = c(t, x)I_n$, where c is a scalar function. Applying theorem 2.1 we can use their technique to extend this analysis to system (20).

Theorem 3.1. *Let $T > 0$ be fixed. For any $t \in (0, T]$ the attainable set $\mathcal{A}(t)$ is of finite perimeter, and there exist $C_1, C_2 > 0$ such that*

$$\mathcal{H}^1(\partial \mathcal{A}(t)) \leq C_1 \left(\mathcal{H}^1(\partial \Omega) + \frac{e^{C_2 t}}{t} \right).$$

In particular, if set $\mathcal{K}$ has the interior sphere property, then the attainable set $\mathcal{A}(t)$ has finite perimeter for any $t \in [0, T]$, and the perimeter of $\mathcal{A}(t)$ is bounded on $[0, T]$.

In order to prove that $\mathcal{A}(t)$ has finite perimeter, we cover the boundary of the attainable set with the following sets

$$\mathcal{B}^{\mathrm{bnd}}(t) := \{x \in \overline{\Omega} : \exists x(\cdot) \in \mathcal{X}(t), \exists s \in [0, t], \text{ with } x(t) = x, \, x(s) \in \partial\Omega\},$$

$$\mathcal{B}^{\mathrm{int}}(t) := \{x \in \Omega : \exists x(\cdot) \in \mathcal{X}(t), \text{ with } x(t) = x \text{ and } x([0, t]) \cap \partial\Omega = \emptyset\}.$$

Note that $\mathcal{B}^{int}(t)$ is the set of points that can be reached by extremal trajectories contained in the interior of Ω, while $\mathcal{B}^{\mathrm{bnd}}(t)$ is the set of points reached by extremal trajectories that touch the boundary $\partial\Omega$. Then

$$\partial\mathcal{A}(t) \subseteq \mathcal{B}^{\mathrm{int}}(t) \cup \mathcal{B}^{\mathrm{bnd}}(t) \cup \partial\Omega.$$

We point out that, in this case, the adjoint system of theorem 2.1 is given by

$$-p'(t) = |p(t)| \left(D_x H(t, x(t), \frac{p(t)}{|p(t)|}) - \overline{\lambda}(t) Dd(x(t)) \right), \qquad (21)$$

where $\overline{\lambda}$ is a bounded positive function.

Consequently, for all $T > 0$, we have that the set $\mathcal{X}(T)$ of the extremal trajectories for system (20) is compact with respect to the $\mathcal{C}^1$ norm.

With an easy adaptation we can recover from Ref. 6 a Lipschitz estimate for the velocities of extremal trajectories on the boundary of Ω.

Lemma 3.1 (Ref. 6, lemma 4.1). *Let T and δ be given positive numbers, let $t_1, t_2 \in (0, T]$, and let*

$$y_i : [0, t_i + \delta] \to \mathbb{R}^2 \qquad (i = 1, 2)$$

be optimal trajectories for (20). There is a positive number σ such that, if $y_i(t_i) \in \partial\Omega$ $(i = 1, 2)$ and $|y_1(t_1) - y_2(t_2)| \leq \sigma$, then

$$\left| \frac{y_1'(t_1)}{|y_1'(t_1)|} - \frac{y_2'(t_2)}{|y_2'(t_2)|} \right| \leq C_\Omega |y_1(t_1) - y_2(t_2)|,$$

where C_Ω is a positive constant depending only on the regularity of $\partial\Omega$.

In order to estimate $\mathcal{H}^1(\mathcal{B}^{\mathrm{bnd}}(t))$ the main idea is to use the perimeter of $\partial\Omega$. At this aim we define the set

$$\mathcal{B}^{\mathrm{bnd}}(t, \delta) := \{x \in \mathcal{B}^{\mathrm{bnd}}(t) : x((t - \delta, t]) \cap \partial\Omega = \emptyset\}.$$

Lemma 3.2. *Let $T > 0$ be fixed, let $t \in (0, T]$, and let $y_1, y_2 \in \mathcal{B}^{\mathrm{bnd}}(t, \delta)$. For $i = 1, 2$, let $y_i(\cdot) \in \mathcal{X}(t)$ and let $s_i \in [0, t - \delta)$ such that*

$$y_i(t) = y_i, \quad y_i(s_i) \in \partial\Omega, \quad y_i((s_i, t]) \cap \partial\Omega = \emptyset.$$

There is a constant $C_0 > 0$ such that, for any $\delta > 0$, a constant $\sigma_\delta > 0$ exists, so that, if

$$|y_1(s_1) - y_2(s_2)| \leq \sigma_\delta,$$

then

$$|y_1 - y_2| \leq C_0 |y_1(s_1) - y_2(s_2)|.$$

Proof. Without loss of generality, we can suppose $s_1 \leq s_2$. There exist adjoint states $p_i(\cdot)$ such that, for $s \geq s_i$ the extremal trajectories $y_i(\cdot)$ solve

$$\begin{cases} y_i'(s) = D_p H(s, y_i(s), p_i(s)) \\ -p_i'(s) = |p_i(s)| D_x H(s, y_i(s), \frac{p_i(s)}{|p_i(s)|}). \end{cases} \qquad (22)$$

We normalize arcs $p_i(\cdot)$ so that

$$H(s_i, y_i(s_i), p_i(s_i)) = |f^*(s_i, y_i(s_i))p_i(s_i)| = 1, \qquad i = 1, 2.$$

Applying Gronwall's lemma we have

$$\frac{1}{M} e^{-L(s-s_i)} \leq |p_i(s)| \leq M e^{L(s-s_i)} \qquad \forall s \in [s_i, t].$$

Moreover, thanks to system (22), for any $s \in [s_2, t]$,

$$\begin{aligned}
|y_1(s) - y_2(s)| &\leq |y_1(s) - y_2(s)| + |p_1(s) - p_2(s)| \\
&\leq C\big(|y_1(s_2) - y_2(s_2)| + |p_1(s_2) - p_2(s_2)|\big).
\end{aligned}$$

Now, we focus our attention on the right-hand side of this inequality. For the first term, we have that

$$\begin{aligned}
|y_1(s_2) - y_2(s_2)| &\leq |y_1(s_1) - y_2(s_2)| + |y_1(s_2) - y_1(s_1)| \\
&\leq |y_1(s_1) - y_2(s_2)| + M|s_1 - s_2| \\
&\leq (1 + ML_0)|y_1(s_1) - y_2(s_2)|,
\end{aligned}$$

invoking the Lipschitz continuity (of rank L_0) of the minimum time function. Similarly, we have that

$$\begin{aligned}
|p_1(s_2) - p_2(s_2)| &\leq |p_1(s_1) - p_2(s_2)| + |p_1(s_2) - p_1(s_1)| \\
&\leq |p_1(s_1) - p_2(s_2)| + C'|y_1(s_1) - y_2(s_2)|.
\end{aligned}$$

Recalling the expression of $D_p H(t, x, p)$ in (19), we have that

$$|p_1(s_1) - p_2(s_2)| = \Bigg| \overbrace{|f^*(s_1, y_1(s_1))p_1(s_1)|}^{=1} \big[f(s_1, y_1(s_1))f^*(s_1, y_1(s_1))\big]^{-1} y_1'(s_1)$$
$$- \overbrace{|f^*(s_2, y_2(s_2))p_2(s_2)|}^{=1} \big[f(s_2, y_2(s_2))f^*(s_2, y_2(s_2))\big]^{-1} y_2'(s_2) \Bigg|.$$

Moreover, we also have that

$$|y_1'(s_1) - y_2'(s_2)| \leq$$
$$\leq |y_1'(s_1)| \left(\left| \frac{y_1'(s_1)}{|y_1'(s_1)|} - \frac{y_2'(s_2)}{|y_2'(s_2)|} \right| + |y_2'(s_2)| \left| \frac{1}{|y_1'(s_1)|} - \frac{1}{|y_2'(s_2)|} \right| \right).$$

Since, for $i = 1, 2$, $|y_i'(s_i)|y_i'(s_i)/|y_i'(s_i)| = f(s_i, y_i(s_i))u_i(s_i)$, and $u_i(s_i) \in \partial B$, we obtain that

$$\left| \frac{1}{|y_1'(s_1)|} - \frac{1}{|y_2'(s_2)|} \right| = \left| f^{-1}(s_1, y_1(s_1)) \frac{y_1'(s_1)}{|y_1'(s_1)|} - f^{-1}(s_2, y_2(s_2)) \frac{y_2'(s_2)}{|y_2'(s_2)|} \right|.$$

As a consequence of lemma 3.1 and of Lipschitz continuity of f we have the conclusion. $\qquad\square$

Thanks to lemma 3.2 we have a bound for the perimeter of $\mathcal{B}^{\mathrm{bnd}}(t)$. We give a proof for the reader's convenience.

Proposition 3.1. *Let $T > 0$ be fixed. There is a constant $C_0 > 0$ such that, for all $t \in [0, T]$,*

$$\mathcal{H}^1(\mathcal{B}^{\mathrm{bnd}}(t)) \leq C_0 \mathcal{H}^1(\partial\Omega).$$

Proof. For any fixed $\delta > 0$, let C_0 and σ_δ the constants of lemma 3.2. Let $\varepsilon > 0$, and let $\{B_n\}_{n \in \mathbb{N}}$ be a family of sets such that

$$\partial\Omega \subseteq \bigcup_{n \in \mathbb{N}} B_n, \quad \mathcal{H}^1(\partial\Omega) + \varepsilon \geq \sum_{n=1}^{\infty} \mathrm{diam}(B_n), \quad \mathrm{diam}(B_n) \leq \inf\{\sigma_\delta, \varepsilon\}.$$

Consider the covering of $\mathcal{B}^{\mathrm{bnd}}(t, \delta)$, given by

$$K_n = \{x \in \mathcal{B}^{\mathrm{bnd}}(t, \delta) : x(s) \in B_n, \, x((s, t]) \cap \partial\Omega = \emptyset\}.$$

In view of lemma 3.2, for all $n \in \mathbb{N}$, we have that

$$\mathrm{diam}(K_n) \leq C_0 \varepsilon,$$
$$\mathrm{diam}(K_n) \leq C_0 \mathrm{diam}(B_n).$$

Thus,

$$\mathcal{H}^1_{C_0\varepsilon}(\mathcal{B}^{\mathrm{bnd}}(t, \delta)) \leq \sum_{n=1}^{\infty} \mathrm{diam}(K_n) \leq C_0(\mathcal{H}^1(\partial\Omega) + \varepsilon).$$

Now, letting $\varepsilon \to 0$ and then $\delta \to 0$, we have the desired result, since C_0 is independent of ε and δ. $\qquad\square$

Now, we turn our attention to set $\mathcal{B}^{\mathrm{int}}(t)$. Following the main ideas of Ref. 3, we can see that this set has the *interior ball property*, and then it has finite perimeter for any $t > 0$.

Remark 3.1. Let $x(\cdot)$ be an extremal solution on the interval $[0, T]$, such that $x([0, T]) \subseteq \Omega$. Then there exists some $\eta = \eta(x) > 0$ such that, for all $s \in [0, T]$, we have that $B_\eta(x(s)) \subseteq \Omega$.

Proposition 3.2. *Let $T > 0$ be fixed. Then there exists a constant $c_T > 0$ such that for any $x \in \mathcal{B}^{\mathrm{int}}(T)$, with $x(\cdot)$ extremal solution so that $x(T) = x$, then there exists some $\eta = \eta(x) > 0$ such that*

$$B\left(x(t) - c_T t \frac{p(t)}{|p(t)|}, c_T t\right) \cap B\left(x(t), \eta \frac{e^{-LT}}{1 + LT}\right) \subseteq \mathcal{A}(t).$$

From Ref. 1 and proposition 3.2, we can derive the following bound for the perimeter.

Proposition 3.3. *Fix $T > 0$. There exists $C > 0$ such that*

$$\mathcal{H}^1(\mathcal{B}^{\mathrm{int}}(t)) \leq \frac{C}{c_T t}.$$

The perimeter of $\mathcal{B}^{\mathrm{int}}(t)$ is in inverse ratio to the time t, since the interior sphere property is proportional to the time. This means that, if we fix a time $\vartheta > 0$, for any $t \geq \vartheta$ we have a uniform estimate (i.e. $C/c_T\vartheta$). In addition, if set $\mathcal{K}$ has the interior sphere property of radius r, then for all $t \in [0, T]$

$$\mathcal{H}^1(\mathcal{B}^{\mathrm{int}}(t)) \leq \frac{C}{r}.$$

References

1. O. Alvarez, P. Cardaliaguet, R. Monneau, *Interf. Free Bound.* **7**, 415 (2005).
2. A.V. Arutyunov and S.M. Aseev, *SIAM J. Control Optim.* **35**, 930 (1997).
3. P. Cannarsa and P. Cardaliaguet, *J. Convex Anal.* **13**, 253 (2006).
4. P. Cannarsa and H. Frankowska, *ESAIM Control Optim. Calc. Var.* **12**, 350 (2006).
5. P. Cannarsa and C. Sinestrari, *Semiconcave functions, Hamilton-Jacobi equations and optimal control* (Birkhäuser, Boston, 2004).
6. P. Cardaliaguet and C. Marchi, *SIAM J. Control Optim.* **45**, 1017 (2006).
7. F.H. Clarke, *Optimization and Nonsmooth analysis* (Wiley-Interscience, New York, 1983).
8. A.Y. Dubovitskii and A.A. Milyutin, *USSR Comput. Math. and Math. Physics* **5**, 1 (1965).
9. H. Frankowska, *J. Convex Anal.* **13**, 299 (2006).
10. R.V. Gamkrelidze, *Izk. Akad. Nauk., USSR Sec. Mat.* **24**, 315 (1960).
11. P. Loewen and R.T. Rockafellar, *Trans. Amer. Math. Soc.* **325**, 39 (1991).
12. L.S. Pontryagin, V.G. Boltyanskii, R.V. Gamkrelidze, E.F. Mishchenko, *The Mathematical theory of otpimal processes* (Interscience Publishers John Wiley & Sons Inc., New York-London, 1962).
13. R.B. Vinter, *Optimal Control* (Birkhäuser, Boston, Basel, Berlin, 2000).

A GENERALIZED HOPF-LAX FORMULA:
ANALYTICAL AND APPROXIMATIONS ASPECTS

I. CAPUZZO DOLCETTA

Dipartimento di Matematica, Università di Roma "La Sapienza",
00185 Roma, Italia
E-mail: capuzzo@mat.uniroma1.it

In this paper we discuss the validity of the Hopf-Lax representation formula for
solutions of evolution Hamilton-Jacobi equations governed by state-dependent
and perhaps non coercive Hamiltonians. Some applications to PDE's on the
Heisenberg group and finite differences approximations are also presented.

Keywords: Hamilton-Jacobi, Hopf-Lax formula, Heisenberg group, finite differences approximations

1. Introduction

The purpose of this Note is to report some recent research on scalar evolution Hamilton-Jacobi equations of the form

$$\begin{cases} u_t + \Phi\left(H_0(x, D_x u)\right) = 0\,, \ (t,x) \in (0, +\infty) \times \mathbb{R}^N \\ u(0, x) = g(x)\,, \ x \in \mathbb{R}^N, \end{cases} \tag{1}$$

where $p \to H_0(x,p)$ is convex and $1-$positively homogeneous and Φ is a
convex, nondecreasing scalar function. The focus is on Hamiltonians H_0
which are state-dependent and perhaps non coercive with respect to the
gradient variable. The main issue addressed here is the validity of a representation formula of Hopf-Lax type (often referred also as Hopf-Lax-Oleinik
formula) for the solution of the above initial value problem. In the language
of optimization theory, these type of formulas represent the solution of problem (1) as a marginal function, parametrized by the time variable t. In the
model state-independent example where $H_0(D_x u) = |Du|$ and $\Phi(r) = \frac{1}{2}r^2$,
is well-known that the Hopf-Lax solution is given by

$$u(t,x) = \inf_{y \in \mathbb{R}^N} \left[g(y) + \frac{|x-y|^2}{2t}\right].$$

Representations formula as the above are of course very useful in the analysis of Hamilton-Jacobi equations. We just mention here applications in optimal control theory[3,23] , large time behaviour of solutions[5] , weak KAM theory[16] , transport theory[29] , geometric front propagation[26] .

The validity of such representation formulas has been widely investigated in the state-independent coercive case starting from the seminal paper Ref. 20, see also Refs. 2,28 for recent general results. Few results are available in the state-dependent case, see however Refs. 24,27 in this respect.

Our formula

$$u(t,x) = \inf_{y \in \mathbb{R}^N} \left[g(y) + t\Phi^* \left(\frac{d(x;y)}{t} \right) \right],$$

where Φ^* is the convex conjugate of Φ and d is a distance function suitably associated with the 1-positively homogeneous function H_0, covers in particular the case of Hamilton-Jacobi equations associated with sub-Riemannian metrics; in this case $H_0(x,p) - |\sigma(x)p|$ where $\sigma(x)$ is a $M \times N$ matrix satisfying the Chow-Hörmander rank condition, see for example Ref. 8 . Let us recall for convenience that this condition amounts to the requirement that there exists an integer $k \leq N$ such that the Lie algebra spanned by the columns of $\sigma(x)$ by means of commutators of length $\leq k$ has rank N at any $x \in \mathbb{R}^N$.

An interesting special case discussed in greater detail in Sections 3 and 4 is that of the Heisenberg Hamiltonian; in that case the matrix $\sigma(x)$ is given, for $x = (x_1, x_2, x_3) \in \mathbb{R}^3$, by

$$\sigma(x) = \begin{pmatrix} 1 & 0 & 2x_2 \\ 0 & 1 & -2x_1 \end{pmatrix}$$

In that case, our Hopf-Lax type formula can be regarded as a simplified expression for the value function of the Mayer optimal control problem for the well-known Brockett's[9] system where * denotes transposition:

$$\inf g\left(x(t;x,c(.))\right), \quad \dot{x}(t) = \sigma^*(x(t))\,c(t), x(0) = x, \quad c(t) \in R^2, \ |c(t)| \leq 1.$$

Some application of our representation formula to the asymptotic analysis of the heat operator on the Heisenberg group and to the error analysis of an approximation scheme for the numerical computation of the solution of problem (1) are outlined in Section 4.

Most of the results reported here are taken from the recent papers or preprints[1,10,11,15] .

2. A generalized eikonal equation

Consider a set-valued mapping $x \to C(x) \subseteq \mathbb{R}^N$ such that

$$C(x) \quad \text{is convex and compact and} \quad 0 \in C(x)\,, \forall\, x \in \mathbb{R}^N\,, \tag{2}$$

$$d_{\mathcal{H}}(C(x), C(y)) \leq L|x - y|\,, \tag{3}$$

where

$$d_{\mathcal{H}}(C(x), C(y)) = \max \left[\max_{a \in C(x)} \text{dist}(a, C(y)); \max_{b \in C(y)} \text{dist}(a, C(x)) \right]$$

is the usual Hausdorff metric.

We assume also that $C(x)$ is "fat" enough in the following sense:

there exist $\delta > 0$, a natural number $M \leq N$ and for each $x \in \mathbb{R}^N$ an $M \times N$ smooth matrix $\sigma(x)$ with bounded Jacobian such that the columns of $\sigma(x)$ satisfy the Chow-Hörmander rank condition of order k at any $x \in \mathbb{R}^N$ and

$$\sigma^*(x) B_{\mathbb{R}^M}(0, \delta) \subseteq C(x)\,, \ \forall\, x \in \mathbb{R}^N\,.$$

The above condition, labelled **(CC)** from now on, is trivially satisfied if $C(x)$ contains an $N-$dimensional ball; observe however that **(CC)** may be fulfilled even when $C(x)$ has empty interior at some x, see examples below. Denoting by $B_1(0)$ the Euclidean unit ball, it is easy to check that condition **(CC)** holds e.g. in the following cases:

$$C(x) = a(x)\, B_{\mathbb{R}^M}(0, \delta)\,, \ a(x) \geq a_0 > 0 \tag{4}$$

$$C(x) = \sqrt{A(x)}\, B_{\mathbb{R}^M}(0, \delta)\,, \ A(x) \text{ positive definite} \tag{5}$$

$$C(x) = \sigma^*(x)\, B_{\mathbb{R}^M}(0, \delta)\,, \ \sigma(x) \text{ satisfying } \textbf{(CC)}. \tag{6}$$

let us consider now the support function of the set $C(x)$, namely

$$H_0(x, p) = \sup_{q \in C(x)} q \cdot p \tag{7}$$

It is not hard to check that under the assumptions made, H_0 satisfies

$$p \mapsto H_0(x, p) \text{ convex}\,, H_0(x, \lambda p) = \lambda H_0(x, p) \text{ for all } \lambda > 0, \tag{8}$$

$$H_0(x, p) \geq 0\,, |H_0(x, p) - H_0(y, p)| \leq L\left(|x - y|(1 + |p|)\right), \tag{9}$$

the last property of H_0 following from the fact that

$$d_{\mathcal{H}}(C(x), C(y)) = \max_{|p| \leq 1} |H_0(x, p) - H_0(y, p)|\,.$$

Furthermore, it is well-known from convex analysis that

$$\partial_p H_0(x,0) = C(x)$$

where $\partial_p H$ denotes the partial subdifferential with respect to the p variable. The Hamiltonians corresponding to the cases (4),(5),(6) are given respectively, by

$$H_0(x,p) = a(x)|p| \,, \; H_0(x,p) = \sqrt{A(x)p \cdot p} \,, \; H_0(x,p) = |\sigma(x)p| \,.$$

This last example includes, in particular, the case of the Grushin Hamiltonian given on $\mathbb{R}^2$ by $H_0(x,p) = \sqrt{p_1^2 + x_1^2 p_2^2}$, as well as that of Heisenberg Hamiltonian given on $\mathbb{R}^3$ by

$$H_0(x,p) = \sqrt{(p_1 + 2x_2 p_3)^2 + (p_2 - 2x_1 p_3)^2} \,. \tag{10}$$

See Ref. 8 for a discussion of the role of these Hamiltonians in sub-Riemannian geometry. Further examples are provided by Hamiltonians of the form $H_0(x,p) = \max \left[H_0^1(x,p), ..., H_0^l(x,p) \right]$, where the H_0^i are of the types above.

Condition **(CC)** and the classical Chow Connectivity Theorem[8] imply that the set $F_{x,y}$ of all trajectories $X(\cdot)$ of the differential inclusion

$$\dot{X}(t) \in \partial_p H_0(X(t),0) \,, \quad X(0) = x, \; X(T) = y$$

for some $T = T(X(\cdot)) > 0$, is non empty for all x and y in $\mathbb{R}^N$ and, consequently the minimum connection time

$$d(x;y) := \inf_{X(\cdot) \in F_{x,y}} T(X(\cdot))$$

is well defined and finite on $\mathbb{R}^N$. Moreover, $d(x,y)$ is a sub-Riemannian metric on $\mathbb{R}^N$ such that

$$\frac{1}{C}|x - y|^{\frac{1}{k}} \le d(x,y) \le C|x - y|^{\frac{1}{k}} \,,$$

and therefore, $x \to d(x,y)$ is a $\frac{1}{k}$−Hölder continuous function, see Ref. 8 for a nice presentation of these topics.

By Dynamic Programming techniques one can verify that $x \to d(x,y)$ solves for each fixed $y \in \mathbb{R}^N$ the generalized eikonal problem

$$\begin{cases} H_0(x, D_x d) = 1 \,, \; x \in \mathbb{R}^N \setminus \{y\} \\ d(y) = 0 \,. \end{cases} \tag{11}$$

in the viscosity sense[3,23] .

3. The generalized Hopf-Lax formula

Let us consider now the evolution problem

$$\begin{cases} u_t + \Phi\left(H_0(x, D_x u)\right) = 0 \,, \ (t,x) \in (0, +\infty) \times \mathbb{R}^N \\ u(0,x) = g(x) \,, \ x \in \mathbb{R}^N \end{cases} \tag{12}$$

where $\Phi : \mathbb{R}^N \to [0, +\infty)$, $\Phi(0) = 0$, is convex, increasing and H_0 as in (7). In this setting, an Hopf-Lax type formula holds for the solution of (12) under the assumption for H_0 made in the previous section.

Theorem 3.1. *Assume $H_0(x, p) = \sup_{q \in C(x)} q \cdot p$ with $C(x)$ satisfying conditions (2),(3) as well as condition (CC). If $g \in C(\mathbb{R}^N)$ is linearly bounded below, i.e.*

$$g(x) \geq -C(1 + |x|) \ \text{for some } C > 0 \,,$$

then the function

$$u(t,x) = \inf_{y \in \mathbb{R}^N} \left[g(y) + t\Phi^* \left(\frac{d(x;y)}{t} \right) \right], \tag{13}$$

where Φ^* is the convex conjugate of Φ, is the unique continuous linearly bounded below viscosity solution of the Cauchy problem (12).

This representation formula generalizes to the present framework the classical Hopf-Lax formula[2,4] for the case of uniformly convex state independent Hamiltonians $H = H(p)$ as well as the one established in Ref. 24 for the special case of the Heisenberg Hamiltonian (10). Comparison and uniqueness results for problem (12) suited to the present setting can be found in Refs. 11,13 .

The proof of the fact that the Hopf-Lax function (13) is a solution of problem (12) is based on viscosity calculus and makes essential use of stability properties of viscosity solutions with respect to inf operations. We briefly sketch it here and refer to Ref. 11 for full details.

The first step is to show that the functions v^y defined by

$$v^y(t,x) = g(y) + t\Phi^* \left(\frac{d(x;y)}{t} \right),$$

where $y \in \mathbb{R}^N$ plays the role of a parameter, do satisfy equation(12) for each fixed y. A formal argument is as follows: assume that (11) has smooth solutions $d(x) = d(x,y)$ for each fixed y and suppose that v^y is smooth.

Then, denoting by v_t^y the partial derivative of v^y with respect to the t variable,

$$v_t^y(t,x) = \Phi^*(\tau) - \tau(\Phi^*)'(\tau) \, , \quad D_x v^y(t,x) = (\Phi^*)'(\tau)D_x d(x,y)$$

where $\tau = \frac{d(x;y)}{t} > 0$. It is not hard to check, using the positive homogeneity of H_0, see assumption (8), that d solves the eikonal equation (11). Assuming for simplicity that Φ^* is smooth, the reciprocity formula of convex analysis

$$\Phi\left((\Phi^*)'(s)\right) + \Phi^*(s) = s\left(\Phi^*\right)'(s) \, ,$$

yields

$$v_t^y(t,x) + \Phi\left(H_0(x, D_x v^y(t,x))\right) = 0 \, , \quad (t,x) \in (0,+\infty) \times \mathbb{R}^N \, .$$

The procedure outlined above can be made rigorous by viscosity calculus.

The second step of the proof is to show, using the stability properties of viscosity solutions with respect to inf operations, that the lower envelope

$$u(t,x) = \inf_{y \in \mathbb{R}^N} v^y(t,x)$$

of this family of special solutions, is lower semicontinuous and solves (12) in the viscosity sense. Observe that, since $p \to \Phi\left(H_0(x,p)\right)$ is convex, it is enough at this purpose to check that

$$\lambda + H(x,\eta) = 0 \quad \forall(\eta,\lambda) \in D^- u(x,t)$$

at any (x,t), where $D^- u(x,t)$ is the subdifferential of u at (x,t), see Ref. 6.

The third step is to check the initial condition $u(0,x) = g(x)$ and the fact that u is bounded below by a function of linear growth. This can be done much in the same way as in Ref. 2.

Let us conclude this section by observing that in the model example

$$\begin{cases} u_t + \frac{1}{2}|\sigma(x)D_x u|^2 = 0 \, , \ (t,x) \in (0,+\infty) \times \mathbb{R}^N \\ u(0,x) = g(x) \, , \ x \in \mathbb{R}^N \end{cases} \tag{14}$$

with σ as in (6), the Hopf-Lax solution given by (13) is

$$u(t,x) = \inf_{y \in \mathbb{R}^N} \left[g(y) + \frac{d(x,y)^2}{2t}\right] \, ,$$

which can be regarded as a generalized form of the classical Yosida-Moreau inf-convolution of function g, see Ref. 14 for more informations on this issue.

142

4. The Hopf-Lax formula for the Heisenberg Hamiltonian

The Heisenberg Hamiltonian

$$H_0(x, p) = \sqrt{(p_1 + 2x_2 p_3)^2 + (p_2 - 2x_1 p_3)^2} = |\sigma(x)p|$$

where $x = (x_1, x_2, x_3)$ and

$$\sigma(x) = \begin{pmatrix} 1 & 0 & 2x_2 \\ 0 & 1 & -2x_1 \end{pmatrix}$$

is strictly related to the dynamics of the famous Brockett's example[9] in control theory

$$\dot{x}(t) = \sigma^*(x(t))c(t), \qquad x(0) = x \tag{15}$$

and to the analysis of some asymptotic properties of the heath kernel on the Heisenberg group, see Ref. 7.

Thanks to the Lie bracket relation

$$[(1, 0, 2x_2), (0, 1, -2x_1)] = -4(0, 0, 1),$$

the matrix $\sigma(x)$ satisfies the Chow-Hörmander rank condition with $k = 2$. Then, as a particular case of the results of the previous section, the minimum time for connecting x to y by system (15) with controls

$$c : [0, +\infty) \to \mathbb{R}^2, \ |c(t)| \le 1,$$

is the appropriate distance $d(x; y)$ occurring in the Hopf-Lax function

$$u(t, x) = \inf_{y \in \mathbb{R}^N} \left[g(y) + t\Phi^* \left(\frac{d(x; y)}{t} \right) \right] \tag{16}$$

solving in the viscosity sense (note that $x \to d(x; y)$ is in the present case just $\frac{1}{2}$- Hölder continuous) the Cauchy problem:

$$\begin{cases} u_t + \Phi\left(|\sigma(x)Du|\right) = 0, \ (t, x) \in (0, +\infty) \times \mathbb{R}^3 \\ u(0, x) = g(x), \ x \in \mathbb{R}^3 \end{cases} \tag{17}$$

The Hopf-Lax formula defines the nonlinear Lax-Oleinik semigroup[16] $S(t)$ on the space $BC(\mathbb{R}^N)$ of bounded continuous function:

$$S(t)g = \inf_{y \in \mathbb{R}^N} \left[g(y) + t\Phi^* \left(\frac{d(x; y)}{t} \right) \right]$$

Using this, it is easy to deduce, see Ref. 1, some useful properties of the solution of problem (17), namely

$$S(t + s)g \le S(t)g, \forall s > 0 \quad, \quad \inf_{\mathbb{R}^3} g \le S(t)g \le \sup_{\mathbb{R}^3} g$$

and

$$\| (S(t)g - S(t)\tilde{g}))^+ \|_\infty \leq \|(g - \tilde{g})^+\|_\infty .$$

Moreover, if $\frac{\Phi^*(s)}{s} \to +\infty$ as $s \to +\infty$, then the expansion of support property holds: if

$$\operatorname{supp} g \subseteq \{x \in \mathbb{R}^3 : d(x;0) \leq R\}$$

then

$$\operatorname{supp} S(t)g \subseteq \{x \in \mathbb{R}^3 : d(x;0) \leq R + R(t)\} ,$$

where $R(.)$ is a nondecreasing function depending on $\|g^-\|_\infty$ and Φ^*.

These and other properties of Hopf-Lax function are useful in the error analysis of finite difference schemes for the numerical computation of the solution of (17) which will be briefly described later on.

4.1. *A singular perturbation problem on the Heisenberg group*

Next we present a result from Ref. 15 , see Ref. 22 for previous result of this type. The proof here exhibits the role played by the Hopf-Lax function in a singular perturbation problem for the heat equation on the Heisenberg group $H^1 = (\mathbb{R}^3, \oplus)$.

The Heisenberg group is a stratified Lie group whose group operation $\oplus$ is defined as

$$y \oplus x = (x_1 + y_1, x_2 + y_2, x_3 + y_3 + 2(x_1 y_2 - x_2 y_1))$$

Note that

$$y \oplus x - x \oplus y = 2 \left(Jx \cdot y \right) e_3$$

where J is the matrix

$$J = \begin{pmatrix} 0 & -2 & 0 \\ 2 & 0 & 0 \\ 0 & 0 & 0 \end{pmatrix}$$

It is immediate to check that the trajectory $x(t; x, y)$ of the Brockett's control system

$$\dot{x}(t) = \sigma^*(x(t))y \qquad x(0) = x$$

with control $y = (y_1, y_2, 0)$ is given by

$$x(t; x, y) = (x_1 + ty_1, x_2 + ty_2, x_3 + 2t(x_1y_2 - x_2y_1)) = x \oplus ty .$$

In particular, $x(1; x, y) = x \oplus y$.

The horizontal gradient $D_{H^1}u$ of a smooth function u is defined as

$$D_{H^1}u(x) = \sigma(x)Du(x) = (u_{x_1} + 2x_2u_{x_3}, u_{x_2} - 2x_1u_{x_3})$$

while its Heisenberg Laplacian[21] $\Delta_{H^1}u$ is given by the second order expression

$$\Delta_{H^1}u = \mathrm{Tr}\left(\sigma^*(x)\sigma(x)D^2\right)u .$$

Consider for $\epsilon > 0$, the heat diffusion problem

$$\begin{cases} w_t^\epsilon(t, x) - \epsilon\,\Delta_{H^1}w^\epsilon(t, x) = 0 , & (t, x) \in (0, +\infty) \times \mathbb{R}^3 \\ w^\epsilon(0, x) = e^{-\frac{g(x)}{\epsilon}} , & x \in \mathbb{R}^3 . \end{cases} \tag{18}$$

The question we address is to identify the limit of functions w^ϵ as the diffusion coefficient ϵ is sent to 0^+. It is natural in this context to look at the log transform[17] of function w^ϵ, namely

$$u^\epsilon = -2\epsilon\,\log w^\epsilon .$$

Function u^ϵ satisfies the quasilinear equation

$$\begin{cases} u_t^\epsilon - \epsilon\,\Delta_{H^1}u^\epsilon + \frac{1}{2}|\sigma(x)D_xu^\epsilon|^2 = 0 , & (t, x) \in (0, +\infty) \times \mathbb{R}^3 \\ u^\epsilon(0, x) = g(x) , & x \in \mathbb{R}^3 . \end{cases} \tag{19}$$

The formal guess is that $u^\epsilon = -2\epsilon\,\log w^\epsilon$ should converge to the viscosity solution $u(t, x)$ of the Hamilton-Jacobi equation

$$\begin{cases} u_t + \frac{1}{2}|D_{H^1}u|^2 = 0 , & (t, x) \in (0, +\infty) \times \mathbb{R}^3 \\ u^\epsilon(0, x) = g(x) , & x \in \mathbb{R}^3 , \end{cases} \tag{20}$$

that is to the Hopf-Lax function

$$u(t, x) = \inf_{y \in \mathbb{R}^3}\left[g(y) + \frac{d^2(x; y)}{2t}\right]$$

where, by the result of the previous section, d is the minimum time needed to connect x to y through the trajectories of the controlled Brockett's system

$$\dot{x}(t) = \sigma^*(x(t))\,c(t) , \quad |c(t)| \leq 1 .$$

Let us sketch briefly the elements of the proof of the formal guess; it makes essential use of the Hopf-Lax's formula and the Large Deviations Principle[19] . The functions w^ϵ have the integral representation

$$w^\epsilon(t, x) = \int_{\mathbb{R}^3} e^{-\frac{g(x)}{2\epsilon}} p(\epsilon t, x, y)\, dy$$

where p is the fundamental solution of $w_t - \epsilon\, \Delta_{H^1} w$. Function p is estimated, see Ref. 22, by

$$\frac{1}{M|B^d(x, \sqrt{t})|} e^{-M\frac{d^2(x,y)}{t}} \leq p(t, x, y) \leq \frac{M}{|B^d(x, \sqrt{t})|} e^{-\frac{d^2(x,y)}{Mt}}$$

for some $M > 1$ and small t, where d solves the eikonal equation

$$\begin{cases} |\sigma(x)D_x d)| = 1 \,, \ x \in \mathbb{R}^N \setminus \{y\} \\ d(y) = 0 \end{cases} \tag{21}$$

Hence, the family of probability measures

$$P_\epsilon(B) = P_\epsilon^{t,x}(B) = \int_B p(\epsilon t, x, y)\, dy$$

satisfy the conditions needed to apply Large Deviation Principle with rate function $I^{t,x}(y) = \frac{d^2(x,y)}{4t}$. This gives the Laplace-Varadhan type result:

$$\lim_{\epsilon \to 0^+} \epsilon \log \left(\int_{\mathbb{R}^3} e^{\frac{-g(y)}{2\epsilon}}\, dP_\epsilon(y) \right) = \sup_{y \in \mathbb{R}^3} [g(y) - I(y)]$$

and, as a consequence,

$$\lim_{\epsilon \to 0^+} -2\epsilon \log w^\epsilon = \inf_{y \in \mathbb{R}^3} \left[g(y) + \frac{d^2(x; y)}{2t} \right]$$

The above result confirms that even in the present sub-riemannian context, the inf-convolution can be regarded as a singular limit of integral convolutions, see Ref. 10 for the standard euclidean heat equation and also Ref. 18 for a general remark in this direction. Similar asymptotic results for more general matrices satisfying the **(CC)** condition can be found in Ref. 14 .

The way of deriving the Hopf-Lax function via the logarithmic Hopf-Cole transform and the Large Deviation Principle is also closely related to the Maslov's approach[25] to Hamilton-Jacobi equations based on idempotent analysis. In that approach, the basefield $\mathbb{R}$ of ordinary calculus is replaced by the semiring $\mathbb{R}^* = \mathbb{R} \cup \{\infty\}$ with operations $a \cap b = \min\{a, b\}$, $a \cup b = a + b$. A more detailed description of this relationship goes beyond the scopes of the present paper; just observe that the nonsmooth operation $a \cap b$ has the smooth approximation $a \cap b = \lim_{\epsilon \to 0^+} -\epsilon \log \left(e^{-\frac{a}{\epsilon}} + e^{-\frac{b}{\epsilon}} \right)$.

146

4.2. *Convergence rate of finite differences approximation*

We address in this section some issues related to the computation of approximate solutions of

$$\begin{cases} u_t + \Phi\left(|\sigma(x)Du|\right) = 0\,, & (t,x) \in (0,+\infty) \times \mathbb{R}^3 \\ u(0,x) = g(x)\,, & x \in \mathbb{R}^3 \end{cases} \tag{22}$$

where

$$\sigma(x)Du(x) = \sqrt{(u_{x_1} + 2x_2 u_{x_3})^2 + (u_{x_2} - 2x_1 u_{x_3})^2}$$

is the Heisenberg Hamiltonian considered in the previous section. Since $\sigma(x)$ has rank 2, (22) is a non coercive equation.

For Hamilton-Jacobi equations $u_t + H(x, D_x u) = 0$ which are coercive, that is $\lim_{|p| \to +\infty} H(x,p) = +\infty$, finite differences schemes with uniform grid size $h > 0$ have been investigated in Ref. 12 and the following optimal error estimate

$$\textbf{(EE)} \qquad ||u - u_{appr}||_\infty \le C\sqrt{h}$$

was established. The validity of this estimate in the coercive case depends, in particular, on the Lipschitz continuity of the viscosity solution.

The purpose in Ref. 1 is to design an approximation scheme for equation (22) (as previously mentioned the solution is just $\frac{1}{2}-$ Hölder continuous in this case) in such a way as to obtain the same rate of convergence proved in Ref. 12 for the coercive case.

The idea is to construct a grid compatible with the Heisenberg group translations

$$y \oplus x = (x_1 + y_1, x_2 + y_2, x_3 + y_3 + 2(x_1 y_2 - x_2 y_1))\,,$$

(note that $\oplus$ is non commutative) and dilations

$$\alpha \cdot x = (\alpha x_1, \alpha x_2, \alpha^2 x_3) \qquad \alpha > 0\,.$$

The grid nodes of discrete lattice are chosen to be

$$\xi_{i,j,k} = (ih, jh, (4k + 2ij)h^2)$$

where $h > 0$ and i, j, k are integers. Observe that

$$\xi_{i,j,k} \oplus \pm h \cdot e_1 = \xi_{i\pm 1,j,k}\,, \qquad\qquad \xi_{i,j,k} \oplus \pm h \cdot e_2 = \xi_{i,j\pm 1,k\mp i}\,.$$

$$\xi_{l,m,n} \oplus \xi_{i,j,k} = \xi_{l+i,m+j,k+n-jl}$$

Formulas above show compatibility between the discrete lattice and the group operations $\oplus$ and $\cdot$

For a time step $\Delta t \leq Ch$ one solves the discrete equations

$$U_{i,j,k}^{n+1} = U_{i,j,k}^n - \Delta t\, H_{num}\left(\frac{1}{h}(\Delta_+^1 U)_{i,j,k}^n\,,\,\frac{1}{h}(\Delta_+^1 U)_{i-1,j,k}^n\,,\right.$$

$$\left.\frac{1}{h}(\Delta_+^2 U)_{i,j,k}^n\,,\,\frac{1}{h}(\Delta_+^2 U)_{i,j-1,k+i}^n\right)$$

where

$$(\Delta_+^1 U)_{i,j,k} = U_{i+1,j,k} - U_{i,j,k}, \qquad (\Delta_+^2 U)_{i,j,k} = U_{i,j+1,k-i} - U_{i,j,k}.$$

The numerical Hamiltonian H_{num} used in the scheme is chosen to be monotone and consistent with (22). For example, one can use the Osher-Sethian[26] upwind scheme is

$$H_{num}(u_1, u_2, v_1, v_2)$$

$$= \Phi\left(\left(\min(u_1,0)^2 + \max(u_2,0)^2 + \min(v_1,0)^2 + \max(v_2,0)^2\right)^{\frac{1}{2}}\right)$$

which is monotone on the interval $[-\Lambda, \Lambda]$ if $1 - \frac{2\Delta t}{h}\Phi'(2\Lambda) \geq 0$.

Under the assumption that $\Phi : \mathbb{R}^N \to [0, +\infty)$, $\Phi(0) = 0$ is convex, increasing and

$$\frac{\Phi^*(s)}{s} \to +\infty \quad \text{as} \quad s \to +\infty$$

and some extra conditions on the initial datum g, including

- g has compact support,
- g is Lipschitz continuous with respect to left translations,
- $\sup_{z \in \mathbb{R}^3} |g(z \oplus \delta e_3) - g(z)| \leq L|\delta|$,

in Ref. 1 it is proved that, for some positive constant C,

$$\textbf{(EE)} \qquad \Theta := \sup_{i,j,k \in \mathbb{Z}} \sup_{n=1,\ldots,\frac{T}{\Delta t}} |U_{i,j,k}^n - u(p\Delta t, \xi_{i,j,k})| \leq C\sqrt{h}$$

for all $0 < h < 1$.

An important role in the proof is played by the fact, which can be checked using the Hopf-Lax formula (16), that the exact solution of problem (22) is Lipschitz continuous with respect to right translations with some explicitly computable constant $L(t)$. Proof makes use in a crucial way of the typical viscosity comparison technique and is based on "doubling variables" and careful analysis of the auxiliary function

$$\Psi : [0, T] \times \mathbb{R}^3 \times \left\{(\xi_{i,j,k}, n\Delta t), i, j, k \in \mathbb{Z}, n = 0, \ldots, \frac{T}{\Delta t}\right\} \to \mathbb{R}$$

148

defined by

$$\Psi(\eta, t, \xi, s) = u(\eta, t) - U^p_{i,j,k} + (5(\|g\|_\infty + 1) + \frac{\Theta}{2})\beta_\epsilon(-\xi \oplus \eta, t - s) - \frac{\Theta(t - s)}{4T}.$$

Here, $\xi = \xi_{i,j,k}$, $s = n\Delta t$ and

$$\beta_\epsilon(x, t) = \beta\left(\left(\left(\frac{x_1^2}{\epsilon^2} + \frac{x_2^2}{\epsilon^2}\right)^2 + \frac{x_3^2}{\epsilon^2}\right)^{\frac{1}{4}}, \frac{t}{\epsilon}\right)$$

ϵ is a positive real number and β a smooth function on $\mathbb{R} \times \mathbb{R}$, satisfying

$$\beta(0,0) = 1, \qquad 0 \leq \beta \leq 1, \qquad \beta(r, t) = 0 \quad \text{if } r^4 + t^4 > 1.$$

Among the main technical steps of the proof we mention the following estimates on the partial derivatives of β_ϵ and on their "discrete increments" at a global maximum point $(\eta_0, t_0, \xi_0, s_0)$ of the auxiliary function Ψ:

$$\left| \frac{1}{h}(\Delta^1_+(\beta_\epsilon(- \cdot \oplus \eta_0, t_0 - s_0 + \Delta t)))_{i_0, j_0, k_0} \right.$$
$$\left. + (D_H \beta_\epsilon)_1(-\xi_0 \oplus \eta_0, t_0 - s_0) \right| \leq Ch^{\frac{1}{2}},$$

$$\left| \frac{1}{h}(\Delta^1_+(\beta_\epsilon(- \cdot \oplus \eta_0, t_0 - s_0 + \Delta t)))_{i_0-1, j_0, k_0} \right.$$
$$\left. + (D_H \beta_\epsilon)_1(-\xi_0 \oplus \eta_0, t_0 - s_0) \right| \leq Ch^{\frac{1}{2}},$$

$$\left| \frac{1}{h}(\Delta^2_+(\beta_\epsilon(- \cdot \oplus \eta_0, t_0 - s_0 + \Delta t)))_{i_0, j_0, k_0} \right.$$
$$\left. + (D_H \beta_\epsilon)_2(-\xi_0 \oplus \eta_0, t_0 - s_0) \right| \leq Ch^{\frac{1}{2}},$$

$$\left| \frac{1}{h}(\Delta^2_+(\beta_\epsilon(- \cdot \oplus \eta_0, t_0 - s_0 + \Delta t)))_{i_0, j_0-1, k_0+i_0} \right.$$
$$\left. + (D_H \beta_\epsilon)_2(-\xi_0 \oplus \eta_0, t_0 - s_0) \right| \leq Ch^{\frac{1}{2}}.$$

We refer to Ref. 1 for full details of the proof and numerical experiments.

References

1. Y. Achdou, I. Capuzzo Dolcetta, Approximation of solutions of Hamilton-Jacobi equations on the Heisenberg group, cpde-preprint yar30073 (June 2006), available at http://cpde.iac.rm. cnr.it/ricerca.php.

2. O. Alvarez, E.N. Barron, H. Ishii, *Indiana University Mathematical Journal*, **48**, 993 (1999).

3. M. Bardi, I. Capuzzo Dolcetta, *Optimal Control and Viscosity Solutions of Hamilton - Jacobi - Bellman Equations*, Systems & Control: Foundations and Applications (Birkhauser Verlag, Boston, MA, 1997).

4. M. Bardi L.C. Evans, *Nonlinear Anal.*, **8**, 1373 (1984).

5. G. Barles, P.E. Souganidis, *SIAM Journal Math. Anal.*, **31**, 925 (2000).

6. E.N Barron, R. Jensen, *Comm. Partial Differential Equations*, **15**, 1713 (1990).

7. R. Beals, B. Gaveau, P.C. Greiner, *J. Math. Pures Appl. (9)*, **79**, 633 (2000).

8. A. Bellaïche, *The tangent space in sub-Riemannian geometry*, in *Sub-Riemannian Geometry*, eds. A. Bellache and J.-J. Risler, Progr. Math., Vol. 144 (Birkhäuser Verlag, Basel, 1996): 1–78.

9. R. W. Brockett, *Control theory and singular Riemannian geometry*, in *New directions in applied mathematics*, eds. P.J. Hilton and G.J. Young (Springer-Verlag, New York, 1982): 11–27.

10. I. Capuzzo Dolcetta, *Representations of solutions of Hamilton-Jacobi equations*, in *Nonlinear equations: methods, models and applications (Bergamo, 2001)*, Progress in Nonlinear Differential Equations and Applications, Vol. 54 (Birkhauser Verlag, Basel, 2003): 79–90.

11. I. Capuzzo Dolcetta, H. Ishii, Hopf-Lax formulas for state dependent Hamilton-Jacobi equations (in preparation, 2007).

12. M.G. Crandall, P.L. Lions, *Math.Comp.*, **43**, 1 (1984).

13. A. Cutri, F. Da Lio, Comparison and existence results for evolutive non-coercive first-order Hamilton-Jacobi equations, *ESAIM Control Optim. Calc. Var.* (to appear), available at http://cpde.iac.rm.cnr.it/ricerca.php (September 2005).

14. F. Dragoni, *Discrete Contin. Dyn. Syst.-A*, **17**, 713 (2007).

15. F. Dragoni, *NoDEA Nonlinear Differential Equations Appl.*, **14**, 429 (2007).

16. A. Fathi, *Weak KAM theorem in Lagrangian dynamics*, Cambridge Studies in Advanced Mathematics (Cambridge University Press, Cambridge, 2007).

17. W. H. Fleming, M. H. Soner, *Controlled Markov Processes and Viscosity Solutions*, Applications of Mathematics, Vol. 25 (Springer-Verlag, New York, 1993).

18. J.B. Hiriart-Urruty, C.Lemaréchal *Fundamentals of Convex Analysis*, Grundlehren Text Edition (Springer-Verlag, Berlin, 2001).

19. F. den Hollander *Large Deviations*, Fields Institute Monographs, Vol. 14 (Amer. Math. Soc., Providence, 2000).

20. E. Hopf, *J. Math. Mech.*, **14**, 951 (1965).

21. D. Jerison, A. Sanchez Calle, *Subelliptic second order differential operators*, in Complex analysis, III (College Park, Md., 1985–86), Lecture Notes in Mathematics, Vol. 1277 (Springer, Berlin, 1987): 46–77.

22. R. Léandre, *J. Funct. Anal.*, **74**, 399 (1987).

23. P.-L. Lions, *Generalized solutions of Hamilton-Jacobi equations*, Research Notes in Mathematics, Vol. 69 (Pitman, Boston, Mass., 1982).

24. J. J. Manfredi, B. Stroffolini, *Comm. Partial Differential Equations*, **27**, 1139 (2002).
25. V.P. Maslov, On a new principle of superposition for optimization problems, *Uspekhi Mat. Nauk (Russian Math. Surveys)*, **42**, 39, 255 (1987).
26. S. Osher, J.A. Sethian, *J. Comput. Phys.*, **79**, 12 (1988).
27. R.T. Rockafellar, P.R.Wolenski, *SIAM J. Control Optim.*, **39**, 1351 (2000).
28. A. Siconolfi, *Trans. Amer. Math. Soc*, **355**, 1987 (2003).
29. C. Villani, *Optimal transport, old and new* (Springer, Berlin, 2007).

REGULARITY OF SOLUTIONS
TO ONE-DIMENSIONAL AND MULTI-DIMENSIONAL
PROBLEMS IN THE CALCULUS OF VARIATIONS

F. H. CLARKE

Institut universitaire de France et Université de Lyon
Institut Camille Jordan UMR 5208
Université Claude Bernard Lyon 1
La Doua, 69622 Villeurbanne, France
E-mail: clarke@math.univ-lyon1.fr

We review the long-standing issue of regularity of solutions to the basic problem in the calculus of variations, in both the one-dimensional and the multi-
dimensional settings. It is shown how certain recent results fit in with the
classical ones, in particular the theories of De Giorgi and Hilbert-Haar.

Keywords: Calculus of variations; regularity; necessary conditions; existence.

1. Introduction

We begin in the middle, with two of the celebrated problems proposed by
Hilbert in Paris in 1900:

The 20th problem: Is it not the case that every *regular* variational problem has a solution, provided certain assumptions on the boundary conditions are satisfied, and provided also, if need be, that the concept of solution
is suitably extended?

The 19th problem: Are the solutions of regular problems in the calculus
of variations always analytic?

These questions bear upon the following *basic problem* in the calculus
of variations: to minimize the functional

$$J(u) := \int_\Omega F(x, u(x), Du(x)) \, dx$$

over the functions $u : \Omega \to \mathbb{R}$ assuming prescribed values on $\Gamma := \partial\Omega$:

$$u(x) = \phi(x), \; x \in \Gamma.$$

152

Here Ω is a *domain* in $\mathbb{R}^n$: an open bounded connected set, and Du denotes the gradient of u. In fact, Hilbert was referring to the case $n = 2$ in his problems, but we shall assume only $n \geq 2$ for now; the case $n = 1$, which is the context in which the subject began (in 1696, arguably, but certainly no later than 1744), has a markedly different character, and will be considered in the final section. Hilbert also took the function $F(x, u, z)$ (the *Lagrangian*) to be analytic; the problem is *regular* if F_{zz} is positive definite everywhere. We do not discuss in this article the so-called *vector case* of the problem in which u is vector-valued (that is, in which there are several unknown functions).

The decade preceding the formulation of Hilbert's problems had been marked by a controversy over the *Dirichlet principle*, which affirms the equivalence between functions u minimizing the Dirichlet functional ($n = 2$)

$$\int_\Omega (u_x^2 + u_y^2) \, dx \, dy$$

and solutions u of Laplace's equation $u_{xx} + u_{yy} = 0$. As Weierstrass and Hilbert pointed out in response to (notably) Riemann's assertions, the existence of a minimum here (and the very class in which to seek one) is problematic. Hilbert went on to give the first rigorous treatment of the issue in 1904, in a context which succeeded in limiting the class of functions u involved to Lipschitz ones. But it became clear that a more general type of function space was needed, and eventually the work of Levi, Tonelli, Morrey, Sobolev and others, led to the theory of *Sobolev spaces*, which provides a suitable context in which to assert the *existence* of a solution to the basic problem.

In the meantime, however, significant progress on Hilbert's regularity question was made:

(1) If u is C^3, then u is analytic [Bernstein 1904];
(2) If u is C^2, then u is C^3 [Lichtenstein 1912];
(3) If u is $C^{1,\alpha}$ (that is, has a gradient which is Hölder continuous of order $\alpha \in (0, 1]$), then u is C^2 [Hopf 1929].

These results lowered the *regularity threshold* to $C^{1,\alpha}$; once this level of regularity is present in the solution, then it is as regular as the Lagrangian permits: C^r ($r \geq 2$), C^∞, or analytic.

Letting $W^{1,1}(\Omega)$ denote the usual Sobolev space, we now consider the following reduced basic problem (P):

$$\text{minimize } J(u) := \int_\Omega F(Du(x)) \, dx : u - \phi \in W_0^{1,1}(\Omega)$$

under the following *Standing Hypotheses*:

(1) F is of class C^2 and $F_{zz} > 0$ everywhere;

(2) F is coercive of order $p > 1$: for certain constants $\sigma > 0$ and μ,

$$F(z) \geq \sigma |z|^p + \mu, \, z \in \mathbb{R}^n;$$

(3) $\phi : \mathbb{R}^n \to \mathbb{R}$ is Lipschitz.

It is a standard exercise in the theory of Sobolev spaces to invoke the *direct method* introduced by Tonelli, in which one exploits the weak sequential compactness of a minimizing sequence and the weak lower semicontinuity of the convex functional J, to deduce the existence of a solution u to problem (P) (u is the unique solution, since J is strictly convex). Note that this is an answer of sorts to Hilbert's 20th problem. The issue now becomes the *regularity* of the solution u, especially since functions in the Sobolev space $W^{1,1}(\Omega)$ are not even continuous necessarily.

The reasons for wanting regularity of the solution u are manifold. For example, *continuity* of u (or more precisely, of one of its representatives) on the closure $\overline{\Omega}$ of Ω would mean that the boundary conditions are assumed in the conventional pointwise manner (rather than in the sense of trace). *Differentiability* of u would mean that Du can be interpreted as the true gradient, and not just the weak distributional derivative. If u is *locally Lipschitz* in Ω (or equivalently, Lipschitz on compact subsets of Ω), then the Euler equation in weak form can be asserted to hold:

$$\int_\Omega \nabla F(Du(x)) \cdot D\psi(x) \, dx = 0, \, \psi \in C_c(\Omega).$$

And finally, in view of the results cited above, local $C^{1,\alpha}$ regularity would imply that the full regularity of the Lagrangian F is inherited by u, up to and including analyticity (as in Hilbert's 19th problem).

There are two major 20th century developments on the regularity issue to report, and both were first obtained in the context of the reduced problem (P). We now examine these in turn.

2. The theorem of De Giorgi

The Lagrangian F is said to be *uniformly elliptic* provided that for some $\epsilon > 0$ we have $F_{zz}(z) \geq \epsilon I$ for all $z \in \mathbb{R}^n$. It is said to be *almost quadratic* if for certain constants $c_0, c_1, d_0 > 0, d_1 > 0$ we have

$$c_0 + d_0 |z|^2 \leq F(z) \leq c_1 + d_1 |z|^2, \, z \in \mathbb{R}^n.$$

154

The Dirichlet integrand, which is precisely quadratic, is the canonical example of such a Lagrangian. As we know from harmonic analysis, its minimizers are analytic in Ω. In 1957, De Giorgi[15] proved the following celebrated result:

Theorem 2.1. *Let F be C^2, uniformly elliptic and almost quadratic. Then the solution u to (P) is locally $C^{1,\alpha}$ in Ω.*

Note that the boundary function ϕ plays no role here, and that the theorem is strictly one of 'interior regularity'. De Giorgi's proof proceeded by obtaining a linearized Euler equation for u, to which a new regularity result on elliptic pde's was then applied to get the required conclusion. This difficult result was introduced in the same article, and other proofs of it were later given by Nash (1958), and by Moser (1960).

The main effect of the theorem, from the point of view of the present discussion, is to reduce the regularity threshold to 'locally Lipschitz'. Let us make this explicit by recording the following simple consequence of the theorem which, surprisingly, is not stated in De Giorgi's article:

Corollary 2.1. *Under merely the hypotheses that F is C^2 and regular, if the solution u to the problem (P) is locally Lipschitz, then it is locally $C^{1,\alpha}$.*

This is proved by replacing the original Lagrangian by one which agrees with F on a bounded set containing the values of $Du(x)$ and which is uniformly elliptic and almost quadratic. Then u is still a solution of the problem (by convexity), and De Giorgi's theorem applies to yield the conclusion.

The theorem above has been extended in a variety of ways. Certain evident limits to such extensions, however, as well as possible grounds for pessimism, arise from certain examples due to Giusti and Giaquinta. We refer here to special cases of (P) in which the Lagrangian is uniformly elliptic and satisfies

$$c_0 + d_0|z|^2 \leq F(z) \leq c_1 + d_1|z|^4, \ z \in \mathbb{R}^n,$$

and yet the solution fails to be continuous in Ω. In such examples, though, the boundary function ϕ is itself discontinuous. This motivates the thought that if some regularity properties were imposed on ϕ, then perhaps this would induce regularity of the solution u.

This idea, which harkens back to Hilbert's successful analysis of the Dirichlet principle, is precisely the one that underlies the other significant approach to the regularity issue; we turn to it now.

3. Hilbert-Haar theory

The classical Hilbert-Haar approach, which in contrast to Theorem 2.1 makes no additional structural assumptions on F, requires instead that ϕ satisfy the *bounded slope condition* (BSC) defined below. The ingredients of the theory come from several sources. Hilbert is responsible for the first version of the comparison (or maximum) principle (in the Dirichlet context), which Haar extended to other Lagrangians. Rado introduced the 'three point condition', a forerunner of the BSC below. The idea of applying comparison to a translated solution is due to von Neumann. The BSC was formulated and studied in its present form by Hartman and Nirenberg, while Stampacchia[21] coined the term BSC and applied it to variational problems.

The bounded slope condition of rank K is the requirement that given any point γ on the boundary, there exist two affine functions

$$y \mapsto \langle \zeta_\gamma^-, y - \gamma \rangle + \phi(\gamma),\ y \mapsto \langle \zeta_\gamma^+, y - \gamma \rangle + \phi(\gamma)$$

agreeing with ϕ at γ whose 'slopes' satisfy $|\zeta_\gamma^-| \leq K, |\zeta_\gamma^+| \leq K$ and such that

$$\langle \zeta_\gamma^-, \gamma' - \gamma \rangle + \phi(\gamma) \leq \phi(\gamma') \leq \langle \zeta_\gamma^+, \gamma' - \gamma \rangle + \phi(\gamma)\ \ \forall \gamma' \in \Gamma.$$

Let $Lip(\Omega)$ denote the class of globally Lipschitz functions on Ω. The Hilbert-Haar theorem (see Chapter 1 of Giusti[16]) is the following:

Theorem 3.1. *Let ϕ satisfy the BSC of rank K. Then there is a solution u of problem (P) when it is restricted to $Lip(\Omega)$, and the solution u is Lipschitz on Ω of rank K.*

It is possible to show[3,14,18] that the solution of (P) relative to $Lip(\Omega)$ is in fact the solution relative to $W^{1,1}(\Omega)$, so we now deduce

Corollary 3.1. *If ϕ satisfies the BSC of rank K, then the solution u of problem (P) is Lipschitz on Ω of rank K.*

It is clear that Theorems 2.1 and 3.1, or rather their corollaries, work in tandem to assert the higher regularity of the solution u to (P), as follows: if ϕ satisfies the BSC, then u is locally $C^{1,\alpha}$ and inherits the full regularity of the Lagrangian up to analyticity.

It is natural to ask now how restrictive the BSC is. It is certainly a serious limitation of the allowable boundary conditions on 'flat parts' of Γ, since it forces ϕ to be affine. But the BSC becomes more interesting when

Ω is sufficiently curved. Ω is said to be *uniformly convex* if, for some $\epsilon > 0$, every point γ on the boundary admits a hyperplane H through γ such that

$$d_H(\gamma') \geq \epsilon|\gamma' - \gamma|^2 \quad \forall \gamma' \in \Gamma.$$

Miranda's Theorem[20] states that when Ω is uniformly convex, then any ϕ of class C^2 satisfies the BSC. Later, Hartman[17] showed that when Ω is uniformly convex and Γ is $C^{1,1}$, then ϕ satisfies the BSC if and only if ϕ is itself $C^{1,1}$. We can say therefore that as a hypothesis, the BSC essentially restricts the boundary data to be smooth. When ϕ is not affine, the BSC also forces Ω to be convex, a hypothesis that we will add to our Standing Hypotheses as we turn now to some new results that center around a weakening of the BSC.

4. New boundary hypotheses

We now turn our attention to the present, or at least the very recent past. We assume throughout this section that Ω is convex. Clarke[9] has introduced a new hypothesis on ϕ, the *lower bounded slope condition* (lower BSC) of rank K: given any point γ on the boundary, there exists an affine function $y \mapsto \langle \zeta_\gamma, y - \gamma \rangle + \phi(\gamma)$ with $|\zeta_\gamma| \leq K$ such that

$$\langle \zeta_\gamma, \gamma' - \gamma \rangle + \phi(\gamma) \leq \phi(\gamma') \quad \forall \gamma' \in \Gamma.$$

Being one-sided, the lower BSC naturally admits a counterpart: an *upper* BSC that is satisfied by ϕ exactly when $-\phi$ satisfies the lower BSC.

4.1. *Interior regularity*

The significance of the 'partial' BSC hypothesis stems from the following result:[9]

Theorem 4.1. *If ϕ satisfies the lower bounded slope condition, then the solution u of (P) is locally Lipschitz in Ω. In fact, there is a constant $\overline{K}$ with the property that for any subdomain Ω' of distance $\delta > 0$ from Γ, we have*

$$|u(x) - u(y)| \leq (\overline{K}/\delta)|x - y| \quad \forall x, y \in \Omega'.$$

Thus the one-sided BSC gives the crucial regularity property: u is locally Lipschitz in Ω. This allows us to assert that u is a weak solution of the Euler equation, in the absence of any restrictive growth conditions on F, and of course allows the application of Theorem 2.1 to deduce higher-order regularity. Informally, it appears that in return for 'half the BSC hypothesis', we obtain considerably more than half the conclusion. Of course the principal thing that has been sacrificed is the continuity of u up to the boundary, but as we shall see below, this can be recovered under a variety of additional hypotheses.

As in the case of the BSC, it behooves us to examine the conditions under which the one-sided BSC can be asserted to hold. In this new context, the property that Ω be curved has less importance than before; flat parts of the boundary do not force ϕ to be affine. Nonetheless, curvature can still serve a purpose: Bousquet[1] has shown that when Ω is uniformly convex, then ϕ satisfies the lower (upper) BSC if and only if it is the restriction to Γ of a function which is semiconvex (semiconcave), a familiar and useful property in pde's (see for example Ref. 5). In the uniformly convex case, therefore, Theorem 4.1 extends Hilbert-Haar theory to boundary data that is semiconvex or semiconcave rather than C^2 (or $C^{1,1}$).

We remark that the proofs of Theorems 3.1 and 4.1 have something in common: both of them construct a new minimizer from u itself with which to compare u. In the classical case, this is done by translation. The principal new idea in the proof of Theorem 4.1 is to construct a new minimizer through dilation rather than translation.

4.2. *Continuity at the boundary*

In a variety of situations, it turns out that the lower or upper BSC *does* imply continuity at the boundary, and even a global Hölder condition in some cases. A counterexample due to Bousquet and Cannarsa (in the Dirichlet context) shows, however, that the gradient of u can become unbounded.

Theorem 4.2. *Suppose that in addition to the hypotheses of Theorem 4.1, one of the following holds:*

(a) Γ is a polyhedron, or
(b) Γ is $C^{1,1}$ and $p > (n+1)/2$, or
(c) Ω is uniformly convex.

Then u is continuous on $\overline{\Omega}$. In cases (a) and (b), u satisfies a Hölder condition on $\overline{\Omega}$.

We remark that we know of no example in which (under the hypotheses of Theorem 4.1) the solution fails to be continuous on $\overline{\Omega}$. Indeed, we know of no example in which ϕ is Lipschitz and u fails to be continuous.

4.3. *More general Lagrangians*

It turns out to be challenging to extend the new results given above to Lagrangians depending on x and u as well as Du, principally because the comparison principle does not hold in that case. However, using the technique of *barrier functions*, Bousquet and Clarke[2] have obtained a result for Lagrangians of the form $F(Du) + G(x, u)$; we describe it now. We assume that F is uniformly elliptic, that $G(x, u)$ is measurable in x and differentiable in u, and that for every bounded interval U in $\mathbb{R}$ there is a constant L_U such that for almost all $x \in \Omega$,

$$|G(x, u) - G(x, u')| \leq L_U |u - u'| \quad \forall u, u' \in U.$$

We also postulate that for some bounded function b, the integral $\int_\Omega G(x, b(x))\, dx$ is well-defined and finite.

Theorem 4.3. *Under these hypotheses, and when ϕ satisfies the Lower or Upper BSC, any bounded solution u of the basic problem is locally Lipschitz in Ω.*

We remark in connection with this result that it is possible to formulate additional hypotheses on the data which imply *a priori* that any solution to the basic problem is bounded, and that additional structural hypotheses lead as before to continuity at the boundary.

5. The one-dimensional case

Finally, let us address the beginning of the subject: the one-dimensional case, for which there is always a different notation. The basic problem (P) now corresponds to the minimization of the functional

$$J(x) := \int_a^b F(t, x(t), x'(t))\, dt$$

over the functions $x : [a, b] \to \mathbb{R}^N$ in some given class X, and subject to prescribed endpoint conditions: $x(a) = A$, $x(b) = B$. Note that we now allow $N > 1$, the case of several unknown functions; the generic name for the variables becomes (t, x, v) rather than (x, u, z).

For Euler and his contemporaries, all functions were smooth, so the issue of the regularity of the solutions did not arise; implicitly, X was a space of very smooth functions. On a more rigorous level, when the degree of smoothness becomes a consideration, we can deduce that if x is C^1 to start with, and if F is regular (which continues to mean $F_{vv} > 0$) and at least C^2, then x inherits the full degree of regularity of F, up to analyticity. This is known as the Hilbert-Weierstrass theorem (*circa* 1890), and is a consequence of the Euler equation together with the implicit function theorem. By the 19th century, however, the possible nonsmoothness of solutions began to be recognized as an important point, in view of such concrete evidence as nonsmooth soap (minimal) surfaces.

Results were obtained for the class PWS of *piecewise smooth* functions, notably. An important breakthrough was duBois-Reymond's proof of the integral form of the Euler equation: if x solves (P) relative to PWS, then there exists a piecewise smooth function p such that

$$(p'(t), p(t)) = \nabla F(t, x(t), x'(t))$$

at all non-corner points; here, ∇F refers to the gradient in the (x, v) variables. This condition subsumes the earlier Erdmann condition, and like it, can sometimes be used to deduce the smoothness of x (in this setting, the absence of corners). Note however that x has to be assumed piecewise smooth *a priori*. This is unfortunate, since, although suitable necessary conditions can be asserted in PWS, the class of piecewise smooth functions is of no help in regard to the existence issue.

It was Tonelli's major contribution to show that existence theory can be developed successfully in the class AC of *absolutely continuous* functions. But within AC, the ability to derive the Euler equation fails in general, so our steps forward (on existence) are accompanied by one step back (on the necessary conditions). A way out of this quandary is to find reasonable supplementary hypotheses on the Lagrangian which will imply that x is *Lipschitz*. The reason for this is that all the classical results for the class PWS carry over to this class (now that Lebesgue has given us his integral). In other words, just as in the multi-dimensional case treated in the previous sections, the regularity threshold is situated at Lipschitz continuity of the solution.

Of course, in the one-dimensional case there is little help to be found from examining the boundary conditions (as in the Hilbert-Haar theory). In fact, the methodology used to obtain regularity theorems has been overwhelmingly based upon analyzing the necessary conditions. This is in stark contrast to the multi-dimensional case, where the necessary conditions don't seem to yield very much directly. We conclude therefore that our best hope to extend regularity theory lies in the possibility of deriving stronger necessary conditions in more general circumstances.

We proceed to report on just such a recent development. For this purpose, we examine a Lagrangian $F(t, x, v)$ which is merely measurable in t and (x, v) (see[10] for the precise meaning of this) and lower semicontinuous in (x, v). No hypotheses of smoothness or convexity are made. Instead, we assume the following *generalized Tonelli-Morrey* condition [GTM]: for every bounded subset S of $\mathbb{R}^n$ there exist a constant c and a summable function d such that for almost every t, for every $(x, v) \in S \times \mathbb{R}^n$, for every $(\zeta, \psi) \in \partial_P F(t, x, v)$, one has

$$\frac{|\zeta|}{1 + |\psi|} \le c \left\{ |F(t, x, v)| + |v| \right\} + d(t).$$

Here, $\partial_P F$ refers to the *proximal subgradient* of F with respect to the (x, v) variables, a basic construct in nonsmooth analysis.

When F is C^2 (or somewhat less), the growth condition of [GTM] is equivalent to

$$|D_x F| \le c_1 \left\{ |F| + |v| + |D_v F| \right\} + d_1(t) + \left\{ c_2 \left(|F| + |v| \right) + d_2(t) \right\} |D_v F|.$$

The special case $c_2 = d_2 = 0$ corresponds to a class of Lagrangians that has been considered by Clarke and Vinter[13] in connection with regularity. In this case, and when in addition the term involving $|D_v F|$ is placed on the left side of the inequality, as follows:

$$|D_x F| + |D_v F| \le c_1 \left\{ |F| + |v| \right\} + d_1(t),$$

(thereby making the condition a more stringent hypothesis), we obtain a growth condition first postulated by Tonelli in order to be able to derive the necessary conditions in the class AC.

In the much more general setting now being considered, the new [GTM] still has the same effect: it allows one to deduce necessary conditions that must be satisfied by any solution to (P). These conditions have the nature of the classical ones, but expressed in such a way as to take account of the nonsmoothness of F. Here is just one such result taken from Clarke:[10]

Theorem 5.1. *Let x be a solution to (P) relative to the class AC, where F satisfies the generalized Tonelli-Morrey growth condition [GTM]. Then there exists an arc p satisfying the Euler inclusion*

$$p'(t) \in \mathrm{co} \left\{ \omega : (\omega, p(t)) \in \partial_L F(t, x(t), x'(t)) \right\} \quad \text{a.e.,}$$

and the Weierstrass condition: for almost all $t \in [a, b]$ we have

$$F(t, x(t), v) - F(t, x(t), x'(t)) \ge \langle p(t), v - x'(t) \rangle \quad \forall v \in \mathbb{R}^n.$$

The Euler inclusion in this statement involves the *limiting subdifferential* $\partial_L F$; it reduces to the integral form of the Euler equation when F is smooth. The Weierstrass condition is the familiar one of the classical theory.

This theorem admits a more general form in which the cost depends on the endpoint values of the arc x, which need be only a *local* minimum in a specified sense. Beyond this, and most significantly, the theorem can be stated for *extended-valued* Lagrangians F, so that problems in optimal control are subsumed by it. But these are chapters in a different story, so let us proceed instead with our quest for regularity consequences.

We say that the Lagrangian F is **coercive** if for any bounded subset S of $\mathbb{R}^n$ there exists a function $\theta : [0, \infty) \to \mathbb{R}$ satisfying

$$\lim_{r \to +\infty} \frac{\theta(r)}{r} = +\infty$$

and such that

$$F(t, x, v) \geq \theta(|v|) \quad \forall\, (t, x, v) \in [a, b] \times S \times \mathbb{R}^n.$$

We remark that coercivity is a familiar ingredient in the theory of existence of solutions. The symbiosis between necessary conditions and regularity is well illustrated by the following new result.

Corollary 5.1. *If x solves (P) relative to the class* AC, *where F is coercive, bounded above on bounded sets, and satisfies the generalized Tonelli-Morrey growth condition* [GTM], *then x is Lipschitz on* $[a, b]$.

Proof. In view of Theorem 5.1, we know that an arc p exists which satisfies the Weierstrass condition. Let M be an upper bound on

$$F\left(t, x(t), \frac{x'(t)}{1 + |x'(t)|}\right)$$

for $t \in [a, b]$, and let θ be a coercivity function for F when the variable x is restricted to the bounded set consisting of the values of $x(t)$ on $[a, b]$. Then, taking $v := x'(t)/(1 + |x'(t)|)$ in the Weierstrass inequality leads to (almost everywhere)

$$\theta(|x'(t)|) \leq F(t, x(t), x'(t)) \leq M + |p(t)|\,|x'(t)|\,.$$

Since $|p(t)|$ is bounded and $\lim_{r \to \infty} \theta(r)/r = +\infty$, it follows from this inequality that $x'(t)$ is essentially bounded on $[a, b]$. $\square$

This corollary even extends (relative to previous results) the class of *smooth* Lagrangians for which the necessary conditions can be asserted and

162

regularity inferred. A simple exampleof this is provided (for $N = 1$) by

$$F(t, x, v) = \exp\left\{(1 + x^2 + t^2)v^2\right\}.$$

This Lagrangian satisfies the hypotheses of the classical Tonelli existence theorem as well as those of the corollary. Thus a solution to (P) over AC exists, and any solution x is Lipschitz. Because F is strictly convex in v, it then follows[13] that x is C^1, and finally we derive all the higher regularity of x from the Hilbert-Weierstrass theorem.

There are other structural hypotheses yielding regularity results that serve to highlight the extremely general Lagrangians that can be treated in the one-dimensional case, as compared to the very special structure that seems to be required in the multi-dimensional case. We end the discussion with one further example,[10] which asserts the regularity of the solution for one-dimensional problems having *autonomous* Lagrangians (F is called autonomous when it has no explicit dependence on the variable t).

Theorem 5.2. *Let x solve (P) relative to* AC, *where the Lagrangian F is coercive, bounded above on bounded sets, and autonomous. Then x is Lipschitz on $[a, b]$.*

This result (among others that we do not touch upon here) was first proved by Clarke and Vinter[13] under the added requirement that F be locally Lipschitz and convex in v. A more direct and simplified proof in that setting can be given.[11]

The literature on the venerable subject of regularity in the calculus of variations, and on the inextricably linked issues of existence and necessary conditions, is huge and still growing. We make no attempt here to be complete, but rather we refer the interested reader to the classic books of Giaquinta, Giusti and Morrey, and to the representative (but not exhaustive) more recent references appearing below, in which detailed bibliographic information appears.

References

1. P. Bousquet, *J. Convex Analysis* **14**, 119 (2007).
2. P. Bousquet and F. Clarke, *J. Differential Equations* **243**, 489 (2007).
3. G. Buttazzo and M. Belloni, *A survey on old and recent results about the gap phenomenon*, in *Recent Developments in Well-Posed Variational Problems*, eds. R. Lucchetti and J. Revalski (Kluwer, Dordrecht, 1995) pp. 1–27.
4. G. Buttazzo, M. Giaquinta and S. Hildebrandt, *One-dimensional Variational Problems*, Oxford Lecture Series in Mathematics and its Applications, Vol. 15 (Clarendon Press, 1998).

5. P. Cannarsa and C. Sinestrari, *Semiconcave Functions, Hamilton-Jacobi Equations, and Optimal Control* (Birkhäuser, Boston, 2004).
6. A. Cellina, *SIAM J. Control Optim.* **40**, 1270 (2001).
7. F. H. Clarke, *Regularity, existence and necessary conditions for the basic problem in the calculus of variations*, in *Contributions to the modern calculus of variations*, ed. L. Cesari (Longman, London, 1987) pp. 80–90.
8. F. H. Clarke, *Trans. Amer. Math. Soc.* **336**, 655 (1993).
9. F. Clarke, *Ann. Scuola Norm. Sup. Pisa Cl. Sci. (5)* **4**, 511 (2005).
10. F. Clarke, *Necessary Conditions in Dynamic Optimization*, Memoirs of the Amer. Math. Soc., No. 816, Vol. 173, 2005.
11. F. Clarke, *Ergodic Theory Dynam. Systems* **27**, 1 (2007).
12. F. H. Clarke and R. B. Vinter, *J. Differential Equations* **59**, 336 (1985).
13. F. H. Clarke and R. B. Vinter, *Trans. Amer. Math. Soc.* **289**, 73 (1985).
14. R. De Arcangelis, *Ann. Univ. Ferrara* **35**, 135 (1989).
15. E. D. Giorgi, *Mem. Accad. Sci. Torino* **3**, 25 (1957).
16. E. Giusti, *Direct Methods in the Calculus of Variations* (World Scientific, Singapore, 2003).
17. P. Hartman, *Pacific J. Math.* **18**, 495 (1966).
18. C. Mariconda and G. Treu, *Proc. Amer. Math. Soc.* **130**, 395 (2001).
19. C. Mariconda and G. Treu, *J. Optim. Theory Appl.* **112**, 167 (2002).
20. M. Miranda, *Ann. Scuola Norm. Sup. Pisa Cl. Sci. (3)* **19**, 233 (1965).
21. G. Stampacchia, *Comm. Pure Appl. Math.* **16**, 383 (1963).

STABILITY ANALYSIS
OF SLIDING MODE CONTOLLERS

F. H. CLARKE

Institut universitaire de France et Université de Lyon
Institut Camille Jordan UMR 5208
Université Claude Bernard Lyon 1
La Doua, 69622 Villeurbanne, France
E-mail: clarke@math.univ-lyon1.fr
http://math.univ-lyon1.fr

R. B. VINTER

Department of Electrical and Electronic Engineering
Imperial College, London
hibition Road, London SW7 2BT, UK
E-mail: r.vinter@imperial.ac.uk
http://www3.imperial.ac.uk

In this paper we announce a new framework for a rigorous stability analysis of sliding-mode controllers. We give unrestrictive conditions under which such feedback controllers are robustly stabilizing. These conditions make allowance for large disturbance signals, for modeling, actuator and observation measurement errors, and also for the effects of digital implementation of the control. The proposed stability analysis techniques involve two Lyapunov-type functions. The first is associated with passage to the sliding surface in finite time; the second, with convergence to the state associated with the desired equilibrium point. Application of the techniques is illustrated with reference to higher-order linear systems in control canonical form.

Keywords: Sliding Mode Control, Discontinuous Control, Lyapunov Functions, Feedback, Stabilization

1. Introduction

Sliding-mode control is a technique for feedback stabilization, in the presence of disturbances. We refer to the books of Utkin,[7] Spurgeon and Edwards[3] and Slotine and Li.[6] The state is driven to a subset Σ of the state space, the *sliding surface*, in finite time.

Subsequently, the state trajectory remains in Σ and moves asymptotically to a value consistent with the desired equilibrium point: See Figure 1.

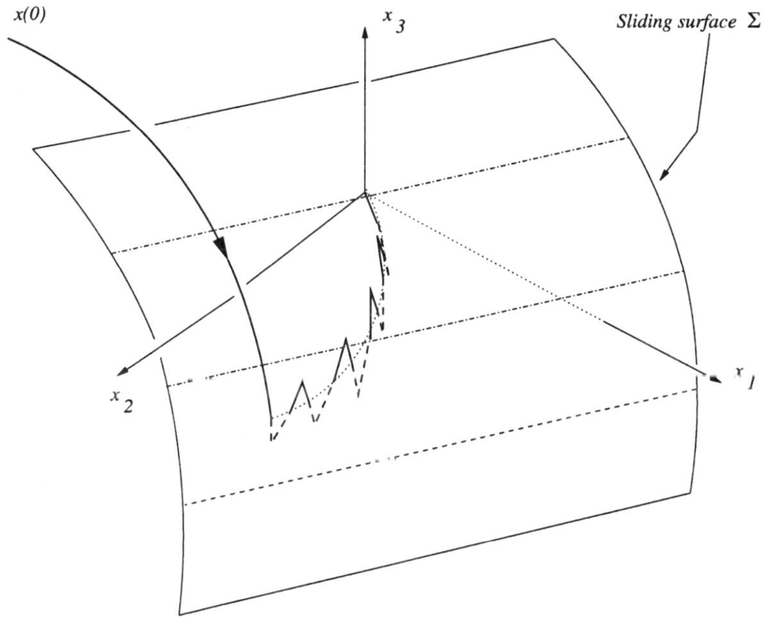

Fig. 1.   A Closed-Loop State Trajectory under Sliding-Mode Control

A notable feature of sliding-mode controls is their stabilizing properties for dynamic systems, in the presence of *large* disturbances.

Sliding-mode controllers, relating the control u to the current state x, commonly take the form

$$u(x) \; = \; g(x) + \chi(x) \tag{1}$$

in which $g(x)$ and $\chi(x)$ are respectively smooth and discontinuous terms. The purpose of the discontinuous term $\chi(x)$ is to force the state to approach Σ at a uniformly positive rate (controllers containing such terms are said to satisfy the 'sliding condition'). The continuous term $g(x)$ is preliminary feedback, configuring the system so that the sliding condition can be achieved. Since, if ever the state trajectory departs from Σ, the controller drives it back towards Σ, we expect that the state trajectory remains in Σ in some sense.

If sliding-mode control is implemented digitally, with a high sample rate, the control values generated by the control law are typically observed to switch rapidly, after the state trajectory first crosses Σ, in such a manner that the state trajectory remains close to Σ, and lies in Σ in the limit, as the sample period tends to zero. Traditional approaches to analysing the stabilizing properties of sliding-mode controllers, expounded for example by Utkin,[7] are along the following lines. Since rapid switching keeps the state close to Σ, we might expect that the state trajectory will be similar to that generated by an 'equivalent' control law,

$$u_{\mathrm{equiv}}(x) \;=\; g(x) + \chi_{\mathrm{equiv}}(x)$$

in which $\chi_{\mathrm{equiv}}(x)$ is a smooth function, chosen to ensure that the time derivative of the evolving state vector is tangent to Σ. Classical techniques (Lyapunov theory or eigenvalue analysis in the linear case) can be used to study the stability properties of the equivalent control law and, by implication (perhaps) those of discrete implementations of the original control, in the limit as the sample period tends to zero.

We present analytic tools for a rigorous stability analysis of sliding-mode controllers. Broad, unrestrictive conditions are given under which these controllers are stabilizing, in the presence not only of large disturbances, but also of modeling, actuator and observation errors. These conditions invoke the existence of two Lyapunov type functions V_1 and V_2, the first associated with passage to the sliding surface in finite time, and the second with convergence to the state associated with the desired set point. The approach takes account, from the beginning, of implementational constraints. The main stability theorem concerns the stabilizing properties of a 'sample-and-hold' implementation of the feedback controller (1), in the presence of modeling, actuator and measurement errors, when the bounds on these errors and the sample period are suitably small. The stability theorem is accompanied by a description of the 'dynamics' of the state trajectory, in the form of a differential inclusion, in the limit as the actuator and measurement errors, and also the upper bound on sample period, tend to zero.

The advantages of our approach over the classical one are as follows. First, by examining the behaviour of the sample-and-hold implementation of the controller directly we avoid the additional hypotheses and the analytical apparatus in Utkin[7] associated with defining Filippov solutions to the continuous time dynamic and controller equations (via equivalent controls), and also the difficulties of choosing the 'correct' equivalent control from among the candidate equivalent controls, when they are not unique.

Second, our approach tells us that the controller is robust: it retains its stabilizing properties when we take account of implementational constraints and small modeling, actuator and measurement errors.

The sliding-mode stability tools introduced in this paper are used to demonstrate the stabilizing properties of a class of sliding-mode controllers applied to a general n^{th}-order linear system in control canonical form, with large disturbances. The controllers employed here are well-known; see, e.g., the books of Spurgeon and Edwards and of Slotine and Li,[3,6] where emphasis is on the second-order case and the 'disturbances' describe, in part, departures from linearity of a nonlinear model such as the rigid pendulum. Our analysis brings to these old controllers new insights, however, regarding their robustness to modeling errors and the effects of digital implementation.

The systematic use of a pair of Lyapunov functions appears to be a novel approach to analysing the stability properties of digitally-implemented sliding-mode control systems, in the presence of modeling and measurement errors. But this paper includes many ingredients from earlier research. In particular, the sliding-mode controller is discontinuous; the use of a single control Lyapunov function to establish the stabilizing properties of sample-and-hold implementations of discontinuous feedbacks has been systematically studied, and is the subject of numerous papers, those of Clarke et al.[1,2] for example. The link between the existence of a single C^2 control Lyapunov function and robustness of the corresponding control system to measurement errors is also very well understood; see for example Refs. 4,5.

A future paper, besides providing full details of the proofs of the stability results announced here, will report on their application to the analysis of new sliding-mode controllers for nonlinear systems, such as the nonholonomic integrator. It will show that, with small changes, the stability analysis presented here can be applied when the 'delay-free sample-and-hold digital control' implementation scheme considered in this paper is replaced by a wide range of alternative implementation schemes (involving, for example, time delays, pre-filtering of the control system, and various forms of 'regularization' of the discontinuous sliding-mode control law).

In Euclidean space, the length of a vector x is denoted by $|x|$, and the closed unit ball $\{x : |x - a| \leq R\}$ by $B(a, R)$. $d_\Sigma(x)$ denotes the Euclidean distance of the point x from the set Σ, namely $\min\{|x - x'| : x' \in \Sigma\}$.

2. System Description

Let $\{t_0 = 0, t_1 \ldots, \}$ be an increasing sequence of numbers such that $t_i \to \infty$ as $i \to \infty$. (We refer to such a sequence as a 'partition of $[0, \infty)$'.) Consider the dynamic system, governing the evolution of the closed-loop state trajectory $x(\cdot)$, described as follows:

$$\left.\begin{array}{lll}
\dot{x}(t) & = & f(x(t), u(t), d(t)) \qquad \text{a.e. } t \in [0, \infty) \\
u(t) & = & a_i + g(x(t_i) + m_i) + v_i \ \text{a.e. } t \in [t_i, t_{i+1}), \ i = 0, 1, 2, \ldots \\
v_i & = & \chi(x(t_i) + m_i) \qquad\qquad\quad i = 0, 1, 2, \ldots \\
d(t) & \in & D \qquad\qquad\qquad\qquad\quad\ \text{a.e. } t \in [0, \infty),
\end{array}\right\} \qquad (2)$$

in which $f : R^n \times R^m \times R^k \to R^n$, $g : R^n \to R^m$ and $\chi : R^n \to R^m$ are given functions satisfying

$$\chi(x) \ \subset \ V \quad \text{for all } x \in R^m \,,$$

and $D \subset R^k$ and $V \subset R^m$ are given sets.

In the above equations, $d(\cdot) : [0, \infty) \to R^k$ is a measurable, essentially bounded function describing the disturbance signal. The sequences $\{a_i\}$ and $\{m_i\}$ describe the n-vector actuator errors and m-vector measurement errors at successive sample instants, respectively.

The following hypotheses will be imposed:

(H1) f and g are continuous and of linear growth: there exist $c_f, c_g > 0$ such that

$$|f(x, u, d)| \ \leq \ c_f(1 + |x| + |u|) \quad \text{for all } (x, u, d) \in R^n \times R^m \times D$$
$$|g(x)| \ \leq \ c_g(1 + |x|) \quad \text{for all } x \in R^n.$$

(H2) For any bounded sets $X \subset R^n$ and $U \subset R^m$, there exists $K = K(X, U) > 0$ such that the following Lipschitz condition holds:

$$|f(x, u, d) - f(x', u, d)| \ \leq \ K|x - x'| \quad \text{for all } x, x' \in X, \ (u, d) \in U \times D \,.$$

(H3) V and D are closed bounded sets.

3. Lyapunov Functions for Sliding Mode Control

We assume that the control feedback design

$$u \ = \ g(x) + \chi(x)$$

has been carried out on the basis of a Lyapunov stability analysis, in ignorance of the measurement and actuator errors and on the basis of a possibly inaccurate dynamic model:

$$\dot{x} = f^0(x, u, d) .$$

To study the effects of sliding-mode control via Lyapunov stability analysis, it is helpful to introduce not one, but two Lyapunov-like functions $V_1 : R^n \to [0, \infty)$ and $V_2 : R^n \to [0, \infty)$, a decrease function $W : R^n \to [0, \infty)$ associated with V_2, and a subset $\Sigma \subset R^n$ of the state space (the 'sliding set').

V_1 will be used to capture the property that the sliding-mode control drives the state into a 'boundary layer' about Σ, in finite time. V_2 is associated with the subsequent motion of the state to a neighbourhood of the origin. V_1, V_2 and W will be required to satisfy the following conditions.

(LF1): V_1 is continuous. V_1 is of class C^1 on $R^n \backslash \Sigma$. Furthermore, there exists $\omega_1 > 0$ such that

$$\nabla V_1(x) \cdot f^0(x, g(x) + \chi(x), d) < -\omega_1 \quad \text{for all } x \in R^n \backslash \Sigma, \, d \in D$$
$$V_1(x) = 0 \quad \text{for} \quad x \in \Sigma$$
$$V_1(x) > 0 \quad \text{for} \quad x \notin \Sigma.$$

Now write

$$F_0(x) := \bigcap_{\rho > 0} \left\{ f^0(x', g(x') + v', d) : x' \in B(x, \rho), \, v' \in \chi(x'), \, d \in D \right\} .$$

We may think of $F_0(x)$ as consisting of all possible velocity values $\dot{x}$ (in limiting terms, and for the nominal dynamics given by f^0) when the state is at x.

(LF2): V_2 and W are continuous. V_2 is of class C^1 on $R^n \backslash \{0\}$. Furthermore

$$\sup_{w \in F_0(x)} \langle \nabla V_2(x), w \rangle \leq -W(x) \quad \text{for all } x \in \Sigma \backslash \{0\}.$$
$$V_2(0) = 0 \quad \text{and} \quad V_2(x) > 0 \text{ for } x \in \Sigma \backslash \{0\}$$
$$W(0) = 0 \quad \text{and} \quad W(x) > 0 \quad \text{for } x \in \Sigma \backslash \{0\}.$$

(LF3): $V_1 + V_2$ is a proper function; i.e., for any $\alpha \geq 0$, the sub-level set $\{x : V_1(x) + V_2(x) \leq \alpha\}$ is bounded.

4. Sufficient Conditions for Stability

In this section we give sufficient conditions for stabilization to the origin, via sliding-mode control. Since the control system description we have adopted takes account of digital implementation and allows for disturbances, we cannot expect state trajectories to converge to the zero state, as t tends to ∞. Instead we give conditions for *practical semiglobal stability*, i.e. conditions under which, for any two balls $B(0,R)$ (the set of initial states) and $B(0,r)$ (the target set) in R^n, $R > r > 0$, a state trajectory that issues from the set of initial states is driven to the target set in bounded time, and remains in the target set thereafter.

Theorem 4.1 (Conditions for Practical Semiglobal Stabilization).
Assume (H1)-(H3). Suppose there exist functions V_1, V_2 and W, and a set Σ satisfying hypotheses (LF1)-(LF3). Choose any numbers $R > 0$, $r > 0$ $(R > r)$, $\bar{\omega} \in (0,\omega_1)$ and $\bar{\epsilon} > 0$. Then there exist positive numbers $R_* > R$ $(R_*$ depends only on $R)$, $e_m, e_a, \delta, e_1, e_2$ and $T > 0$, with the following properties:
Take any sequences $\{m_i\}$ and $\{a_i\}$ in R^m and R^n respectively, partition $\{t_i\}$, measurable function $d : [0,\infty) \to D$ and $x_0 \in B(0,R)$ satisfying

$$|m_i| \le e_m, \quad |a_i| \le e_a \quad and \quad |t_{i+1} - t_i| \le \delta \quad for \ all \ i. \tag{3}$$

Suppose in addition that we have

$$|\langle \nabla V_1(x), f(x,g(x)+v,d) - f^0(x,g(x)+v,d)\rangle| \le e_1$$
$$for \ all \ v \in \chi(x), \ x \in B(0,R_*)\backslash\Sigma, \ d \in D \tag{4}$$

and

$$|\langle \nabla V_2(x), f(x,g(x)+v,d) - f^0(x,g(x)+v,d)\rangle| \le e_2$$
$$for \ all \ v \in \chi(x), \ x \in B(0,R_*)\backslash\{0\}, \ d \in D. \tag{5}$$

Let $x(\cdot) : [0,\infty) \to R^n$ be any solution to eqn.(2) with $x(0) = x_0$. (One such solution exists). Then

$$x(t) \in B(0,R_*) \quad for \ all \ t \ge 0$$

and

$$x(t) \in B(0,r) \quad for \ all \ t \ge T.$$

Furthermore,

$$d_\Sigma(x(t)) \le \bar{\epsilon} \quad for \ all \ t \in [V_1(x(0)/\bar{\omega},\infty).$$

A proof of the theorem will be given in a future publication. We remark that the way in which the modeling error $f - f^0$ is constrained in (4) and (5) (relative to inner products with the gradients of V_1 and V_2) leads to a new interpretation of 'matched' errors.

A standard 'sequential compactness' analysis, combined with an application of the results of the preceding theorem, provides information, summarized as Prop. 4.1 below, about the set of solutions to eqn.(2), in the limit as the sample period tends to zero and modeling, actuator and measurement errors vanish. Full details will be reported elsewhere.

Let $\Phi : R^n \rightsquigarrow R^n$ be the set-valued function

$$\Phi(x) := \bigcap_{\rho>0} \mathrm{co}\,\{f(x,g(x)+v,d)\,|\,v=\chi(x'),\,x' \in B(x,\rho), d \in D\} \quad . \quad (6)$$

Here, $\mathrm{co}\,S$ denotes 'closed convex hull of S', and $T_S(\mathrm{x})$ is the tangent cone to the closed set S at a point $x \in S$:

$$T_S(x) := \{v \in R^n : \exists\, v_i \to v \text{ and } \epsilon_i \downarrow 0 \text{ s.t. } x + \epsilon_i v_i \in S \text{ for all } i\}\,.$$

Proposition 4.1. *Assume (H1)-(H3). Suppose that there exist functions V_1, V_2 and W satisfying (LF1)-(LF3) (when f replaces f_0). Take a measurable function $d(\cdot) : [0,\infty) \to D$ and families of sequences of n-vectors $\{m_{j0}, m_{j1}, \ldots\}_{j=1}^{\infty}$ and of m-vectors $\{a_{j0}, a_{j1}, \ldots\}_{j=1}^{\infty}$ satisfying*

$$\lim_{j\to\infty} \sup_i |m_{ji}| \to 0, \quad \lim_{j\to\infty} \sup_i |a_{ji}| \to 0\,.$$

Let $\{t_{j0}, t_{j1}, \ldots\}_{j=1}^{\infty}$ be a family of partitions satisfying

$$\lim_{j\to\infty} \sup_i |t_{j(i+1)} - t_{ji}| = 0\,.$$

Fix $x_0 \in R^n$. For each j, let $x_j(.)$ be a solution to eqn.(2) (with initial value x_0), when $\{t_{j0}, t_{j1}, \ldots\}$, $\{a_{j0}, a_{j1}, \ldots\}$ and $\{m_{j0}, m_{j1}, \ldots\}$ replace $\{t_0, t_1, \ldots\}$, $\{a_0, a_1, \ldots\}$ and $\{m_0, m_1, \ldots\}$ respectively.

Then there exists a locally Lipschitz continuous arc $x(\cdot) : [0,\infty) \to R^n$ and $T_1 \geq 0$ such that

$$T_1 \leq \omega_1^{-1} V_1(x_0)$$

and, for some subsequence (we do not relabel),

$$x_j(\cdot) \to x(\cdot) \quad \text{as } j \to \infty$$

uniformly on bounded subsets of $[0,\infty)$. We have

$$x(t) \in \begin{cases} R^n \backslash \Sigma & \text{for } t < T_1 \\ \Sigma & \text{for } t \geq T_1\,, \end{cases}$$

$$x(t) \to 0 \quad \text{as } t \to \infty$$

and, for a.e. $t \in [0, \infty)$,

$$\dot{x}(t) \in \begin{cases} \Phi(x(t)) & \text{if } t \leq T_1 \\ \Phi(x(t)) \cap T_\Sigma(x(t)) & \text{if } t > T_1 . \end{cases}$$

This proposition describes the behaviour of state trajectories under 'continuous time' control feedback, when the response to continuous time control feedback is interpreted as the limit of a sample-and-hold implementation, as the sample period goes to zero. It asserts that, if there exist a pair of Lyapunov functions with the stated properties, the state trajectories reach the switching surface in finite time. They then move asymptotically towards the origin, while remaining in Σ. The proposition also provides details of a differential inclusion that the state trajectory satisfies, after it has entered Σ.

5. An Example

Consider the linear system in control canonical form, relating the n-vector state $x(t)$ to the scalar control $u(t)$:

$$\frac{d}{dt}x(t) = Ax(t) + b(d(t) + u(t)) \tag{7}$$

in which

$$A = \begin{bmatrix} 0 & 1 & 0 & . & . & 0 & 0 \\ 0 & 0 & 1 & 0 & . & 0 & 0 \\ . & & . & & . & . & . \\ 0 & 0 & 0 & . & . & 0 & 1 \\ -a_0 & -a_1 & . & . & . & . & -a_{n-1} \end{bmatrix} \quad \text{and} \quad b = \begin{bmatrix} 0 \\ . \\ . \\ . \\ 0 \\ 1 \end{bmatrix} .$$

Here $a_0, \ldots, a_{n-1}$ are known parameters, d is an unmeasured scalar disturbance signal, assumed to satisfy, for some given $d_{\max}(> 0)$, the condition

$$|d(t)| \leq d_{\max} .$$

We seek state feedback

$$u = \phi(x),$$

to achieve closed loop asymptotic stabilization, for arbitrary disturbance signals $d(\cdot)$.

This design problem is addressed in the sliding-mode literature as follows (see, e.g., Ref. 3). Fix coefficients $\lambda_0, \ldots, \lambda_{n-1}$ of a stable polynomial, of degree $n - 1$,

$$\lambda(\sigma) = \lambda_0 + \lambda_1\sigma + \ldots + \lambda_{n-2}\sigma^{n-2} + \sigma^{n-1}$$

and define the scalar-valued function of the state

$$s(x) = \lambda_0 x_1 + \lambda_1 x_2 + \ldots + \lambda_{n-2} x_{n-1} + x_n . \tag{8}$$

Define also the k vector

$$k = \text{col}\{a_0, a_1 - \lambda_0, \ldots, a_{n-1} - \lambda_{n-2}\} .$$

Fix $K > d_{\max}$. The following control law has been proposed:

$$u = \phi(x) , \tag{9}$$

where

$$\phi(x) = k^T x - K \, \text{sgn}\{s(x)\} . \tag{10}$$

Here, $\text{sgn}(\cdot)$ is the 'signum function'

$$\text{sgn}(s) = \begin{cases} +1 \text{ if } s > 0 \\ -1 \text{ if } s \leq 0 \end{cases}$$

The rationale here is that, if we substitute control law (9) into (7) and take account of the fact that $\frac{d}{dt} x_i = x_{i+1}$, for $i = 1, \ldots, n-1$, there results (in the case $s(x(t)) \neq 0$)

$$\begin{aligned}
\frac{d}{dt}|s(x(t))| &= \text{sgn}\{s(x(t))\}[\frac{d}{dt} x_n + \lambda_{n-2} x_n + \ldots + \lambda_0 x_2] \\
&= \text{sgn}\{s(x(t))\}[-a_0 x_1 + \ldots - a_{n-1} x_n + a_0 x_1 \\
&\quad + (a_1 - \lambda_0) x_2 + \ldots + (a_{n-1} - \lambda_{n-2}) x_n \\
&\quad - K \, \text{sgn}\{s(x(t)\} + d(t) + \lambda_{n-2} x_n + \ldots + \lambda_0 x_2] \\
&\leq -K \text{sgn}\{s(x(t)\}\text{sgn}\{s(x(t)\} + |d(t)| \\
&\leq -\omega ,
\end{aligned}$$

where $\omega = (K - d_{\max}) \, (> 0)$. These calculations suggest that, if $s(x(0)) \neq 0$, then $x(t)$ arrives at the set

$$\Sigma = \{x : s(x) = 0\} \tag{11}$$

at a positive time $\bar{T}$ ($\bar{T} \leq |s(x(0))|/(K - d_{\max})$), where it remains thereafter. Furthermore, for $t > \bar{T}$ we have that

$$s(x(t)) = \frac{d^{n-1}}{dt^{n-1}} x_1 + \lambda_{n-2} \frac{d^{n-2}}{dt^{n-2}} x_1 + \ldots + \lambda_0 x_1 = 0 . \tag{12}$$

From this we can surmise that $x_1(t)$ (and hence also $x_2(t), \ldots, , x_n$) converge to zero, as $t \to 0$.

The methods of this paper can be used to justify these conclusions, taking account of implementational effects and the presence of modeling and measurement errors. We show:

174

Proposition 5.1. *Take $R > 0$ and $r > 0$ $(R > r)$. Take also $d_{max} > 0$ and $K > d_{max}$. Then there exist positive numbers $\bar{a} > 0$, $\bar{m} > 0$, $\bar{\delta} > 0$ and $T > 0$ with the following properties:*

Take any partition $\{t_i\}$, measurable function $d(\cdot) : [0, \infty) \to R$, and sequences $\{a_i\}$ and $\{m_i\}$ satisfying

$$|t_{i+1} - t_i| < \bar{\delta}, \ |a_i| \leq \bar{a}, \ |m_i| \leq \bar{m} \ \text{for all } i, \ ||d(\cdot)||_{L^\infty} \leq d_{max} .$$

Then, for any $x_0 \in B(0, R)$, the solution $x : [0, \infty) \to R^n$ to

$$\left.\begin{array}{l} \dot{x}(t) = Ax(t) + b(d(t) + u(t))) \quad a.e. \ t \in [0, \infty) \\ u(t) = a_i + \phi(x(t_i) + m_i) \quad a.e. \ t \in [t_i, t_{i+1}) \\ x(0) = x_0 \end{array}\right\} \tag{13}$$

satisfies

$$x(t) \in B(0, r) \quad \text{for all } t \geq T .$$

Here, $\phi(\cdot)$ is the mapping (10).

Proof. We apply Thm. 4.1, making the following identifications:

$$f(x, u, d) = Ax + b(u + d), \ g(x) = k^T x, \ \chi(x) = K \operatorname{sgn}\{s(x)\}$$
$$D = [-d_{\max}, +d_{\max}] \text{ and } V = [-K, +K] .$$

The hypotheses (H1)-H(3) on the dynamics are satisfied for this choice of data. The assertions of the propositon will have been proved, if we can construct functions V_1, V_2 and W satisfying the conditions $(LF1)$–$(LF3)$.
    Take $V_1 : R^n \to R^+$ to be the function

$$V_1 = |s(x)|,$$

where $s(\cdot)$ is the function (8). The sliding set is

$$\Sigma = \{x \,|\, V_1(x) = 0\} = \{x \,|\, s(x) = 0\}.$$

V_1 is continuous, of class C^1 on $R^n \backslash \Sigma$. It can, furthermore, be shown that, for any $x \in R^n \backslash \Sigma$:

$$\nabla V_1(x) \cdot f(x, g(x) + \chi(x), d) \leq -\omega_1 ,$$

where ω_1 is the positive number $\omega = K - d_{\max}$. Thus, V_1 satisfies (LF1).

With a view to constructing V_2 and W, we introduce the matrices

$$\bar{A} = \begin{bmatrix} 0 & 1 & 0 & . & . & 0 \\ 0 & 0 & 1 & 0 & . & 0 \\ . & & . & . & . & . \\ 0 & . & . & . & . & 1 \\ -\lambda_0 & . & . & . & . & -\lambda_{n-2} \end{bmatrix} \quad \text{and} \quad J = \begin{bmatrix} I_{(n-1)\times(n-1)} \\ -\underline{\lambda}^T \end{bmatrix}$$

in which $\underline{\lambda}$ is the vector

$$\underline{\lambda} = \text{col}\{\lambda_0, \ldots, \lambda_{n-2}\} \ .$$

Note that J has full column rank and that $\Sigma = \text{range}\{J\}$. It follows that $J^T J$ is invertible and, given any $x \in R^n$, there exists a unique $\xi \in R^{n-1}$ such that $x = J\xi$. Since $\bar{A}$ is a 'stable' matrix, there exists a symmetric matrix $\bar{P}$ and $\gamma > 0$ and $c > 0$ such that

$$\xi^T \bar{P} \bar{A} \xi \ \leq \ -\gamma|\xi|^2 \text{ and } \xi^T \bar{P} \xi \ \geq c|\xi|^2 \quad \text{for all } \xi \in R^{n-1}.$$

(See Ref. 8.) Define

$$P = \begin{bmatrix} \bar{P} & \underline{0} \\ \underline{0}^T & 0 \end{bmatrix}$$

in which $\underline{0}$ is the $(n-1)$ vector $\text{col}\{0, \ldots, 0\}$. Finally we set

$$V_2(x) = \frac{1}{2}x^T P x \quad \text{for all } x \in R^n \ .$$

The function W is taken to be

$$W(x) = \gamma |[J^T J]^{-1} J^T x|^2 \ ,$$

The functions V_2 and W constructed in this way are continuous, vanish at the origin, and are of class C^2. Note also that, if $x \in \Sigma\backslash\{0\}$, then $x = J\xi$ for some $\xi \neq 0$. It follows that

$$W(x) = \gamma|\xi|^2 > 0 \quad \text{and} \quad V_2(x) = \xi^T \bar{P} \xi > 0 \ ,$$

i.e. V_2 and W are positive on $\Sigma\backslash\{0\}$. It can also be shown that, for any $x \in \Sigma\backslash\{0\}$, the decrease property of (LF2) holds. It is a straightforward matter to show that $V_1 + V_2$ is proper (condition (LF3)). All the hypotheses of Thm. 4.1 are satisfied. The assertions of the proposition follow.

References

1. F. Clarke, Y. S. Ledyaev, L. Rifford and R. Stern, *SIAM J. Control Optim.* **39**, 25 (2000).
2. F. H. Clarke, Y. S. Ledyaev, E. D. Sontag and A. I. Subbotin, *IEEE Trans. Automat. Control* **42**, 1394 (1997).
3. C. Edwards and S. K. Spurgeon, *Sliding Mode Control* (Taylor and Francis, London, 1998).
4. Y. S. Ledyaev and E. D. Sontag, *Nonlinear Analysis* **37**, 813 (1999).
5. Y. Lin, E. D. Sontag and Y. Wang, *SIAM J. Control Optim.* **34**, 124 (1996).
6. J. J. E. Slotine and W. Li, *Applied Nonlinear Control* (Prentice Hall, Englewood Cliffs NJ, 1991).
7. V. I. Utkin, *Sliding Modes in Control Optimization* (Springer Verlag, Berlin, 1992).
8. J. L. Willems, *Stability Theory of Dynamical Systems* (Thomas Nelson and Sons, London, 1970).

GENERALIZED DIFFERENTIATION OF PARAMETERIZED FAMILIES OF TRAJECTORIES

M. GARAVELLO

Dipartimento di Matematica e Applicazioni, Università di Milano-Bicocca,
20125 Milano, Italy,
Present address: Di.S.T.A., Università del Piemonte Orientale,
15100 Alessandria, Italy
E-mail: mauro.garavello@mfn.unipmn.it

E. GIREJKO

Bialystok Technical University,
ul. Wiejska 45a, 15-351, Bialystok, Poland.
E-mail: ewagir@wp.pl

B. PICCOLI

I.A.C., C.N.R.,
00161 Roma, Italy
E-mail: b.piccoli@iac.cnr.it

We consider the problem of weak differentiability for a parameterized family of trajectories. The study is motivated by various problems in ordinary differential equations and control theory.
Some results are presented and their applicability is shown by means of few examples. The methods are based on generalized differentiation of flows or, more generally, of set-valued maps.

Keywords: Parameterized families of trajectories; generalized differentials; necessary and sufficient conditions for optimality.

1. Introduction

This paper deals with weak differentiability of parameterized families of trajectories of vector fields. This topic is extensively studied in the literature, since various problems in mathematical analysis can be formulated in such a way. First let us describe precisely the setting.

Let P be a normed space and consider the following family of Cauchy problems

$$\begin{cases} \dot{x}(t) = f_p(x(t), t), \\ x(0) = \bar{x}_p, \end{cases} \tag{1}$$

where $p \in P$ is the parameter, $f_p : \mathbb{R}^n \times [0, +\infty[\to \mathbb{R}^n$ is a parameterized family of vector fields and $\bar{x}_p \in \mathbb{R}^n$ a parameterized family of initial conditions. We indicate by x_p a solution to (1), thus we get a parameterized family of trajectories (in case of non uniqueness of solutions to (1) one has to take a selection).

We list some examples of problems which can be stated in terms of differentiability of the family x_p, at some time in the domain of definition, with respect to the parameters.

Differentiability with respect to initial data. The most classical example is the differentiability of a trajectory of a $\mathcal{C}^1$ vector field with respect to the initial data. In this case $x, p \in \mathbb{R}^n$, and for fixed $p_0 \in \mathbb{R}^n$ we set

$$\begin{cases} f_p = f, \\ \bar{x}_{p_0+v} = x_{p_0}(0) + v, \end{cases}$$

where $v \in \mathbb{R}^n$. It is well known that the family is differentiable at every time (at which solutions are defined) and is expressed by the solution of the adjoint matrix equation:

$$\dot{V}(t) = f_x(x_{p_0}(t), t) \cdot V(t), \qquad V(0) = Id,$$

where f_x is the Jacobian matrix of f, in the sense that $d/dp \, x_p(t)|_{p=p_0} = V(t)$.

When the vector field f is not smooth, especially when f is not continuous or it has not the uniqueness property for trajectories, then this result does not hold in general. Hence the necessity to introduce weaker concepts of differentiability. For this, various results on generalized differentials of maps may be used, see Refs. 2,3,11,13,19,22–25,28.

Differentiability with respect to the vector field. The second example is the investigation of how the solution to a Cauchy problem for the same initial data depends on the vector field. Let us consider directly discontinuous vector fields. For instance, fix a time dependent family of discontinuous vector fields $f_\varepsilon : \mathbb{R} \times \mathbb{R} \to \mathbb{R}^+$ ($\varepsilon \geq 0$) with exactly one discontinuity at $\bar{x} > 0$, so that each f_ε can be written in the form

$$f_\varepsilon(x, t) = \begin{cases} f_\varepsilon^-(x, t), & \text{if } x \leq \bar{x}, \\ f_\varepsilon^+(x, t), & \text{if } x > \bar{x}, \end{cases} \tag{2}$$

with f_ε^- and f_ε^+ C^1 vector fields. This means that we take parameter $p = \varepsilon \in \mathbb{R}$. For every $\varepsilon \geq 0$, consider the Cauchy problem

$$\begin{cases} \dot{x}(t) = f_\varepsilon(x(t), t), \\ x(0) = 0, \end{cases} \tag{3}$$

and solutions $x_\varepsilon(\cdot)$ on $[0, T]$ crossing $\bar{x}$ at a certain time $0 < t_\varepsilon < T$. It means that each $x_\varepsilon(\cdot)$ solves

$$\begin{cases} \dot{x}_\varepsilon(t) = f_\varepsilon^-(x_\varepsilon(t), t), & \text{if } 0 < t < t_\varepsilon, \\ \dot{x}_\varepsilon(t) = f_\varepsilon^+(x_\varepsilon(t), t), & \text{if } t_\varepsilon < t \leq T, \\ x_\varepsilon(0) = 0, \\ x_\varepsilon(t_\varepsilon) = \bar{x}. \end{cases} \tag{4}$$

Our aim is to compute the differential of $x_\varepsilon(t)$ for $t > t_0$ with respect to ε.

Variations of trajectories of control systems. The main ingredient to prove necessary conditions for optimality in optimal control problems is that of considering variations of trajectories. The most known are "needle" variations used in the proof of Pontryagin Maximum Principle; see Ref. 21. Consider for example the optimal control problem in Bolza form:

$$\dot{x} = f(t, x, u), \qquad x \in \mathbb{R}^n, \ u \in U, \tag{5}$$

$$\min \int_0^T L(t, x(t), u(t))\, dt, \qquad x(0) = \bar{x}, \ x(T) \in S, \tag{6}$$

where $f : [0, T] \times \mathbb{R}^n \times U \to \mathbb{R}^n$ is the controlled dynamics, U is the set of controls, L is the Lagrangian cost, $\bar{x} \in \mathbb{R}^n$ is the initial condition and $S \subset \mathbb{R}^n$ is the target. Assume f regular enough and U compact so to have a solution to (5) for every control $u(\cdot)$ defined on $[0, T]$. Given a candidate optimal control u^*, a control variation is a parameterized family of controls u_ε, $u_0 = u^*$, so that the corresponding trajectories x_ε (satisfying the initial condition) give a variation of the trajectory x_0. Needle variations consist in fixing $\tau \in [0, T]$ (Lebesgue point for both $f(\cdot, x_0(\cdot), u_0(\cdot))$ and $L(\cdot, x_0(\cdot), u_0(\cdot)))$, $\omega \in U$, and setting $u_\varepsilon = \omega$ if $t \in [\tau - \varepsilon, \tau]$ and $u_\varepsilon = u_0$ otherwise. Computing differentials of trajectory variations amounts to consider the above problem with $p = \varepsilon \in \mathbb{R}$, $f_\varepsilon(x, t) = f(t, x, u_\varepsilon)$ and $\bar{x}_p \equiv \bar{x}$.

Needle variations give rise to trajectory variations differentiable in classical sense only after time τ. A setting for which generalized differentiability holds, for every time, was introduced in Refs. 12,20. Other possible variations can be stated in terms of switching times; see Example 5.2 below. Other works in this direction are Refs. 1,4,5,7,10,15,16,26,27.

Optimal Syntheses. A technique to provide solutions to problems as (5)–(6) is to look for an optimal synthesis. Roughly speaking an optimal synthesis consists of a family of trajectories $z_x(\cdot)$, one for every initial condition $z(0) = x$, thus embedding (5)–(6) in a family of problems. To construct an optimal synthesis, one usually finds a synthesis formed by extremal trajectories, i.e. satisfying Pontryagin Maximum Principle. Then, if the synthesis is regular enough, it follows optimality, in the sense that each z_x solves the corresponding optimal control problem. Here regularity can be expressed in terms of weak differentiability of z_x with respect to initial conditions (plus some other conditions); see Refs. 6,9,17,18,20.

The aim of this paper is thus to provide general results on weak differentiability of x_p with respect to p, so to address all examples above. The approach we use is quite natural. We start from the incremental ratio, then we split it in two different ways and consider two different methods depending on how the incremental ratio was split. The first approach is inspired by the results of weak differentiability for the flow obtained in Refs. 12,20 and on the use of additional families of trajectories, to apply some generalizations of the Theorem 2.9 of Ref. 8. In the second approach, we consider the (possibly multivalued) flow of an ordinary differential equation and we apply generalized differentiation theories for multivalued maps.

The paper is organized as follows. In Section 2 we describe the problem and the methods used. Sections 3 and 4 develop respectively the first and the second approach and contain the main results with the proofs. Finally Section 5 presents some examples of applicability of the theory developed.

2. Basic definitions

Let P be a normed space and fix $p_0, q_0 \in P$. Consider the family of Cauchy problems (1), fix $T > 0$ and assume

(C-1) for every $p \in P$ the vector field f_p is measurable w.r.t. t and continuous w.r.t. x;

(C-2) for p, q in a neighborhood of p_0 there exists a solution defined on $[0, T]$ to

$$\begin{cases} \dot{x}(t) = f_p(x(t), t), \\ x(0) = \bar{x}_q. \end{cases} \tag{7}$$

Remark 2.1. Notice that the assumption (C-1) with an additional growth condition on f_p is sufficient to ensure local forward existence of a Caratheodory solution to the Cauchy problem (1). By a Caratheodory solution to (1) we mean an absolutely continuous function $x : [0, h] \to \mathbb{R}^n$, $(h > 0)$, such that

$$x(t) = \bar{x}_p + \int_0^t f_p(x(s), s)ds$$

for every $t \in [0, h]$.

Notation 2.1. We denote by $x_p(\cdot)$ a solution to the Cauchy problem (1) and by $x_p^q(\cdot)$ a solution to the Cauchy problem (7). If the Cauchy problem (7) admits a unique solution, then the function $x_p^q(\cdot)$ is clearly fixed. In the other case we will select $x_p^q(\cdot)$ from the family of solutions so to satisfy some additional conditions. We denote by $y_{p_0,v}^q(t)$, $y_p^{q_0,v}(t)$ and $y_{p_0,v}(t)$, respectively, the directional derivative in the direction $v \in P$ of $x_p^q(t)$ w.r.t. p at $p = p_0$, the directional derivative in the direction $v \subset P$ of $x_p^q(t)$ w.r.t. q at $q = q_0$ and the directional derivative in the direction $v \in P$ of $x_p(t)$ w.r.t. p at $p = p_0$.

Our aim is to compute $y_{p_0,v}(t)$ under suitable assumptions. There are at least two natural ways to evaluate it:

(a) the first one consists in writing

$$y_{p_0,v}(t) = \lim_{\varepsilon \to 0} \frac{x_{p_0+\varepsilon v}(t) - x_{p_0}(t)}{\varepsilon}$$

$$= \lim_{\varepsilon \to 0} \frac{1}{\varepsilon} \left[(x_{p_0+\varepsilon v}^{p_0+\varepsilon v}(t) - x_{p_0}^{p_0+\varepsilon v}(t)) + (x_{p_0}^{p_0+\varepsilon v}(t) - x_{p_0}^{p_0}(t)) \right]$$

$$= \lim_{\varepsilon \to 0} \frac{(x_{p_0+\varepsilon v}^{p_0+\varepsilon v}(t) - x_{p_0}^{p_0+\varepsilon v}(t))}{\varepsilon} + \lim_{\varepsilon \to 0} \frac{(x_{p_0}^{p_0+\varepsilon v}(t) - x_{p_0}^{p_0}(t))}{\varepsilon}; \qquad (8)$$

(b) the second one consists in writing

$$y_{p_0,v}(t) = \lim_{\varepsilon \to 0} \frac{x_{p_0+\varepsilon v}(t) - x_{p_0}(t)}{\varepsilon}$$

$$= \lim_{\varepsilon \to 0} \frac{1}{\varepsilon} \left[(x_{p_0+\varepsilon v}^{p_0+\varepsilon v}(t) - x_{p_0+\varepsilon v}^{p_0}(t)) + (x_{p_0+\varepsilon v}^{p_0}(t) - x_{p_0}^{p_0}(t)) \right]$$

$$= \lim_{\varepsilon \to 0} \frac{(x_{p_0+\varepsilon v}^{p_0+\varepsilon v}(t) - x_{p_0+\varepsilon v}^{p_0}(t))}{\varepsilon} + \lim_{\varepsilon \to 0} \frac{(x_{p_0+\varepsilon v}^{p_0}(t) - x_{p_0}^{p_0}(t))}{\varepsilon}. \qquad (9)$$

Remark 2.2. Note that the first addendum of the equation (9) and the second one of the equation (8) look like generalized differentials of the flow with respect to initial data (see for example Ref. 2,3,11,13,19,24–26). More precisely, the first addendum of (9) is similar to $\lim_{\varepsilon \to 0} y_{p_0+\varepsilon v}^{p_0,v}$, i.e. the limit

182

on ε of the differential of the flow, which can be granted by uniformity assumptions on such differentials.

On the other hand, the remaining addenda in (9) and (8) look like weak differentiations with respect to the dynamics. Indeed, the second term of (9) is exactly $y_{p_0,v}^{p_0}$, i.e. a weak differential with respect to the dynamics at x_{p_0} along the direction v, whilst the first term in (8) is similar to $\lim_{\varepsilon \to 0} y_{p_0,v}^{p_0+\varepsilon v}$, i.e. the limit as $\varepsilon \to 0$ of a weak differentiation with respect to the dynamics at $x_{p_0+\varepsilon v}$.

Before considering the general case, we propose an example where the dynamics f_p are smooth vector fields.

Example 2.1. Consider the following Cauchy problem

$$\begin{cases} \dot{x}(t) = f_p(x(t), t), \\ x(0) = \bar{x}_q. \end{cases} \tag{10}$$

where $p, q \in P$ and the vector fields f_p are of class C^2. In particular (10) admits a unique local solution for every initial point.

Let $v \in P$ and $\varepsilon > 0$. Consider the trajectory $x_{p_0+\varepsilon v}^{p_0+\varepsilon v}$, solution to (10) on $[0, T]$ where $p = p_0 + \varepsilon v$ and $q = p_0 + \varepsilon v$, and $\gamma_t : [t, T] \to \mathbb{R}^n$ the solution to

$$\begin{cases} \frac{d}{ds} \gamma_t(s) = f_{p_0}(\gamma_t(s), s), \\ \gamma_t(t) = x_{p_0+\varepsilon v}^{p_0+\varepsilon v}(t). \end{cases} \tag{11}$$

Using Taylor's expansion and classical differential with respect to the initial point, we have

$$\gamma_{t+h}(T) - \gamma_t(T) = [\gamma_{t+h}(t+h) - \gamma_t(t+h)]$$

$$+ \int_{t+h}^{T} [f_{p_0}(\gamma_{t+h}(s), s) - f_{p_0}(\gamma_t(s), s)]\, ds$$

$$= \gamma_{t+h}(t+h) - \gamma_t(t+h) + h \int_{t+h}^{T} f_{p_0,x}(\gamma_t(s), s) \frac{d}{dt}\gamma_t(s)\, ds + o(h)$$

$$= \gamma_{t+h}(t+h) - \gamma_t(t+h)$$

$$+ h \int_{t+h}^{T} f_{p_0,x}(\gamma_t(s), s) M(s, t+h) v(t+h)\, ds + o(h)$$

$$= \gamma_{t+h}(t+h) - \gamma_t(t+h) + h M(T, t+h) v(t+h) - h v(t+h) + o(h),$$

where $f_{p_0,x}$ denotes the Jacobian matrix with respect to x of the flow f_{p_0}, $M(s, t)$ is the fundamental matrix solution to the linear system

$$\frac{\partial M}{\partial s}(s, t) = f_{p_0,x}(\gamma_t(s), s) M(s, t),$$

and

$$v(t+h) = \lim_{\eta \to 0^+} \frac{\gamma_{t+\eta}(t+h) - \gamma_t(t+h)}{\eta}.$$

Since the vector fields are of class $\mathcal{C}^2$, then

$$v(t+h) = \frac{\gamma_{t+h}(t+h) - \gamma_t(t+h)}{h} + o(h)$$

and so

$$\gamma_{t+h}(T) - \gamma_t(T) = M(T, t+h)\left[\gamma_{t+h}(t+h) - \gamma_t(t+h)\right] + o(h).$$

Define the function $\psi(t) := \gamma_t(T)$. Clearly ψ is differentiable and

$$\psi(T) - \psi(0) = \int_0^T \dot{\psi}(t)dt = \lim_{h \to 0^+} \int_0^T \frac{\psi(t+h) - \psi(t)}{h} dt$$

$$= \lim_{h \to 0^+} \int_0^T \frac{\gamma_{t+h}(T) - \gamma_t(T)}{h} dt$$

$$= \lim_{h \to 0^+} \frac{1}{h} \int_0^T M(T, t+h)\left[\gamma_{t+h}(t+h) - \gamma_t(t+h)\right] dt.$$

Observing that

$$\lim_{h \to 0} \frac{\gamma_{t+h}(t+h) - \gamma_t(t)}{h} = \frac{d}{dt} x_{p_0 + \varepsilon v}^{p_0 + \varepsilon v}(t),$$

we have

$$x_{p_0 + \varepsilon v}^{p_0 + \varepsilon v}(T) - x_{p_0}^{p_0 + \varepsilon v}(T) = \psi(T) - \psi(0)$$

$$= \lim_{h \to 0^+} \int_0^T M(T, t+h) \left[\frac{\gamma_{t+h}(t+h) - \gamma_t(t+h)}{h}\right] dt$$

$$= \int_0^T M(T, t) \left[f_{p_0 + \varepsilon v}(x_{p_0 + \varepsilon v}^{p_0 + \varepsilon v}(t), t) - f_{p_0}(x_{p_0 + \varepsilon v}^{p_0 + \varepsilon v}(t), t)\right] dt.$$

Notice that the final equation of the example expresses the numerator of the first addendum of equation (8). We consider now weaker assumptions on the vector fields in order to have similar expressions.

3. Approach (a)

This section deals with the estimate of the directional derivative $y_{p_0, v}(t)$ using the decomposition given in (8). The first part of the Section is dedicated to the statement of the main results (Theorems 3.1 and 3.2), while the second one to the proofs. In order to do this, we prove Proposition 3.1 and some technical lemmas.

184

Theorem 3.1. *Fix $T > 0$. Suppose that the function*

$$p \mapsto x_{p_0}^p(T)$$

is Gateaux differentiable at $p_0 \in P$.

(1) Fix $v \in P$. For every $\varepsilon > 0$ and $t \in [0, T]$ assume that the Cauchy problem

$$\begin{cases} \dot{\gamma}(s) = f_{p_0}(\gamma(s), s), \\ \gamma(t) = x_{p_0 + \varepsilon v}^{p_0 + \varepsilon v}(t). \end{cases} \tag{12}$$

admits a unique solution $\gamma_t : [t, T] \to \mathbb{R}^n$. Define

$$\bar{t} = \sup \left\{ t \in [0, T] : \forall s \in [0, t] \, \forall \tau \in [s, t] \, \gamma_s(\tau) = x_{p_0 + \varepsilon v}^{p_0 + \varepsilon v}(\tau) \right\}.$$

Assume that f_{p_0} is a Lipschitz continuous function and for p in a neighborhood of p_0, f_p is bounded on $\mathcal{T} := \{(\gamma_t(s), s) : t \in [\bar{t}, T], \, s \in [t, T]\}$. Then, there exists a constant $C > 0$ such that

$$\limsup_{\varepsilon \to 0^+} \left| \frac{x_{p_0 + \varepsilon v}(T) - x_{p_0}(T)}{\varepsilon} \right| \leq \left| y_{p_0}^{p_0, v}(T) \right|$$

$$+ \liminf_{\varepsilon \to 0} \frac{C}{\varepsilon} \int_0^T \left| f_{p_0 + \varepsilon v}(x_{p_0 + \varepsilon v}^{p_0 + \varepsilon v}(t), t) - f_{p_0}(x_{p_0 + \varepsilon v}^{p_0 + \varepsilon v}(t), t) \right| dt.$$

Moreover, if

$$\liminf_{\varepsilon \to 0^+} \left| \frac{x_{p_0 + \varepsilon v}(T) - x_{p_0}(T)}{\varepsilon} \right| = \limsup_{\varepsilon \to 0^+} \left| \frac{x_{p_0 + \varepsilon v}(T) - x_{p_0}(T)}{\varepsilon} \right|,$$

then

$$\left| y_{p_0, v}(T) \right| \leq \liminf_{\varepsilon \to 0} \frac{C}{\varepsilon} \int_0^T \left| f_{p_0 + \varepsilon v}(x_{p_0 + \varepsilon v}^{p_0 + \varepsilon v}(t), t) - f_{p_0}(x_{p_0 + \varepsilon v}^{p_0 + \varepsilon v}(t), t) \right| dt$$

$$+ \left| y_{p_0}^{p_0, v}(T) \right|.$$

(2) Fix $v \in P$. Assume that, for every $p \in P$, the vector field f_p is of class C^2 and the limit for $\varepsilon \to 0^+$ of

$$\frac{1}{\varepsilon} \int_0^T M(T, t) \left[f_{p_0 + \varepsilon v}(x_{p_0 + \varepsilon v}^{p_0 + \varepsilon v}(t), t) - f_{p_0}(x_{p_0 + \varepsilon v}^{p_0 + \varepsilon v}(t), t) \right] dt,$$

exists, where $M(s, t)$ is the fundamental matrix solution to the system

$$\frac{\partial M}{\partial s}(s, t) = \frac{\partial}{\partial x} f_{p_0}(\gamma_t(s), s) M(s, t),$$

and $\gamma_t(s)$ is the solution to

$$\begin{cases} \dot{\gamma}_t(s) = f_{p_0}(\gamma(s), s), \\ \gamma_t(t) = x_{p_0 + \varepsilon v}^{p_0 + \varepsilon v}(t). \end{cases}$$

Then

$$y_{p_0,v}(T) = \lim_{\varepsilon \to 0} \frac{1}{\varepsilon} \int_0^T M(T,t) \left[f_{p_0+\varepsilon v}(x_{p_0+\varepsilon v}^{p_0+\varepsilon v}(t), t) - f_{p_0}(x_{p_0+\varepsilon v}^{p_0+\varepsilon v}(t), t) \right] dt$$
$$+ y_{p_0}^{p_0,v}(T).$$

Remark 3.1. Notice that the fundamental matrix solution $M(t,s)$ depends also on ε.

Theorem 3.2. *Fix $T > 0$, $p_0 \in P$ and $v \in P$. For every $\varepsilon > 0$ and $t \in [0,T]$, consider $x_{p_0+\varepsilon v}^{p_0+\varepsilon v}(\cdot)$ and assume that the Cauchy problem*

$$\begin{cases} \dot{\gamma}(s) = f_{p_0}(\gamma(s), s), \\ \gamma(t) = x_{p_0+\varepsilon v}^{p_0+\varepsilon v}(t), \end{cases}$$

admits a unique solution $\gamma_t : [t,T] \to \mathbb{R}^n$. Suppose that the function

$$p \mapsto x_{p_0}^p(T)$$

is Gateaux differentiable at p_0.

(1) Assume that the functions $w(t) = x_{p_0+\varepsilon v}^{p_0+\varepsilon v}(t)$ and $\gamma_t(s)$ satisfy for some $C > 0$

    (A1) for every $0 \le t_1 \le t_2 \le T$

$$\left| \gamma_{t_2}(T) - \gamma_{t_1}(T) \right| \le C \left| \gamma_{t_2}(t_2) - \gamma_{t_1}(t_2) \right| + o(|t_2 - t_1|); \qquad (13)$$

    (A2) there exists $h > 0$ s.t. for every $t \in [0, T[$

$$\left| w(t+h) - \gamma_t(t+h) \right| \le Ch. \qquad (14)$$

Then

$$\limsup_{\varepsilon \to 0^+} \left| \frac{x_{p_0+\varepsilon v}(T) - x_{p_0}(T)}{\varepsilon} \right| \le \left| y_{p_0}^{p_0,v}(T) \right|$$
$$+ \liminf_{\varepsilon \to 0} \frac{C}{\varepsilon} \int_0^T \left| f_{p_0+\varepsilon v}(x_{p_0+\varepsilon v}^{p_0+\varepsilon v}(t), t) - f_{p_0}(x_{p_0+\varepsilon v}^{p_0+\varepsilon v}(t), t) \right| dt.$$

Moreover, if

$$\liminf_{\varepsilon \to 0^+} \left| \frac{x_{p_0+\varepsilon v}(T) - x_{p_0}(T)}{\varepsilon} \right| = \limsup_{\varepsilon \to 0^+} \left| \frac{x_{p_0+\varepsilon v}(T) - x_{p_0}(T)}{\varepsilon} \right|,$$

then

$$\left| y_{p_0,v}(T) \right| \le \liminf_{\varepsilon \to 0} \frac{C}{\varepsilon} \int_0^T \left| f_{p_0+\varepsilon v}(x_{p_0+\varepsilon v}^{p_0+\varepsilon v}(t), t) - f_{p_0}(x_{p_0+\varepsilon v}^{p_0+\varepsilon v}(t), t) \right| dt$$
$$+ \left| y_{p_0}^{p_0,v}(T) \right|.$$

(2) *Assume that, for every $\varepsilon > 0$, the functions $w(t) = x_{p_0+\varepsilon v}^{p_0+\varepsilon v}(t)$ and $\gamma_t(s)$ satisfy*

   (B1) there exists $\bar{h} > 0$ and a positive constant C such that

$$\sup_{0<h<\bar{h}} \int_0^T \frac{|w(t+h) - \gamma_t(t+h)|}{h} \leq C$$

   (B2) there exists a Radon measure μ_ε on $[0,T]$ such that the function

$$t \mapsto \frac{w(t+h) - \gamma_t(t+h)}{h}$$

   weakly converges, as $h \to 0^+$, to μ_ε in the space of Radon measures, seen as the dual of C_0;*

   (B3) for every h there exists a continuous function $\alpha_h : [0,T] \to \mathbb{R}^{n\times n}$ such that

$$\alpha_h \to \alpha_\varepsilon$$

   uniformly on $[0,T]$ as $h \to 0$ and

$$\gamma_{t+h}(T) - \gamma_t(T) = \alpha_h(t)\left(\gamma_{t+h}(t+h) - \gamma_t(t+h)\right) + o(h).$$

If the limit for $\varepsilon \to 0^+$ of $\frac{1}{\varepsilon} \int_0^T \alpha_\varepsilon(t)d\mu_\varepsilon(t)$ exists, then

$$y_{p_0,v}(T) = \lim_{\varepsilon \to 0} \frac{1}{\varepsilon} \int_0^T \alpha_\varepsilon(t)d\mu_\varepsilon(t) + y_{p_0}^{p_0,v}(T).$$

(3) *Assume that, for every $\varepsilon > 0$, the functions $w(t) = x_{p_0+\varepsilon v}^{p_0+\varepsilon v}(t)$ and $\gamma_t(s)$ satisfy*

   (B1') the function $\psi(t) := \gamma_t(T)$ is approximately continuous at 0, i.e.

$$\lim_{h\to 0^+} \frac{1}{h} \int_0^h \psi(s)ds = \psi(0);$$

   (B2') the function

$$t \mapsto \frac{w(t+h) - \gamma_t(t+h)}{h}$$

   is continuous in $[0,T]$ for h sufficiently small and converges uniformly to a continuous function m_ε on $[0,T]$ as $h \to 0^+$;

   (B3') for every $h > 0$ there exists in $L^1(0,T)$ a function $\alpha_h : [0,T] \to \mathbb{R}^{n\times n}$ such that

$$\alpha_h \rightharpoonup^* \alpha_\varepsilon, \quad \text{as } h \to 0$$

   where α_ε is a Radon measure on $[0,T]$ and the convergence is in the weak topology. Moreover,*

$$\gamma_{t+h}(T) - \gamma_t(T) = \alpha_h(t)\left(\gamma_{t+h}(t+h) - \gamma_t(t+h)\right) + o(h).$$

If the limit for $\varepsilon \to 0^+$ of $\frac{1}{\varepsilon} \int_0^T m_\varepsilon(t) d\alpha_\varepsilon(t)$ exists, then

$$y_{p_0,v}(T) = \lim_{\varepsilon \to 0} \frac{1}{\varepsilon} \int_0^T m_\varepsilon(t) d\alpha_\varepsilon(t) + y_{p_0}^{p_0,v}(T).$$

3.1. *Technical proofs*

This subsection deals with the proofs of Theorems 3.1 and 3.2. We now proceed with technical lemmas giving boundedness and estimates of trajectories produced by variations.

Lemma 3.1. *Fix $T > 0$ and a continuous function $w : [0,T] \to \mathbb{R}^n$. Assume that, for every $t \in [0,T]$, there exists a continuous curve $\gamma_t : [t,T] \to \mathbb{R}^n$ such that $\gamma_t(t) = w(t)$ and the function $(s,t) \mapsto \gamma_t(s)$ is measurable. Moreover, suppose that there exists a positive constant $C > 0$ such that (A1) and (A2) of Theorem 3.2 hold. Then*

$$\big|w(T) - \gamma_0(T)\big| \le C \int_0^T \liminf_{h \to 0^+} \frac{\big|w(t+h) - \gamma_t(t+h)\big|}{h} dt. \tag{15}$$

Proof. The assumptions on w and γ imply that the function

$$\phi(t) := \liminf_{h \to 0^+} \frac{\big|w(t+h) - \gamma_t(t+h)\big|}{h}$$

is measurable. Define

$$\psi(t) := \big|\gamma_t(T) - \gamma_0(T)\big|.$$

Clearly, $\psi(t)$ is a measurable function. Moreover, for every $t \in [t,T]$

$$\begin{aligned}
\big|\psi(t+h) - \psi(t)\big| &\le \big|\gamma_{t+h}(T) - \gamma_t(T)\big| \\
&\le C\big|\gamma_{t+h}(t+h) - \gamma_t(t+h)\big| + o(h) \\
&= C\big|w(t+h) - \gamma_t(t+h)\big| + o(h) \\
&\le C^2 h + o(h).
\end{aligned}$$

Therefore ψ is a Lipschitz continuous function and consequently ψ is differentiable almost everywhere. Moreover,

$$\big|\dot\psi(t)\big| \le C\phi(t)$$

for almost every $t \in [0,T]$. Define the Lipschitz continuous function

$$x(t) := \psi(t) - \int_0^t \phi(s) ds.$$

It is clear that $\dot{x}(t) \leq (C-1)\phi(t)$ for almost every $t \in [0,T]$ and $x(0) = 0$. Thus

$$x(T) \leq (C-1) \int_0^T \phi(t)dt,$$

and so

$$\psi(T) \leq C \int_0^T \phi(t)dt,$$

which is the thesis. $\qquad\square$

Lemma 3.2. *Fix $T > 0$ and a measurable function $w : [0,T] \to \mathbb{R}^n$. Assume that, for every $t \in [0,T]$ there exists a curve $\gamma_t : [t,T] \to \mathbb{R}^n$ such that $\gamma_t(t) = w(t)$ and the function $(s,t) \mapsto \gamma_t(s)$ is measurable. Moreover, suppose that, for the functions w and γ_t, hypotheses (B1), (B2) and (B3) of Theorem 3.2 hold, with α and μ in the place, respectively, of α_ε and μ_ε. Then*

$$w(T) - \gamma_0(T) = \int_0^T \alpha(t)d\mu.$$

Proof. We prolong $\gamma_s(t)$ and $w(t)$ in the following way. We assume that

(1) $w(t) = w(T)$ for every $t \geq T$;
(2) $\gamma_s(t) = \gamma_T(T)$ for every $s, t \geq T$;
(3) $\gamma_s(t) = \gamma_s(T)$ for every $s \in [0,T]$ and $t \geq T$.

Define the function

$$\psi(t) = \gamma_t(T) - \gamma_0(T).$$

We have

$$\frac{\psi(t+h) - \psi(t)}{h} = \alpha_h(t)\left(\frac{w(t+h) - \gamma_t(t+h)}{h}\right) + \frac{o(h)}{h}.$$

It thus holds

$$\int_0^T \frac{\psi(t+h) - \psi(t)}{h}dt = \int_0^T \alpha_h(t)\frac{w(t+h) - \gamma_t(t+h)}{h}dt + \frac{o(h)}{h}T. \quad (16)$$

In particular, ψ is a BV function. Indeed, for some $L > 0$

$$\sup_{0<h<\bar{h}} \int_0^T \frac{|\psi(t+h) - \psi(t)|}{h}dt \leq L \sup_{0<h<\bar{h}} \int_0^T \frac{|w(t+h) - \gamma_t(t+h)|}{h} + T\frac{o(h)}{h}$$

$$\leq LC + T\frac{o(h)}{h}.$$

Passing to the limit when $h \to 0^+$ in (16) we get

$$\int_0^T d\psi(t) = \int_0^T \alpha(t)d\mu,$$

where we use the BV property of ψ, and the hypotheses (B2) and (B3). Hence

$$\psi(T) = \int_0^T \alpha(t)d\mu(t).$$

The proof is completed. $\qquad\square$

The next lemma is dual with respect to the previous one, since the convergence required in hypotheses (B2') and (B3') are in the dual topologies with respect to the convergences required in hypotheses (B2) and (B3) of the previous lemma.

Lemma 3.3. *Fix $T > 0$ and a measurable function $w : [0,T] \to \mathbb{R}^n$. Assume that, for every $t \in [0,T]$ there exists a curve $\gamma_t : [t,T] \to \mathbb{R}^n$ such that $\gamma_t(t) = w(t)$ and the function $(s,t) \mapsto \gamma_t(s)$ is measurable. Moreover, suppose that, for the functions w and γ_t, hypotheses (B1'), (B2') and (B3') of Theorem 3.2 hold, with α and m in the place, respectively, of α_ε and m_ε. Then*

$$w(T) - \gamma_0(T) = \int_0^T md\alpha.$$

Proof. We follow the same proof of Lemma 3.2. We prolong the functions $w(t)$ and $\gamma_s(t)$ as before. One can write

$$\int_0^T \frac{\psi(t+h) - \psi(t)}{h} dt = \frac{1}{h}\left[\int_T^{T+h} \psi(s)ds - \int_0^h \psi(s)ds\right].$$

By assumptions we have that

$$\lim_{h \to 0^+} \frac{1}{h}\left[\int_T^{T+h} \psi(s)ds - \int_0^h \psi(s)ds\right] = \psi(T) - \psi(0)$$

and the limit as $h \to 0^+$ of

$$\int_0^T \alpha_h(t)\frac{w(t+h) - \gamma_t(t+h)}{h} dt$$

exists and is finite. So the conclusion easily follows. $\qquad\square$

The next corollary is an easy consequence of the previous lemma.

Corollary 3.1. *Assume the same hypotheses as in Lemma 3.2. If for every $t \in [0, T]$ there exists $\Lambda_{T,t}$, a collection of linear maps $L_{T,t} : \mathbb{R}^n \to \mathbb{R}^n$, such that $\alpha_h(s) \in \Lambda_{T,s+h}$ for $s \in [0, T]$ and h sufficiently small, then*

$$w(T) \in \gamma_0(T) + \int_0^T \bar{\Lambda}\, d\mu,$$

where

$$\int_0^T \bar{\Lambda}\, d\mu(t) = \left\{ \int_0^T \alpha(t) d\mu : \alpha \in \bar{\Lambda},\ \alpha \text{ continuous} \right\}$$

and $\bar{\Lambda}$ is the closure of the set $\cup_{t \in [0,T]} \Lambda_{T,t}$.

Proposition 3.1. *Fix $v \in P$ and $\varepsilon > 0$. Consider a trajectory $x_{p_0+\varepsilon v}^{p_0+\varepsilon v}$. For every $t \in [0, T]$ assume that the Cauchy problem*

$$\begin{cases} \dot{\gamma}(s) = f_{p_0}(\gamma(s), s), \\ \gamma(t) = x_{p_0+\varepsilon v}^{p_0+\varepsilon v}(t). \end{cases} \tag{17}$$

admits a unique solution $\gamma_t : [t, T] \to \mathbb{R}^n$. Define

$$\bar{t} = \sup \left\{ t \in [0, T] : \forall s \in [0, t]\, \forall \tau \in [s, t]\, \gamma_s(\tau) = x_{p_0+\varepsilon v}^{p_0+\varepsilon v}(\tau) \right\}.$$

Assume that f_{p_0} is a Lipschitz continuous function and for p in a neighborhood of p_0, f_p is bounded on

$$\mathcal{T} := \{(\gamma_t(s), s) : t \in [\bar{t}, T],\ s \in [t, T]\}.$$

Then there exists $C > 0$ such that

$$\left| x_{p_0+\varepsilon v}^{p_0+\varepsilon v}(T) - x_{p_0}^{p_0+\varepsilon v}(T) \right| \leq C \int_0^T \left| f_{p_0+\varepsilon v}(x_{p_0+\varepsilon v}^{p_0+\varepsilon v}(t), t) - f_{p_0}(x_{p_0+\varepsilon v}^{p_0+\varepsilon v}(t), t) \right| dt.$$

Proof. We apply Lemma 3.1 in the case where the function w is equal to $x_{p_0+\varepsilon v}^{p_0+\varepsilon v}$. We have

$$\left| \gamma_{t_2}(T) - \gamma_{t_1}(T) \right| \leq \left| \gamma_{t_2}(t_2) - \gamma_{t_1}(t_2) \right| + \int_{t_2}^T \left| f_{p_0}(\gamma_{t_2}(s), s) - f_{p_0}(\gamma_{t_1}(s), s) \right| ds$$

$$\leq \left| \gamma_{t_2}(t_2) - \gamma_{t_1}(t_2) \right| + L \int_{t_2}^T \left| \gamma_{t_2}(s) - \gamma_{t_1}(s) \right| ds,$$

where we use the Lipschitzianity of f_{p_0}. It follows from Gronwall's Lemma that (A1) is fulfilled. Moreover we get

$$\left| w(t+h) - \gamma_t(t+h) \right| \leq \int_t^{t+h} \left| f_{p_0+\varepsilon v}(w(s), s) - f_{p_0}(\gamma_t(s), s) \right| ds \leq 2Ch,$$

where C is defined by

$$C = \sup \left\{ f_{p_0 + \varepsilon v}(T), f_{p_0}(T) \right\},$$

which is finite from the assumptions. So also (A2) is satisfied.

Therefore

$$\left| x_{p_0 + \varepsilon v}^{p_0 + \varepsilon v}(T) - x_{p_0}^{p_0 + \varepsilon v}(T) \right| \leq C \int_0^T \liminf_{h \to 0} \frac{\left| w(t+h) - \gamma_t(t+h) \right|}{h} dt$$

$$\leq C \int_0^T \liminf_{h \to 0} \int_t^{t+h} \frac{\left| f_{p_0 + \varepsilon v}(w(s), s) - f_{p_0}(\gamma_t(s), s) \right|}{h} ds$$

$$\leq C \int_0^T \left| f_{p_0 + \varepsilon v}(w(s), s) - f_{p_0}(w(s), s) \right| ds$$

$$+ L \int_0^T \liminf_{h \to 0} \frac{1}{h} \int_t^{t+h} \left| w(s) - \gamma_t(s) \right| ds$$

$$= C \int_0^T \left| f_{p_0 + \varepsilon v}(w(s), s) - f_{p_0}(w(s), s) \right| ds,$$

and the proof is completed. $\qquad\qquad\qquad\qquad\qquad\qquad\qquad\qquad\square$

Remark 3.2. If the Cauchy problem (17) has not a unique solution for every $t \in [0, T]$, then Proposition 3.1 can also be applied with $\gamma_0(T)$ instead of $x_{p_0}^{p_0 + \varepsilon v}(T)$. So, in order to evaluate $\left| x_{p_0 + \varepsilon v}^{p_0 + \varepsilon v}(T) - x_{p_0}^{p_0 + \varepsilon v}(T) \right|$ one has to consider also the additional term $\left| \gamma_0(T) - x_{p_0}^{p_0 + \varepsilon v}(T) \right|$.

We can now proof the main results about the estimate of $y_{p_0, v}(t)$.

Proof of Theorem 3.1. Consider first point 1. By Proposition 3.1 we have

$$\left| x_{p_0 + \varepsilon v}^{p_0 + \varepsilon v}(T) - x_{p_0}^{p_0 + \varepsilon v} \right|(T) \leq C \int_0^T \left| f_{p_0 + \varepsilon v}(x_{p_0 + \varepsilon v}^{p_0 + \varepsilon v}(t), t) - f_{p_0}(x_{p_0 + \varepsilon v}^{p_0 + \varepsilon v}(t), t) \right| dt$$

and so, since $\varepsilon > 0$,

$$\frac{\left| x_{p_0 + \varepsilon v}^{p_0 + \varepsilon v}(T) - x_{p_0}^{p_0 + \varepsilon v}(T) \right|}{\varepsilon} \leq \frac{C}{\varepsilon} \int_0^T \left| f_{p_0 + \varepsilon v}(x_{p_0 + \varepsilon v}^{p_0 + \varepsilon v}(t), t) - f_{p_0}(x_{p_0 + \varepsilon v}^{p_0 + \varepsilon v}(t), t) \right| dt.$$

On the other hand, by assumption of Gateaux differentiability, we get

$$\lim_{\varepsilon \to 0^+} \frac{x_{p_0}^{p_0 + \varepsilon v}(T) - x_{p_0}^{p_0}(T)}{\varepsilon} = y_{p_0}^{p_0, v}(T).$$

So the first part of the theorem clearly holds.

Consider now point 2. Example 2.1 shows that

$$x_{p_0+\varepsilon v}^{p_0+\varepsilon v}(T) - x_{p_0}^{p_0+\varepsilon v}(T)$$

$$= \int_0^T M(T,t)\left[f_{p_0+\varepsilon v}(x_{p_0+\varepsilon v}^{p_0+\varepsilon v}(t),t) - f_{p_0}(x_{p_0+\varepsilon v}^{p_0+\varepsilon v}(t),t)\right]dt.$$

The assumption of Gateaux differentiability permits to conclude. $\square$

Proof of Theorem 3.2. Consider first point 1. By Lemma 3.1 we have

$$\left|x_{p_0+\varepsilon v}^{p_0+\varepsilon v}(T) - x_{p_0}^{p_0+\varepsilon v}(T)\right| \leq C\int_0^T \left|f_{p_0+\varepsilon v}(x_{p_0+\varepsilon v}^{p_0+\varepsilon v}(t),t) - f_{p_0}(x_{p_0+\varepsilon v}^{p_0+\varepsilon v}(t),t)\right|dt.$$

The assumption of Gateaux differentiability permits to conclude in the same way as in the proof of the previous theorem.

Consider now point 2. By Lemma 3.2 there exist a continuous function α_ε and a Radon measure μ_ε such that

$$x_{p_0+\varepsilon v}^{p_0+\varepsilon v}(T) - x_{p_0}^{p_0+\varepsilon v}(T) = \int_0^T \alpha_\varepsilon(t)d\mu_\varepsilon(t).$$

The proof of the point 3 is similar to that of the point 2. $\square$

Remark 3.3. If in point 2 of Theorem 3.2 we use Lemma 3.3 we come to the conclusion

$$y_{p_0,v} = \int_0^T md\alpha(t) + y_{p_0}^{p_0,v}(T).$$

Remark 3.4. The convergence as $\varepsilon \to 0^+$ of the integrals

$$\frac{1}{\varepsilon}\int_0^T \alpha_\varepsilon(t)d\mu_\varepsilon(t)$$

is assured for example by the following condition. There exists a real number $\beta \in [0,1]$ such that $\alpha_\varepsilon\varepsilon^{-\beta}$ converges uniformly in $\mathcal{C}^0(0,T)$ and $\mu_\varepsilon\varepsilon^{\beta-1}$ weakly* converges to a Radon measure.

4. Approach (b)

To deal with approach (b) we need to consider generalized differentiation theories: we indicate by $\mathcal{D}$ a generalized differentiation theory and by $\mathcal{D}$-differential the corresponding generalized differential. We write $\Lambda \in \mathcal{D}(F;x,y;C)$ which means that Λ is a $\mathcal{D}$-differential of a set-valued map $F:\mathbb{R}^n \twoheadrightarrow \mathbb{R}^n$ at the point (x,y) of the graph of F in the direction C, where C is a closed convex cone in $\mathbb{R}^n$. The object $\mathcal{D}(F;x,y;C)$ is required to be a set of nonempty compact subsets of $\mathrm{Lin}(\mathbb{R}^n, \mathbb{R}^n)$, the space of linear maps from $\mathbb{R}^n$ to $\mathbb{R}^n$. For examples of generalized differentiation theories see Refs. 11,14,23,28.

For every $p \in P$, define the possibly multivalued map $\Phi_p^{t_0,t_1}(\cdot) = \Phi_p(t_0, t_1, \cdot)$ as the flow map of the vector field f_p, where $t_0 \leq t_1$ are respectively the initial and final time.

Definition 4.1. Fix $p \in P$. We say that Λ_p is a minimal generalized differential of the flow map $\Phi_p^{t_0,t_1}(\cdot)$ at the point (x_1, x_2) where $x_2 \in \Phi_p^{t_0,t_1}(x_1)$, if for every γ solution to

$$\begin{cases} \dot{\gamma}(t) = f_p(\gamma(t), t), \\ \gamma(t_0) = x, \end{cases}$$

there exists $L \in \Lambda_p$ such that

$$\gamma(t_1) - x_2 = L(x - x_1) + o(|x - x_1|).$$

We indicate by $min\mathcal{D}(\Phi_p^{t_0,t_1}(\cdot); x_1, x_2)$ the set of all minimal $\mathcal{D}$-differentials.

Theorem 4.1. *Fix $p_0 \in P$, $v \in P$, and $v_1 \in \mathbb{R}^n$. Assume that, for every $\varepsilon > 0$, there exists a trajectory $x_{p_0+\varepsilon v}^{p_0}$ such that the followings hold.*

(1) $\bar{x}_{p_0+\varepsilon v} = \bar{x}_{p_0} + \varepsilon v_1 + o(\varepsilon)$.
(2) There exists $\Lambda_{p_0+\varepsilon v} \in min\mathcal{D}(\Phi_{p_0+\varepsilon v}^{0,T}; \bar{x}_{p_0}, x_{p_0+\varepsilon v}^{p_0}(T))$.
(3) There exists a compact set Λ of linear functions from $\mathbb{R}^n$ to $\mathbb{R}^n$ such that $\cup_{\varepsilon > 0} \Lambda_{p_0+\varepsilon v} \subseteq \Lambda$.
(4) For every $t \in [0, T]$, the Cauchy problem

$$\begin{cases} \dot{\gamma}(s) = f_{p_0}(\gamma(s), s), \\ \gamma(t) = x_{p_0+\varepsilon v}^{p_0}(t), \end{cases}$$

admits a unique solution $\gamma_t : [t, T] \to \mathbb{R}^n$ such that:

(a) there exist $K > 0$ and $\bar{h} > 0$ such that, for every $t \in [0, T]$

$$\sup_{0 < h < \bar{h}} \int_0^T \frac{\left| x_{p_0+\varepsilon v}^{p_0}(t + h) - \gamma_t(t + h) \right|}{h} \leq K;$$

(b) there exists a Radon measure μ_ε on $[0, T]$ such that the function

$$t \mapsto \frac{x_{p_0+\varepsilon v}^{p_0}(t + h) - \gamma_t(t + h)}{h}$$

weakly converges to μ_ε in the sense of measures as $h \to 0^+$;*

194

*(c) for every h there exists a continuous function $\alpha_h : [0,T] \to \mathbb{R}^{n \times n}$
such that*

$$\alpha_h \to \alpha_\varepsilon$$

uniformly on $[0,T]$ as $h \to 0$ and

$$\gamma_{t+h}(T) - \gamma_t(T) = \alpha_h(t)\left(\gamma_{t+h}(t+h) - \gamma_t(t+h)\right) + o(h).$$

(5) The real number $\frac{1}{\varepsilon}\int_0^T \alpha_\varepsilon(t)d\mu_\varepsilon$ converges as $\varepsilon \to 0^+$.

Then there exists $L \in \Lambda$ and a sequence $\varepsilon_n \to 0$ as $n \to +\infty$ such that

$$\lim_{n \to +\infty} \frac{x_{p_0+\varepsilon_n v}(T) - x_{p_0}(T)}{\varepsilon_n} = Lv + \lim_{n \to +\infty} \frac{1}{\varepsilon_n}\int_0^T \alpha_{\varepsilon_n}d\mu_{\varepsilon_n}.$$

Proof. By the definition of a minimal generalized differential we have that,
for every $\varepsilon > 0$,

$$x_{p_0+\varepsilon v}^{p_0+\varepsilon v}(T) - x_{p_0+\varepsilon v}^{p_0}(T) = L_\varepsilon(\bar{x}_{p_0+\varepsilon v} - \bar{x}_{p_0}) + o(|\bar{x}_{p_0+\varepsilon v} - \bar{x}_{p_0}|)$$
$$= L_\varepsilon(\varepsilon v_1 + o(\varepsilon)) + o(|\bar{x}_{p_0+\varepsilon v} - \bar{x}_{p_0}|).$$

Dividing by $\varepsilon > 0$ we get

$$\frac{x_{p_0+\varepsilon v}^{p_0+\varepsilon v}(T) - x_{p_0+\varepsilon v}^{p_0}(T)}{\varepsilon} = L_\varepsilon v_1 + \frac{o(\varepsilon)}{\varepsilon}.$$

By assumptions, L_ε belongs to the compact set Λ and hence there exists
a sequence ε_n and $L \in \Lambda$ such that $L_{\varepsilon_n} \to L$. Therefore

$$\lim_{n \to +\infty} \frac{x_{p_0+\varepsilon_n v}^{p_0+\varepsilon_n v}(T) - x_{p_0+\varepsilon_n v}^{p_0}(T)}{\varepsilon_n} = Lv_1.$$

Applying Lemma 3.2 with $w = x_{p_0+\varepsilon_n v}^{p_0}$, we obtain that

$$x_{p_0+\varepsilon_n v}^{p_0}(T) - x_{p_0}^{p_0}(T) = \int_0^T \alpha_{\varepsilon_n}(t)d\mu_{\varepsilon_n}(t)$$

and so

$$\lim_{n \to +\infty} \frac{x_{p_0+\varepsilon_n v}(T) - x_{p_0}(T)}{\varepsilon_n} = Lv_1 + \lim_{n \to +\infty} \frac{1}{\varepsilon_n}\int_0^T \alpha_{\varepsilon_n}d\mu_{\varepsilon_n}.$$

This concludes the proof. $\qquad\square$

Remark 4.1. Notice that the previous theorem holds even if we substitute
assumption 4 according to the hypotheses of Theorems 3.1 and 3.2.

Moreover we can substitute assumption 5 of the theorem with a hypothesis on the strong convergence of α_ε in a topological space and on the weak convergence of measures μ_ε in the dual space.

5. Applications of the main results

In this section we present some scalar examples that clarify how to apply the results of previous sections. This choice follows from the fact that generalizations to the vector-valued case are trivial, but would render this part more involved and hard to be followed.

Example 5.1. (Differentiability with respect to the vector field). Fix a time dependent family of discontinuous vector fields $f_\varepsilon : \mathbb{R} \times \mathbb{R} \to \mathbb{R}^+$ ($\varepsilon \geq 0$) with exactly one discontinuity at $\bar{x} > 0$ as in (2) and $\lim_{x \to \bar{x}^+} f_\varepsilon(x,t) > 0$ for every $\varepsilon \geq 0$ and $t \geq 0$. Define x_ε as in (3), (4). Fix $\varepsilon > 0$, set $w(\cdot) = x_\varepsilon(\cdot)$ and, for every $t \in [0,T]$ consider a solution $\gamma_t(\cdot)$ to

$$\begin{cases} \frac{d}{dt}\gamma_t(s) = f_0(\gamma_t(s), s), & \text{if } t < s \leq T, \\ \gamma_t(t) = w(t) = x_\varepsilon(t). \end{cases}$$

We have that

$$\frac{|w(t+h) - \gamma_t(t+h)|}{h} = \left| \frac{1}{h} \int_t^{t+h} [f_\varepsilon(w(s), s) - f_0(\gamma_t(s), s)]\, ds \right|$$

which converges pointwise to

$$\left| f_\varepsilon(w(t), t) - f_0(w(t), t) \right|$$

as $h \to 0^+$. More precisely we have that the convergence is uniform on the intervals $[0, \bar{t}]$ and $[\bar{t}, T]$, where $\bar{t}$ satisfies $w(\bar{t}) = \bar{x}$.

Moreover the function $\alpha_h(t)$, defined in Theorem 3.2, is given by

$$1 + \frac{1}{\gamma_{t+h}(t+h) - \gamma_t(t+h)} \cdot$$

$$\cdot \left[\chi_{[0,t_\varepsilon^h]}(t+h) \int_{t+h}^{t_\varepsilon^h} f_0^-(\gamma_{t+h}(s), s)ds - \chi_{[0,t_\varepsilon^0]}(t+h) \int_{t+h}^{t_\varepsilon^0} f_0^-(\gamma_t(s), s)ds \right]$$

$$+ \frac{1}{\gamma_{t+h}(t+h) - \gamma_t(t+h)} \cdot$$

$$\cdot \left[\int_{\max\{t_\varepsilon^h, t+h\}}^T f_0^+(\gamma_{t+h}(s), s)ds - \int_{\max\{t_\varepsilon^0, t+h\}}^T f_0^+(\gamma_t(s), s)ds \right],$$

where t_ε^h and t_ε^0 are respectively the times satisfying $\gamma_{t+h}(t_\varepsilon^h) = \bar{x}$ and $\gamma_t(t_\varepsilon^0) = \bar{x}$. Therefore, the boundness of vector fields implies the boundness of $\alpha_h(t)$ in the space of Radon measures and so we have compactness of $\alpha_h(t)$ with respect to the weak* convergence. Hence point 3 of Theorem 3.2 can be applied.

Example 5.2. (Switching times). Fix f_1 and f_2, two bounded $\mathcal{C}^1$ vector fields on $\mathbb{R}$, and $0 < \bar{t} < T$. Let $x(t)$ be defined as follows. On $[0, \bar{t}]$, $x(t)$ satisfies

$$\begin{cases} \dot{x}(t) = f_1(x(t), t), \\ x(0) = x_0, \end{cases}$$

where $x_0 \in \mathbb{R}$, while on $[\bar{t}, T]$, $x(t)$ satisfies

$$\dot{x}(t) = f_2(x(t), t).$$

We perform a variation in this way. For $\varepsilon > 0$, $x_\varepsilon(t)$ is equal to $x(t)$ if $0 \leq t \leq \bar{t}$. On $[\bar{t}, \bar{t} + \varepsilon]$, $x_\varepsilon(t)$ solves

$$\begin{cases} \dot{x}_\varepsilon(t) = f_1(x_\varepsilon(t), t), \\ x_\varepsilon(\bar{t}) = x(\bar{t}), \end{cases}$$

while on $[\bar{t} + \varepsilon, T]$, $x_\varepsilon(t)$ satisfies

$$\dot{x}_\varepsilon(t) = f_2(x_\varepsilon(t), t).$$

Define the functions $\gamma_t(s)$ in the following way. If $t \leq \bar{t}$, then $\gamma_t(s) = x(s)$ for every $s \in [t, T]$. If $t \geq \bar{t} + \varepsilon$, then $\gamma_t(s) = x_\varepsilon(s)$ for every $s \in [t, T]$. Finally, if $\bar{t} < t < \bar{t} + \varepsilon$, then $\gamma_t(s)$ is the solution to

$$\begin{cases} \dot{\gamma}_t(s) = f_2(\gamma_t(s), s), \\ \gamma_t(t) = x_\varepsilon(t); \end{cases}$$

see Figure 1.

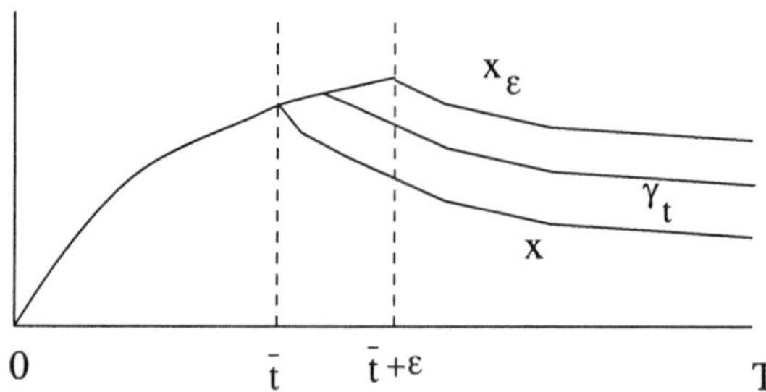

Fig. 1.   The trajectory x_ε and the functions γ_t for a variation on the switching time.

It is easy to check that (B1') and (B2') of Theorem 3.2 hold. More precisely, the term

$$\frac{x_\varepsilon(t + h) - \gamma_t(t + h)}{h}$$

converges uniformly to $f_\varepsilon(x_\varepsilon(t), t) - f_0(x_\varepsilon(t), t)$.

In the same way as in Example 2.1, we deduce also (B3'). Thus we can apply Theorem 3.2.

Example 5.3. (Control theory) Consider the following Cauchy problem

$$\begin{cases} \dot{x}(t) = \sqrt{|x(t)|} + \eta(t)u(t), \\ x(0) = 0, \end{cases} \tag{18}$$

where $u \in [0,1]$ is the control, $t \in [0,T]$ and $\eta \geq 0$ is a $\mathcal{C}^1$ function. Let $u_0 \equiv 0$ and $x_0 \equiv 0$ be a corresponding solution. We perform a variation of the trajectory x_0 and we apply Lemma 3.3. Notice that in this case the differential of the final point of the trajectory with respect to the initial condition is zero, since we take the initial condition always equal to 0.

Fix a function $\alpha_1(\varepsilon) \in \,]0, T[$ and a function $\alpha_2(\varepsilon) \in \,]0, T - \alpha_1(\varepsilon)]$. Define the family of controls

$$u_\varepsilon(t) := \begin{cases} 0, & \text{if } 0 \leq t \leq T - \alpha_1(\varepsilon), \\ \varphi(\varepsilon), & \text{if } T - \alpha_1(\varepsilon) < t \leq T, \end{cases} \tag{19}$$

with $\varphi(\varepsilon) \in [0,1]$ and a corresponding family of trajectories

$$w(t) = x_\varepsilon(t) = \begin{cases} 0, & \text{if } 0 \leq t \leq T - \alpha_1(\varepsilon) - \alpha_2(\varepsilon), \\ \left(\frac{t - T + \alpha_1(\varepsilon) + \alpha_2(\varepsilon)}{2}\right)^2, & \text{if } T - \alpha_1(\varepsilon) - \alpha_2(\varepsilon) \leq t \leq T - \alpha_1(\varepsilon), \\ z(t), & \text{if } T - \alpha_1(\varepsilon) \leq t \leq T, \end{cases} \tag{20}$$

where the function z satisfies

$$\begin{cases} \dot{z}(t) = \sqrt{|z(t)|} + \eta(t)\varphi(\varepsilon), \\ z(T - \alpha_1(\varepsilon)) = \left(\frac{\alpha_2(\varepsilon)}{2}\right)^2; \end{cases}$$

see Figure 2.

Further, for every $t \in [0, T - \alpha_1(\varepsilon)]$, define $\gamma_t : [t, T] \to \mathbb{R}$ by

$$\gamma_t(s) = \begin{cases} w(s), & \text{if } t \leq s \leq T - \alpha_1(\varepsilon), \\ \left(\frac{s - T + \alpha_1(\varepsilon) + \alpha_2(\varepsilon)}{2}\right)^2, & \text{if } T - \alpha_1(\varepsilon) \leq s \leq T, \end{cases} \tag{21}$$

while, for every $t \in [T - \alpha_1(\varepsilon), T]$, define $\gamma_t : [t, T] \to \mathbb{R}$ by the solution to the Cauchy problem

$$\begin{cases} \dot{\gamma}_t(s) = \sqrt{\gamma_t(s)}, \\ \gamma_t(t) = w(t); \end{cases}$$

see Figure 3.

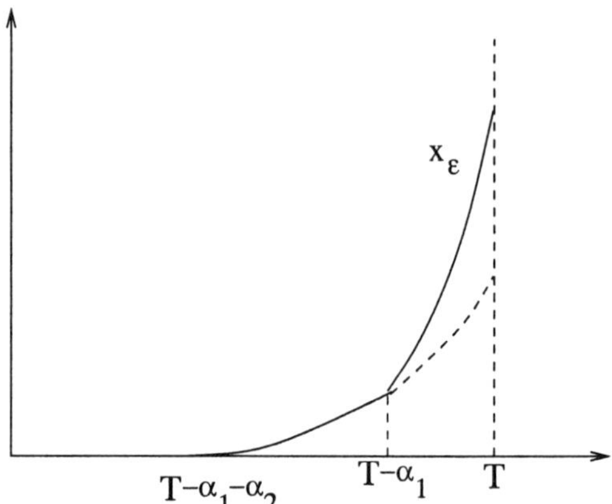

Fig. 2.   The trajectory x_ε.

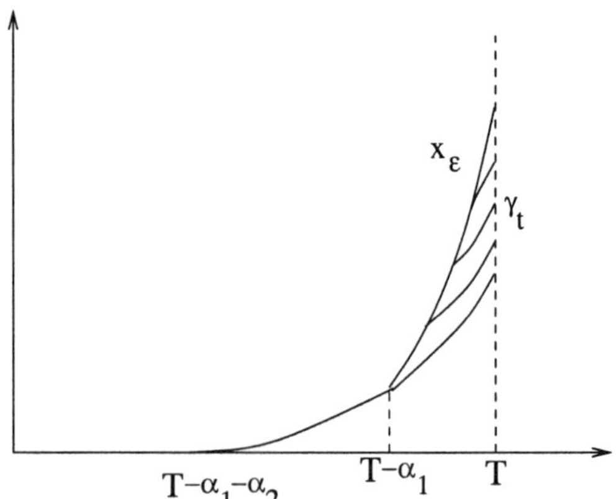

Fig. 3.   The curves γ.

We check the hypotheses (B1'), (B2') and (B3'). In order to simplify the notation we omit the dependence on ε. Note that (B1') is granted by definition. Let us focus on (B2'). Notice that

$$\left| w(t+h) - \gamma_t(t+h) \right| = 0$$

for $t < T - \alpha_1$ and h sufficiently small. If $t \geq T - \alpha_1$, then

$$\frac{w(t+h) - \gamma_t(t+h)}{h} = \frac{1}{h} \int_t^{t+h} \left[\sqrt{w(s)} - \sqrt{\gamma_t(s)} \right] ds + \frac{\varphi}{h} \int_t^{t+h} \eta(s) ds.$$

Passing to the limit with $h \to 0$, the first term of the right hand side of the above equation tends to 0, while the second one tends to $\varphi \eta(t)$ pointwise. Moreover,

$$\frac{d}{dt} \frac{w(t+h) - \gamma_t(t+h)}{h} = \frac{1}{h} \left[\sqrt{w(t+h)} - \sqrt{\gamma_t(t+h)} \right]$$
$$+ \frac{\varphi}{h} \left[\eta(t+h) - \exp\left\{ \int_t^{t+h} \frac{1}{2\sqrt{\gamma_t(s)}} ds \right\} \eta(t) \right].$$

which, for every $t \geq T - \alpha_1$, is bounded when h is sufficiently small. By Ascoli-Arzelà theorem, we conclude that

$$\frac{w(t+h) - \gamma_t(t+h)}{h}$$

converges when $h \to 0^+$ uniformly, and hence also in the sense of measures, to $\varphi \eta(t)$. Thus we get (B2').

Let us pass to (B3'). We have

$$\gamma_{t+h}(T) - \gamma_t(T) = \left(\sqrt{w(t+h)} + \frac{T - t - h}{2} \right)^2$$
$$- \left(\sqrt{w(t)} + \frac{h}{2} + \frac{T - t - h}{2} \right)^2.$$

Thus

$$\alpha_h(t) = \left[1 + (T - t - h) \frac{\sqrt{w(t+h)} - \sqrt{w(t)} - \frac{h}{2}}{w(t+h) - (\sqrt{w(t)} + \frac{h}{2})^2} \right] \chi_{[T-\alpha_1, T]}(t)$$
$$= \left[(1 + (T - t - h) \frac{1}{2\sqrt{w(t)}} + o(h)) \right] \chi_{[T-\alpha_1, T]}(t),$$

where χ denotes the characteristic function of a set. Passing to the limit with $h \to 0$ we conclude that

$$\alpha_h \to \alpha_0,$$

in $L^1(0, T)$, where

$$\alpha_0(t) = \left(1 + \frac{T - t}{2\sqrt{w(t)}} \right) \chi_{[T-\alpha_1, T]}(t).$$

Hence (B3') is satisfied. Lemma 3.3 permits to conclude that

$$w(T) - \gamma_0(T) = \int_0^T \left(1 + \frac{T-t}{2\sqrt{w(t)}}\right) \chi_{[T-\alpha_1,T]}(t)\varphi\eta(t)dt$$

$$= \varphi \int_{T-\alpha_1}^T \left(1 + \frac{T-t}{2\sqrt{w(t)}}\right) \eta(t)dt.$$

We have

$$\frac{w(T) - x_0(T)}{\varepsilon} = \frac{w(T) - \gamma_0(T)}{\varepsilon} + \frac{\gamma_0(T)}{\varepsilon}$$

$$= \frac{\varphi}{\varepsilon} \int_{T-\alpha_1}^T \eta(t)dt + \frac{\varphi}{\varepsilon} \int_{T-\alpha_1}^T \eta(t)\frac{T-t}{2\sqrt{w(t)}}dt + \frac{(\alpha_1+\alpha_2)^2}{4\varepsilon}$$

$$= I_1 + I_2 + I_3.$$

Since $w(t) \geq \frac{\alpha_2^2}{4}$, then

$$I_2 \leq \frac{\varphi}{\varepsilon\alpha_2} \int_{T-\alpha_1}^T \eta(t)(T-t)dt$$

$$\leq \frac{\varphi\alpha_1}{\varepsilon\alpha_2} \int_{T-\alpha_1}^T \eta(t)dt.$$

Consider first the special case when $\eta \equiv 1$, $\varphi \equiv 1$, $\alpha_1 = \varepsilon^{1/2+\beta_1}$ and $\alpha_2 = \varepsilon^{1/2+\beta_2}$, with $\beta_1 \geq 0$ and $\beta_2 \geq 0$. We get

$$I_1 = \varepsilon^{\beta_1-\frac{1}{2}},$$

$$I_2 \leq \varepsilon^{2\beta_1-\beta_2-\frac{1}{2}},$$

$$I_3 = \frac{(\varepsilon^{\beta_1} + \varepsilon^{\beta_2})^2}{4}.$$

Choosing $\beta_1 = \frac{1}{2}$ and $\beta_2 \in]0, \frac{1}{2}[$, we conclude that $I_1 = 1$, and I_2, I_3 tend to 0 as $\varepsilon \to 0^+$; thus we have produced a first order non zero variation.

Let us take now $\eta(t) = (T-t)^{\beta_3}$, $\varphi = \varepsilon^{\beta_4}$, $\alpha_1 = \varepsilon^{1/2+\beta_1}$ and $\alpha_2 = \varepsilon^{1/2+\beta_2}$, with $\beta_1 \geq 0$, $\beta_2 \geq 0$, $\beta_3 \geq 0$ and $\beta_4 \geq 0$. In this case

$$I_1 = \frac{1}{\beta_3 + 1}\varepsilon^{\beta_1+\beta_1\beta_3+\beta_3/2+\beta_4-1/2},$$

$$I_2 \leq \frac{1}{\beta_3 + 2}\varepsilon^{2\beta_1-\beta_2+\beta_1\beta_3+\beta_3/2+\beta_4-1/2},$$

$$I_3 = \frac{(\varepsilon^{\beta_1} + \varepsilon^{\beta_2})^2}{4}.$$

In order to get I_1 constant and I_2, I_3 going to 0 we need

$$\begin{cases} \beta_1 + \beta_1\beta_3 + \beta_3/2 + \beta_4 - 1/2 = 0, \\ 2\beta_1 - \beta_2 + \beta_1\beta_3 + \beta_3/2 + \beta_4 - 1/2 > 0, \\ \beta_1 > 0, \\ \beta_2 > 0, \end{cases} \tag{22}$$

which is equivalent to

$$\begin{cases} \beta_1 = \frac{1-\beta_3-2\beta_4}{2(1+\beta_3)}, \\ 2\beta_2(1+\beta_3) + \beta_3 + 2\beta_4 < 1, \\ \beta_1 > 0, \\ \beta_2 > 0. \end{cases} \tag{23}$$

The first equation implies $\beta_1 \leq \frac{1}{2}$. If $\beta_1 = \frac{1}{2}$, then the first equation of the system implies that $\beta_3 = \beta_4 = 0$, thus $\varphi \equiv 1$ and $\eta \equiv 1$ as before. Finally, choosing $\beta_1 < \frac{1}{2}$ we get infinitely many solutions to (22).

Remark 5.1. If we consider the control system (18) and we want to maximize the cost $x(T)$, then the previous example shows that the trajectory $x \equiv 0$ is not optimal, since the first order variation $x_\varepsilon(t)$ of Example 5.3 satisfies

$$\lim_{\varepsilon \to 0^+} \frac{x_\varepsilon(T)}{\varepsilon} > 0.$$

Remark 5.2. Notice that the classical needle variations are not useful in the above example. Indeed, consider a C^1 function η such that $\eta(t) > 0$ for every $t \in [0, T[$ and $\eta(T) = 0$ (for example $\eta(t) = (T - t)^{\beta_3}$ with $\beta_3 > 0$). Two cases happen.

(1) A needle variation is originated at $t = T$. It means that controls u_ε are defined by

$$u_\varepsilon(t) = \begin{cases} 0, & \text{if } 0 \leq t \leq T - \varepsilon, \\ \varphi, & \text{if } T - \varepsilon < t \leq T. \end{cases}$$

In this case the corresponding trajectories take on the form

$$x_\varepsilon(T) = \varphi\eta(T - \varepsilon)\varepsilon + o(\varepsilon).$$

Thus

$$\frac{x_\varepsilon(T) - x_0(0)}{\varepsilon} = \frac{\varphi\eta(T - \varepsilon)\varepsilon + o(\varepsilon)}{\varepsilon} \to 0,$$

as $\varepsilon \to 0^+$.

(2) A needle variation is originated at $t = \tau < T$. It means that controls u_ε are defined by

$$u_\varepsilon(t) = \begin{cases} 0, & \text{if } 0 \le t \le \tau - \varepsilon, \\ \varphi, & \text{if } \tau - \varepsilon < t \le \tau, \\ 0, & \text{if } \tau - \varepsilon < t \le T. \end{cases}$$

As before we have

$$x_\varepsilon(\tau) = \varphi\eta(\tau - \varepsilon)\varepsilon + o(\varepsilon),$$

while

$$x_\varepsilon(T) = \left(\frac{T - \tau}{2}\right)^2 + \sqrt{\varepsilon}(T - \tau)\sqrt{\varphi\eta(\tau - \varepsilon) + \frac{o(\varepsilon)}{\varepsilon}} + O(\varepsilon).$$

So

$$\frac{x_\varepsilon(T) - x_0(0)}{\varepsilon}$$

is unbounded for $\varepsilon \to 0^+$.

Therefore the needle variation in the first case is not a first order variation, while in the second case is not an admissible variation.

Example 5.4. (Control theory) Consider the following Cauchy problem

$$\begin{cases} \dot{x}(t) = -\sqrt{|x(t)|} + u(t), \\ x(0) = 0, \end{cases}$$

where $u > 0$ is the control, $t \in [0, T]$. Let $u_0 \equiv 0$ and $x_0 \equiv 0$ be a corresponding solution. We perform a variation of this trajectory using the controls

$$u_\varepsilon(t) = \sqrt{\varepsilon}.$$

Thus the trajectories $x_\varepsilon(\cdot)$ satisfy

$$\begin{cases} \dot{x}_\varepsilon(t) = -\sqrt{|x_\varepsilon(t)|} + \sqrt{\varepsilon}, \\ x_\varepsilon(0) = 0, \end{cases}$$

while the functions $\gamma_t(\cdot)$ satisfy

$$\begin{cases} \dot{\gamma}_t(s) = -\sqrt{|\gamma_t(s)|}, \\ \gamma_t(t) = x_\varepsilon(t); \end{cases}$$

see Figure 4.

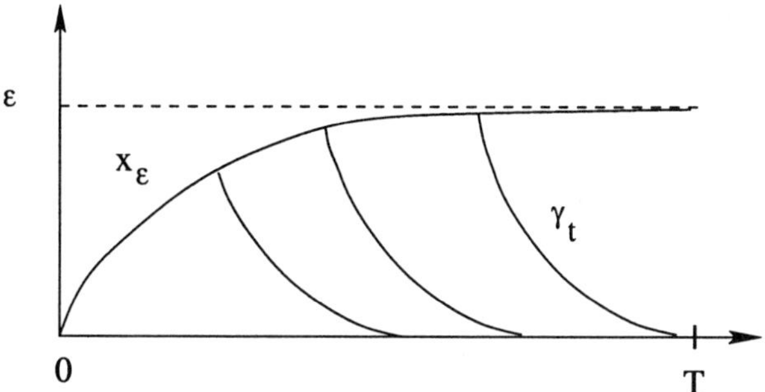

Fig. 4. The trajectory x_ε and the functions γ_t.

We want to apply Lemma 3.3. As in the previous example the differential of the trajectory with respect to the initial condition is zero. The hypothesis (B1') is trivially satisfied. We have

$$x_\varepsilon(t+h) - \gamma_t(t+h) = \int_t^{t+h} \left[-\sqrt{x_\varepsilon(s)} + \sqrt{\gamma_t(s)} \right] ds + \int_t^{t+h} \sqrt{\varepsilon}\, ds$$

and so

$$\frac{x_\varepsilon(t+h) - \gamma_t(t+h)}{h} \to \sqrt{\varepsilon}$$

as $h \to 0^+$. This implies (B2').

We have $x_\varepsilon(t) \leq \varepsilon$ for every $t \geq 0$ and setting $\bar{t} = 2\sqrt{\varepsilon}$

$$\gamma_t(t + \bar{t}) = 0$$

if $\gamma_t(t) \in [0, \sqrt{\varepsilon}]$ Then it is clear that

$$\gamma_t(T) = 0$$

for every $t \in [0, T - 2\sqrt{\varepsilon}]$ and hence $\alpha_h(t)$ is equal to 0 for every $t \in [0, T - 2\sqrt{\varepsilon}[$ and h sufficiently small. Moreover, if T is sufficiently big, then

$$\gamma_{t+h}(T) \sim \left[\sqrt{\varepsilon} - \frac{T - t - h}{2} \right]^2,$$

$$\gamma_t(T) \sim \left[\sqrt{\varepsilon} - \frac{T - t}{2} \right]^2,$$

$$\gamma_{t+h}(t + h) \sim \varepsilon,$$

$$\gamma_t(t + h) \sim \left[\sqrt{\varepsilon} - \frac{h}{2} \right]^2.$$

Therefore we get

$$\alpha_h(t) = \begin{cases} \frac{h+4\sqrt{\varepsilon}-2(T-t)}{4\sqrt{\varepsilon}-h}, & \text{if } t \in [T-2\sqrt{\varepsilon},T], \\ 0, & \text{if } t \in [0,T-2\sqrt{\varepsilon}[, \end{cases}$$

and so $\alpha_h(t) \to \alpha_0(t)$ when $h \to 0^+$ in $L^1(0,T)$, where

$$\alpha_0(t) = \begin{cases} \frac{4\sqrt{\varepsilon}-2(T-t)}{4\sqrt{\varepsilon}}, & \text{if } t \in [T-2\sqrt{\varepsilon},T], \\ 0, & \text{if } t \in [0,T-2\sqrt{\varepsilon}[. \end{cases}$$

Thus (B3') is satisfied. Lemma 3.3 permits to conclude that

$$x_\varepsilon(T) = \int_{T-2\sqrt{\varepsilon}}^{T} \frac{4\sqrt{\varepsilon}-2(T-t)}{4}\,dt = \varepsilon.$$

Remark 5.3. One can deal classical needle variations as before (possibly adding a function η).

Acknowledgments

The second author was partially supported through a European Community Marie Curie Fellowship and in the framework of the CTS, and by Bialystok Technical University under the grant W/WI/2/07.

References

1. A. A. Agrachev, G. Stefani, P. Zezza, *SIAM J. Control Optim.* **41**, 991 (2002).
2. J. P. Aubin, Contingent Derivatives of Set-Valued Maps and Existence of Solutions to Nonlinear Inclusions and Differential Inclusions, in *Mathematical analysis and applications, Part A*, Adv. in Math. Suppl. Stud., Vol. 7, eds. L. Nachbin (Academic Press, Orlando, 1981) pp. 160–232.
3. J. P. Aubin, H. Frankowska, *Set-Valued Analysis*, Systems & Control: Foundations & Applications, Vol. 2 (Birkhäuser, Boston, MA, 1990).
4. R. M. Bianchini, M. Kawski, *SIAM J. Control Optim.* **42**, 218 (2003).
5. R. M. Bianchini, G. Stefani, *SIAM J. Control Optim.* **31**, 900 (1993).
6. V. G. Boltyanskii, *SIAM J. Control* **4**, 326 (1966).
7. A. Bressan, *SIAM J. Control Optim.* **23**, 38 (1985).
8. A. Bressan, *Hyperbolic systems of conservation laws* (Oxford University Press, Oxford, 2000).
9. P. Brunovský, *J. Differential Equations* **38**, 317 (1980).
10. F. H. Clarke, *SIAM J. Control Optim.* **14**, 1078 (1976).
11. F. Clarke, *Optimization and Nonsmooth Analysis*, Classics in Applied Mathematics, Vol. 5, second edition (Society for Industrial and Applied Mathematics (SIAM), Philadelphia, PA, 1990).
12. M. Garavello, B. Piccoli, *SIAM J. Control Optim.* **43**, 1867 (2005).
13. E. Girejko, *Rend. Semin. Mat. Univ. Politec. Torino* **63**, 357 (2005).

14. E. Girejko, B. Piccoli, *Set-Valued Anal.* **15**, 163 (2007).

15. H. Halkin, Necessary conditions for optimal control problems with differentiable or nondifferentiable data, in *Mathematical control theory (Proc. Conf., Australian Nat. Univ., Canberra, 1977)*, Lect. Notes in Math. 680 (Springer-Verlar, Berlin , 1978), pp. 77–118.

16. B. Kaškosz, S. Lojasiewicz Jr., *Nonlinear Anal.* **9**, 109 (1985).

17. M. Kiefer, H. Schätler, *SIAM J. Control Optim.* **37**, 1346 (1999).

18. U. Ledzewicz, A. Nowakowski, H. Schätler, *J. Optim. Theory Appl.* **122**, 345 (2004).

19. B. S. Mordukhovich, *Variational Analysis and Generalized Differentiation I & II*, Grundlehren der mathematischen Wissenschaften, Vol. 330 and 331 (Springer-Verlag, Berlin, 2006).

20. B. Piccoli, H. J. Sussmann, *SIAM J. Control Optim.* **39**, 359 (2000).

21. L. S. Pontryagin, V. G. Boltyanskii, R. V. Gamkrelidze, E. F. Mishchenko, *The mathematical theory of optimal processes*, Translated by D. E. Brown (A Pergamon Press Book. The Macmillan Co., New York 1964).

22. R. T. Rockafollar, R. J. B. Wcts, *Variational analysis*, Grundlehren der Mathematischen Wissenschaften, Vol. 317 (Springer-Verlag, Berlin, 1998).

23. H. J. Sussmann, *New theories of set-valued differentials and new version of the maximum principle of optimal control theory*, in *Nonlinear Control in the Year 2000, Vol. 2*, Lecture Notes in Control and Inform. Sci., 259, eds. A. Isidori, F. Lamnabhi-Lagarrigue and W. Respondek (Springer-Verlag, London, 2000), pp. 487–526.

24. H. J. Sussmann, Path-integral generalized differentials, in *Proceedings of the 41st IEEE 2002 Conference on Decision and Control, Las Vegas, Nevada* (Las Vegas, Nevada, December 10–13, 2002): 1101–1106.

25. H. J. Sussmann, Warga derivate containers and other generalized differentials, in *Proceedings of the 41st IEEE 2002 Conference on Decision and Control* (Las Vegas, Nevada, December 10–13, 2002): 4728–4732.

26. H. J. Sussmann, Generalized differentials, variational generators, and the maximum principle with state contraints, in *Nonlinear and Optimal Control Theory* (Lectures given at the C.I.M.E. Summer School held in Cetraro (Cosenza), June 21–29, 2004), Lecture Notes in Mathematics, eds. G. Stefani and P. Nistri (Springer-Verlag (Fondazione C.I.M.E.), to appear).

27. R. Vinter, *Optimal control*, Systems & Control: Foundations & Applications (Birkhäuser Boston, Inc., Boston, MA, 2000).

28. J. Warga, *SIAM J. Control Optim.* **21**, 837 (1983).

SAMPLED-DATA REDESIGN FOR NONLINEAR MULTI-INPUT SYSTEMS

L. GRÜNE[†] and K. WORTHMANN[‡]

Mathematisches Institut, Fakultät für Mathematik, Physik und Informatik
Universität Bayreuth, 95440 Bayreuth, Germany
[†] *E-mail: lars.gruene@uni-bayreuth.de*
[‡] *E-mail: karl.worthmann@uni.bayreuth.de*

We investigate the sampled-data redesign problem for nonlinear control affine multi-input systems and consider sampled-data feedback laws for which the trajectories of the sampled-data closed loop system converge to the continuous time trajectories with a prescribed rate of convergence as sampling time vanishes. We analyze geometric conditions for the existence of such sampled-data feedback laws and give formulae and algorithms for their computation.

Keywords: sampled-data system, controller design, Lie brackets, Fliess expansion, Taylor expansion, nonlinear control affine system, convergence rate

1. Introduction

Feedback controllers are nowadays typically implemented using digital devices. In contrast to analog implementations, these devices are not able to evaluate the feedback law continuously in time but only at discrete sampling time instances. Thus, the controller must be designed as a sampled-data controller, whose simplest (and most widely used) implementation is a zero order hold, i.e., the feedback law is evaluated at each sampling time and the resulting control value is kept constant and applied on the sampling interval until the next sampling time.

A popular design method for sampled-data controllers is the design of a controller based on the continuous-time plant model, followed by a discretization of the controller. In other words, the continuous control function generated by the continuous-time controller is replaced by a piecewise constant and thus nonsmooth control function generated by the sampled-data controller. If the sampling interval is sufficiently small, then the choice of an appropriate sampled-data controller can be done in a very straightforward way, however, hardware or communication constraints may prohibit the use of small sampling intervals, in which case more sophisticated techniques have to be used. A good introduction to this subject in the nonlinear setting considered here can be found in the survey paper Ref. 11. An important class of such techniques is the *sampled-data redesign*, in which a sampled-data controller is constructed which inherits certain properties of a previously designed feedback law for the continuous-time system. The survey papers Refs. 5,12 summarize a couple of such redesign techniques. The analytical approaches in these papers are restricted to single-input systems, i.e., for systems with a one dimensional control variable, a condition which we relax in this paper.

More precisely, in this paper we extend the redesign technique presented in Ref. 4 to multi-input control affine nonlinear systems. This technique solves the redesign problem by designing a controller which is asymptotically optimal in the sense that we maximize the order at which the difference between the trajectories of the continuous and the sampled-data system converges to zero as the sampling time tends to zero. This amounts to investigating whether a sampled-data feedback law for a desired order exists and, in case the answer is positive, how it can be computed.

Concerning the conditions for the existence of higher order sampled-data feedback laws, it turns out that like in the single-input case the answer lies in the geometry of the system, expressed via the possible directions of the solution trajectories, which in turn are determined by the Lie brackets of the vector fields. Compared to the single-input case, the main difference of our multi-input results lies in the fact that the presence of more control variables typically facilitates the construction of a higher order sampled-data feedback law, an effect we illustrate in our numerical examples. In particular, it turns out that the design of sampled-data feedback laws of arbitrary order is always feasible if the control dimension equals the state dimension and the matrix composed of the control vector fields has full rank.

Since for higher orders the existence conditions and formulae for the sampled-data feedback laws become fairly complicated, we restrict our analytical results to low orders in order to illustrate the geometric nature of the conditions. For general orders we provide a Maple code which checks the respective conditions and computes the resulting sampled-data feedback, if possible. Here the second main difference to the single-input case appears: while in the single input case this computation was based on the successive solution of several one dimensional linear equations, the multi-input case can be tackled algorithmically via the solution of a suitable least squares

208

problem.

2. Problem formulation

We consider a nonlinear plant model

$$\dot{x}(t) = f(x(t), u(t)) \tag{1}$$

with vector field $f : \mathbb{R}^n \times \mathbb{U} \to \mathbb{R}^n$ which is continuous and locally Lipschitz in x, state $x(t) \in \mathbb{R}^n$ and control $u(t) \in \mathbb{U} \subset \mathbb{R}^m$. Throughout the paper we assume that a smooth static state feedback $u_0 : \mathbb{R}^n \to \mathbb{R}^m$ has been designed which solves some given control problem for the continuous-time closed-loop system

$$\dot{x}(t) = f(x(t), u_0(x(t))) \qquad x(0) = x_0. \tag{2}$$

Our goal is now to design $u_T(x)$ such that the corresponding sampled-data solution $\phi_T(t, x_0, u_T)$ of the closed-loop system using a sampler and zero order hold

$$\dot{x}(t) = f(x(t), u_T(x(kT))), \quad t \in [kT, (k+1)T) \tag{3}$$

$k = 0, 1, \ldots$, reproduces the behavior of the continuous-time system and thus improves the performance of the sampled-data closed loop system.

Our approach uses an asymptotic analysis in order to study the difference between the continuous-time model (2) and the sampled-data model (3). To this end, for a function $a : \mathbb{R} \times \mathbb{R}^n \to \mathbb{R}$ we write $a(T, x) = \mathcal{O}(T^q)$, if for any compact set $K \subset \mathbb{R}^n$ there exists a constant $C > 0$ (which may depend on K) such that the inequality $a(T, x) \leq CT^q$ holds for all elements $x \in K$. If we consider a specific set K we explicitly write $a(T, x) = \mathcal{O}(T^q)$ on K.

In order to obtain asymptotic estimates, we consider an "output" function $h : \mathbb{R}^n \to \mathbb{R}$ and derive series expansions for the difference

$$\Delta h(T, x_0, u_T) := |h(\phi(T, x_0)) - h(\phi_T(T, x_0, u_T))|, \tag{4}$$

where $\phi(T, x_0)$ denotes the solution of the continuous-time system (2). Note that h here is not a physical output of the system but rather a scalar auxiliary function which can be chosen arbitrarily. In particular, we will use $h_i(x) = x_i$, $i = 1, \ldots, n$, in order to establish $\Delta h_i(T, x_0, u_T) = \mathcal{O}(T^q)$ which then implies

$$\Delta \phi(T, x_0, u_T) := \|\phi(T, x_0) - \phi_T(T, x_0, u_T)\|_\infty = \mathcal{O}(T^q) \tag{5}$$

measured in the maximum norm $\| \cdot \|_\infty$. From this estimate it follows by a standard induction argument that on each compact interval $[0, t^*]$ we obtain $\Delta\phi(t, x_0, u_T) \leq \mathcal{O}(T^{q-1})$ for all times $t = kT$, $k \in \mathbb{N}$ with $t \in [0, t^*]$ which in particular allows to carry over stability properties from ϕ to ϕ_T, see Refs. 17,18.

In order to facilitate this analysis we restrict ourselves to control affine systems where the ordinary differential equations in (1)–(3) take the form

$$\dot{x}(t) = g_0(x(t)) + \sum_{i=1}^{m} g_i(x(t))u_{0,i}(x(t)), \tag{6}$$

with smooth vector fields $g_0, g_1, \ldots, g_m : \mathbb{R}^n \to \mathbb{R}^n$ and controls $u_{0,1}, \ldots, u_{0,m} : \mathbb{R}^n \to \mathbb{R}$. Note that the continuous-time feedback $u_0(x) = (u_{0,1}(x), \ldots, u_{0,m}(x))^T$ is represented in a vectorial form. In Refs. 4,14 we investigated single input systems, i.e., $u(t) \in \mathbb{R}$. In this paper we extend these results to the multi-input case, i.e., $m > 1$.

We look at sampled-data feedback laws meeting the following definition.

Definition 2.1. An admissible sampled-data feedback law u_T is a family of maps $u_T : \mathbb{R}^n \to \mathbb{R}^m$, parameterized by the sampling period $T \in (0, T^*]$ for some maximal sampling period T^*, such that for each compact set $K \subset \mathbb{R}^n$ the inequality

$$\sup_{x \in K,\, T \in (0,T^*]} \| u_T(x) \|_\infty < \infty$$

holds.

Note that for existence and uniqueness of the solutions of (3), we do not need any continuity assumptions on u_T. Local boundedness is, however, imposed, because unbounded feedback laws are physically impossible to implement and often lead to closed-loop systems which are very sensitive to modelling or approximation errors, cfr., e.g., the examples in Refs. 3,15,17.

A special class of sampled-data feedback laws is given by

$$u_T(x) = \sum_{j=0}^{M} T^j u_j(x), \tag{7}$$

where the $u_j(x)$ are vectors $(u_{j,1}(x), \ldots, u_{j,m}(x))^T$. We will see later that this is exactly the form needed for our purpose. Inserting the sampled-data feedback (7) into our affine control system (6) leads to

$$\dot{x} = g_0(x) + \sum_{i=1}^{m} g_i(x)u_{T,i}(x)$$

$$= g_0(x) + \sum_{i=1}^{m} g_i(x) \sum_{j=0}^{M} T^j u_{j,i}(x). \tag{8}$$

Here the second index of $u_{j,i}(x)$ denotes the i-th component of the vector $u_j(x)$. Another way of writing (8) is

$$\dot{x} = g_0(x) + G(x)u_T(x) \quad \text{with} \quad G(x) = \begin{pmatrix} g_{1,1}(x) & \cdots & g_{m,1}(x) \\ \vdots & \ddots & \vdots \\ g_{1,n}(x) & \cdots & g_{m,n}(x) \end{pmatrix}. \tag{9}$$

In the sequel we use the following notation: for subsets $D \subset \mathbb{R}^n$ we write $\operatorname{cl} D$, $\operatorname{int} D$ for the closure and the interior of D. The notation $|\cdot|$ stands for the Euclidean norm while $\|x\|_\infty = \max_{i=1,\ldots,n} |x_i|$ denotes the maximum norm in $\mathbb{R}^n$. Furthermore, cfr. Ref. 6, we denote the directional derivative of a function $h : \mathbb{R}^n \to \mathbb{R}$ in the direction of $g : \mathbb{R}^n \to \mathbb{R}^n$ by

$$L_g h(x) := \frac{d}{dx} h(x) \cdot g(x)$$

and the Lie bracket of vector fields g_i, $g_j : \mathbb{R}^n \to \mathbb{R}^n$ by

$$[g_i, g_j] = \frac{d}{dx} g_j \cdot g_i - \frac{d}{dx} g_i \cdot g_j.$$

3. Fliess series expansion

In this section we provide the basic series expansion used for the redesign of u_T. Although the admissible sampled-data feedback u_T according to Definition 2.1 may in principle be completely unrelated to $u_0 = (u_{0,1}, \ldots, u_{0,m})^T$, in the sequel it will turn out that a certain relation between u_0 and u_T must hold. More precisely, we will see that the resulting sampled-data feedback (if existing) will be of the form (7) with $u_{0,1}(x), \ldots, u_{0,m}(x)$ from (2) and $u_{1,1}(x), \ldots, u_{M,m}(x) : \mathbb{R}^n \to \mathbb{R}$ being locally bounded functions. This structure appears to be rather natural and was also obtained as the outcome of the design procedure in several other papers, cfr. Refs. 1,10,16 and also for our problem in the single input case.[4] Thus, we develop our series expansion for these feedback laws.

In order to formulate our results we define multinomial coefficients $\binom{n}{n_0 \ \cdots \ n_M} := \frac{n!}{n_0! n_1! \ldots n_M!}$ as well as multi-indices $\nu := (n_0, n_1, \ldots, n_M)$ and use the notations $|\nu| := n_0 + n_1 + \ldots + n_M$ and $\|\nu\| = \sum_{i=0}^{M} i \, n_i$. Our analytical considerations are based on the following theorem which is a generalization of [14, Theorem 3.1] to the multi-input case.

Theorem 3.1. *Consider the control affine system* (6)*, a smooth function* $h : \mathbb{R}^n \to \mathbb{R}$*, the continuous-time closed-loop system* (2) *and the sampled-*

data closed-loop system (3) with controller u_T given by (7). Then, for sufficiently small T, we can write:

$$h(\phi_T(T, x, u_T)) = h(x) + \sum_{s=0}^{M} T^{s+1}\left[\sum_{i=1}^{m} L_{g_i}h(x)u_{s,i} + p_s(x, u_0, \ldots, u_{s-1})\right]$$
$$+ \mathcal{O}(T^{M+2}) \tag{10}$$

where $p_0(x) = L_{g_0}h(x)$ and $p_s(x, u_0, \ldots, u_{s-1})$, $s = 1, \ldots, M$, is given by

$$\sum_{k=1}^{s} \sum_{i_0=0,\ldots,i_k=0}^{m} \frac{L_{g_{i0}} \cdots L_{g_{ik}}h(x)}{(k+1)!} \tag{11}$$

$$\cdot \sum_{\substack{v\in\mathbb{N}_0^m: \\ \sum_{i=1}^{m} v_i = s-k}} \prod_{j=1}^{m}\left(\sum_{\substack{|\nu_j|=c_j \\ \|\nu_j\|=v_j}} \binom{c_j}{n_{0,j}\, n_{1,j}\, \cdots\, n_{M,j}}\prod_{l=0}^{M} u_{l,j}^{n_{l,j}}\right)$$

with $u_i = (u_{i,1}, \ldots, u_{i,m})^T$. Here c_j denotes $\#\{i_l \mid l = 1, \ldots, k : i_l = j\}$.

For the proof, we need the following result, which can be found, e.g., in [8, Theorem 4.2]

Proposition 3.1. *For $a_i \in \mathbb{R}, i = 0, 1, 2, \ldots, M$ and $n \in \mathbb{N}$ we have the equality*

$$(a_0 + a_1 + \ldots + a_M)^n = \sum_{\substack{|\nu|=n \\ n_0,n_1\ldots n_M \geq 0}} \binom{n}{n_0\ \ldots\ n_M}a_0^{n_0} \cdots a_M^{n_M}.$$

Proof. (Theorem 3.1) Using the Fliess series expansion, see [6, Theorem 3.1.5], we can write

$$h(\phi(t, x)) - h(x) = \sum_{k=0}^{\infty} \sum_{i_0=0,\ldots,i_k=0}^{m} L_{g_{i_0}} \ldots L_{g_{i_k}}h(x)\int_0^t d\xi_{i_k} \ldots d\xi_{i_0}.$$

The expressions $\int_0^t d\xi_{i_k} \ldots d\xi_{i_0}$ denote iterated integrals as defined in Ref. 2. Next we consider a single summand of the inner sum of this expression. We assign $i_k, \ldots, i_0$ to a vector $(c_1, \ldots, c_m)$, where $c_j := \#\{i_l \mid l = 0, \ldots, k : i_l = j\}$. Since the values $u_0, \ldots, u_m$ are independent of t, the order of integration may be changed arbitrarily. That means that the value of the integral is independent of the order of the $L_{g_{i_j}}$-operators and it follows

$$\int_0^t d\xi_{i_k} \ldots d\xi_{i_0} = \frac{T^{k+1}}{(k+1)!}\prod_{j=1}^{m} u_{T,j}^{c_j}. \tag{12}$$

212

Using Equation (12) we can write

$$\frac{h(\phi_T(T,x,u_T)) - h(x)}{T} = \sum_{k=0}^{\infty} \sum_{i_0=0,\dots,i_k=0}^{m} L_{g_{i_0}} \dots L_{g_{i_k}} h(x) \frac{T^k}{(k+1)!} \prod_{j=1}^{m} u_{T,j}^{c_j}.$$

Like in the single-input case[14] we use Lemma 3.1 in order to transform the components of (7). This leads to the expression

$$u_{T,j}^{c_j} = \sum_{\substack{|\nu_j|=c_j \\ n_{0,j},\dots,n_{M,j} \geq 0}} \binom{c_j}{n_{0,j}\, n_{1,j} \cdots n_{M,j}} u_{0,j}^{n_{0,j}} \cdots u_{M,j}^{n_{M,j}} T^{\|\nu_j\|}$$

with $|\nu_j| = \sum_{l=0}^{M} n_{l,j}$ and $\|\nu_j\| = \sum_{l=0}^{M} l n_{l,j}$. Hence it follows

$$\frac{h(\phi_T(T,x,u_T)) - h(x)}{T} = H_1 + \sum_{k=0}^{M} \sum_{i_0=0,\dots,i_k=0}^{m} L_{g_{i_0}} \dots L_{g_{i_k}} h(x) \frac{T^k}{(k+1)!}$$

$$\cdot \prod_{j=1}^{m} \left(\sum_{|\nu_j|=c_j} \binom{c_j}{n_{0,j}\, n_{1,j} \cdots n_{M,j}} \prod_{l=0}^{M} u_{l,j}^{n_{l,j}} \cdot T^{\|\nu_j\|} \right),$$

where H_1 denotes an $\mathcal{O}(M+1)$ term. Define $s := k + \sum_{j=1}^{m} \|\nu_j\| = k + \sum_{j=1}^{m} \sum_{l=0}^{M} l n_{l,j}$ and sum over all terms of order $\leq M$. Collecting all terms of order strictly greater than M in H_1 we can rewrite the last equation as

$$\frac{h(\phi_T(T,x,u_T)) - h(x)}{T} = H_1 + \sum_{s=0}^{M} T^s \sum_{k=0}^{s} \sum_{i_0,\dots,i_k=0}^{m} \frac{L_{g_{i_0}} \dots L_{g_{i_k}} h(x)}{(k+1)!}$$

$$\cdot \sum_{\substack{v \in \mathbb{N}_0^m: \\ \sum_{i=1}^{m} v_i = s-k}} \prod_{j=1}^{m} \left(\sum_{\substack{|\nu_j|=c_j \\ \|\nu_j\|=v_j}} \binom{c_j}{n_{0,j}\, n_{1,j} \cdots n_{M,j}} \prod_{l=0}^{M} u_{l,j}^{n_{l,j}} \right).$$

Observe that for each $s > 0$ the sum for $k = 1 \dots, s$ is exactly (11). Thus, in order to complete the proof it remains to show that the summands for $k = 0$ equals the remaining terms in (10).

To this end first consider $s = 0$ and $k = 0$. This leads to $s - k = 0$ and the only vector $v \in \mathbb{N}_0^m$ satisfying $\sum_{i=1}^{m} v_i = s - k$ is the zero vector. Consequently we obtain

$$\sum_{i_0=0}^{m} L_{g_{i_0}} h(x) \underbrace{\sum_{0 \in \mathbb{N}_0^m} \prod_{j=1}^{m} \sum_{n_{0,j}=c_j} u_{0,j}^{n_{0,j}}}_{=\prod_{j=1}^{m} u_{0,j}^{c_j}} = L_{g_0} h(x) + \sum_{i=1}^{m} L_{g_i} h(x) u_{0,i}. \qquad (13)$$

For $s = 1, \ldots, M$ and $k = 0$ one computes $\sum_{i=1}^{m} v_i = s$ and differs between the cases $i_0 = 0$ and $i_0 > 0$. For $i_0 = 0$ this provides $L_{g_0} h(x)$ multiplied with a sum with respect to the empty set because it holds $c_j = 0$ for all $j = 1, \ldots, m$ and thus $|v_j| = 0$ for all $j = 1, \ldots, m$. This causes $\|v_j\| = 0$ for all $j = 1, \ldots, m$ and leads to a contradiction due to $\sum_{i=1}^{m} \|v_j\| = \sum_{i=1}^{m} v_i = s > 0$. Accordingly $i_0 = 0$ doesn't provide any additional terms. Continuing we consider $i_0 = i \in \{1, \ldots, m\}$. It holds $c_i = 1$ and $c_j = 0$ for all $j = 1, \ldots, m$ with $j \neq i$. This leads to $|v_j| = 0$ and $\|v_j\| = 0$ for $j = 1, \ldots, m$ with $j \neq i$. Consequently we have to consider $v = s e_i$ where e_i denotes the i-th unit vector. Hence we multiply $L_{g_i} h(x)$ with $u_{s,i}$ due to

$$c_j = \begin{cases} 1 & \text{for } j = i, \\ 0 & \text{otherwise}. \end{cases}$$

Overall we obtain $\sum_{s=1}^{M} T^s \sum_{i=1}^{m} L_{g_i} h(x) u_{s,i}$ for $s > 0$ and $k = 0$, which finishes the proof. $\qquad\square$

Remark 3.1. While (10) is a straightforward extension of the single-input case, the p_s differ in our multi-input case in terms of an additional combinatorial condition. One has to choose all vectors $v \in \mathbb{N}_0^m$ whose components add up to $s - k$.

Remark 3.2. For later reference we will explicitly compute the components of p_1 and p_2 according to (11). For $p_1(x, u_0)$ we have $s - k = 0$ and thus the combinatorial condition $v \in \mathbb{N}_0^m : \sum_{i=1}^{m} v_i = s - k$ is only satisfied for $v = 0_{\mathbb{N}_0^m}$. Now we have to distinguish three cases, the first one is $i_0 = i_1 = 0$. Here $c_j = 0$ for all $j \in \{1, \ldots, m\}$ and it results $\frac{1}{2} L_{g_0} L_{g_0} h(x)$. The second case is $i_0 = 0$, $i_1 \neq 0$ and respectively $i_0 \neq 0$, $i_1 = 0$. Here it exists exactly one $c_j > 0$ and due to $\|v_j\| = 0$ it follows

$$\sum_{i=1}^{m} \frac{1}{2}(L_{g_0} L_{g_i} h(x) + L_{g_i} L_{g_0} h(x)) u_{0,i}(x).$$

The last case, $s = 1$, $i_0 > 0$, $i_1 > 0$, provides

$$\sum_{i=1, j=1}^{m} \frac{L_{g_i} L_{g_j} h(x)}{2} u_{0,j}(x) u_{0,i}(x).$$

For $p_2(x, u_0, u_1)$ we need to distinguish between $k = 1$ and $k = 2$. For $k = 1$ we have $s - k = 1$ and thus $v = e_1, \ldots, e_m$. Furthermore, we have to discern three cases. The first is $i_0 = i_1 = 0$ which leads to the empty set. The second, $i_0 = 0$, $i_1 > 0$ respectively $i_0 > 0$, $i_1 = 0$, provides

$$\sum_{i=1}^{m} \frac{1}{2}[L_{g_0}L_{g_i}h(x) + L_{g_i}L_{g_0}h(x)]u_{1,i}(x).$$

Finally, in the case $i_0 > 0$ and $i_1 > 0$ we obtain

$$\sum_{\substack{i=1,j=1 \\ i \neq j}}^{m} \frac{1}{2}L_{g_i}L_{g_j}h(x)u_{1,i}(x)u_{0,j}(x) + \sum_{i=1}^{m} L_{g_i}^2 h(x)u_{1,i}(x)u_{0,i}(x).$$

For $k = 2$ it holds $s - k = 0$ and thus we have only to regard $v = 0_{\mathbb{N}_0^m}$. The computations take course analogously to the case $s = k = 1$. Again we obtain terms dependent on components of u_0. Hence, altogether $p_2(x, u_0, u_1)$ provides

$$\frac{1}{2}\sum_{i=1}^{m} \left([L_{g_0}L_{g_i} + L_{g_i}L_{g_0}]h(x) + \sum_{j=1}^{m} L_{g_i}L_{g_j}h(x)u_{0,j}(x) \right) u_{1,i}(x)$$

$$+ \frac{1}{6}\sum_{i=1}^{m} [L_{g_i}L_{g_0}^2 + L_{g_0}L_{g_i}L_{g_0} + L_{g_0}^2 L_{g_i}]h(x)u_{0,i}(x) + \frac{1}{6}L_{g_0}^3 h(x)$$

$$+ \frac{1}{6}\sum_{i=1,j=1}^{m} [L_{g_i}L_{g_j}L_{g_0} + L_{g_i}L_{g_0}L_{g_j} + L_{g_0}L_{g_i}L_{g_j}]h(x)u_{0,i}(x)u_{0,j}(x)$$

$$+ \frac{1}{6}\sum_{i=1,j=1,k=1}^{m} L_{g_i}L_{g_j}L_{g_k}h(x)u_{0,i}(x)u_{0,j}(x)u_{0,k}(x). \tag{14}$$

Remark 3.3. Computer algebra systems, such as Maple, can be used to compute expansions of the difference (4) for particular examples, cfr. Remark 4.5.

4. Necessary and sufficient conditions

In this section we investigate necessary and sufficient conditions for the existence of an admissible feedback law u_T which achieves

$$\Delta h(T, x, u_T) = \mathcal{O}(T^q) \tag{15}$$

or

$$\Delta \phi(T, x, u_T) = \mathcal{O}(T^q) \tag{16}$$

and provide formulae for these feedback laws. Since the computations with respect to the sufficient condition turn out to be fairly involved we restrict our analytical computations to the case $q \leq 4$ and provide a Maple procedure for the general case. As we will see, $q = 4$ is the first nontrivial case in the sense that that (15) and (16) for $q \leq 3$ can always be satisfied without any further conditions. For the case $q = 4$ it turns out that the cases (15) and (16) require different conditions. In particular, for (16) we obtain a much stronger necessary condition than for (15). Thus, we state them in two separate theorems starting with (15).

The next theorem is a consequence from Theorem 3.1 performing a careful evaluation of the p_i-terms. It generalizes corresponding results for the single input-case $m = 1$, see [14, Theorem 4.11] for the cases (i) and (ii), [4, Theorem 3.1] for case (iii), and also [5, Theorem 3.2]. For the formulation of the theorem we use the notation

$$
u^{(i)}(x) = \begin{pmatrix} u_1^{(i)}(x) \\ \vdots \\ u_m^{(i)}(x) \end{pmatrix} = \frac{1}{(i+1)!} \frac{d^i}{dt^i} \begin{pmatrix} u_{0,1}(\phi(t,x)) \\ \vdots \\ u_{0,m}(\phi(t,x)) \end{pmatrix} \Bigg|_{t=0}
$$
$$
= \frac{1}{(i+1)!} \frac{d^i u_0(\phi(t,x))}{dt^i} \Bigg|_{t=0}. \tag{17}
$$

Note that this definition coincides with the continuous-time controller for $i = 0$. Furthermore we abbreviate

$$
\dot{u}_j(x) := \frac{\partial u_{0,j}(x)}{\partial x} \left[g_0 + \sum_{i=1}^{m} g_i u_{0,i} \right] = 2 u_j^{(1)}(x). \tag{18}
$$

Theorem 4.1. *Consider the vector field (6), the continuous-time closed-loop system (2), the sampled-data closed-loop system (3), a smooth function $h : \mathbb{R}^n \to \mathbb{R}$ and a compact set $K \subset \mathbb{R}^n$. Then the following assertions hold for $u^{(i)}$ from (17):*

(i) $\Delta h(T, x_0, u_T) = \mathcal{O}(T^2)$ holds on K for

$$
u_T(x) = u^{(0)}(x).
$$

(ii) $\Delta h(T, x_0, u_T) = \mathcal{O}(T^3)$ holds on K for

$$
u_T(x) = u^{(0)}(x) + T u^{(1)}(x).
$$

(iii) If there exists a bounded function $\alpha_h : K \to \mathbb{R}^m$ satisfying

$$
\sum_{i=1}^{m} L_{g_i} h(x) \alpha_{h,i}(x) = \sum_{i=1}^{m} \left[L_{[g_0,g_i]} h(x) + \sum_{\substack{j=1 \\ j \neq i}}^{m} L_{[g_j,g_i]} h(x) u_{0,j}(x) \right] \dot{u}_i(x), \tag{19}
$$

216

then there exists u_T such that

$$\Delta h(T, x_0, u_T) = \mathcal{O}(T^4) \tag{20}$$

holds on K with

$$u_T(x) = \begin{cases} u^{(0)}(x) + Tu^{(1)}(x) + T^2 u^{(2)}(x) + \frac{T^2}{12}\alpha_h(x), & x \in \mathrm{cl}\,\widetilde{K} \\ u^{(0)}(x) + Tu^{(1)}(x), & x \notin \mathrm{cl}\,\widetilde{K}, \end{cases} \tag{21}$$

where $\widetilde{K} := \{x \in K \mid \exists i : L_{g_i} h(x) \neq 0\}$.

Conversely, if an admissible sampled-data feedback law $\tilde{u}_T = u_T + \mathcal{O}(T^3)$ for u_T from (21) satisfies (20) on a set $\widehat{K} \subseteq \widetilde{K}$, then there exists a bounded function α satisfying (19) on $\mathrm{cl}\,\widehat{K}$.

Proof. Our smoothness assumptions enable us to use the Taylor series expansion for the solution of the ordinary differential equation (6). To this end we define the differential operator $\tilde{L} := L_{g_0 + \sum_{i=1}^m L_{g_i} u_{0,i}}$ and apply the Taylor series expansion to our output-function h, i.e.,

$$h(\phi(T, x)) - h(x) = \sum_{i=1}^{q-1} \frac{T^i}{i!} \tilde{L}^i h(x) + \mathcal{O}(T^q). \tag{22}$$

Hence, from the Taylor expansion of $h(\phi(t, x))$ in $t = 0$ we obtain the identity

$$\begin{aligned}
h(\phi(T, x)) &= h(x) + T\tilde{L}h(x) + \mathcal{O}(T^2) \\
&= h(x) + T\left[L_{g_0} h(x) + \sum_{i=1}^m L_{g_i} h(x) \cdot u_{0,i}(x) \right] + \mathcal{O}(T^2) \\
&= h(x) + T\left[\sum_{i=1}^m L_{g_i} h(x) u_i^{(0)}(x) + p_0(x) \right] + \mathcal{O}(T^2) \tag{23} \\
&= h(\phi_T(T, x, u_T)) + \mathcal{O}(T^2) \tag{24}
\end{aligned}$$

using Theorem 3.1 in the last step. Thus (i) holds.

For the proof of (ii) we apply $\tilde{L}$ twice to the output-function h and exploit the shape of $p_1(x, u_0)$ outlined in Remark 3.2:

$$
\begin{aligned}
h(\phi(t,x)) &= h(x) + T\tilde{L}h(x) + \frac{T^2}{2}\tilde{L}^2 h(x) + \mathcal{O}(T^3) \\
&= h(x) + T\left[\sum_{i=1}^{m} L_{g_i}h(x)u_{0,i}(x) + p_0(x)\right] + T^2 p_1(x, u_0) \\
&\quad + \frac{T^2}{2}\sum_{i=1}^{m} L_{g_i}h(x)\frac{\partial u_{0,i}(x)}{\partial x}\left[g_0 + \sum_{i=1}^{m} g_i u_{0,i}\right] + \mathcal{O}(T^3) \\
&= h(x) + \sum_{s=0}^{1} T^{s+1}\left[\sum_{i=1}^{m} L_{g_i}h(x)u_i^{(s)}(x) + p_s(x, u^{(0)}, \ldots, u^{(s-1)})\right] \\
&\quad + \mathcal{O}(T^3) \\
&= h(\phi_T(T, x, u_T)) + \mathcal{O}(T^3)
\end{aligned}
$$
(25)
(26)

where we used Theorem 3.1 and $u_T = u^{(0)} + Tu^{(1)}$ in the last step. This shows (ii).

In order to prove (iii) we have to examine the Taylor series expansion up to order four. To this aim we consider the threefold application of $\tilde{L}$ and use the identity $\dot{u}_j(x) = \tilde{L}u_{0,j}(x)$:

$$
\begin{aligned}
\tilde{L}^3 h(x) &= L_{g_0}^3 h(x) + \sum_{i=1}^{m}\left[L_{g_i}L_{g_0}^2 + L_{g_0}L_{g_i}L_{g_0} + L_{g_0}^2 L_{g_i}\right]h(x)u_{0,i}(x) + \\
&\quad + \sum_{j=1}^{m}\sum_{i=1}^{m}\left[L_{g_j}L_{g_i}L_{g_0} + L_{g_j}L_{g_0}L_{g_i} + L_{g_0}L_{g_j}L_{g_i}\right]h(x)u_{0,i}(x)u_{0,j}(x) \\
&\quad + \sum_{i,j,k=1}^{m} L_{g_k}L_{g_j}L_{g_i}h(x)u_{0,i}(x)u_{0,j}(x)u_{0,k}(x) + \\
&\quad + \sum_{i=1}^{m}\left[L_{g_i}L_{g_0} + 2L_{g_0}L_{g_i}\right]h(x)\dot{u}_i(x) + \sum_{i=1}^{m} L_{g_i}h(x)\tilde{L}\dot{u}_i(x) + \\
&\quad + \sum_{i=1,j=1}^{m} L_{g_i}L_{g_j}h(x)\left[2\dot{u}_j(x)u_{0,i}(x) + \dot{u}_i(x)u_{0,j}(x)\right].
\end{aligned}
$$
(27)

A comparison between (27) and the terms resulting from (14) using u_T

218

from (21) yields

$$h(\phi(T,x)) = \sum_{s=0}^{2} T^{s+1}\left[\sum_{i=1}^{m} L_{g_i}h(x)u_i^{(s)}(x) + p_s(x,u^{(0)},\ldots,u^{(s-1)})\right]$$

$$+ \frac{1}{12}\sum_{i=1}^{m}\left[L_{g_0}L_{g_i}h(x) + \sum_{\substack{j=1\\j\neq i}}^{m} L_{g_j}L_{g_i}h(x)u_{0,j}(x) - \right.$$

$$\left. - L_{g_i}L_{g_0}h(x) - \sum_{\substack{j=1\\j\neq i}}^{m} L_{g_i}L_{g_j}h(x)u_{0,j}(x)\right]\dot{u}_i(x)$$

$$+ h(x) + \mathcal{O}(T^4).$$

With $L_{g_0}L_{g_i}h(x) - L_{g_i}L_{g_0}h(x) = L_{[g_0,g_i]}h(x)$ one obtains

$$h(\phi(T,x)) = \sum_{s=0}^{2} T^{s+1}\left[\sum_{i=1}^{m} L_{g_i}h(x)u_i^{(s)}(x) + p_s(x,u^{(0)},\ldots,u^{(s-1)})\right]$$

$$+ \frac{1}{12}\sum_{i=1}^{m}\left[L_{[g_0,g_i]}h(x) + \sum_{\substack{j=1\\j\neq i}}^{m} L_{[g_j,g_i]}h(x)u_{0,j}(x)\right]\dot{u}_i(x)$$

$$+ h(x) + \mathcal{O}(T^4) \;=\; h(\phi_T(T,x,u_T)) + \mathcal{O}(T^4), \tag{28}$$

where we used Theorem 3.1 and the definition of u_T and α_h in the last step. Thus, the choice (21) ensures a forth order approximation. Note that the function α_h is bounded on K by assumption, which in particular implies that the control law (21) is admissible in the sense of Definition 2.1 on $\widetilde{K}$.

For the converse statement, if u_T from (21) satisfies (20) on $\widehat{K} \subseteq \widetilde{K}$ and Definition 2.1, then the function α in (21) must be bounded. Hence, for each boundary point $x \in \partial\widehat{K}$ we can find a sequence $x_k \to x$ such that $\alpha(x_k)$ is convergent and define $\alpha(x) = \lim_{k\to\infty} \alpha(x_k)$. Then, since all coefficients in (19) are continuous, we obtain that (19) also holds for $x \in \partial\widehat{K}$, i.e., on $\mathrm{cl}\,\widehat{K}$ and the boundedness follows immediately. $\qquad\square$

Remark 4.1. Note that the converse part of statement (iii) is rather weak, as it only provides a necessary condition for the existence of feedback laws of the specific form (21) but not for arbitrary admissible sampled-data feedback laws satisfying (20). It is however, an important building block for the much stronger necessary condition for $\Delta\phi(T,x,u_T) = \mathcal{O}(T^4)$ given in Theorem 4.2, below.

Remark 4.2. In (21) we distinguish between $x \in \operatorname{cl} \widetilde{K}$ and $x \notin \operatorname{cl} \widetilde{K}$. This case differentiation can be interpreted in terms of relative degree (see Ref. 7 for a definition and Ref. 12 for the role of the relative degree in sampled-data feedback design). System (6) with output function h has relative degree one on $\operatorname{cl} \widetilde{K}$ while the relative degree is strictly larger on $K \backslash \operatorname{cl} \widetilde{K}$. This explains why the feedback law (21) has different structure inside and outside $\operatorname{cl} \widetilde{K}$.

Remark 4.3. For driftless systems, i.e., $g_0(x) \equiv 0$, the lie-bracket $[g_0, g_i]$ in (19) is equal to zero for all $i = 1, \ldots, m$. Hence, condition (19) is always satisfied for $m = 1$ and easier to evaluate otherwise.

Now we turn to (16) and deduce assertions for the full state trajectory from Theorem 4.1 by choosing $h_i(x) = x_i$, $i = 1, \ldots, n$.

Theorem 4.2. *Consider the control affine system* (6), *the continuous-time closed-loop system* (2), *the sampled-data closed-loop system* (3) *and a compact set* $K \subset \mathbb{R}^n$ *satisfying* $K = \operatorname{cl} \operatorname{int} K$. *Then the following assertions hold for* $u^{(i)}$ *from* (17)*:*

(i) $\Delta\phi(T, x_0, u_T) = \mathcal{O}(T^2)$ *holds on* K *for*

$$u_T(x) = u^{(0)}(x).$$

(ii) $\Delta\phi(T, x_0, u_T) = \mathcal{O}(T^3)$ *holds on* K *for*

$$u_T(x) = u^{(0)}(x) + T u^{(1)}(x).$$

(iii) If there exists a bounded function $\alpha : K \to \mathbb{R}^m$ *satisfying*

$$\sum_{i=1}^{m} \left[[g_0, g_i](x) + \sum_{\substack{j=1 \\ j \neq i}}^{m} [g_j, g_i](x) u_{0,j}(x) \right] \dot{u}_i(x) = \sum_{i=1}^{m} \alpha_i(x) g_i(x) \tag{29}$$

then

$$\Delta\phi(T, x_0, u_T) = \mathcal{O}(T^4)$$

holds on K *for*

$$u_T(x) = \begin{cases} u^{(0)}(x) + T u^{(1)}(x) + T^2 u^{(2)}(x) + \frac{T^2}{12} \alpha(x), & x \in \operatorname{cl} \widetilde{K} \\ arbitrary, & x \notin \operatorname{cl} \widetilde{K} \end{cases} \tag{30}$$

with $\widetilde{K} := \{x \in K \mid \exists i : g_i(x) \neq 0\}$. *Furthermore, on*

$$K^* = \{x \in K \mid G(x) \text{ from } (9) \text{ has full column rank}\},$$

any feedback law $\tilde{u}_T$ *satisfying* $\Delta\phi(T, x_0, \tilde{u}_T) = \mathcal{O}(T^q)$, $q = 2, 3, 4$, *is of the form* $\tilde{u}_T(x) = u_T(x) + \mathcal{O}(T^{q-1})$ *for* u_T *from (i), (ii) or (iii), respectively,* *and the function α in (29) is unique if it exists. On $\operatorname{cl} K^*$ the sufficient condition (29) is also necessary for the existence of u_T in (iii).*

Proof. Note that (16) is equivalent to (15) for $h_i(x) = x_i$, $i = 1, \ldots, n$. Hence, assertions (i) and (ii) follow immediately from Theorem 4.1 applied to $h_i(x) = x_i$, $i = 1, \ldots, n$.

For the proof of (iii), we first show that under condition (29) any feedback of the form (30) satisfies the assertion.

First note that for $x \notin \operatorname{cl} \widetilde{K}$ the feedback value $u_T(x)$ is indeed arbitrary. This follows since on $K \setminus \operatorname{cl} \widetilde{K}$ the control system is given by $\dot{x} = g_0(x)$. Thus, on the open set $\operatorname{int}(K \setminus \operatorname{cl} \widetilde{K})$ the Taylor series expansions of $\phi(t, x)$ and $\phi_T(t, x, u_T)$ coincide for any order, regardless of the values of u_0 and u_T, i.e., we obtain (16) for any $M > 0$ for arbitrary u_T. By continuity of the expressions in the Taylor series expansion this property carries over to $\operatorname{cl} \operatorname{int}(K \setminus \operatorname{cl} \widetilde{K})$ which contains $K \setminus \operatorname{cl} \widetilde{K}$ because we have assumed $K = \operatorname{cl} \operatorname{int} K$.

It is, hence, sufficient to show that u_T satisfies the assertion for $x \in \operatorname{cl} \widetilde{K}$. Assume that the function α exists and is bounded. Fix $i \in \{1, \ldots, n\}$ and consider the function $h_i(x) = x_i$. A simple computation using the identities

$$L_{g_j} h_i(x) = g_{j,i}(x) \quad \text{and} \quad L_{[g_k, g_j]} h_i(x) = [g_k, g_j](x)_i$$

shows that the function α from (29) satisfies

$$\sum_{i=1}^{m} \alpha_i(x) g_i(x) = \sum_{i=1}^{m} \left[[g_0, g_i](x) + \sum_{\substack{j=1 \\ j \neq i}}^{m} [g_j, g_i](x) u_{0,j}(x) \right] \dot{u}_i(x)$$

$$= \sum_{i=1}^{m} \left[L_{[g_0, g_i]} h(x) + \sum_{\substack{j=1 \\ j \neq i}}^{m} L_{[g_j, g_i]} h(x) u_{0,j}(x) \right] \dot{u}_i(x).$$

Thus, the feedback is of the form (21) for $h = h_i$ and we can use Theorem 4.1 to conclude (15) for $q = 4$ and $i = 1, \ldots, n$ and thus (16).

Now we show the claimed form of the $\tilde{u}_T$ on K^*: From Theorem 3.1 for $M = 0$ it follows that any $\tilde{u}_T$ satisfying $\Delta\phi(T, x_0, \tilde{u}_T) = \mathcal{O}(T^2)$ must fulfill

$$\sum_{i=1}^{m} L_{g_i} h_k(x) u_{0,i}(x) = \sum_{i=1}^{m} L_{g_i} h_k(x) \tilde{u}_{T,i}(x) + \mathcal{O}(T)$$

for $k = 1, \ldots, n$ in order to get the equality "(23) = (24)" (for $\tilde{u}_T$ instead of u_T) for all h_k. Using again $L_{g_j} h_i(x) = g_{j,i}(x)$ one sees that this is equivalent to

$$G(x)u_0(x) = G(x)\tilde{u}_T(x) + \mathcal{O}(T)$$

and since $G(x)$ has full column rank this implies $\tilde{u}_T(x) = u_0 + \mathcal{O}(T)$. The statements for (ii) and (iii) now follow analogously by induction using the equalities "(25) = (26)" and (28). The uniqueness of α on K^* follows again from the full column rank of $G(x)$ because the right hand side of (29) equals $G(x)\alpha(x)$.

Finally, using the uniqueness of u_T in (iii) up to higher order terms, the necessity of (29) on $\operatorname{cl} K^*$ follows from the converse statement in Theorem 4.1(iii) for $\widehat{K} = K^*$. $\qquad\square$

Remark 4.4. Theorem 4.2 has a nice geometric interpretation if we consider the possible directions of the system trajectories. To this end, consider the expansions

$$\phi(T, x) = v_0 + Tv_1 + T^2 v_2 + T^3 v_3 + \dots$$
$$\phi_T(T, x, u_T) = w_0 + Tw_1 + T^2 w_2 + T^3 w_3 + \dots$$

in which the vectors v_i and w_i determine the directions of the respective solution trajectories. While the control value in ϕ may vary in time, the control value in ϕ_T is constant on the sampling interval $[0, T)$. Thus, for each $i = 0, 1, \dots$ the set of possible directions v_i which can be generated by different choices of u_0 is larger or equal than the corresponding set of possible directions w_i generated by different u_T.

The cases (i) and (ii) now show that the sets of possible directions v_i and w_i are indeed identical for $i = 0, 1$ and 2, because (i) and (ii) are unconditionally feasible provided u_T is chosen appropriately. Note that the T-dependence of u_T is crucial in (ii) because it gives us the additional flexibility needed for achieving $w_2 = v_2$.

This is no longer possible for the directions v_3 and w_3 which affect the trajectories with order $\mathcal{O}(T^3)$. Indeed, our analysis shows that the direction v_3 can be decomposed as $v_3 = v_3^1 + v_3^2$, such that $w_3 = v_3^1$ can always be achieved via the $u^{(2)}$ term in u_T while v_3^2 cannot in general be reproduced by w_3. This direction v_3^2 is exactly the expression appearing on the left hand side of (29) which depends on the Lie brackets of the vector fields and on the continuous-time feedback law u_0. Condition (29) now demands that v_3^2 lies in $\operatorname{span}\langle g_1, \dots, g_m\rangle$ such that it can be compensated by the α-term of the sampled-data feedback law u_T.

Remark 4.5. While the formulation of condition (29) is suitable for the geometric interpretation, it is difficult to generalize it to orders $\mathcal{O}(T^q)$, $q \geq 5$. However, using Theorem 3.1 directly we can obtain a simple recursive procedure for computing u_T for arbitrary orders: Assuming that $u_0, \ldots, u_{M-1}$ in (7) are already determined and realize the order $\mathcal{O}(T^{M+1})$. Then, comparing the summand for $s = M$ in (10) with the summand for $s = M$ in the Taylor expansion of $\phi(T, x)$ leads to a (in general overdetermined) linear system

$$G(x)u_M(x) = b(x). \tag{31}$$

If (31) admits a solution, then this defines the M-th component of u_T in (7) which then realizes the order $\mathcal{O}(T^{M+2})$. If (31) does not admit a solution, then the order $\mathcal{O}(T^{M+2})$ cannot be achieved by a sampled data feedback law.

This procedure can be efficiently implemented in MAPLE using the least squares solver in order to solve (31) and checking the residual in order to decide whether (31) is solvable. The MAPLE implementation is available on www.math.uni-bayreuth.de/~lgruene/publ/redesign_multiinput.html.

Furhermore, this procedure shows that we can always achieve any desired order if the matrix G is square, i.e., the control dimension m equals the space dimension n, and invertible.

Remark 4.6. In Ref. 13 it was shown for single-input systems, i.e., $m = 1$, that the condition $[g_0, g_1] \in \mathrm{span}\langle g_1 \rangle$ is necessary and sufficient for the existence of sampled-data feedback laws u_T realising $\Delta\phi(T, x) = \mathcal{O}(T^q)$ for all $q \geq 2$ and all continuous-time feedback laws u_0. We conjecture that the generalization of this condition to the multi-input case is $[g_i, g_j] \in \mathrm{span}\langle g_1, \ldots, g_m \rangle$ for all $i, j = 0, \ldots, m$. Note that the sampled-data feedback laws considered in Ref. 13 are not necessarily locally bounded and thus may not fulfill our Definition 2.1.

5. Examples

We illustrate our results by two examples. We first consider the second order version of the Moore-Greitzer jet engine model

$$\begin{pmatrix} \dot{x}_1 \\ \dot{x}_2 \end{pmatrix} = \begin{pmatrix} -x_2 - 3x_1^2/2 - x_1^3/2 \\ 0 \end{pmatrix} + \begin{pmatrix} 0 \\ -1 \end{pmatrix} u_{0,1} \tag{32}$$

with the continuous-time stabilising backstepping feedback law $u_{0,1}(x) = -7x_1 + 5x_2$ derived in [9, Section 2.4.3]. Here the condition (29) shows that no admissible sampled-data feedback u_T satisfies $\Delta\phi(T, x, u_T) \leq \mathcal{O}(T^4)$, cfr., [4, Section 4]. Now we examine this system with an additional control $u_{0,2} \equiv 0$, i.e.,

$$\begin{pmatrix} \dot{x}_1 \\ \dot{x}_2 \end{pmatrix} = \begin{pmatrix} -x_2 - 3x_1^2/2 - x_1^3/2 \\ 0 \end{pmatrix} - \begin{pmatrix} 0 \\ 1 \end{pmatrix} u_{0,1} + \begin{pmatrix} g_{2,1}(x) \\ g_{2,2}(x) \end{pmatrix} u_{0,2}. \quad (33)$$

Note that this is now an academical example because the vector field $g_2 = (g_{2,1}, g_{2,2})^T$ and its control $u_{0,2}$ do not have any physical meaning. Nevertheless, the additional input allows for the design of higher order sampled-data feedback laws. Indeed, while $u_{0,2} \equiv 0$ implies that the left hand side of our condition (29) coincides for (32) and (33) and evaluates to

$$\begin{pmatrix} -1 \\ 0 \end{pmatrix} \dot{u}_1(x) = \begin{pmatrix} -35x_1 + 18x_2 - \frac{21}{2}x_1^2 - \frac{7}{2}x_1^3 \\ 0 \end{pmatrix} \quad (34)$$

with $\dot{u}_1(x)$ from (18), on the right hand side of condition (29) the coefficients of g_2 yield additional degrees of freedom and (29) becomes

$$\begin{pmatrix} -1 \\ 0 \end{pmatrix} \dot{u}_1(x) = \begin{pmatrix} \alpha_2(x)g_{2,1}(x) \\ -\alpha_1(x) + \alpha_2(x)g_{2,2}(x) \end{pmatrix}. \quad (35)$$

It is easily seen that this equality is satisfied, e.g., for $g_2(x) = (1,0)^T$ and $\alpha(x) = (0, -\dot{u}_1(x))^T$. The performance for this choice of g_2, α, and the resulting feedback law

$$u_T(x) = u^{(0)}(x) + Tu^{(1)}(x) + T^2 u^{(2)}(x) + \frac{T}{12}\alpha(x)$$

is shown in Figure 1.

By means of our Maple-procedure, we may compute feedbacks of even higher order. We took this approach to compute the trajectory for $q = 6$ in Figure 2. Remarkable is that neither the sampled-data feedback for $q = 2$ (i.e., $u_T = u_0$) nor the feedback for $q = 3$ preserve the asymptotic stability of the continuous-time system. In contrast to that the fourth order feedback preserves asymptotic stability and the feedback for $q = 6$ provides an even better performance despite the large sampling period.

For further investigation of our analytically constructed control laws we analyze the three dimensional Moore-Greitzer model. Adding an additional control input $u_{0,2} \equiv 0$ analogously to (33) we obtain the system

$$\begin{pmatrix} \dot{x}_1 \\ \dot{x}_2 \\ \dot{x}_3 \end{pmatrix} = \begin{pmatrix} -x_2 - \frac{3}{2}x_1^2 - \frac{1}{2}x_1^3 - 3x_3 x_1 - 3x_3 \\ 0 \\ -\sigma x_3(x_3 + 2x_1 + x_1^2) \end{pmatrix} + \begin{pmatrix} 0 \\ -1 \\ 0 \end{pmatrix} u_{0,1} + \begin{pmatrix} 1 \\ 0 \\ 0 \end{pmatrix} u_{0,2} \quad (36)$$

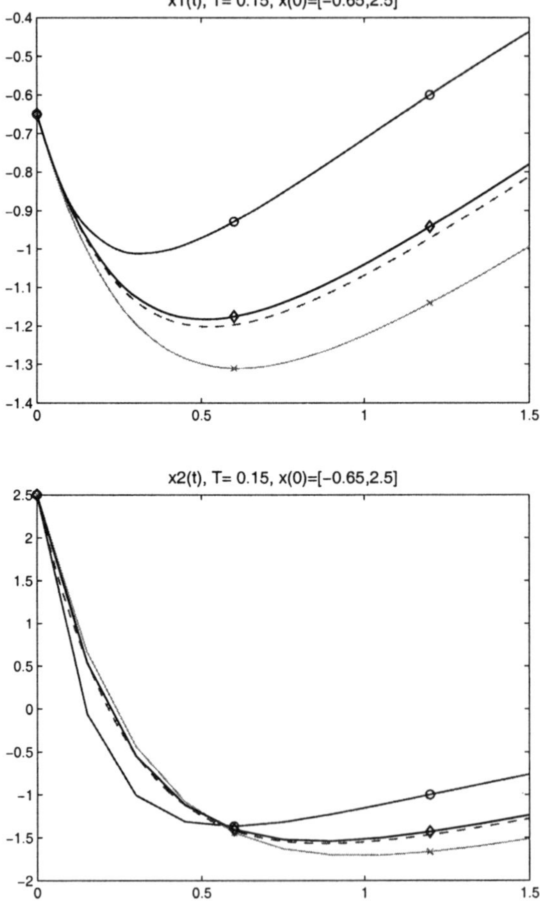

Fig. 1. $x(t)$ for example (33), $T = 0.15$. Continuous-time solution (- -); sampled–data solution for order $q = 2$ (o), $q = 3$ (x) and $q = 4$ ($\Diamond$).

with $\sigma = 2$ and continuous-time controller (cfr. [9, Section 2.4.2])

$$u_{0,1}(x) = -(c_1 - 3x_1)\left(-x_2 - \frac{3}{2}x_1^2 - \frac{1}{2}x_1^3 - 3x_1x_3 - 3x_3\right)$$

$$+c_2\left(x_2 - c_1x_1 + \frac{3}{2}x_1^2 + 3x_3\right) - x_1 - 3\sigma x_3(x_3 + 2x_1 + x_1^2)$$

$$u_{0,2}(x) = 0$$

using the parameters $c_1 = 1$ and $c_2 = 50$. Again, the Maple-routine provides

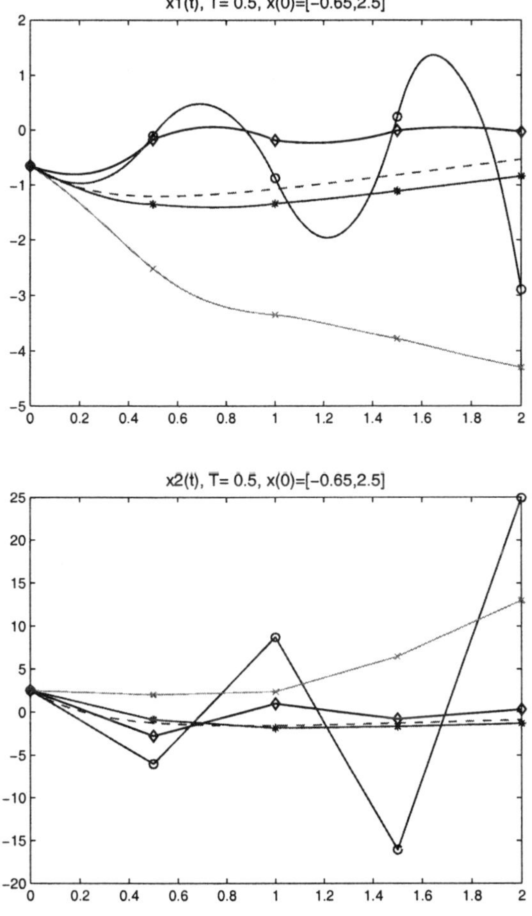

Fig. 2. $x(t)$ for example (33), $T = 0.5$. Continuous-time solution (- -); sampled–data solution for order $q = 2$ (o), $q = 3$ (x), $q = 4$ ($\diamond$) and $q = 6$ (*).

sampled-data feedbacks for $q = 4, 5$, but reveals that there does not exist a control law for order $q = 6$. However, the order $q = 6$ becomes feasible if one adds $g_3(x) = (0, 0, 1)^T u_{0,3}(x)$ with $u_{0,3} \equiv 0$ as a second additional control term. Figure 3 presents the numerical simulations of this design procedure.

The sampled continuous-time feedback $u_T = u_0$ does not retain the asymptotic stability of the continuous-time solution. Instead, it exhibits an

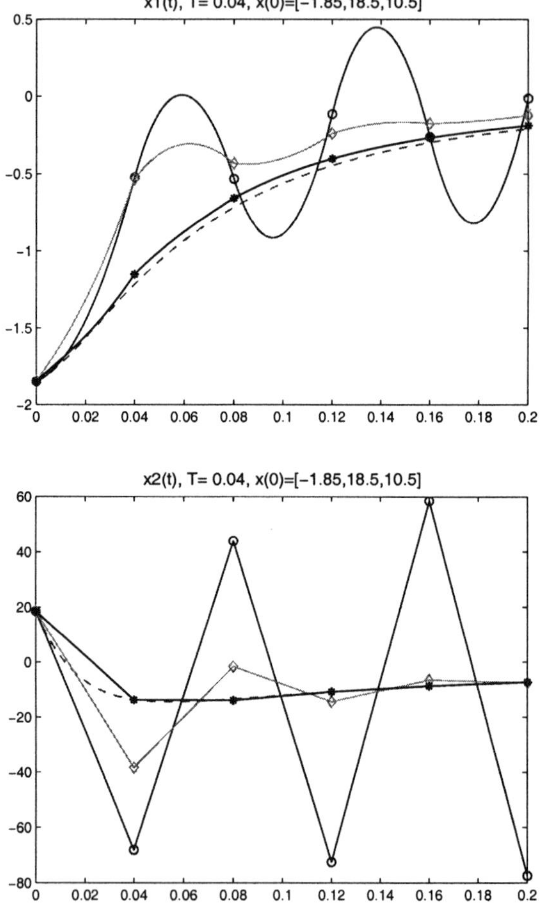

Fig. 3. $x(t)$ for example (36), $T = 0.04$. Continuous-time solution (- -); sampled–data solution for orders $q = 2$ (o), $q = 4$ ($\diamond$) and $q = 6$ (*).

asymptotically stable periodic trajectory and even divergence for sampling periods $T \geq 0.052$. In contrast to that u_T for $q = 4$ and $q = 6$ preserve the asymptotic stability for $T \leq 0.05$ for $q = 4$ and $T \leq 0.064$ for $q = 6$, respectively, while for larger sampling intervals the solutions become first periodic and eventually divergent, too.

References

1. A. Arapostathis, B. Jakubczyk, H.-G. Lee, S. Marcus and E. D. Sontag, *Syst. Contr. Lett.* **13**, 373 (1989).
2. L. Grüne and P. E. Kloeden, *Numer. Math.* **89**, 669 (2001).
3. L. Grüne and D. Nešić, *SIAM J. Control Optim.* **42**, 98 (2003).
4. L. Grüne, D. Nešić and K. Worthmann, *Automatica* (to appear).
5. L. Grüne, D. Nešić, J. Pannek and K. Worthmann, Redesign techniques for nonlinear sapled-data systems, *at-Automatisierungstechnik* (to appear).
6. A. Isidori, *Nonlinear control systems*, 3rd edn. (Prentice Hall, Upper Saddle River, New Jersey, 2002).
7. H. K. Khalil, *Nonlinear systems*, 3rd edition (Prentice Hall, Upper Saddle River, New Jersey, 2002).
8. V. Krishnamurthy, *Combinatorics: theory and applications* (Affiliated East-West Press, Madras, 1985).
9. M. Krstić, I. Kanellakopoulos and P. V. Kokotović, *Nonlinear and adaptive control design* (John Wiley & Sons, New York, 1995).
10. D. S. Laila and D. Nešić, *Automatica* **39**, 821 (2003).
11. D. S. Laila, D. Nešić and A. Astolfi, Sampled-data control of nonlinear systems, in *Advanced Topics in Control Systems Theory: Lecture Notes from FAP 2005*, eds. A. Loría, F. Lamnabhi-Lagarrigue and E. Panteley, Lecture Notes in Control and Information Sciences, Vol. 328 (Springer–Verlag, Berlin, Heidelberg, 2006) pp. 91–137.
12. S. Monaco and D. Normand-Cyrot, *Europ. J. Control* **7**, 160 (2001).
13. S. Monaco and D. Normand-Cyrot, Input-state matching under digital control, in *Proceedings of the 45th Conference on Decision and Control (San Diego, CA, 2006)*: 1806-1811.
14. D. Nešić and L. Grüne, *Automatica* **41**, 1143 (2005).
15. D. Nešić and A. R. Teel, *IEEE Trans. Automat. Control* **49 (7)**, 1103 (2004).
16. D. Nešić and A. R. Teel, *Automatica* **42**, 1801 (2006).
17. D. Nešić, A. R. Teel and P. V. Kokotović, *Syst. Contr. Lett.* **38**, 259 (1999).
18. D. Nešić, A. R. Teel and E. D. Sontag, *Syst. Contr. Lett.* **38**, 49 (1999).

ON THE DEFINITION OF TRAJECTORIES CORRESPONDING TO GENERALIZED CONTROLS ON THE HEISENBERG GROUP

P. MASON

*Institut Elie Cartan UMR 7502, Nancy-Université/CNRS/INRIA,
POB 239, Vandoeuvre-lès-Nancy 54506, France
E-mail: Paolo.Mason@iecn.u-nancy.fr*

In this paper we consider control-affine systems driven by controls that are not absolutely continuous, so that a solution in the classical sense is not defined. We discuss the possibility of defining a generalized notion of solution. Focusing on the Heisenberg example, we state some conjectures that would justify an extension of the definition of solution, and we analyze them by means of examples and counterexamples.

Keywords: Generalized Controls; Motion Planning; Stochastic Differential Equations.

1. Introduction

The aim of this paper is to discuss the possibility of generalizing the notion of solution of an affine control system where the control is supposed to belong to a functional space such that the classical results about local existence and uniqueness do not apply.

This problem has been studied extensively by control theorists for its applications to impulsive systems[1-3] and it is also interesting for probabilists since, as we will precise later, it is connected to the field of stochastic processes.[5,6,8]

The control systems under consideration here have the form

$$\dot{x} = f_0(x) + \sum_{i=1}^{m} \dot{v}_i f_i(x) \tag{1}$$

where $x \in \mathbb{R}^n$ and the vector fields f_i are smooth for every i, and satisfy some suitable growth condition (for instance we can assume that they are sublinear or bounded). We will not insist on the regularity properties of the fields f_i, which actually could be sharpened, since this is not the target of this paper. In fact our aim is to discuss the minimal requirements on $v(\cdot)$ that enable us to give a meaningful definition of solution for (1). Notice that the equation (1) is affine with respect to the derivative of the control and therefore, under the previous assumptions on f_i, it is quite natural to expect that the possible solutions have the same regularity as $v(\cdot)$.

We now define the main objects used throughout this paper. An *admissible control* for (1) is an absolutely continuous function $v(\cdot) = (v_1(\cdot), ..., v_m(\cdot)) : [0, T] \to \mathbb{R}^m$. Notice that, if v is an admissible control, then, setting $u = \dot{v}$ in equation (1), we obtain the affine control system

$$\dot{x} = f_0(x) + \sum_{i=1}^{m} u_i f_i(x), \quad u(\cdot) \in L^1([0, T], \mathbb{R}^m).$$

An *admissible trajectory* is an absolutely continuous curve $x(\cdot)$ satisfying (1) a.e. for some admissible control v. Observe that in general, if m is smaller than the dimension n of the state space, the set of the admissible trajectories for (1) is a proper subset of the class of absolutely continuous curves on $\mathbb{R}^n$.

Let $\mathcal{F} = \text{span}\{f_0 + \sum_{i=1}^{m} u_i f_i, \ u \in \mathbb{R}^m\}$ and consider the vector space generated by the Lie brackets of elements of $\mathcal{F}$

$$Lie_x(\mathcal{F}) = \text{span} \ \big\{ [g_1, [g_2, \ldots [g_{k-1}, g_k]..]](x) \ : \\ g_i \in \mathcal{F} \ \forall i = 1, \ldots, k, \ k \geq 1 \big\}. \tag{2}$$

It is well-known that the condition $Lie_x(\mathcal{F}) = T_x \mathbb{R}^n \simeq \mathbb{R}^n$ for every $x \in \mathbb{R}^n$, i.e. $\mathcal{F}$ is a *bracket-generating* family of vector fields, is a genericity condition (with respect to the $\mathcal{C}^2$ topology) for families of vector fields generated by at least two vector fields. If it is the case, then, by Krener's Theorem, the attainable set for (1), starting from any given point x_0

$$\mathcal{A}_{x_0} := \{x(T) \in \mathbb{R}^n \ : \ x(\cdot) \text{ admissible trajectory for (1)}, \ T > 0\}$$

is "full dimensional" in the sense that $clos(\overset{\circ}{\mathcal{A}}_{x_0}) \supset \mathcal{A}_{x_0}$ (here $\overset{\circ}{\mathcal{A}}_{x_0}$ denotes the set of interior points of $\mathcal{A}_{x_0}$). In particular, in the driftless case $f_0 = 0$, the attainable set coincides with $\mathbb{R}^n$, as a consequence of Chow's Theorem. Moreover every smooth trajectory can be uniformly approximated by admissible trajectories, and therefore the Motion Planning Problem is meaningful in this case (see for instance Refs. 4,9) and it is clearly strictly connected to the problem of generalizing the notion of solution for (1).

1.1. *Previous approaches and results*

In order to give a meaning to (1) in the case in which v is non-admissible one can introduce the following equation

$$dx = f_0(x)dt + \sum_{i=1}^{m} f_i(x)dv_i \qquad (3)$$

whose solutions are defined by means of the Stieltjes integration. It is clear that this new formulation allows one to give a meaning to the solution in the more general case in which $v(\cdot)$ has bounded variation. This solution can be seen as a *generalized solution* to the equation (1). Nevertheless, motivated by mechanical models (see for instance Refs. 1,2), a weaker notion of generalized solution, that applies to the case of discontinuous controls with bounded variation, has been proposed in Ref. 3. In this framework, a generalized solution can be simply regarded as the limit of a converging sequence of admissible solutions with uniformly bounded variation (see also Ref. 7). In particular, with this new definition and in the case in which the vector fields f_i, $i = 1, \ldots, m$ do not commute, the uniqueness of the solution is no more guaranteed.

Now we call $\mathcal{V}_{adm}$, $\mathcal{X}_{adm}$ the spaces of the admissible controls and the admissible trajectories and we consider the *input-output map* associated to (1), i.e.

$$\Phi^{x_0}_{adm} : v(\cdot) \in \mathcal{V}_{adm} \mapsto x(\cdot) \in \mathcal{X}_{adm} \quad \text{where} \quad \begin{cases} \dot{x} = f_0(x) + \sum_{i=1}^{m} \dot{v}_i f_i(x) \\ x(0) = x_0 \end{cases} \qquad (4)$$

A natural way to define a notion of generalized solution of the equation (1) consists in fixing some topological spaces $\mathcal{V} \supset \mathcal{V}_{adm}$, $\mathcal{X} \supset \mathcal{X}_{adm}$ in such a way that the input-output map $\Phi^{x_0}_{adm}$ can be extended to a continuous map $\Phi^{x_0} : \mathcal{V} \to \mathcal{X}$. Then a generalized solution associated to some $v(\cdot)$ belonging to the closure of $\mathcal{V}_{adm}$ with respect to the topology of $\mathcal{V}$ can be simply defined as $\Phi^{x_0}(v)(\cdot)$. For instance it is well-known[8] that in the case in which $m = 1$ the input-output map is continuous in the topology of uniform convergence and therefore in this case, associated to each continuous control $v(\cdot)$, one can define a generalized solution. This is no more true if $m > 1$ and, as one could expect, the vector fields f_i, $i > 0$ do not commute. In this case one needs to consider more appropriate topologies.

The notion of generalized solution is also useful in order to exploit the relation between stochastic differential equations (SDE) and ordinary differential equations of the form (1). In this case $v(\cdot)$ and $x(\cdot)$ represent paths of the stochastic processes that correspond respectively to the input and the output of the SDE. Notice that a path of a stochastic process is in general (with probability 1) a continuous path, so that the case of continuous (but not regular) paths turns out to be of particular importance in the literature.

One important result in this direction has been obtained by T. Lyons in Ref. 5, where the spaces $\mathcal{V}$ and $\mathcal{X}$ are identified with the space of continuous functions with bounded p-variation, with $p > 2$ (for precise definitions, see the next section). More precisely in Ref. 5 the Picard iteration method has been used to prove the existence and uniqueness of the solutions of the equation (3). Such results are inspired by the work of L.C. Young,[10] who essentially proved that the Stieltjes integrals of the form $\int f\,dg$, where f, g have finite p and q variation and $\frac{1}{p} + \frac{1}{q} > 1$, are well-defined.

In Ref. 6 the problem of generalizing the notion of solution to the case in which $v(\cdot)$ has finite p-variation, with $p \geq 2$, has been treated. Even if the results obtained are powerful, they are very difficult to apply for several reasons. Indeed such approach does not guarantee the existence of a generalized solution for any v in the class of functions with bounded p-variation. Moreover, if a solution exists, in general it is not unique. Finally the very theoretical approach precludes the possibility of direct applications of the results.

Our aim is to discuss some alternative ways of defining generalized solutions which are easier to handle. In particular we analyze some conjectures related to the Heisenberg system with the help of some examples and counterexamples.

2. Functional spaces and topologies

In this section we introduce some functional spaces that can be seen as generalizations of the classical spaces of Lipschitz, BV and absolutely continuous functions. The first interesting functional space is the space of the Hölder-α functions $\mathcal{C}^{0,\alpha}([0,T])$, with $\alpha \in (0,1)$, i.e. the space of real-valued (or vector-valued) functions satisfying $\sup_{t_1,t_2 \in [0,T]} \frac{|f(t_1)-f(t_2)|}{|t_1-t_2|^\alpha} < +\infty$. It is easy to verify that it is a Banach space with norm

$$\|f\|_\alpha := |f(0)| + \sup_{t_1,t_2 \in [0,T]} \frac{|f(t_1) - f(t_2)|}{|t_1 - t_2|^\alpha}. \tag{5}$$

It is interesting to investigate the properties of $\mathcal{C}^{0,\alpha}([0,T])$ as α varies. For this purpose the following simple result is useful.

Proposition 2.1. *If $f \in \mathcal{C}^{0,\alpha'}([0,T])$ with $\alpha' > \alpha$, $\lim_{n\to\infty} f_n = f$ uniformly and $\|f_n\|_{\alpha'}$ is uniformly bounded, then $\lim_{n\to\infty} f_n = f$ in the topology* *of* $\| \cdot \|_\alpha$.

Proof. Assume by contradiction that f_n does not converge to f in the norm $\| \cdot \|_\alpha$. Then there must exist a subsequence f_{n_k} and two sequences $x_k, y_k \in [0, T]$ such that

$$\frac{|f_{n_k}(x_k) - f_{n_k}(y_k) - f(x_k) + f(y_k)|}{|x_k - y_k|^\alpha} > C \tag{6}$$

for some $C > 0$. Therefore we can suppose that, up to subsequences, $\lim_{k \to \infty} x_k = \bar{x}$, $\lim_{k \to \infty} y_k = \bar{y}$, and we deduce, from the uniform convergence of f_n to f, that the inequality (6) may hold only if $\bar{x} = \bar{y}$. This clearly contradicts our hypotheses, since the right-hand side of (6) is bounded by $\|f_{n_k} - f\|_{\alpha'} |x_k - y_k|^{\alpha' - \alpha}$, which tends to 0. $\square$

Since for every $f \in \mathcal{C}^{0,\alpha'}([0, T])$ it is easy to find a sequence of smooth functions approximating f as in the hypotheses of the previous result, we get the following corollary.

Corollary 2.1. *The closure of* $\mathcal{C}^\infty([0, T])$ *with respect to* $\| \cdot \|_\alpha$ *contains the set* $\bigcup_{\alpha' > \alpha} \mathcal{C}^{0,\alpha'}([0, T])$.

Notice that the closure of $\mathcal{C}^\infty([0, T])$ with respect to $\| \cdot \|_\alpha$ doesn't coincide with $\mathcal{C}^{0,\alpha}([0, T])$ since, for instance, the function $f(t) = t^\alpha$ cannot be approximated by a smooth function in this norm.

Observe also that $\mathcal{C}^{0,\alpha}([0, T])$ contains properly the space of Lipschitz functions, but it does not contain the space of absolutely continuous functions, since t^β is not Hölder-α for every $\beta < \alpha$. This means that each possible definition of solution to the equation (1) that applies to the case in which $v(\cdot)$ is Hölder cannot be considered as a generalization of the classical notion of Carathéodory solution. Therefore we introduce new spaces that are more general.

The space BV_p of functions with finite p-variation, introduced by Wiener, is the set of functions f satisfying the following condition

$$\sup_{0 = t_0 < t_1 < \ldots < t_N = T} \sum_i |f(t_{i+1}) - f(t_i)|^p < +\infty. \tag{7}$$

In particular by the Minkowski inequality it follows trivially that the map

$$\|f\|_{BV_p} := |f(0)| + \sup_{0 = t_0 < t_1 < \ldots < t_N = T} \left(\sum_i |f(t_{i+1}) - f(t_i)|^p \right)^{1/p}$$

defines a norm on BV_p and the space BV_p turns out to be a Banach space. Moreover it is clear that $\mathcal{C}^{0,1/p} \subset BV_p$.

We give a particular importance to the BV_p functions that are also continuous. In particular we mention two functional spaces that are contained in the class of continuous and BV_p functions. The first one is the set of *regular* BV_p functions[5] (that we denote by BV_p^{reg}), i.e. the functions f satisfying the following condition

$$\limsup_{\substack{0=t_0<t_1<\ldots<t_N=T \\ \max|t_{i+1}-t_i|\to 0}} \sum_i |f(t_{i+1}) - f(t_i)|^p = 0 . \tag{8}$$

It is not difficult to see that the closure of $\mathcal{C}^\infty$ with respect to the norm $\|\cdot\|_{BVp}$ is contained inside the space of BV_p^{reg} functions.

The other functional space that we introduce is a natural generalization of the space of absolutely continuous functions: we say that $f \in AC_p$ if it satisfies the following property:

$$\forall \varepsilon > 0 \quad \exists \delta \text{ s.t. } \sum_i |t_i - s_i| < \delta \Rightarrow \sum_i |f(t_i) - f(s_i)|^p < \varepsilon . \tag{9}$$

Notice that for every $p \geq 1$ we have $\mathcal{C}^{0,1/p} \subset AC_p$. Moreover it is easy to see that we have the following inclusions for every $1 \leq p < p'$:

$$BV_p \cap \mathcal{C} \subset BV_{p'}^{reg} \subset AC_{p'} \subset BV_{p'} \cap \mathcal{C} \tag{10}$$

Indeed these inclusion can be easily derived from the definitions and the uniform continuity (we are working with continuous functions defined on an interval).

In some sense this series of inclusions shows that the spaces that we have introduced are "almost" equivalent, since infinitesimal changes of the parameter p can reverse the inclusions.

3. The Heisenberg example

We consider now the simplest driftless completely non holonomic system, namely a system of the form

$$\dot{x} = \dot{v}_1 F_1(x) + \dot{v}_2 F_2(x), \tag{11}$$

where $x \in \mathbb{R}^3$ and F_1 and F_2 generates the Heisenberg algebra i.e.

$$F_1 = \begin{pmatrix} 1 \\ 0 \\ -x_2/2 \end{pmatrix}, F_2 = \begin{pmatrix} 0 \\ 1 \\ x_1/2 \end{pmatrix} \tag{12}$$

234

Notice that $[F_1, F_2] = (0, 0, 1)^T$ and that the system is nilpotent, in the sense that the brackets of order greater than two annihilate. For every admissible control $v(\cdot)$ the first two components of the corresponding admissible trajectory coincide with $(v_1(\cdot), v_2(\cdot))$ up to additive constants. Moreover the third component is characterized by the equation

$$\dot{x}_3 = \frac{1}{2}(x_1\dot{x}_2 - \dot{x}_1 x_2) \tag{13}$$

that means that, if we set

$$\mathcal{A}[x(\cdot)](t) := \int_0^t \frac{1}{2}(x_1(s)\dot{x}_2(s) - \dot{x}_1(s)x_2(s))\ ds, \tag{14}$$

that is the area computed in counterclockwise sense of the region spanned

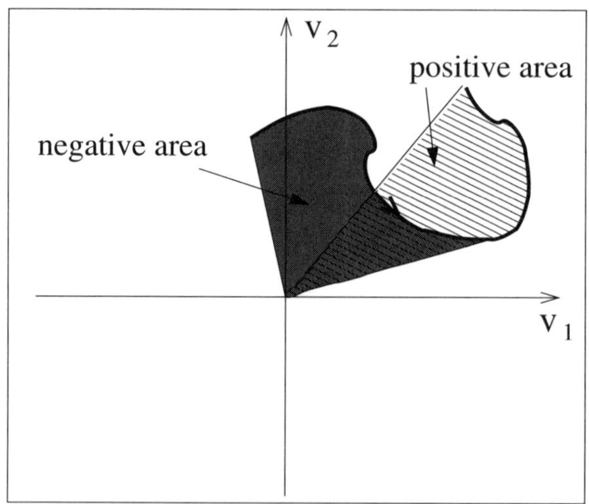

Fig. 1.   The area spanned by a curve

by the planar curve $(x_1(\cdot), x_2(\cdot))$ (see Figure 1), then $x_3(t) = \mathcal{A}[x(\cdot)](t)$. Consider now a non admissible control $v(\cdot)$ in equation (11), and suppose that it is possible to define a generalized solution for v by means of a continuous extension of the input-output map. Then the first two components must coincide with the corresponding controls up to constants. Therefore the problem of generalizing the notion of solution reduces to the following:

Problem: *Given a continuous, but not absolutely continuous, curve $x(\cdot) = (x_1(\cdot), x_2(\cdot))$, is it possible to give a definition of the area generalizing (14)?*

We focus in particular on curves belonging to the spaces $\mathcal{C}^{0,\alpha}([0,T])$. We start with a simple example after which we will discuss some possible conjectures. Fix $\alpha \in]0,1[$ and consider the following sequence of admissible controls:

$$v_\alpha^n(t) = \left(\frac{1}{n^\alpha}\cos(n\,t), \frac{1}{n^\alpha}\sin(n\,t)\right). \tag{15}$$

It is clear that $\|v_\alpha^n(\cdot)\|_\infty \to 0$ and therefore $v_\alpha^n(\cdot)$ is an approximating sequence of the zero function in the $\mathcal{C}^0$ topology. The corresponding area is the following:

$$\mathcal{A}[v_\alpha^n(\cdot)](t) = n^{(1-2\alpha)}t. \tag{16}$$

Therefore we can distinguish three cases:

- If $\alpha > 1/2$ then $\mathcal{A}[v_\alpha^n(\cdot)](t)$ converges uniformly to zero as n goes to ∞,
- If $\alpha = 1/2$ then $\mathcal{A}[v_\alpha^n(\cdot)](t)$ converges uniformly to the function $v(t) = t$,
- If $\alpha < 1/2$ then $\mathcal{A}[v_\alpha^n(\cdot)](t)$ diverges.

This example with $\alpha \leq 1/2$ is enough to conclude that the $\mathcal{C}^0$ norm does not render the input-output map continuous. Therefore we should look for a stronger norm. For this purpose, it is interesting to notice that the functions $v_\alpha^n(\cdot)$ are such that the sequence $\|v_\alpha^n\|_\alpha$ is uniformly bounded. More precisely

$$\lim_{n\to\infty} \|v_\alpha^n\|_{\alpha'} = \begin{cases} 0 & \text{if } \alpha' > \alpha \\ \frac{2}{\pi^\alpha} & \text{if } \alpha' = \alpha \\ \infty & \text{if } \alpha' < \alpha \end{cases}$$

This suggests some natural conjectures. Let $\mathcal{V}_{adm}^{\alpha,K} = \{v \in \mathcal{C}^\infty,\ \|v\|_\alpha \leq K\}$. Then:

C1 If $\alpha > 1/2$ then the input-output map can be extended continuously to the set $\{v \in \mathcal{C}^{0,\alpha},\ \|v\|_\alpha \leq K\}$, seen as the closure of $\mathcal{V}_{adm}^{\alpha,K}$ with respect to the norm $\|.\|_\infty$.

In[5] it was proved that, if we consider the topology given by $\|\cdot\|_{BV_p}$ with $p < 2$, the input-output map can be extended continuously to a map Φ^{x_0} defined on $BV_p \cap \mathcal{C}$. Combining this result with Proposition 2.1 we obtain easily that the conjecture **C1** is true:

$$v_n \stackrel{\|\cdot\|_\infty}{\to} v, \quad \|v_n\|_\alpha \le K \quad \Longrightarrow \quad v_n \stackrel{\|\cdot\|_{\alpha'}}{\to} v \quad \forall \alpha' \in (1/2, \alpha)$$

$$\Longrightarrow v_n \stackrel{\|\cdot\|_{BV^p}}{\to} v, \quad p := 1/\alpha' \quad \Longrightarrow \quad \Phi^{x_0}(v_n) \stackrel{\|\cdot\|_{BV^p}}{\to} \Phi^{x_0}(v)$$

$$\Longrightarrow \quad \Phi^{x_0}(v_n) \stackrel{\|\cdot\|_\infty}{\to} \Phi^{x_0}(v)$$

As a consequence, we obtain that it is possible to generalize the definition of the area $\mathcal{A}[v(\cdot)](t)$ for $v \in \mathcal{C}^{0,\alpha}$ and $\alpha > 1/2$. For $\alpha \le 1/2$ the previous example shows that the analogous conjecture is not true. However for $\alpha = 1/2$ one could still expect the possibility of defining the area on subsequences.

C2 Given $v \in \mathcal{C}^{0,\frac{1}{2}}$, for every sequence of smooth functions v_n converging uniformly to v and with $\|v_n\|_{1/2}$ uniformly bounded, there exists a subsequence such that the corresponding admissible solutions of (11) (or, more in general, of (1)) form a Cauchy sequence in the uniform topology.

If this conjecture was false than one could still try to consider a stronger norm on the space of controls, getting the following conjecture.

C3 If the space of the admissible controls is endowed with the topology given by $\| \cdot \|_{1/2}$, then it is possible to extend continuously the input-output map.

In the previous example the functions were constructed in order to converge to 0, but it is clear that this is far from being a natural way of approximating the function $v = 0$. One can also try to overcome this problem by defining directly a notion of convergence of sequences of functions. For instance, given a function v and a sequence of times $0 = t_0 < t_1 < \ldots < t_{k-1} < t_k = T$, one can consider the piecewise affine approximations obtained joining successively by segments the points $v(t_h)$ and $v(t_{h+1})$, for $h = 0, \ldots, k$. Therefore another possible conjecture is the following

C4 Let $v \in \mathcal{C}^{0,\frac{1}{2}}$ and $v^{(k)}$ be a sequence of piecewise affine approximating functions defined as before and characterized by the times $t_h^{(k)}$, and assume that $\lim_{k \to \infty} \sup_h |t_{h+1}^{(k)} - t_h^{(k)}| = 0$. Then the sequence $\mathcal{A}[v^{(k)}](T)$ is a Cauchy sequence.

We discuss the above conjectures with the help of some examples.

Example A. Consider the following function

$$v : [0, 2\pi] \to \mathbb{C} \qquad v(t) = \sum_{k=0}^{+\infty} \frac{1}{2^k} e^{4^k i t}, \tag{17}$$

which represents a closed curve in the complex plane. We want to prove the following facts.

- $v(\cdot) \in C^{0,\frac{1}{2}}$.
- Consider the sequence of times $t_h = \frac{2\pi h}{4^k}$ $\quad h = 0, \ldots, 4^k - 1$ and define v_k as the piecewise affine function obtained joining the point $v(t_h)$ to $v(t_{h+1})$ for every $h = 0, \ldots, 4^k - 1$, i.e

$$v^{(k)}(t) := v(t_h) + \frac{v(t_{h+1}) - v(t_h)}{t_{h+1} - t_h}(t - t_h) \quad \text{if} \quad t \in [t_h, t_{h+1}].$$

Then $\lim_{k \to +\infty} \mathcal{A}[v^{(k)}(\cdot)](2\pi) = +\infty$.

For the first issue we first fix two real numbers x and y. Then we define an integer number Q depending on x and y (in the computations that follow, $[\cdot]$ denotes the integer part):

$$Q := \left[\log_2 \left(\frac{1}{\sqrt{|x - y|}} \right) \right] \iff 2^Q \leq \frac{1}{\sqrt{|x - y|}} < 2^{Q+1}$$

$$\iff 2^{-Q} \geq \sqrt{|x - y|} > 2^{-Q-1}.$$

We have that

$$|v(x) - v(y)| \leq \left| \sum_{k=0}^{Q} \frac{1}{2^k} \left(e^{4^k i x} - e^{4^k i y} \right) \right| + \left| \sum_{k=Q+1}^{+\infty} \frac{1}{2^k} \left(e^{4^k i x} - e^{4^k i y} \right) \right|.$$

For the second term we have

$$\left| \sum_{k=Q+1}^{+\infty} \frac{1}{2^k} \left(e^{4^k i x} - e^{4^k i y} \right) \right| < 2 \sum_{k=Q+1}^{+\infty} \frac{1}{2^k} = 2^{-Q+1} < 4\sqrt{|x - y|}$$

while we can estimate the first one using the inequality $|e^{i\alpha} - e^{i\beta}| < |\alpha - \beta|$:

$$\left| \sum_{k=0}^{Q} \frac{1}{2^k} \left(e^{4^k i x} - e^{4^k i y} \right) \right| \leq \sum_{k=0}^{Q} \frac{1}{2^k} 4^k |x - y| = \sum_{k=0}^{Q} 2^k |x - y|$$

$$< 2^{Q+1} |x - y| \leq 2\sqrt{|x - y|}.$$

Finally we have found that the inequality $|v(x) - v(y)| \leq 6\sqrt{|x - y|}$ holds for every x and y in the interval $[0, 2\pi]$ and therefore $v(\cdot) \in C^{0,\frac{1}{2}}$.

Remark 3.1. Observe that each term in the Fourier series of $v(\cdot)$ is Hölder-$\frac{1}{2}$ with the same optimal constant.

Now we want to see that the function $v(\cdot)$ defined above, as a function from $[0, 2\pi]$ to $\mathbb{R}^2$, disproves the conjectures **C2** and **C4**. Consider the 4^N times $t_h = \frac{2\pi h}{4^N}$ $h = 0, \ldots, 4^N - 1$. Then the area corresponding to the segment connecting $v(t_h)$ to $v(t_{h+1})$ is given by $\frac{1}{2}\left(v_2(t_h)v_1(t_{h+1}) - v_1(t_h)v_2(t_{h+1})\right)$, where v_1, v_2 denote the first and second components of v.

In particular we have

$$v_1(t_h) = \sum_{k=0}^{N-1} \frac{1}{2^k} \cos(4^{k-N}2\pi h) + \sum_{k=N}^{\infty} \frac{1}{2^k} = \sum_{k=0}^{N-1} \frac{1}{2^k} \cos(4^{k-N}2\pi h) + 2^{-N+1},$$

$$v_2(t_h) = \sum_{k=0}^{N-1} \frac{1}{2^k} \sin(4^{k-N}2\pi h).$$

Therefore

$$v_1(t_h)v_2(t_{h+1}) - v_2(t_h)v_1(t_{h+1}) = \sum_{k=0}^{N-1} \frac{2^{-N+1}}{2^k}\left(\sin(4^k t_{h+1}) - \sin(4^k t_h)\right)$$
$$+ \sum_{n,m=0}^{N-1} \frac{1}{2^{n+m}}\left(\cos(4^n t_h) \sin(4^m t_{h+1}) - \cos(4^m t_{h+1}) \sin(4^n t_h)\right).$$

If we consider the sum over $h = 0, \ldots 4^N - 1$ of these terms and we exchange the order of summation we obtain the area

$$\mathcal{A}[v^{(N)}(\cdot)](2\pi) = \sum_{n,m=0}^{N-1} \frac{1}{2^{n+m}} \sum_{h=0}^{4^N-1} \sin(4^{m-N}2\pi(h+1) - 4^{n-N}2\pi h) \,.\tag{18}$$

Now we want to see that

$$\sum_{h=0}^{4^N-1} \sin(4^{m-N}2\pi(h+1) - 4^{n-N}2\pi h) = 0 \tag{19}$$

if $n \neq m$ $(n, m < N)$. We rewrite $4^m(h+1) - 4^n h = h(4^m - 4^n) + 4^m = hp + q$.

Let M be the greatest common divisor of p and 4^N, and define $D := \frac{4^N}{2M}$. If we prove that

$$(h + D)p + q \equiv hp + q + \frac{4^N}{2} \pmod{4^N}, \tag{20}$$

then $\frac{2\pi}{4^N}((h + D)p + q) \equiv \frac{2\pi}{4^N}(hp + q) + \pi$ and it is clear that the sum of $2D$ successive terms in (19) is 0. Therefore, since $2D$ divides 4^N, the whole series (19) is 0.

To prove (20) notice that $2Dp = 4^N \frac{p}{M} \equiv 0 \pmod{4^N}$, so that either $Dp \equiv 0$

$(\text{mod } 4^N)$ or $Dp \equiv \frac{4^N}{2} \; (\text{mod } 4^N)$. In the first case we would have $4^N \frac{p}{2M} \equiv 0 \; (\text{mod } 4^N)$ which means that $2M$ divides p, which is impossible, since otherwise $2M$ would be a common divisor of p and 4^N. Therefore we obtain $Dp \equiv \frac{4^N}{2} \; (\text{mod } 4^N)$ that immediately gives (20).

So the only non-vanishing terms in (18) correspond to $n = m$

$$\mathcal{A}[v^{(N)}(\cdot)](2\pi) = \sum_{n=0}^{N-1} \frac{1}{4^n} \sum_{h=0}^{4^N-1} \sin(4^{n-N}2\pi)$$

$$= \sum_{n=0}^{N-1} 4^{N-n} \sin(4^{n-N}2\pi) = \sum_{m=1}^{N} 4^m \sin(4^{-m}2\pi).$$

Since $\lim_{m\to\infty} 4^m \sin(4^{-m}2\pi) = 1$ we have that $\lim_{N\to+\infty} \mathcal{A}[v^{(N)}(\cdot)](2\pi) = +\infty$.

Remark 3.2. There is a easy geometric interpretation of this example. Indeed, in (17) each term in the summation corresponds to 4^k turns on a circle centered at the origin with radius $\frac{1}{2^k}$. The area corresponding to such path is exactly π and it is easy to see that the area of a finite sum of l elements in (17) is exactly $l\pi$. As suggested by Figure 2, where the image of $v(\cdot)$ is drawn, the fact that $v(\cdot)$ is given by the sum of an infinite number of circles with decreasing radius and increasing speed renders this function very irregular.

Remark 3.3. A similar example has been discussed by L.C. Young,[10] in order to prove that a particular inequality, that would justify the existence of the Stieltjes integral $\int f \, dg$, does not hold in general if f and g have bounded 2-variation.

Remark 3.4. We proved that there is a sequence of piecewise affine approximations of v such that the corresponding area goes to infinity, but we don't know if this remains true for all possible converging sequences of piecewise affine approximations. However, the geometric interpretion given in Remark 3.2, suggests that this is true. On the other hand, in general, the area does not converge to infinity for smooth approximations (also with uniformly bounded Hölder constant), since

$$v^{(N)}(t) := \sum_{k=0}^{N-1} \frac{1}{2^k} e^{4^k i t} + \sum_{k=N}^{2N-1} \frac{1}{2^k} e^{-4^k i t}$$

is such that $\mathcal{A}[v^{(N)}](2\pi) = 0 \quad \forall N > 0$.

240

Example B. We consider now a slightly modified example:

$$\tilde{v}(t) = \sum_{k=0}^{+\infty} \frac{1}{2^k} e^{(-4)^k i t}$$

In this case, exactly as before, one can prove the following.

- $\tilde{v}(\cdot)$ is Hölder-$\frac{1}{2}$.
- The sequence of times $t_h = \frac{2\pi h}{4^k}$ $h = 0, \dots 4^k - 1$ is such that the area of the corresponding piecewise affine approximation of $\tilde{v}(\cdot)$ does not converge as k goes to infinity even if it is uniformly bounded. More precisely the area of such approximation is

$$\mathcal{A}[\tilde{v}^{(k)}(\cdot)](2\pi) = (-1)^k \sum_{m=1}^{k} (-4)^m \sin(4^{-m}2\pi)$$

and it is easy to see that $\mathcal{A}[\tilde{v}^{(2k)}(\cdot)](2\pi)$ and $\mathcal{A}[\tilde{v}^{(2k+1)}(\cdot)](2\pi)$ converge to two different values as k goes to infinity.

The image of $\tilde{v}$ is depicted in Figure 2.

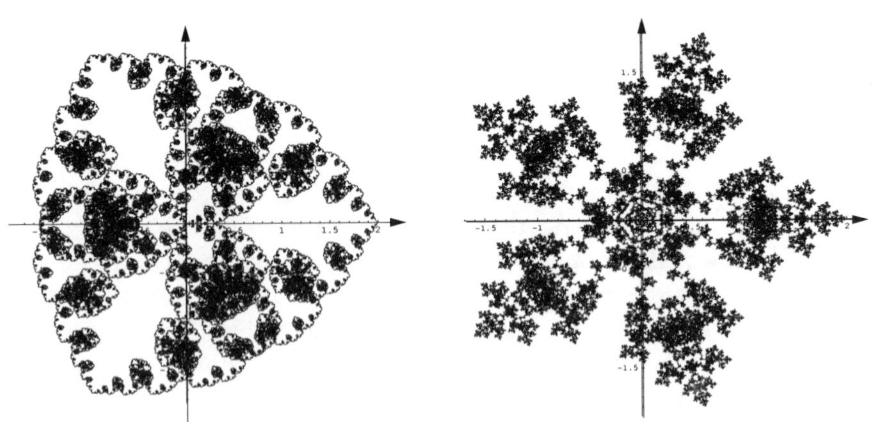

Fig. 2. The images of v and $\tilde{v}$ in the complex plane

Example C. Consider now the following function.

$$\hat{v} : [0, 2\pi] \to \mathbb{C} \qquad \hat{v}(t) = \sum_{k=0}^{+\infty} \frac{1}{\sqrt{k}2^k} e^{4^k i t}.$$

This function satisfies the same properties as the one of the first example, and such properties can be proved exactly in the same way. The most interesting feature of this example is that the approximating sequence $\hat{v}^{(N)} = \sum_{k=0}^{N} \frac{1}{\sqrt{k}2^k} e^{4^k i t}$ converges to $\hat{v}$ in the norm $\| \cdot \|_{1/2}$, while we still have $\lim_{k \to +\infty} \mathcal{A}[\hat{v}^{(N)}](2\pi) = +\infty$. Moreover, similarly to Remark 3.4 it is easy to construct an approximating sequence converging to $\hat{v}$ in the norm $\| \cdot \|_{1/2}$ and such that the corresponding area is uniformly bounded.

Therefore also conjecture **C3** turns out to be false.

4. Conclusion

We have seen that it is not easy to find a definition of generalized solution that applies to the case of Hölder-$\frac{1}{2}$ controls. However notice that, for instance, the first example we have constructed is very particular, since the components v_1 and v_2 seem to "cooperate" in order to increase the area. Therefore it seems quite natural to look for some particular "non-resonance" condition on the space of controls, that could also represent the case of two paths of independent stochastic processes.

Then one could try to define a generalized solution of (11) extending continuously the input-output map relatively to the class of controls satisfying this non-resonance condition. Our future research on this field will be based on this new approach.

Acknowledgements

The author wishes to thank Prof. Andrei A. Agrachev, Dr. Ugo Boscain and Prof. Alberto Bressan for their suggestions and comments on the subject of this paper.

References

1. Aldo Bressan, *Atti Accad. Naz. dei Lincei, Memorie classe di Scienze Mat. Fis. Nat. (9)* **1**, 149 (1989).
2. Aldo Bressan, *Atti Accad. Naz. dei Lincei, Memorie classe di Scienze Mat. Fis. Nat. (8)* **19**, 249 (1989).
3. Alberto Bressan, F. Rampazzo, *Boll. Un. Mat. Ital. B (7)* **2**, 641 (1988).
4. W. Liu, *SIAM J. Control Optim.* **35**, 1989 (1997).

242

5. T. Lyons , *Math. Res. Lett.* **1**, 451 (1994).

6. T. Lyons, *Rev. Mat. Iberoamericana* **14**, 215 (1998).

7. P. Mason, Sistemi di Controllo Impulsivi su Varietà, Tesi di Laurea, Università degli studi di Padova, (Padova, 2002).

8. H.J. Sussmann, *Ann. Probability* **6**, 19 (1978).

9. H. J. Sussmann and W. Liu, Limits of highly oscillatory controls and the approximation of general paths by admissible trajectories, in *Proceedings of the 30th IEEE 1991 Conference on Decision and Control, Brighton, UK* (Brighton, UK, December 11–13, 1991): 437–442.

10. L.C. Young, *Acta Math.* **67**, 251 (1936).

CHARACTERIZATION OF THE STATE CONSTRAINED BILATERAL MINIMAL TIME FUNCTION

C. NOUR

Computer Science and Mathematics Division
Lebanese American University
Byblos Campus, P.O. Box 36
Byblos, Lebanon
E-mail: cnour@lau.edu.lb

We characterize the state constrained bilateral minimal time function as the unique proximal solution of a partial Hamilton-Jacobi equations with certain boundary conditions. This generalizes [13, Theorem 3.4] where Stern studied the unilateral case.

Keywords: Bilateral and unilateral minimal time function; State constraint, Hamilton-Jacobi equations; Proximal analysis; Nonsmooth analysis

1. Introduction

Let F be a multifunction mapping points x in $\mathbb{R}^n$ to subsets $F(x)$ of $\mathbb{R}^n$ and let $S \subset \mathbb{R}^n$ a closed set. Associated with F is the differential inclusion

$$\dot{x}(t) \in F(x(t)) \;\; a.e.\; t \in [0, T], \;\; x(0) = x_0. \tag{1}$$

A solution to (1) is an absolutely continuous function $x(\cdot)$ defined on the interval $[0, T]$ with initial value $x(0) = x_0$, in which case we say that $x(\cdot)$ is a trajectory of F that originates from x_0. The notation $\dot{x}(t)$ refers to the derivative of $x(\cdot)$ at t and is the right derivative if $t = 0$.

We assume throughout this paper that F satisfies the *standing hypotheses*; that is, F takes nonempty compact convex values, has closed graph, and satisfies a linear growth condition: for some positive constants γ and c, and for all $x \in \mathbb{R}^n$,

$$v \in F(x) \implies \|v\| \leq \gamma\|x\| + c.$$

The multivalued function F is also taken to be locally Lipschitz: every $x \in \mathbb{R}^n$ admits a neighborhood $U = U(x)$ and a positive constant $K = K(x)$

such that

$$x_1, x_2 \in U \implies F(x_2) \subseteq F(x_1) + K\|x_1 - x_2\|\overline{B}.$$

We associate with F the following function h, the *lower Hamiltonian*:

$$h(x, p) := \min\{\langle p, v \rangle : v \in F(x)\}.$$

Now let $S \subset \mathbb{R}^n$ be a closed set. A trajectory $x(t)$ of F over $[0, T]$ which satisfies the state constraint $x(t) \in S$ for all $t \in [0, T]$ is called an S-trajectory of F. For $\Sigma \subset S$, a closed set, the S-constrained (unilateral) minimal time function (associated with Σ) $T_S(\cdot, \Sigma) : S \longrightarrow \mathbb{R}^n \cup \{+\infty\}$ is defined as follows: For $\alpha \in S$, $T_S(\alpha, \Sigma)$ is the minimum time taken by an S-trajectory to go from α to Σ (when no such trajectory exists, $T_S(\alpha, \Sigma)$ is taken to be $+\infty$). We assume throughout this note that $T_S(\cdot, \Sigma)$ is extended to be $+\infty$ on S^c, the complement of S. When $S = \mathbb{R}^n$ (unconstrained case) then $T_S(\cdot, \Sigma)$ coincides with the well known minimal time function (denoted by $T(\cdot, \Sigma)$) which has a large literature; see for example Refs. 4,10,11,14,15 for the regularity study of $T(\cdot, \Sigma)$, and Refs. 1,3,6,7,12,15 for its Hamilton-Jacobi characterization. We also invite the reader to see [2, Chapter 4] for a thorough history of such results.

The Hamilton-Jacobi characterization of the state constrained (unilateral) minimal time function is studied (apparently for the first time) in Stern[13] where a characterization of this function is given as the unique proximal solution of Hamilton-Jacobi equations with certain boundary conditions. Let us recall the main result of this paper. We begin with some definitions. Our reference will be the book of Clarke *et al.*[5] and the paper of Stern[13]. For a lower semicontinuous function $f : \mathbb{R}^n \longrightarrow \mathbb{R} \cup \{+\infty\}$ and a point $x \in \text{dom} f := \{x' : f(x') < +\infty\}$, we denote by $\partial_P f(x)$ the *proximal subdifferential* of f at x. We recall that $\zeta \in \partial_P f(x)$ if and only if there exists $\sigma = \sigma(x, \zeta) \geq 0$ such that

$$f(y) - f(x) + \sigma\|y - x\|^2 \geq \langle \zeta, y - x \rangle,$$

for all y in a neighborhood of x.

We say that S is *wedged* provided that at each boundary point x one has pointedness of $N_S^C(x)$; that is, $N_S^C(x) \cap \{-N_S^C(x)\} = \{0\}$, where $N_S^C(x)$ denotes the *Clarke normal cone* to S at x. We note that wedgedness implies S is the closure of its interior, and that any closed convex body is wedged. We also say that *strict inwardness* condition is satisfied if

$$h(x, \zeta) < 0 \quad \forall \zeta \in N_S^C(x), \quad \forall x \in \text{bdry } S.$$

A function $\varphi : \mathbb{R}^n \longrightarrow]-\infty, +\infty]$ is said to be (Σ, S)-*continuous* if there exist $\gamma_\varphi > 0$ and a function $\omega_\varphi : [0, \gamma] \longrightarrow [0, +\infty[$ such that

$$\lim_{s\downarrow 0} \omega(s) = 0 \text{ and } \varphi(x) \leq \omega_\varphi(d_\Sigma(x)) \ \forall x \in S \cup \{\Sigma + \gamma_\varphi B\},$$

where $d_\Sigma(\cdot)$ is the Euclidean distance function and B is the unit open ball. If $T_S(\cdot, \Sigma)$ is (Σ, S)-continuous then we say that F is S-*constrained small time controllable*. It is well known that the S-constrained small time controllability condition is equivalent to the continuity of $T_S(\cdot, \Sigma)$ on $S \cap \{\Sigma + \nu B\}$ for some $\nu > 0$. Now we can state the main result of Stern[13].

Theorem 1.1. *Assume that S is compact and wedged, and that the strict inwardness condition holds. Assume also that F is S-constrained small time controllable. Then the function $T_S(\cdot, \Sigma)$ is the unique lower semicontinuous function $\varphi : \mathbb{R}^n \longrightarrow]-\infty, +\infty]$ which is (Σ, S)-continuous, bounded below on $\mathbb{R}^n$, identically 0 on Σ, identically $+\infty$ on S^c and satisfies*

- $h(x, \partial_P \varphi(x)) + 1 = 0$ *for all* $x \in \{\text{int } S\} \setminus \Sigma$,
- $h(x, \partial_P \varphi(x)) + 1 \geq 0$ *for all* $x \in \{\text{int } S\} \cap \Sigma$, *and*
- $h(x, \partial_P \varphi(x)) + 1 \leq 0$ *for all* $x \in \{\text{bdry } S\} \setminus \Sigma$,

The purpose of this note is to generalize the preceding result to the bilateral case, that is, find a Hamilton-Jacobi characterization for the S-*constrained bilateral minimal time function* $T_S : S \times S \longrightarrow [0, +\infty]$ defined by the following: For $(\alpha, \beta) \in S \times S$, $T_S(\alpha, \beta)$ is the minimum time taken by a S-trajectory to go from α to β (when no such trajectory exists, $T_S(\alpha, \beta)$ is taken to be $+\infty$). We note that the unconstrained bilateral case is studied in Nour[8] where we give a Hamilton-Jacobi characterization and some regularity results.

In the next section we present our main result and then we sketch its proof. More details can be found in the forthcoming paper of Nour & Stern[9].

2. Main result

We begin with the following definition. We say that a function $\varphi : \mathbb{R}^n \times \mathbb{R}^n \longrightarrow]-\infty, +\infty]$ is *bilateral S-continuous* if there exist $\gamma_\varphi > 0$ and a function $\omega_\varphi : [0, \gamma_\varphi] \longrightarrow [0, +\infty[$ such that $\lim_{s\downarrow 0} \omega_\varphi(s) = 0$ and

$$\varphi(x, y) \leq \omega_\varphi(\|x - y\|) \ \forall \alpha \in S, \ \forall x, y \in (\alpha + \gamma_\varphi B) \cap S.$$

If the function $T_S(\cdot, \cdot)$ is bilateral S-continuous then we say that F is *bilateral S-constrained small time controllable*.

Let $D : \{(\alpha, \alpha) : \alpha \in \mathbb{R}^n\}$ be the diagonal set. The following is our main result (here the function $T_S(\cdot, \cdot)$ is also assumed to be $+\infty$ outside $S \times S$).

Theorem 2.1. *Assume that S is compact and wedged, and that the strict inwardness condition holds. Assume also that F is bilateral S-constrained small time controllable. Then the function $T_S(\cdot, \cdot)$ is the unique lower semicontinuous function $\varphi : \mathbb{R}^n \times \mathbb{R}^n \longrightarrow] - \infty, +\infty]$ which is bilateral S-continuous, bounded below on $\mathbb{R}^n \times \mathbb{R}^n$, identically 0 on $D \cap (S \times S)$, identically $+\infty$ on $(S \times S)^c$ and satisfies*

- $\forall \alpha \neq \beta \in \text{int } S$ *and* $\forall(\zeta, \theta) \in \partial_P \varphi(\alpha, \beta)$ *we have:*

$$1 + h(\alpha, \zeta) = 0,$$

- $\forall \alpha \in \text{int } S$ *and* $\forall(\zeta, \theta) \in \partial_P \varphi(\alpha, \alpha)$ *we have:*

$$1 + h(\alpha, \zeta) \geq 0,$$

- $\forall \alpha \in \text{bdry } S$ *and* $\forall \beta \in S$ *such that* $\alpha \neq \beta$ *we have:*

$$\forall(\zeta, \theta) \in \partial_P \varphi(\alpha, \beta), \quad 1 + h(\alpha, \zeta) \leq 0, \text{ and}$$
$$\forall(\zeta, \theta) \in \partial_P \varphi(\beta, \alpha), \quad 1 + h(\beta, \zeta) \leq 0.$$

Sketch of the proof. We consider the multifunction $\tilde{F}$ defined over $\mathbb{R}^n \times \mathbb{R}^n$ by $\tilde{F}(x, y) := F(x) \times \{0\}$. For $\Sigma := D \cap (S \times S)$ we can verify that the $S \times S$-constrained (unilateral) minimal time function for the new dynamic $\tilde{F}$ associated to Σ (denoted by $\tilde{T}_{S \times S}(\cdot, \Sigma)$) coincides with the S-constrained bilateral minimal time function $T_S(\cdot, \cdot)$. Then the result of the theorem follows by applying Theorem 1.1 to $\tilde{T}_{S \times S}(\cdot, \Sigma)$ (which is $T_S(\cdot, \cdot)$) after verifying that:

- We have an $S \times S$-constrained trajectory tracking result for $\tilde{F}$ (this follows since we already have an S-constrained trajectory tracking result for F and since a trajectory of $\tilde{F}$ is of the form $(x(t), \beta)$ where $x(t)$ is a trajectory of F).

- A function $\varphi : \mathbb{R}^n \times \mathbb{R}^n \longrightarrow] - \infty, +\infty]$ is bilateral S-continuous if and only if it is (Σ, S)-continuous (for $\tilde{F}$). $\square$

References

1. M. Bardi, *SIAM J. Control Optim.* **27**, 776 (1989).
2. M. Bardi and I. Capuzzo-Dolcetta, *Optimal Control and Viscosity Solutions of Hamilton-Jacobi-Bellman Equations* (Birkhauser, Boston, 1997).

3. M. Bardi and V. Staicu, *Acta Appl. Math.* **31**, 201 (1993).
4. P. Cannarsa and C. Sinestrari, *Calc. Var. Partial Differential Equations* **3**, 273 (1995),.
5. F. H. Clarke, Y. S. Ledyaev, R. J. Stern and P. R. Wolenski, *Nonsmooth analysis and control theory*, Graduate Texts in Mathematics, Vol. 178 (Springer-Verlag, New York, 1998).
6. F. H. Clarke and C. Nour, *J. Convex Anal.*, **11**, 413 (2004).
7. C. Nour, *ESAIM Control Optim. Calc. Var.* **12**, 120 (2006).
8. C. Nour, *J. Convex Anal.* **3**, 61 (2006).
9. C. Nour and R. J. Stern, The state constrained bilateral minimal time function (in progress).
10. N. N. Petrov, *J. Appl. Math. Mech.* **34**, 785 (1970).
11. P. Soravia, *SIAM J. Control Optim.* **31**, 604 (1993).
12. P. Soravia, *Comm. Partial Differential Equations* **18**, 1493 (1993).
13. R. J. Stern, *SIAM J. Control Optim.* **43**, 697 (2004).
14. V. M. Veliov, *Journal of Optimization Theory and Applications* **94**, 335 (1997).
15. P. Wolenski, Y. Zhuang, *SIAM J. Control Optim.* **36**, 1048 (1998).

EXISTENCE AND A DECOUPLING TECHNIQUE FOR THE NEUTRAL PROBLEM OF BOLZA WITH MULTIPLE VARYING DELAYS

N. L. ORTIZ

*Department of Mathematical Sciences, Virginia Commonwealth University,
Richmond, VA 23284, USA
E-mail: nlortiz@vcu.edu*

In this article we consider the generalized problem of Bolza with different time
varying delays appearing in the state and velocity variables. We derive existence
of solutions and prove a natural extension of a decoupling theorem, which was
originally introduced by Clarke in the non delay case[2] and was previously
extended[7] to the neutral case when both of the delays appearing are given by
one function.

Keywords: Generalized problem of Bolza; Multiple delays; Non-smooth analysis; Neutral systems; Optimal control.

1. Introduction

In this article we study the generalized problem of Bolza involving two
different time-varying delays that appear in the state and velocity variables. We will refer to this problem as the multiple delays neutral problem
of Bolza, or multiple delays NPB. We seek to extend previous results[8] to
guarantee existence of solutions and use a decoupling technique introduced
by Clarke[2] in the case without delays as an important tool to derive necessary conditions analogous to Ref. 7. In this note we consider the functional

$$\Pi(x) := \ell\big(x(T)\big) + \int_0^T L\big(t, x(t), x(t - \Delta_1(t)), \dot{x}(t), \dot{x}(t - \Delta_2(t))\big)\, dt. \quad (1)$$

Here $L : [0, T] \times \mathbb{R}^{4n} \to (-\infty, \infty]$ and $\ell : \mathbb{R}^n \to (-\infty, \infty]$ are given and
are allowed to reach ∞ in order to incorporate constraints. The given delay
functions $\Delta_1 : [0, T] \to [0, \Delta_0]$ and $\Delta_2 : [0, T] \to [0, \Delta_0]$, with $\Delta_0 > 0$
a constant, are assumed to be continuous. In addition, we assume that
$\Delta_1(0) = 0$, $\Delta_1(\cdot)$ is Lipschitz with Lipschitz constant $K_{\Delta_1} = 1$, and extend

$\dot{\Delta}_1(s) := 0$ for each $s \in [-\Delta_0, 0) \cup [T, T + \Delta_0]$. We are also given the initial data $c : [-\Delta_0, 0] \to \mathbb{R}^n$, which we assume to be in $\mathbf{L}^2[-\Delta_0, 0]$.

We now state the main problem involved:

$$\text{minimize } \Pi(x) \tag{2}$$

over all absolutely continuous functions $x : [0, T] \to \mathbb{R}^n$ with initial conditions $x(0) = x_0$, $x(t) = c(t)$ for each $t \leq 0$.

The generalized problem of Bolza was introduced by Rockafellar[11] as a means to study a broad class of optimal control problems. In Ref. 11 he proved existence of solutions introducing a method of proof now referred to as the "direct method" in the calculus of variations. We previously extended this method to prove existence of solutions of NPB when the delay present was given by a single constant.[8] In this note we offer a natural extension to the multiple delay NPB (2). A number of results involving necessary conditions for the generalized problem of Bolza can be obtained in the literature in the case without delay[1,2,5] and various cases involving delays.[6,7,9] In Ref. 2, Clarke introduced a decoupling technique and a decoupling theorem that served as a powerful tool to derive necessary conditions. This "decoupling" theorem was extended to NPB with varying delay given by the same function[7] and here we provide a proof for the more general case with multiple time delay functions. Further, we show that our results subsume the ones previously obtained.

2. Main Assumptions and Preliminaries

Our assumptions are the following:

(i) ℓ is lower semi-continuous and bounded below;

(ii) $L(t, x, y, v, w)$ is lower semi-continuous in (x, y, v, w), is $\mathcal{L} \times \mathcal{B}$-measurable on $[0, T] \times \mathbb{R}^{4n}$, and is (jointly) convex in (v, w).

(iii) There exists a nondecreasing function $\theta : [0, \infty) \to \mathbb{R}$ satisfying $\lim_{r \to \infty} \theta(r)/r = \infty$ so that

$$L(t, x, y, v, w) \geq \theta(|v|) + \theta(|w|) \qquad \text{for all } v, w \in \mathbb{R}^n.$$

In what follows, we use the natural modification of the maximized Hamiltonian:

$$H : \mathbb{R} \times \mathbb{R}^{4n} \to \mathbb{R}$$
$$H(t, x, y, p, q) := \sup_{(v,w) \in \mathbb{R}^{2n}} \{ \langle p, v \rangle + \langle q, w \rangle - L(t, x, y, v, w) \}.$$

As in non delay problems (see Ref. 11), H is upper semi-continuous in (x, y, v, w), $\mathcal{L} \times \mathcal{B}$-measurable on $[0, T] \times \mathbb{R}^{4n}$, and is (jointly) convex in (p, q).

3. Existence of Solutions

In order to prove our main existence theorem for (2), we rely on the following preliminary lemmas and proposition.

Lemma 3.1. *Suppose* $\mathcal{X} \subseteq \mathcal{AC}[0, T]$ *and* $\mathcal{V} \subseteq \mathbf{L}^1[0, T]$ *are nonempty and*

$$\sup_{x(\cdot) \in \mathcal{X}, v(\cdot) \in \mathcal{V}} \int_0^T L(t, x(t), x(t - \Delta_1(t)), v(t), v(t - \Delta_2(t))) \, dt < K < \infty. \quad (3)$$

Then $\mathcal{V}$ *is weakly sequentially precompact.*

Proof. By the Dunford-Pettis criterion[4] it is enough to show that

(a) $\sup_{v(\cdot) \in \mathcal{V}} \|v(\cdot)\|_1 < \infty$, and
(b) for all $\varepsilon > 0$ there exists a $\delta > 0$ such that $m(I) < \delta$ implies $\int_I |v(s)| \, ds < \varepsilon$ for all $v(\cdot) \in \mathcal{V}$.

Given hypothesis (iii) and our assumed estimate above (3) we have that for each $v(\cdot) \in \mathcal{V}$,

$$\int_0^T \theta(|v(t)|) + \theta(|v(t - \Delta_2(t))|) \, dt < K.$$

Thus, $\int_0^T \theta(|v(t)|) \, dt \leq K - T\theta(0)$ since $\theta(0)$ is the minimum value of θ. Let $I \subseteq [0, T]$ be any measurable set, $R > 0$ so that $\theta(R) > 0$, and define $A := I \cap \{t : |v(t)| \leq R\}$ and $B := I \cap \{t : |v(t)| > R\}$. Then

$$\int_I |v(t)| \, dt = \int_A |v(t)| \, dt + \int_B |v(t)| \, dt$$

$$\leq Rm(I) + \int_B \frac{|v(t)|}{\theta(|v(t)|)} \theta(|v(t)|) \, dt$$

$$\leq Rm(I) + k \sup_{r \geq R} \frac{r}{\theta(r)},$$

where $k := (K - T\theta(0))$. Since $r/\theta(r) \to 0$ as $r \to \infty$, (a) follows for large R by letting $I = [0, T]$. If we let $\varepsilon > 0$, choose $R > 0$ such that $\sup_{r \geq R} r/\theta(r) < \varepsilon/2k$, and set $\delta = \varepsilon/2R$ then (2) follows as well. $\qquad\square$

Next we show the weak lower semicontinuity of the integral part of (1)

$$\Lambda(x) := \int_0^T L(t, x(t), x(t - \Delta_1(t)), v(t), v(t - \Delta_2(t)))\, dt$$

and rely on the following proposition, a proof of which can be found in Theorem 1 of Ref. 10.

Proposition 3.1. *Suppose $x(\cdot)$, $y(\cdot)$, $v(\cdot)$, and $w(\cdot)$ are measurable satisfying $L\big(x(\cdot), y(\cdot), v(\cdot), w(\cdot)\big) \in \mathbf{L}^1[0, T]$. Then $\int_0^T L(t, x(t), y(t), v(t), w(t))\, dt$ is equal to the supremum of*

$$\int_0^T \big[\langle p(t), v(t)\rangle + \langle q(t), w(t)\rangle - H(t, x(t), y(t), p(t), q(t))\big]\, dt$$

taken over $\big(p(\cdot), q(\cdot)\big)$ in $\mathbf{L}^\infty$ (= the bounded measurable functions defined from $[0, T]$ into $\mathbb{R}^{2n}$).

Lemma 3.2. *Suppose sequences $\{x_i(\cdot)\} \subseteq \mathbf{L}^2[0, T]$ and $\{v_i(\cdot)\} \subseteq \mathbf{L}^1[0, T]$ are such that $x_i(t) \to \bar{x}(t)$ for almost all t in $[0, T]$ and $v_i(t) \xrightarrow{w} \bar{v}(t)$ (weak convergence in $\mathbf{L}^1$) where each $x_i(s) := c(s)$ and $v_i(s) := 0$ for each $s \in [-\Delta_0, 0)$. Then,*

$$\int_0^T L\big(t, \bar{x}(t), \bar{x}(t - \Delta_1(t)), \bar{v}(t), \bar{v}(t - \Delta_2(t))\big)\, dt$$

$$\leq \liminf_{i \to \infty} \int_0^T L\big(t, x_i(t), x_i(t - \Delta_1(t)), v_i(t), v_i(t - \Delta_2(t))\big)\, dt\,.$$

252

Proof. It follows from Proposition 3.1 that

$$
\liminf_{i\to\infty} \int_0^T L(t, x_i(t), x_i(t - \Delta_1(t)), v_i(t), v_i(t - \Delta_2(t)))\, dt
$$

$$
= \liminf_{i\to\infty} \sup_{(p(\cdot),q(\cdot))\in \mathbf{L}^\infty} \left\{ \int_0^T \langle p(t), v_i(t)\rangle + \langle q(t), v_i(t - \Delta_2(t))\rangle \right.
$$
$$
\left. - H\left(t, x_i(t), x_i(t - \Delta_1(t)), p(t), q(t)\right)\, dt \right\}
$$

$$
\geq \sup_{(p(\cdot),q(\cdot))\in \mathbf{L}^\infty} \liminf_{i\to\infty} \left\{ \int_0^T \langle p(t), v_i(t)\rangle + \langle q(t), v_i(t - \Delta_2(t))\rangle \right.
$$
$$
\left. - H\left(t, x_i(t), x_i(t - \Delta_1(t)), p(t), q(t)\right)\, dt \right\}
$$

$$
\geq \sup_{(p(\cdot),q(\cdot))\in \mathbf{L}^\infty} \left\{ \int_0^T \langle p(t), \bar{v}(t)\rangle + \langle q(t), \bar{v}(t - \Delta_2(t))\rangle \right.
$$
$$
\left. - \limsup_{i\to\infty} H\left(t, x_i(t), x_i(t - \Delta_1(t)), p(t), q(t)\right)\, dt \right\}
$$

$$
\geq \sup_{(p(\cdot),q(\cdot))\in \mathbf{L}^\infty} \left\{ \int_0^T \langle p(t), \bar{v}(t)\rangle + \langle q(t), \bar{v}(t - \Delta_2(t))\rangle \right.
$$
$$
\left. - H\left(t, \bar{x}(t), \bar{x}(t - \Delta_1(t)), p(t), q(t)\right)\, dt \right\}
$$

$$
= \int_0^T L\left(t, \bar{x}(t), \bar{x}(t - \Delta_1(t)), \bar{v}(t), \bar{v}(t - \Delta_2(t))\right)\, dt.
$$

The second inequality is justified by Fatou's Lemma and $v_i(t) \xrightarrow{w} \bar{v}(t)$, the last inequality since H is upper semi-continuous, and the final equality by Proposition 3.1. $\qquad\square$

We now proceed to our main theorem.

Theorem 3.1. *If there exists $x(\cdot) \in \mathcal{AC}[0, T]$ that is feasible for (2), then there exists an arc $\bar{x}(\cdot) \in \mathcal{AC}[0, T]$ that solves (2).*

Proof. Suppose that $x(\cdot)$ is feasible. Select a minimizing feasible sequence $\{x_i(\cdot)\} \subset \mathcal{AC}[0, T]$ so that for each i, $\ell(x_i(T)) + \Lambda(x_i) \leq \ell(x(T)) + \Lambda(x) < \infty$. Since $\ell(\cdot)$ is bounded below it follows that $\Lambda(x_i)$ is bounded above, and so by Lemma 3.1 there exists $\bar{v}(\cdot) \in \mathbf{L}^1[0, T]$ and a subsequence (which we do not relabel) satisfying $\dot{x}_i(\cdot) \xrightarrow{w} \bar{v}(\cdot)$. Define $\bar{x}(\cdot) \in \mathcal{AC}[0, T]$ by

$$
\bar{x}(t) = c(0) + \int_0^t \bar{v}(s)\, ds
$$

and set equal to $c(t)$ for $t \in [-\Delta_0, 0]$. Then $x_i(t) \to \bar{x}(t)$ for all $t \in [-\Delta_0, T]$. It follows from Lemma 3.2 and our hypothesis (i) that

$$\ell(\bar{x}(T)) + \Lambda(\bar{x}) \leq \liminf_{i \to \infty} \ell(x_i(T)) + \Lambda(x_i)$$

Since $\{x_i(\cdot)\}$ is a minimizing sequence, it follows that $\bar{x}(\cdot)$ solves (2). $\square$

4. The Decoupling Technique

Let $\mathcal{X} := \mathbb{R}^n \times \mathbf{L}^2[0, T] \times \mathbf{L}^2[0, T] \times \mathbf{L}^1[0, T] \times \mathbf{L}^1[0, T]$. We define a new Bolza-type functional $\Gamma : \mathcal{X} \to (-\infty, \infty]$, similar to (1) by

$$\Gamma\left(\gamma, u(\cdot), w(\cdot), v_1(\cdot), v_2(\cdot)\right) := \ell(\gamma) + \int_0^T L(t, u(t), w(t), v_1(t), v_2(t)) \, dt. \tag{4}$$

Our original problem (2) is equivalent to minimizing Γ over $\mathcal{X}$ subject to the following constraints

$$u(t) = x(0) + \int_0^t v_1(s) \, ds \tag{5}$$

$$\gamma = x(0) + \int_0^T v_1(t) \, dt \tag{6}$$

$$v_2(t) = v_1(t - \Delta_2(t)) \tag{7}$$

$$w(t) = \xi(t)[c(t - \Delta_1(t)) - x_0] + x_0 + \int_0^t v_1(s - \Delta_1(s))[1 - \dot{\Delta}_1(s)] \, ds$$

$$= \xi(t)[c(t - \Delta_1(t)) - x_0] + x_0 + \int_0^t v_2(s - \Delta(s))[1 - \dot{\Delta}_1(s)] \, ds, \tag{8}$$

where $\Delta(s) := \Delta_1(s) - \Delta_2(s)$, $v_1(t) := 0$ for each $t \in [-\Delta_0, 0)$, and $\xi(t)$ is defined by,

$$\xi(t) := \begin{cases} 1, & \text{if } t - \Delta_1(t) < 0; \\ 0, & \text{else}. \end{cases} \tag{9}$$

Indeed, (5) says that $x(t) := u(t)$ is absolutely continuous, (6) implies that γ is the endpoint $x(T)$, and (7)-(8) together say that $w(t) = x(t - \Delta_1(t))$ for $t \in [0, T]$. We introduce the following function $\mathcal{D} : \mathcal{X} \to \mathbb{R}$ to measure how far an element $(\gamma, u(\cdot), w(\cdot), v_1(\cdot), v_2(\cdot)) \in \mathcal{X}$ is from satisfying

254

(5)-(8):

$$\mathcal{D}\big(\gamma, u(\cdot), w(\cdot), v_1(\cdot), v_2(\cdot)\big) =$$

$$\left| \gamma - x_0 - \int_0^T v_1(t)\, dt \right| + \int_0^T \left| u(t) - x_0 - \int_0^t v_1(s)\, ds \right|^2 dt$$

$$+ \int_0^T \left| w(t) - \xi(t)[c(t - \Delta_1(t)) - x_0] - x_0 - \int_0^t v_2(s - \Delta(s))[1 - \dot{\Delta}_1(s)]\, ds \right|^2 dt$$

$$+ \int_0^T |v_2(t) - v_1(t - \Delta_2(t))|\, dt.$$

It is clear that (5)-(8) hold if and only if $\mathcal{D}\big(\gamma, u(\cdot), w(\cdot), v_1(\cdot), v_2(\cdot)\big) = 0$. We now extend the decoupling technique to the multiple delays NPB.

Theorem 4.1 (Decoupling Theorem). *Suppose the main assumptions (i)-(iii) hold and $\epsilon > 0$. Then there exist a constant $\sigma > 0$, an element $\big(\bar{\gamma}, \bar{u}(\cdot), \bar{w}(\cdot), \bar{v}_1(\cdot), \bar{v}_2(\cdot)\big) \in \mathcal{X}$, and absolutely continuous arcs $p(\cdot)$ and $q(\cdot)$ defined on $[0, T]$ so that*

(a) $\mathcal{D}\big(\bar{\gamma}, \bar{u}(\cdot), \bar{w}(\cdot), \bar{v}_1(\cdot), \bar{v}_2(\cdot)\big) < \epsilon$;
(b) $\Gamma\big(\bar{\gamma}, \bar{u}(\cdot), \bar{w}(\cdot), \bar{v}_1(\cdot), \bar{v}_2(\cdot)\big)$ *is within ϵ of the minimum value in (2);*
(c) for almost all $t \in [0, T]$, the map

$$(u, w, v_1, v_2) \mapsto L(t, u, w, v_1, v_2) - \langle \dot{p}(t), u \rangle - \langle \dot{q}(t), w \rangle - \langle p(t), v_1 \rangle$$

$$- \langle q(t + \Delta(t))[1 - \dot{\Delta}_1(t + \Delta(t))], v_2 \rangle$$

$$+ \sigma \Big\{ |u - \bar{u}(t)|^2 + |w - \bar{w}(t)|^2 + |v_1 - \bar{v}_1(t)|^2 + |v_2 - \bar{v}_2(t)|^2 \Big\}$$

where $\Delta(t) := \Delta_1(t) - \Delta_2(t)$ is minimized at $(u, w, v_1, v_2) = \big(\bar{u}(t), \bar{w}(t), \bar{v}_1(t), \bar{v}_2(t)\big)$; and
(d) the map

$$\gamma \mapsto \ell(\gamma) + \langle \gamma, p(T) \rangle + \sigma |\gamma - \bar{\gamma}|^2$$

is minimized at $\gamma = \bar{\gamma}$.

Proof. We introduce a new problem $P_{\eta, \alpha, \beta}$ and we seek to minimize, over all arcs $x(\cdot)$, the functional $\quad \Lambda_{\eta, \alpha(\cdot), \beta(\cdot)}(x) :=$

$$\ell\big(x(T) + \eta\big) + \int_0^T L\big(t, x(t) + \alpha(t), x(t - \Delta_1(t)) + \beta(t), \dot{x}(t), \dot{x}(t - \Delta_2(t))\big)\, dt.$$

$$\tag{10}$$

Here $x(0) = x_0$, $x(t)$ is set equal to $c(t)$ for $t \in [-\Delta_0, 0)$, where $c(0) = x_0$, $\big(\eta, \alpha(\cdot), \beta(\cdot)\big) \in \mathbb{R}^n \times \mathbf{L}^2[0, T] \times \mathbf{L}^2[0, T]$, and $\dot{x}(t - \Delta_2(t)) := 0$ when

$t - \Delta_2 < 0$ and $\beta(t) := 0$ when $t - \Delta_1(t) < 0$. Define a value function $V : \mathbb{R}^n \times \mathbf{L}^2[0,T] \times \mathbf{L}^2[0,T] \to (-\infty, \infty]$ by setting $V(\eta, \alpha(\cdot), \beta(\cdot))$ as the minimum value in (10). Here $V(\eta, \alpha(\cdot), \beta(\cdot)) = \infty$ if there are no feasible arcs for $P_{\eta,\alpha,\beta}$, V is lower semi-continuous and if $V(\eta, \alpha(\cdot), \beta(\cdot)) < \infty$ then a solution to $P_{\eta,\alpha,\beta}$ exists.[7]

Fix $\epsilon > 0$. From the Density Theorem of proximal analysis[3] we select $(\bar\eta, \bar\alpha(\cdot), \bar\beta(\cdot)) \in \mathbb{R}^n \times \mathbf{L}^2[0,T] \times \mathbf{L}^2[0,T]$ with $|\bar\eta| + \|\bar\alpha\|_2 + \|\bar\beta\|_2 < \epsilon$ and with $|V(\bar\eta, \bar\alpha(\cdot), \bar\beta(\cdot)) - V(0,0,0)| < \epsilon$. The theorem also guarantees that the proximal subgradient of V is not empty and so there exists $(\zeta, \phi(\cdot), \psi(\cdot)) \in \mathbb{R}^n \times \mathbf{L}^2[0,T] \times \mathbf{L}^2[0,T]$ with $(\zeta, \phi(\cdot), \psi(\cdot)) \in \partial_P V(\bar\eta, \bar\alpha(\cdot), \bar\beta(\cdot))$. We note also that $\psi(t) = 0$ for all $t \in [0, \Delta_0]$ satisfying $t - \Delta_1(t) < 0$. By the proximal subgradient inequality there exist $\delta' > 0$ and $\sigma' > 0$ such that for all $(\eta, \alpha, \beta) \in B((\bar\eta, \bar\alpha, \bar\beta), \delta')$,

$$V(\eta, \alpha, \beta) - V(\bar\eta, \bar\alpha, \bar\beta) + \sigma'\{|\eta - \bar\eta|^2 + \|\alpha - \bar\alpha\|_2^2 + \|\beta - \bar\beta\|_2^2\}$$
$$\geq \langle \zeta, \eta - \bar\eta \rangle + \langle \phi, \alpha - \bar\alpha \rangle + \langle \psi, \beta - \bar\beta \rangle. \quad (11)$$

Let $\bar{x}(\cdot)$ be an optimal solution to $P_{\bar\eta,\bar\alpha,\bar\beta}$, which implies that $V(\bar\eta, \bar\alpha, \bar\beta) = \Lambda_{\bar\eta,\bar\alpha,\bar\beta}(\bar{x})$. Since, for any arc $x(\cdot)$ one has $V(\eta, \alpha, \beta) \leq \Lambda_{\eta,\alpha,\beta}(x)$, (11) may be rewritten to obtain:

$$\Lambda_{\eta,\alpha,\beta}(x) - \langle \zeta, \eta \rangle - \langle \phi, \alpha \rangle - \langle \psi, \beta \rangle + \sigma'\{|\bar\eta - \eta|^2 + \|\bar\alpha - \alpha\|_2^2 + \|\bar\beta - \beta\|_2^2\}$$
$$\geq \text{ibid } (\bar\eta, \bar\alpha, \bar\beta, \bar{x}). \quad (12)$$

Here ibid $(\bar\eta, \bar\alpha, \bar\beta, \bar{x})$ represents the left hand side of the inequality but with the variables (η, α, β, x) substituted by $(\bar\eta, \bar\alpha, \bar\beta, \bar{x})$. We now apply the following change of notation:

$$
\begin{aligned}
x(\cdot) + \alpha(\cdot) &:= u(\cdot) & \dot{x}(\cdot) &:= v_1(\cdot) \\
x(T) + \eta &:= \gamma & & \\
\bar{x}(\cdot) + \bar\alpha(\cdot) &:= \bar{u}(\cdot) & \dot{\bar{x}}(\cdot) &:= \bar{v}_1(\cdot) \\
\bar{x}(T) + \bar\eta &:= \bar\gamma & & \\
x(\cdot - \Delta_1(\cdot)) + \beta(\cdot) &:= w(\cdot) & v_2(\cdot) &:= v_1(\cdot - \Delta_2(\cdot)) \\
\bar{x}(\cdot - \Delta_1(\cdot)) + \bar\beta(\cdot) &:= \bar{w}(\cdot) & \bar{v}_2(\cdot) &:= \bar{v}_1(\cdot - \Delta_2(\cdot))
\end{aligned}
$$

Now, using the new variables and separating square terms (since $(a + b)^2 \leq$

256

$2a^2 + 2b^2)$ in (12) we obtain,

$$\ell(\gamma) \;+\; \int_0^T L\big(t, u(t), w(t), v_1(t), v_2(t)\big)\, dt - \langle \zeta, \gamma - x(T)\rangle$$

$$-\int_0^T \{\langle \phi(t), u(t) - x(t)\rangle + \langle \psi(t), w(t) - x(t - \Delta_1(t))\rangle\}\, dt$$

$$+\sigma\{\|u - \bar{u}\|_2^2 + \|w - \bar{w}\|_2^2 + \|v_1 - \bar{v}_1\|_1^2 + \|v_2 - \bar{v}_2\|_1^2 + |\gamma - \bar{\gamma}|^2\}$$

$$\geq \text{ibid}\,(\bar{\gamma}, \bar{u}, \bar{w}, \bar{x}, \bar{v}_1, \bar{v}_2), \quad (13)$$

where estimations for $x(\cdot)$ in terms of v_1 and v_2 have been made and where $\sigma := 4\sigma'(T + T^2)$. Define arcs $p(\cdot)$ and $q(\cdot)$ in $\mathbf{L}^2[0, T]$ by

$$p(t) = -\zeta - \int_t^T \phi(s)\, ds \tag{14}$$

$$q(t) = -\int_t^T \psi(s)\, ds. \tag{15}$$

$$\int_0^T \langle \phi(t), x(t)\rangle\, dt = \int_0^T \langle \dot{p}(t), x(t)\rangle\, dt$$

$$= \langle p(T), x(T)\rangle - \langle p(0), x(0)\rangle - \int_0^T \langle p(t), v_1(t)\rangle\, dt$$

and

$$\int_0^T \langle \psi(t), x(t - \Delta_1(t))\rangle\, dt = \int_0^T \langle \dot{q}(t), x(t - \Delta_1(t))\rangle\, dt$$

$$= 0 - \langle q(0), x(-\Delta_1(0))\rangle - \int_0^T \langle q(t), v_1(t - \Delta_1(t))[1 - \dot{\Delta}_1(t)]\rangle\, dt$$

$$= -\langle q(0), x(0)\rangle - \int_0^T \langle q(t + \Delta(t)), v_2(t)[1 - \dot{\Delta}_1(t + \Delta(t))]\rangle\, dt.$$

The last equality is reached using our definition of v_1 and v_2 and since $q(s) := 0$ for each $s \geq T$ and $v_2(s) = 0$ when $s - \Delta_2(s) < 0$. Substitution of these into (13) and since $x(s) = \bar{x}(s)$ for $s \in [-\Delta_0, 0]$ we obtain,

$$\ell(\gamma) + \langle p(T), \gamma\rangle + \sigma|\gamma - \bar{\gamma}|^2$$

$$+\int_0^T \{L(t, u(t), w(t), v_1(t), v_2(t)) - \langle \dot{p}(t), u(t)\rangle - \langle \dot{q}(t), w(t)\rangle - \langle p(t), v_1(t)\rangle$$

$$- \langle q(t + \Delta(t))[1 - \dot{\Delta}_1(t + \Delta(t))], v_2(t)\rangle\}\, dt$$

$$+\sigma\{\|u - \bar{u}\|_2^2 + \|w - \bar{w}\|_2^2 + \|v_1 - \bar{v}_1\|_1^2 + \|v_2 - \bar{v}_2\|_1^2\} \geq \text{ibid}\,(\bar{\gamma}, \bar{u}, \bar{w}, \bar{v}_1, \bar{v}_2).$$

$$(16)$$

We know that (16) holds as long as $\|\alpha - \bar{\alpha}\|_2$, $|\eta - \bar{\eta}|$, and $\|\beta - \bar{\beta}\|_2$ are each less than δ'. But by definition and since x can be bound in terms of v_1 and v_2, we simply need to require that $\|u - \bar{u}\|_2, \|w - \bar{w}\|_2, \|v_1 - \bar{v}_1\|_1$, $\|v_2 - \bar{v}_2\|_1$, and $|\gamma - \bar{\gamma}|$ be small in order that (16) holds.

We now prove the statements of the theorem. Part (a) of the theorem is satisfied since, by our change of variables definitions

$$\left| \bar{\gamma} - x_0 - \int_0^T \bar{v}_1(t)\, dt \right| = |\bar{\gamma} - \bar{x}(T)| = |\bar{\eta}|,$$

$$\int_0^T \left| \bar{u}(t) - x_0 - \int_0^t \bar{v}_1(s)\, ds \right|^2 dt = \|\bar{x} - \bar{u}\|_2^2 = \|\bar{\alpha}\|_2^2,$$

$$\int_0^T |v_2(t) - v_1(t - \Delta_2(t))| \, dt = 0,$$

and

$$\int_0^T \left| \bar{w}(t) - \xi(t)\left[c(t - \Delta_1(t)) - x_0\right] - x_0 - \int_0^t \bar{v}_2(s - \Delta(t))[1 - \dot{\Delta}_1(s)]\, ds \right|^2 dt$$

$$= \|\bar{w}(t) - \bar{x}(t - \Delta_1(t))\|_2 = \|\bar{\beta}\|_2^2.$$

Since $\|\bar{\alpha}\|_2 + |\bar{\eta}| + \|\bar{\beta}\|_2 < \epsilon$ statement (a) is satisfied. Part (b) of the theorem is satisfied since $|V(\bar{\eta}, \bar{\alpha}, \bar{\beta}) - V(0,0,0)| < \epsilon$, and by setting $u = \bar{u}, v_1 = \bar{v}_1, v_2 = \bar{v}_2, w = \bar{w}$ in (16) we can see that statement (d) is satisfied as well. Finally, to verify that part (c) holds we let f_t denote the function

$$f_t(u, w, v_1, v_2) = L(t, u, w, v_1, v_2) - \langle \dot{p}(t), u \rangle - \langle \dot{q}(t), w \rangle - \langle p(t), v_1 \rangle$$

$$- \langle q(t + \Delta(t))[1 - \dot{\Delta}_1(t + \Delta(t))], v_2 \rangle$$

$$+ \sigma\{|u - \bar{u}(t)|^2 + |w - \bar{w}(t)|^2 + |v_1 - \bar{v}_1(t)|^2 + |v_2 - \bar{v}_2(t)|^2\}. \quad (17)$$

Assuming hypotheses (ii) and (iii), f_t attains a minimum for almost every t. We will prove part (c) by establishing that for each $r > 0$ the set

$$A(r) := \left\{ t \in [0, T] : f_t(\bar{u}(t), \bar{w}(t), \bar{v}_1(t), \bar{v}_2(t)) \geq \min_{(u, w, v_1, v_2) \in \mathbb{R}^{4n}} f_t(u, w, v_1, v_2) + r \right\}$$

has measure zero. Suppose not, then there exists $r > 0$ with $m(A(r)) > 0$. Let B_i be a decreasing sequence of subsets of $A(r)$ such that for each i one has

$$\frac{2}{i} m(A(r)) < m(B_i) < \frac{3}{i} m(A(r)). \quad (18)$$

Choose $(u'(\cdot), w'(\cdot), v_1'(\cdot), v_2'(\cdot))$ measurable such that for almost all t, $(u'(t), w'(t), v_1'(t), v_2'(t))$ minimizes f_t. To complete our proof we must first show that u', w', $v_1' - \bar{v}_1$, and $v_2' - \bar{v}_2$ are in $\mathbf{L}^2[0,T]$. By our choice of $(u'(\cdot), w'(\cdot), v_1'(\cdot), v_2'(\cdot))$ the inequality

$$f_t(u'(t), w'(t), v_1'(t), v_2'(t)) \le f_t(\bar{u}(t), \bar{w}(t), \bar{v}_1(t), \bar{v}_2(t)) \tag{19}$$

holds for almost all t. Let b be the lower bound for the function L. After simplifying and rearranging we obtain from our definition of f_t and (19),

$$\sigma\{|u'(t) - \bar{u}(t)|^2 + |w'(t) - \bar{w}(t)|^2 + |v_1'(t) - \bar{v}_1(t)|^2 + |v_2'(t) - \bar{v}_2(t)|^2\}$$
$$\le \langle \dot{p}(t), u'(t) - \bar{u}(t) \rangle + \langle \dot{q}(t), w'(t) - \bar{w}(t) \rangle + \langle p(t), v_1'(t) - \bar{v}_1(t) \rangle$$
$$+\langle q(t+\Delta(t))[1-\dot{\Delta}_1(t+\Delta(t))], v_2'(t)-\bar{v}_2(t)\rangle + L(t, \bar{u}(t), \bar{w}(t), \bar{v}_1(t), \bar{v}_2(t))-b. \tag{20}$$

Set

$$W(t) = (u'(t) - \bar{u}(t), w'(t) - \bar{w}(t), v_1'(t) - \bar{v}_1(t), v_2'(t) - \bar{v}_2(t)),$$
$$g(t) = \Big(\dot{p}(t), \dot{q}(t), p(t), q(t + \Delta(t))[1 - \dot{\Delta}_1(t + \Delta(t))]\Big),$$
$$k(t) = L(t, \bar{u}(t), \bar{w}(t), \bar{v}_1(t), \bar{v}_2(t)) - b.$$

Then inequality (20) becomes $\sigma |W|^2 - \langle g(t), W(t) \rangle - k(t) \le 0$ and by the quadratic formula $(2\sigma \|W\|)^2 \le 4\|g\|^2 + 8\sigma\|k\|$. Furthermore, since $g(\cdot) \in \mathbf{L}^2$ and $k(\cdot) \in \mathbf{L}^1$, it follows that the integral of $(2\sigma \|W\|)^2$ is bounded and thus, $W \in \mathbf{L}^2$, proving that u', w', $v_1' - \bar{v}_1$, and $v_2' - \bar{v}_2$ are in $\mathbf{L}^2[0,T]$ and that each v_1' and v_2' is in $\mathbf{L}^1[0,T]$.

To finish our proof of (c) we define a family of functions $(u_i, w_i, v_{1i}, v_{2i}) \in \mathbf{L}^2[0,T] \times \mathbf{L}^2[0,T] \times \mathbf{L}^1[0,T] \times \mathbf{L}^1[0,T]$ as:

$$(u_i(t), w_i(t), v_{1i}(t), v_{2i}(t)) = \begin{cases} (u'(t), w'(t), v_1'(t), v_2'(t)), & \text{if } t \in B_i; \\ (\bar{u}(t), \bar{w}(t), \bar{v}_1(t), \bar{v}_2(t)), & \text{else}. \end{cases} \tag{21}$$

Then,

$$\lim_{i \to \infty} \|(u_i, w_i, v_{1i}, v_{2i}) - (\bar{u}, \bar{w}, \bar{v}_1, \bar{v}_2)\|_2 = 0,$$

since as $i \to \infty$, $m(B_i) \to 0$. From the definition of (u', w', v_1', v_2') and since $m(A(r)) > 0$, it follows that

$$\int_0^T f_t(u_i, w_i, v_{1i}, v_{2i}) \, dt < \int_0^T f_t(\bar{u}, \bar{w}, \bar{v}_1, \bar{v}_2) \, dt.$$

But setting $\gamma = \bar{\gamma}, u = u_i, w = w_i, v_1 = v_{1i}$, and $v_2 = v_{2i}$, this contradicts (16) for i sufficiently large. Hence, it must be that $m(A(r)) = 0$, which proves (c) and completes our proof. $\qquad\square$

5. Conclusion

We expect that this more general version of the decoupling theorem will prove to be a precursor for necessary conditions to our problem (2). If we let $\tilde{q}(t) = q(t + \Delta(t))[1 - \dot{\Delta}_1(t + \Delta(t))]$ we can see that for almost all $t \in [0, T]$, the inclusions

$$\big(\dot{p}(t), \dot{q}(t), p(t), \tilde{q}(t)\big) \in \partial_P L\big(t, \cdot, \cdot, \cdot, \cdot\big)\big(\bar{u}(t), \bar{w}(t), \bar{v}_1(t), \bar{v}_2(t)\big), \qquad (22)$$

$$-p(T) \in \partial_P \ell(\bar{\gamma}), \qquad (23)$$

which follow immediately from statements (c) and (d) of the Decoupling Theorem 4.1 are forms of the Euler-Lagrange and transversality conditions. In addition, our results are consistent and subsume our previous results[7,8] which are obtained by setting $\Delta_1 := \Delta_2$ in Ref. 7 and $\Delta_2 := 0$ in Ref. 8.

References

1. F. H. Clarke, *Methods of Dynamic and Nonsmooth Optimization*, CBMS-NSF Regional Conf. Series in Applied Mathematics, Vol. 57 (Society for Industrial and Applied Mathematics (SIAM), Philadelphia, PA, 1989).
2. F. H. Clarke, *J. Math.Anal.Appl.* **172**, 92 (1993).
3. F. H. Clarke, Y. S. Ledyaev, R. J. Stern and P. R. Wolenski, *Nonsmooth analysis and control theory*, Graduate Texts in Mathematics, Vol. 178 (Springer-Verlag, New York, 1998).
4. R. E. Edwards, *Functional Analysis. Theory and applications*, Corrected reprint of the 1965 original (Dover, New York, 1995).
5. P. D. Lowen, R. T. Rockafellar, *SIAM J. Control Optim.* **34**, 1496 (1996).
6. B. S. Mordukhovich, R. Trubnik, *Ann. Oper. Res.* **101**, 149 (2001).
7. N. Ortiz, *J. Math.Anal.Appl.* **305**, 513 (2005).
8. N. L. Ortiz and P.R. Wolenski, *J. Math.Anal.Appl.* **289**, 260 (2004).
9. N. L. Ortiz and P. R. Wolenski, *Set Valued Anal.* **12**, 225 (2004).
10. R. T. Rockafellar, *Pacific J. Math.*, **39**, 439 (1971).
11. R. T. Rockafellar, *Adv. in Math.*, **15**, 312 (1975).

STABILIZATION PROBLEM FOR NONHOLONOMIC CONTROL SYSTEMS

L. RIFFORD

Laboratoire J.A. Dieudonné, Université de Nice-Sophia Antipolis,
Parc Valrose, 06108 Nice Cedex 02, France
E-mail: rifford@math.unice.fr

We present the stabilization problem in the context of nonholonomic control systems and nonholonomic distributions. Then, we introduce the notions of smooth repulsive stabilizing feedbacks and sections, and review recent results on the existence of such objects.

Keywords: Stabilization problem, nonholonomic control systems

1. Introduction

Throughout this paper, M denotes a smooth connected manifold of dimension n and $\bar{x}$ a point of M.

1.1. *Stabilization of nonholonomic control systems*

Let $f_1, \cdots, f_m$ be a family of m smooth vector fields on M. We say that the *control system* defined as,

$$\dot{x} = \sum_{i=1}^{m} u_i f_i(x), \tag{1}$$

is *nonholonomic* on M (also called totally nonholonomic in Ref. 2) if the following property is satisfied*:

$$\text{Lie}\,\{f_1, \cdots, f_m\}\,(x) = T_x M, \quad \forall x \in M.$$

---

*We recall that $\text{Lie}\,\{f_1, \cdots, f_m\}$ denotes the Lie algebra of vector fields generated by the family $\{f_1, \cdots, f_m\}$. It is the smallest vector subspace S of $\mathcal{X}^{\infty}(M)$ (the set of smooth vector fields on M) containing the f_i's and such that $[f_i, g] \in S$ for any $i = 1, \cdots, m$ and any $g \in S$.

Recall that, for every $x \in M$ and every control $u(\cdot) \in L^1([0,\infty); \mathbb{R}^m)$, there is a unique maximal solution $x(\cdot) = x(\cdot; u, x) : [0, T_u) \to M$ (with $T_u > 0$) to the Cauchy problem

$$\dot{x}(t) = \sum_{i=1}^{m} u_i(t) f_i(x(t)), \quad \text{for a.e.} \quad t \in [0, T_u), \quad x(0) = x.$$

As the next result shows, any nonholonomic control system is controllable on M (see Refs. 4 or 10).

Theorem 1.1 (Chow-Rashevsky Theorem). *Let (1) be a nonholonomic control system on M. Then, for every pair $x, y \in M$, there exists a smooth control $u(\cdot) : [0, 1] \to \mathbb{R}^m$ such that $x(1; u, x) = y$.*

Before presenting the stabilization problem, we need to recall the notion of globally asymptotically stable dynamical system. Let X be a smooth vector field on M, the dynamical system $\dot{x} = X(x)$ is said to be *globally asymptotically stable* at the point $\bar{x}$ (abreviated $\text{GAS}_{\bar{x}}$ in the sequel), if the two following properties are satisfied:

(i) *Lyapunov stability:* for every neighborhood $\mathcal{V}$ of $\bar{x}$, there exists a neighborhood $\mathcal{W}$ of $\bar{x}$ such that, for every $x \in \mathcal{W}$, the solution of $\dot{x}(t) = X(x(t)), x(0) = x$, satisfies $x(t) \in \mathcal{V}$, for every $t \geq 0$.

(ii) *Attractivity:* for every $x \in M$, the solution of $\dot{x}(t) = X(x(t))$, $x(0) = x$, tends to $\bar{x}$ as t tends to $+\infty$.

The purpose of the stabilization problem is the following:
Let (1) be a given nonholonomic control system, does there exist a smooth[†] mapping $k : M \to \mathbb{R}^m$ (called *stabilizing feedback*) such that the dynamical system (called *closed-loop system*) defined as,

$$\dot{x} = \sum_{i=1}^{m} k(x)_i f_i(x) \tag{2}$$

is $\text{GAS}_{\bar{x}}$?

1.2. *Stabilization problem for nonholonomic distributions*

Let Δ be a (totally) *nonholonomic distribution* of rank $m \leq n$ on M. This means that for every $x \in M$, there is a neighborhood $\mathcal{V}_x$ of x in M and a

---

[†]We restrict here our attention to smooth stabilizing feedbacks. In fact, it can be shown that if the control system (1) admits a continuous stabilizing feedback then this feedback can be regularized into a smooth stabilizing feedback.

m-tuple $(f_1^x, \cdots, f_m^x)$ of smooth vector fields on $\mathcal{V}_x$ such that

$$\Delta(y) = \operatorname{Span}\left\{f_1^x(y), \cdots, f_m^x(y)\right\}, \quad \forall y \in \mathcal{V}_x,$$

and moreover,

$$\operatorname{Lie}\left\{f_1^x, \cdots, f_m^x\right\}(y) = T_y M, \quad \forall y \in \mathcal{V}_x.$$

We call *horizontal path* between x to y, any absolutely continuous curve $\gamma(\cdot) : [0,1] \to M$ with $\gamma(0) = x, \gamma(1) = y$ which satisfies

$$\dot{\gamma}(t) \in \Delta(\gamma(t)), \quad \text{for a.e.} \quad t \in [0,1].$$

In the context of nonholonomic distribution, the Chow-Rashevsky Theorem takes the following form.

Theorem 1.2. *Let Δ be a nonholonomic distribution on M. Then, any two points of M can be joined by a smooth horizontal path.*

The stabilization problem for nonholonomic distributions consists in finding, if possible, a *smooth stabilizing section* of Δ at $\bar{x}$, that is, a smooth vector field X on M satisfying $X(x) \in \Delta(x)$ for every $x \in M$ and such that the dynamical system $\dot{x} = X(x)$ is $\mathrm{GAS}_{\bar{x}}$.

1.3. *Two obstructions*

Given a nonholonomic control system (resp. a nonholonomic distribution), there are two obstructions to the existence of stabilizing feedbacks (resp. stabilizing sections). The first one is global while the second one is purely local.

(i) Global obstruction: If fact, if the manifold M admits a smooth dynamical system which is $\mathrm{GAS}_{\bar{x}}$, then it must be homeomorphic to $\mathbb{R}^n$ (see Ref. 12 for further details).

(ii) Local obstruction: Since this obstruction is local, we can assume that we work in $\mathbb{R}^n$, that is, in an open neighborhood $\mathcal{U} \subset \mathbb{R}^n$ of $\bar{x}$. If there is a smooth vector field X on $\mathcal{U}$ which is locally asymptotically stable at $\bar{x}$, then for all ϵ small enough,

$$\exists \delta > 0 \quad \text{such that} \quad \delta B \subset X(\bar{x} + \epsilon B),$$

where B denotes the open unit ball in $\mathbb{R}^n$ (see Ref. 12 for further details). In consequence, if there is a smooth feedback $k : \mathcal{U} \to \mathbb{R}^m$ such that the closed-loop system (2) is locally asymptotically stable at $\bar{x}$, then the result above applies to the dynamics $X(x) = \sum_{i=1}^{m} k(x)_i f_i(x)$. Thus, for all ϵ small enough,

$$\exists \delta > 0 \quad \text{such that} \quad \delta B \subset \left\{ \sum_{i=1}^{m} u_i f_i(\bar{x} + \epsilon B) \mid u \in \mathbb{R}^m \right\},$$

which is the *Brockett necessary condition* (see Ref. 5). In particular, we deduce that any distribution Δ of rank $m < n$ cannot admit a smooth locally stabilizing section at $\bar{x}$.

The obstructions above make it impossible to prove the existence of smooth stabilizing feedbacks (resp. sections) for nonholonomic control systems (resp. distributions). Actually, they motivate the design of new kinds of stabilizing feedbacks. The main contributions in that direction have been: Sussmann,[17] Coron,[8] Clarke, Ledyaev, Sontag and Subbotin,[7] and Ancona and Bressan.[1] The aim of the present paper is to highlight the notion of smooth repulsive stabilizing feedbacks (or sections) and to show that it permits to stabilize most of the nonholonomic control systems (or distributions).

2. Examples

2.1. *The Nonholonomic integrator*

Define in $\mathbb{R}^3$ the two smooth vector fields f_1, f_2 by,

$$f_1 = \frac{\partial}{\partial x_1} + x_2 \frac{\partial}{\partial x_3} \quad \text{and} \quad f_2 = \frac{\partial}{\partial x_2} - x_1 \frac{\partial}{\partial x_3}.$$

Note that at any point $x \in \mathbb{R}^3$, the three vectors $f_1(x), f_2(x), [f_1, f_2](x)$ are linearly independent. Hence the associated so-called nonholonomic integrator (or Brockett integrator)

$$\dot{x} = u_1 f_1(x) + u_2 f_2(x),$$

is nonholonomic on $\mathbb{R}^3$. As it is well-known, this control system can be stabilized at the origin by a feedback which is smooth outside the vertical line[‡]. Denote by $\mathcal{S}$ the vertical line of equation $x_1 = x_2 = 0$ in $\mathbb{R}^3$. Using an adapted control-Lyapunov function (see Refs. 11 or 9 for further details), we can construct a mapping $k = (k_1, k_2) : \mathbb{R}^3 \mapsto \mathbb{R}^2$ in such a way that the following properties are satisfied:

[‡]We refer the reader to Ref. 3 for a detailed study of the stabilization of the nonholonomic integrator by discontinuous feedbacks.

(i) The mapping k is locally bounded and smooth on $M \setminus \mathcal{S}$.

(ii) The set $\mathbb{R}^3 \setminus \mathcal{S}$ is invariant with respect to the dynamical system

$$\dot{x} = k_1(x)f_1 + k_2(x)f_2(x),$$

and steers asymptotically all its trajectories to the origin.

In fact, the feedback above can be extended to the whole space $\mathbb{R}^3$. In this way, we can construct a feedback which is smooth outside $\mathcal{S}$, discontinuous at the points of $\mathcal{S}$, and for which the closed-loop system is $\text{GAS}_{\bar{x}}$ in the sense of Carathéodory (see below). Such a feedback is an example of what we call a smooth repulsive stabilizing feedback.

2.2. *The Riemannian case*

Assume through this paragraph that the nonholonomic distribution Δ is given by $\Delta(x) = T_x M$ for every $x \in M$, our aim is to show how to construct a stabilizing section for Δ. For that, consider a smooth and complete Riemannian metric g on M and denote by d_g the *Riemannian distance* associated with g. We recall that, for any $x, y \in M$, the Riemannian distance between x and y is defined as

$$d_g(x, y) = \inf \left\{ \text{length}_g(\gamma(\cdot)) \right\}, \quad \forall x, y \in M,$$

where the infimum is taken over all the $\mathcal{C}^1$ paths $\gamma(\cdot) : [0, 1] \to M$ joining x to y, and where

$$\text{length}_g(\gamma(\cdot)) = \int_0^1 \sqrt{g_{\gamma(t)}(\dot{\gamma}(t), \dot{\gamma}(t))}\, dt.$$

Let $V_g : M \to \mathbb{R}$ be the function defined by

$$V_g(x) = d_g(\bar{x}, x)^2, \quad \forall x \in M.$$

It is easy to show that V_g is Lipschitz continuous on M and smooth outside the set $\mathcal{S}$ defined as the cut-locus[§] from the point $\bar{x}$. Define the vector field X on M by

$$X(x) = -\nabla_g V_g(x), \quad \forall x \in M \setminus \mathcal{S},$$

[§]The cut-locus from $\bar{x}$ is defined as the closure of the set where V_g is not differentiable. We refer the reader to Ref. 16 for a detail study of the distance function and the cut-locus from $\bar{x}$.

where $\nabla_g V_g(x)$ denotes the gradient of V_g at x with respect to the metric g. By construction, any trajectory of $\dot{x}(t) = X(x(t))$ tends to $\bar{x}$ as $t \to \infty$ and satisfies the following property:

$$\forall t \geq 0, \quad x(t) \notin \mathcal{S}.$$

In fact, X can be extended into a global section of Δ on M which is smooth outside the cut-locus from $\bar{x}$, discontinuous at the points of $\mathcal{S}$ and whose the associated dynamics drives all its Carathéodory trajectories asymptotically to $\bar{x}$. Such a stabilizing section corresponds to what we call a smooth repulsive stabilizing section of Δ on M.

3. Smooth repulsive stabilization

Our aim is now to make precise the notions of smooth repulsive stabilizing feedbacks or sections, and to show what kind of results we are able to prove.

3.1. $SRS_{\bar{x},\mathcal{S}}$ vector fields

Let $\mathcal{S}$ be a closed subset of M and X be a smooth vector field on M. The dynamical system $\dot{x} = X(x)$ is said to be *smooth repulsive globally asymptotically stable at $\bar{x}$ with respect to $\mathcal{S}$* (denoted in short $SRS_{\bar{x},\mathcal{S}}$) if the following properties are satisfied:

(i) The vector field X is locally bounded on M and smooth on $M \setminus \mathcal{S}$.

(ii) The dynamical system $\dot{x} = X(x)$ is globally asymptotically stable at $\bar{x}$ in the sense of Carathéodory, namely, for every $x \in M$, there exists a solution of

$$\dot{x}(t) = X(x(t)), \quad \text{for almost every } t \in [0, \infty), \quad x(0) = x, \qquad (3)$$

and, for any $\epsilon > 0$, every solution of $\dot{x}(t) = X(x(t))$ almost everywhere in $[0, \epsilon)$ (called Carathéodory solution of $\dot{x} = X(x)$) can be extended to $[0, \infty)$ and tends to $\bar{x}$ as t tends to ∞. Moreover, for every neighborhood $\mathcal{V}$ of $\bar{x}$, there exists a neighborhood $\mathcal{W}$ of $\bar{x}$ such that, for $x \in \mathcal{W}$, the solutions of (3) satisfy $x(t) \in \mathcal{V}$, for every $t \geq 0$.

(iii) For every $x \in M$, the solutions of (3) satisfy $x(t) \notin \mathcal{S}$, for every $t > 0$.

Given the nonholonomic control system (1), we shall say that a mapping $k : M \to \mathbb{R}^m$ is a *smooth repulsive stabilizing feedback at $\bar{x}$* (denoted in short $SRS_{\bar{x}}$ feedback) for (1), if it is locally bounded on M and if there is a closed set $\mathcal{S} \subset M$ such that k is smooth on $M \setminus \mathcal{S}$ and such that its associated closed-loop system is $SRS_{\bar{x},\mathcal{S}}$. In the same way, given a nonholonomic distribution Δ and a vector field X on M, we shall say that X is a *smooth repulsive stabilizing section at $\bar{x}$* (denoted in short SRS$_{\bar{x}}$ section) for Δ, if X is a section of Δ on M and if the dynamical system $\dot{x} = X(x)$ is SRS$_{\bar{x},\mathcal{S}}$ for some closed set $\mathcal{S} \subset M$.

3.2. *Existence results of SRS$_{\bar{x}}$ feedbacks*

Using a technique of cancellation of the bifurcation points of a "discontinuous" stabilizing feedback, we proved in Ref. 13 the following result:

Theorem 3.1. *If M has dimension two, any nonholonomic control system of the form (1) admits a SRS$_{\bar{x}}$ feedback.*

Using the classical method of local approximation of a nonholonomic control systems by an homogeneous one together with the technique of cancellation developed on surfaces, we were able in Ref. 14 to demonstrate the following result:

Theorem 3.2. *If M has dimension three, any nonholonomic control system of the form (1) admits a SRS$_{\bar{x}}$ feedback defined on a neighborhood of $\bar{x}$.*

We refer the interested reader to Refs. 13 and 14 for more details on these results.

3.3. *Existence results of SRS$_{\bar{x}}$ sections*

The method presented above in the Riemannian case can also be developed in the sub-Riemannian setting; we need for that to introduce material of sub-Riemannian geometry. Let Δ be a nonholonomic distribution of rank $m \leq n$ on M, the set of horizontal paths $\gamma(\cdot) : [0,1] \to M$ such that $\gamma(0) = \bar{x}$, denoted by $\Omega_\Delta(\bar{x})$, endowed with the $W^{1,1}$-topology, inherits of a Banach manifold structure (see Ref. 15 for further details). The *end-point mapping from $\bar{x}$* is the mapping $E_{\bar{x}} : \Omega_\Delta(\bar{x}) \to M$ defined by

$$E_{\bar{x}}(\gamma(\cdot)) = \gamma(1), \quad \forall \gamma(\cdot) \in \Omega_\Delta(\bar{x});$$

it is a smooth mapping on $\Omega_\Delta(\bar{x})$. A path $\gamma(\cdot)$ is said to be *singular* if it is horizontal and if it is a critical point of the end-point mapping $E_{\bar{x}}$, that is, if the differential of $E_{\bar{x}}$ at $\gamma(\cdot)$ is not a submersion. Let g be a smooth Riemannian metric on M, the *sub-Riemannian distance $d_{SR}(x,y)$* between two points x, y of M is defined by

$$d_{SR}(x,y) = \inf \left\{ \text{length}_g(\gamma(\cdot)) \mid \gamma(\cdot) \in \Omega_\Delta(\bar{x}) \right\}.$$

According to the Chow-Rashevsky Theorem, since Δ is nonholonomic on M, the sub-Riemannian distance is well-defined and continuous on $M \times M$. Moreover, if the manifold M is a complete metric space[¶] for the sub-Riamannian distance d_{SR}, then, since M is connected, for every pair x, y of points of M there exists an horizontal path $\gamma(\cdot) : [0,1] \to M$ joining x to y such that

$$d_{SR}(x, y) = \text{length}_g(\gamma(\cdot)).$$

Such a horizontal path is said to be *minimizing*. The following result has been obtained recently with Trélat (see Ref. 15).

Theorem 3.3. *Let Δ be a smooth nonholonomic distribution of rank $m \leq n$ on M. Assume that there exists a smooth Riemannian metric g on Δ for which M is complete and no nontrivial singular path is minimizing. Then, there exist a section X of Δ on M, and a closed nonempty subset S of M, of Hausdorff dimension lower than or equal to $n - 1$, such that X is $SRS_{\bar{x},S}$.*

Note that if $m = n$, then obviously there exists no singular path (it is the Riemannian case). In fact, the main assumption of Theorem 3.3 (the absence of nontrivial singular minimizing paths) is automatically satisfied for a large class of sub-Riemannian structures such as the fat distributions or the medium-fat distributions associated with a generic metric (we refer the reader to Ref. 15 for further details). Here, we just want to emphasize the fact that the main assumption of Theorem 3.3 is satisfied generically for distributions with rank greater than two.

Let $m \geq 3$ be a positive integer and $\mathcal{G}_m$ be the set of pairs (Δ, g), where Δ is a rank m distribution on M and g is a smooth Riemannian metric on Δ, endowed with the Whitney $\mathcal{C}^\infty$ topology. It is shown in Ref. 6 that there exists an open dense subset W_m of $\mathcal{G}_m$ such that every element of W_m does not admit nontrivial minimizing singular paths. This means that, for $m \geq 3$, generically, the main assumption of Theorem 3.1 is satisfied. Therefore, as a by-product of the Chitour-Jean-Trélat Theorem, we have the following result:

Corollary 3.1. *A generic nonholonomic distribution of rank ≥ 3 admits a $SRS_{\bar{x}}$ section.*

[¶]Note that, since the distribution Δ is nonholonomic on M, the topology defined by the sub-Riemannian distance d_{SR} coincides with the original topology of M (see Refs 4 or 10).

268

We notice that in Ref. 15, we are able to remove, in the compact and orientable three-dimension case, the assumption on the absence of singular minimizing paths. We refer the interested reader to Ref. 15 for further details on that result.

3.4. *A Nonholonomic dream*

In view of the results presented here, one might hope that the following conjecture is true.

Conjecture. Any nonholonomic control system admits a SRS feedback.

References

1. F. Ancona and A. Bressan, *ESAIM Control Optim. Calc. Var.*, **4**, 445 (1999).
2. A. A. Agrachev and Y. L. Sachkov, *Control theory from the geometric viewpoint*, Encyclopaedia of Mathematical Sciences, Vol. 87. Control Theory and Optimization, II. (Springer-Verlag, Berlin, 2004).
3. A. Astolfi. *Eur. J. Control*, **4** (1), 49 (1998).
4. A. Bellaïche, The tangent space in sub-Riemannian geometry, in *Sub-Riemannian Geometry*, eds. A. Bellache and J.-J. Risler, Progr. Math., Vol. 144, (Birkhäuser Verlag, Basel, 1996), pp. 1–78.
5. R. W. Brockett, Asymptotic stability and feedback stabilization, in *Differential geometric control theory (Houghton, Mich., 1982)*, eds. R. W. Brockett, R. S. Millman and H. J. Sussmann, Progr. Math., Vol. 27 (Birkhäuser Boston, Boston, MA, 1983), pp. 181–191.
6. Y. Chitour, F. Jean and E. Trélat, *J. Diff. Geom.* **73** (1), 45 (2006).
7. F. H. Clarke, Y. S. Ledyaev, E. D. Sontag and A. I. Subbotin, *IEEE Trans. Automat. Control* **42**, 1394 (1997).
8. J.-M. Coron, *Math. Control Signals Systems*, **5** (3), 295 (1992).
9. Y. S. Ledyaev and L. Rifford (unpublished, 2007).
10. R. Montgomery, *A tour of subriemannian geometries, their geodesics and applications*, Mathematical Surveys and Monographs, Vol. 91 (American Mathematical Society, Providence, RI, 2002).
11. L. Rifford, Problèmes de stabilisation en théorie du contrôle, PhD thesis, Université Lyon I (Lyon, France, 2000).
12. L. Rifford, The Stabilization Problem: AGAS and SRS Feedbacks, in *Optimal control, stabilization and nonsmooth analysis*, eds. M. M. M.S. de Queiroz and P. Wolenski, Lecture Notes in Control and Inform. Sci., Vol. 301 (Springer, Berlin, 2004), pp. 173–184.
13. L. Rifford, *Rend. Semin. Mat. Torino* **64** (1), 55 (2006).
14. L. Rifford, *J. Differential Equations*, **226** (2), 429 (2006).
15. L. Rifford and E. Trélat, On the stabilization problem for nonholonomic distributions, *J. Eur. Math. Soc. (JEMS)* (to appear), available at http://arxiv.org/abs/math/0610363v1 (October 2006).

16. T. Sakai, *Riemannian geometry*, Translations of Mathematical Monographs, Vol. 149 (American Mathematical Society, Providence, RI, 1996). Translated from the 1992 Japanese original by the author.
17. H. J. Sussmann, *J. Differential Equations* **31**, 31 (1979).

PROXIMAL CHARACTERIZATION
OF THE REACHABLE SET
FOR A DISCONTINUOUS DIFFERENTIAL INCLUSION

V. R. RIOS

Departamento de Matemática, Facultad Experimental de Ciencias,
Universidad del Zulia, Maracaibo, Edo. Zulia, Venezuela
E-mail: vrios@luz.edu.ve
http://www.fec.luz.edu.ve

P. R. WOLENSKI

Mathematics Department, College of Arts and Sciences,
Louisiana State University, Baton Rouge, LA 70803, USA
E-mail: wolenski@math.lsu.edu
http://www.math.lsu.edu

The graph of the reachable set is characterized via limiting Hamilton-Jacobi inequalities when the system dynamics satisfy a discontinuous dissipative Lipschitz condition.

Keywords: Reachable Set; Dissipative Lipschitz Maps; Hamilton-Jacobi Inequality; Differential Inclusion.

1. Introduction

This paper considers a dynamical control system governed by the differential inclusion

$$\dot{x}(t) \in F(x(t)) \quad \text{a.e. } t \in I := [0, \infty), \tag{DI}$$

where F is a multifunction mapping $\mathbb{R}^n$ into the subsets of $\mathbb{R}^n$ that satisfies assumptions that will be stated below. A solution (or trajectory) to (DI) is an absolutely continuous (AC) function $x(\cdot) : I \to \mathbb{R}^n$ satisfying (DI). Given a compact set $M \subset \mathbb{R}^n$, the *reachable set of F at time* $t \geq 0$ is defined as

$$\mathcal{R}_F(t) := \Big\{ x(t) : x(\cdot) \text{ solves (DI) on } [0, t], \text{ with } x(0) \in M \Big\},$$

and the *graph of* $\mathcal{R}_F(\cdot)$ is given by

$$\mathcal{G}(\mathcal{R}_F) := \Big\{(t, x) : t \geq 0, \text{ and } x \in \mathcal{R}_F(t)\Big\}.$$

We are interested in characterising $\mathcal{G}(\mathcal{R}_F)$ in terms of Hamiltonian inequalities for possibly discontinuous F. Throughout this paper, we will assume the multifunction F is endowed with the following structural properties **(SP)**:

- $F(x)$ is a nonempty, compact, and convex set, for all $x \in \mathbb{R}^n$;
- There is a constant $c > 0$ so that $\sup\{\|v\| : v \in F(x)\} \leq c(1 + \|x\|)$ for all $x \in \mathbb{R}^n$;
- $F(\cdot)$ is upper semicontinuous **(US)**, which in presence of the previous assumptions, is equivalent to $\mathcal{G}(F) := \{(x, v) : v \in F(x)\}$ being closed.

The fact that $\mathcal{R}_F(t)$ is nonempty is well known, see for instance the existence theory exposed in Refs. 1,5,6. Moreover, using compactness of trajectories (see Theorem 4.1.11 of Ref. 5) it is easy to check that $\mathcal{R}_F(t)$ and $\mathcal{G}(\mathcal{R}_F)$ are compact and closed, respectively.

The motif of this note arises from Clarke[2], where the author considers a time-dependent multifunction $G : I \times \mathbb{R}^n \rightrightarrows \mathbb{R}^n$ satisfying (SP) and a Lipschitz condition, both jointly in (t, x). The main result in Ref. 2 states that $\mathcal{G}(\mathcal{R}_G)$ can be characterized as the unique closed subset S of $I \times \mathbb{R}^n$ for which the following conditions hold, with $S_T := \{x \in \mathbb{R}^n : (T, x) \in S\}$:

$$\theta + H_G(t, x, \zeta) = 0, \text{ for all } (\theta, \zeta) \in N_S^P(t, x), \text{ and all } (t, x) \in S; \tag{1}$$

$$\lim_{T \downarrow 0} S_T = M. \tag{2}$$

Equation (1) involves the *upper Hamiltonian* of G, $H_G : I \times \mathbb{R}^n \times \mathbb{R}^n \rightarrow \mathbb{R}$, defined by

$$H_G(t, x, p) := \sup\{\langle v, p \rangle : v \in G(t, x)\},$$

(analogously, the *lower Hamiltonian* of G is $h_G(t, x, p) := \inf\{\langle v, p \rangle : v \in G(t, x)\}$). For autonomous G the t-variable will be dropped in the previous definitions. Another key ingredient in the above result is the *proximal normal cone* $N_S^P(x)$ of a closed set $S \subset \mathbb{R}^m$ at $x \in S$, defined as the set of elements $\zeta \in \mathbb{R}^m$ for which there exists $\sigma = \sigma(\zeta, x) \geq 0$ such that

$$\langle \zeta, x' - x \rangle \leq \sigma \|x' - x\|^2 \quad \text{for all } x' \in S.$$

In equation (2) the limit is interpreted via the Hausdorff metric d_H. This means precisely that, given $\varepsilon > 0$, there exists $\delta > 0$ such that

$$0 \leq T < \delta \implies d_H(S_T, M) := \max\left\{ \sup_{\alpha \in S_T} d(\alpha, M), \sup_{\beta \in M} d(\beta, S_T) \right\} < \varepsilon,$$

where $d(\alpha, K) := \inf\{\|\alpha - v\| : v \in K\}$ is the distance from a point $\alpha \in \mathbb{R}^n$ to a closed set $K \subset \mathbb{R}^n$.

The goal of this article is to extend the above characterization to an important class of discontinuous dynamics F that we discuss below (see section 2). Following Ref. 2, our main result (Theorem 3.1) shows that under milder assumptions than heretofore imposed, a characterization of $\mathcal{G}(\mathcal{R}_F)$ can be provided by replacing (1) with an appropriate pair of Hamiltonian inequalities, one of which must reflect the possible discontinuity of F.

It is noteworthy that equation (1) is obtained in Ref. 2 as a consequence of the invariant properties satisfied by the auxiliary pairs $(\mathcal{R}_G, \pm\widetilde{G})$, where

$$\widetilde{G}(t, x) := \big\{(1, v) : v \in G(t, x)\big\}.$$

These invariant properties translate into $\mathcal{G}(\mathcal{R}_G)$ being simultaneously a subsolution and a supersolution of the exact *Hamilton-Jacobi* **(HJ)** equation (1) in the following sense (c.f. Ref. 2): a closed set $S \subseteq I \times \mathbb{R}^n$ is a *subsolution* of (1) if

$$\theta + H_G(t, x, \zeta) \geq 0, \text{ for all } (\theta, \zeta) \in N_S^P(t, x), \text{ and all } (t, x) \in S, \quad (3)$$

and a *supersolution* of (1) if

$$\theta + H_G(t, x, \zeta) \leq 0, \text{ for all } (\theta, \zeta) \in N_S^P(t, x), \text{ and all } (t, x) \in S. \quad (4)$$

Property (4) is not necessarily enjoyed by $\mathcal{G}(\mathcal{R}_F)$ under discontinuous (DI), for the Lipschitz invariant result that is used in Ref. 2 to obtain (4) does not cover, for example, the type of discontinuous dynamics considered in this paper. This situation is illustrated in the next section (see also Refs. 8,9).

The rest of this note is organized as follows. Section 2 contains precise assumptions on the data, and an invariant prerequisite that holds under these assumptions. The main result and its proof are presented in Section 3.

2. DL dynamics and invariance

Invariance properties of differential equations have been studied extensively. More recently, weak and strong invariance properties of differential inclusions have also received considerable attention due mainly to their significant applications in Hamilton-Jacobi theory. We suggest Refs. 4,9 (and references therein) for concise histories on flow invariance theory and its repercussions. Regarding the terminology, let us recall that for a given closed set $S \subseteq \mathbb{R}^n$ and F as above, the pair (S, F) is said *weakly invariant* if for each $x_0 \in S$, there is a solution $x(\cdot)$ to (DI), with $x(0) = x_0$, and satisfying $x(t) \in S$ for all $t \in I$. Similarly, (S, F) is *strongly invariant* if for each $x_0 \in S$ every solution $x(\cdot)$ to (DI), with $x(t_0) = x_0$, also satisfies $x(t) \in S$ for all $t \in I$. Weak invariance characterizations hold under general hypotheses similar to (SP) (see the works of Aubin[1], Frankowska et al.[10,11] and Veliov[16]), and they have contributed to establishing subsolutions of the HJ equation in different approaches (minimax, proximal, viscosity: the history is sketched in Ref. 4). In this same sense, we will benefit from the following time-dependent result, which appeared as Theorem 1 in Ref. 9.

Theorem 2.1. *Let $\widetilde{I} \subseteq I$ be a subinterval. Suppose a nonautonomous multifunction $G : \widetilde{I} \times \mathbb{R}^n \rightrightarrows \mathbb{R}^n$ satisfies (SP), with the upper semicontinuity requirement replaced by the weakened Scorza–Dragoni property, and let $S \subset \mathbb{R}^n$ be closed. Then (S, G) is weakly invariant if and only if there exists a null set $A \subset \widetilde{I}$ such that*

$$h_G(t, x, \zeta) \le 0, \tag{5}$$

for all $t \in \widetilde{I} \setminus A$, $x \in S$, and $\zeta \in N_S^P(x)$. In this case, (5) holds at all points of density of a certain countable family of pairwise disjoint closed sets $I_k \subset \widetilde{I}$ (k=1,2,...,) for which $G(\cdot, \cdot)$ is upper semicontinuous on each $I_k \times \mathbb{R}^n$.

On the other hand, supersolutions to the HJ equation (1) have generally been obtained by invoking classical criteria for strong invariance[2,11,17]. These criteria have typically required a Lipschitz assumption on the multifunction $G(\cdot, \cdot)$ (see Theorems 4.3.8 in Ref 5, 5.2 in Ref. 3, 4.10 in Ref. 11, and 2.1 in Ref. 13), which can be stated as follows (we shall restrict ourselves to global definitions to simplify the exposition, but local versions hold as well): a convex-valued multifunction G is *Lipschitz* if there exists a constant λ such that, for all (t, x), $(s, y) \in I \times \mathbb{R}^n$, and all $\zeta \in \mathbb{R}^n$ we have

$$\left| H_G(t, y, \zeta) - H_G(s, x, \zeta) \right| \le \lambda \|\zeta\| \, \|(t - s, y - x)\|. \tag{6}$$

Within a non-Lipschitz framework, and consequently beyond the scope of the standard theory, a class of *Dissipative* dynamics has been used in Refs. 6, 14 to model dry friction forces in physical phenomena. An autonomous convex-valued multifunction $D : \mathbb{R}^n \rightrightarrows \mathbb{R}^n$ is Dissipative if

$$H_D(y, y - x) + H_D(x, x - y) \le 0, \quad \text{for all } x, y \in \mathbb{R}^n. \tag{7}$$

The main result of this paper will hold under a generalization of the previous dissipativity concept. The aforementioned notion was introduced by T. Donchev[7] and consists of a quasi-quadratic perturbation on the right-hand side of (7). More precisely, a convex-valued multifunction F is called *Dissipative Lipschitz* (**DL**) if there is a constant $\mathcal{K}$ such that

$$H_F(y, y - x) - H_F(x, y - x) \leq \mathcal{K}\|y - x\|^2, \quad \text{for all } x, y \in \mathbb{R}^n. \tag{8}$$

Property (8) is essentially weaker than properties (7) and (6): taking the multifunction G autonomous, it is clear that (8) follows from (6) by setting $\zeta = y - x$, and obviously (7)$\Rightarrow$(8). Nevertheless, the converse of these implications may fail, as can be appreciated by considering the multifunctions $D(x) := -\partial\|x\|$, and $F(x) := D(x) + \{x\}$, where ∂ denotes the subdifferential of convex functions, and $x \in \mathbb{R}$. Notice that D and F satisfy (SP), D is dissipative, and F is dissipative Lipschitz (with $\mathcal{K} = 1$). On the other hand, D and F are discontinuous at $x = 0$, which implies they are not Lipschitz. Moreover, F does not satisfy the dissipativity condition (7).

In case of discontinuous DL dynamics the exact Hamilton-Jacobi equation (1) may not hold. In fact, letting $M := \{0\}$ and $S := \mathcal{G}(\mathcal{R}_D) = I \times \{0\}$ we have the strict inequality

$$0 + H_D(0, \zeta) = H_{\widetilde{D}}((0,0),(0,\zeta)) > 0, \text{ for all } (0, \zeta) \in N_S^P(0,0)\backslash\{(0,0)\}.$$

However, the pairs $(S, -\widetilde{D})$ and $(S, \widetilde{D})$ are weakly and strongly invariant respectively, since any trajectory $z(\cdot)$ of the dissipative multimap $\widetilde{D}(z) := \{(1, v) : v \in D(x)\}$, with $z(0) \in S$, is necessarily of the form $z(t) = (t+t_0, 0)$ for some $t_0 \in I$ (here, by state augmentation, we view t as a component of the state $z = (t, x) \in I \times \mathbb{R}^n$).

The previous example suggests that, under assumption (8), finding an appropriate complementary condition to (3) that helps characterize $\mathcal{G}(\mathcal{R}_F)$ in terms of Hamilton-Jacobi inequalities, can be accomplished with a more general criterion for strong invariance. The following result, which appeared in Ref. 9, meets such a requirement. We include here an autonomous version of the proof for completeness and since it is somewhat simpler than the one given there.

Proposition 2.1. (*Corollary 5 of Ref. 9*) *Let $S \subseteq \mathbb{R}^n$ be closed and suppose F satisfies* (SP) *and* (8)*. Then the system (S, F) is strongly invariant if and only if*

$$\limsup_{y \to_\zeta x} H_F(y, \zeta) \leq 0, \tag{9}$$

for all $x \in S$, and all $\zeta \in N_S^P(x)$.

275

Remark 2.1. Given a nonzero vector $\zeta \in \mathbb{R}^n$, the notation $y \to_\zeta x$ in (9) signifies the limit of y approaching x along the vector ζ; in other words, $y \to_\zeta x$ if and only if $y \to x$ and $\frac{y-x}{\|y-x\|} \to \frac{\zeta}{\|\zeta\|}$.

Proof. First, we show that (9) is sufficient for strong invariance. Given any trajectory $x(\cdot)$ of (DI), with $x(0) = x_0 \in S$, the multifunction

$$G(t,x) := \left\{ v \in F(x) : \langle \dot{x}(t) - v, x(t) - x \rangle \leq \mathcal{K}\|x(t) - x\|^2 \right\}$$

satisfies (SP), with the weakened Scorza-Dragoni property (see page 70 of Ref. 15) replacing upper semicontinuity. Let $T > 0$, and set $\widetilde{I} := [0, T] \subset I$. The last property means that for any $\varepsilon > 0$ there is a closed set $I_\varepsilon \subset \widetilde{I}$ with Lebesgue measure $\mu(\widetilde{I} \setminus I_\varepsilon) < \varepsilon$, such that the restriction of G to $I_\varepsilon \times \mathbb{R}^n$ is US in (t, x). Let J_ε denote the points of density of I_ε, and define $A := \widetilde{I} \setminus (\cup_{\varepsilon > 0} J_\varepsilon)$ which has null measure in $\widetilde{I}$ (see page 274 in Ref. 12). Let $x \in S$, and $\zeta \in N_S^P(x)$. Given $t \in \widetilde{I} \setminus A$, there is some $\varepsilon > 0$ for which $t \in J_\varepsilon$. By taking the $\liminf$ all over the sequences $(\rho, z) \to (t, x)$ in $I_\varepsilon \times \mathbb{R}^n$, the lower semicontinuity of $h_G(\cdot, \cdot, \zeta)$ yields

$$h_G(t, x, \zeta) \leq \liminf_{\rho \to t,\ z \to x} h_G(\rho, z, \zeta) \leq \liminf_{\rho \to t,\ y \to_\zeta x} h_G(\rho, y, \zeta). \qquad (10)$$

Inequalities (9) and (10), and the fact that $h_G \leq H_F$ imply

$$h_G(t, x, \zeta) \leq \limsup_{y \to_\zeta x} H_F(y, \zeta) \leq 0,$$

and this inequality actually holds for almost all $t \in I$, as can be readily seen by repeating the previous argument over each member of the sequence of intervals $[T, 2T], [2T, 3T], \ldots$ This in turn implies the weak invariance of G according to Theorem 2.1. If $y(\cdot)$ is an invariant trajectory of G with $y(0) = x_0$, the very definition of G guarantees that

$$\frac{d}{dt}\|x(t) - y(t)\|^2 \leq 2\mathcal{K}\|x(t) - y(t)\|^2,$$

which implies $x(t) = y(t) \in S$, for all $t \in I$ via Gronwall's lemma. This establishes the strong invariance property of (S, F).

Conversely, assume that the system (S, F) is strongly invariant. Let $x \in S$, $\zeta \in N_S^P(x)$, and a sequence $y_i \to_\zeta x$ be given. For each i, consider $v_i \in F(y_i)$ such that $H_F(y_i, \zeta) = \langle v_i, \zeta \rangle$, and define $G_i : \mathbb{R}^n \rightrightarrows \mathbb{R}^n$ by

$$G_i(y) := \{w \in F(y) : \langle v_i - w, y_i - y \rangle \leq \mathcal{K}\|y_i - y\|^2\}.$$

It is easy to check that G_i satisfies (SP), and since $G_i \subseteq F$ the strong invariance of (S, F) yields, in particular, the weak invariance of (S, G_i), for $i = 1, 2, \ldots$ If $w_i \in G_i(x)$ satisfies $\langle w_i, \zeta \rangle = h_{G_i}(x, \zeta)$, then Theorem 2.1

implies $\langle w_i, \zeta \rangle \leq 0$ (see also Theorem 4.2.10 of Ref. 5). Furthermore, due to (SP) the sequences v_i and w_i are bounded. Rearranging terms from the definition of $G_i(x)$, and using the properties of v_i, and w_i we have

$$\limsup_{i \to \infty} H_F(y_i, \zeta) = \limsup_{i \to \infty} \langle v_i, \zeta \rangle$$

$$= \limsup_{i \to \infty} \|\zeta\| \left\langle v_i, \frac{y_i - x}{\|y_i - x\|} \right\rangle$$

$$\leq \limsup_{i \to \infty} \|\zeta\| \left(\mathcal{K}\|y_i - x\| + \left\langle w_i, \frac{y_i - x}{\|y_i - x\|} \right\rangle \right)$$

$$= \limsup_{i \to \infty} \langle w_i, \zeta \rangle \leq 0.$$

Hence condition (9) is satisfied with the limsup taken all over the sequences $y_i \to_\zeta x$. Therefore, (9) holds as stated. $\qquad \square$

3. Main result

We now proceed to establish a proximal characterization of $\mathcal{G}(\mathcal{R}_F)$ in terms of Hamilton-Jacobi inequalities. The proof given below follows the lines of Theorem 1 of Ref. 2, but extends the same result to the US–DL framework by incorporating the novel Hamiltonian condition provided in Proposition 2.1.

Remark 3.1. We must clarify at this point that the limsup over $y \to_\zeta x$ in (9) can be replaced by the a priori weaker condition of taking this limit over $\delta \to 0^+$, with $y = x + \delta\zeta$, without changing the equivalence with strong invariance. This simplification will be in effect when estimating inequality (b) below.

Theorem 3.1. *Assume F satisfies (SP) and property (8). Then the graph of its reachable set $\mathcal{G}(\mathcal{R}_F)$, is the unique closed subset S of $I \times \mathbb{R}^n$ satisfying the following for all $(\theta, \zeta) \in N_S^P(t, x)$, and all $(t, x) \in S$:*
  (a) $\theta + H_F(x, \zeta) \geq 0$,
  (b) $\theta + \limsup\limits_{y \to_\zeta x} H_F(y, \zeta) \leq 0$,
  (c) $\lim_{T \downarrow 0} S_T = M$.

Proof. We start by recalling how to establish property (a). Let $\tilde{F}(t, x) := \{1\} \times F(x)$. By definition of reachable set, for any $(\tau, \alpha) \in \mathcal{G}(\mathcal{R}_F)$ there is a trajectory $y(\cdot)$ of (DI) defined on $[0, \tau]$, with $y(\tau) = \alpha$ and $y(0) \in M$. The previous also implies that $(t, y(t)) \in \mathcal{G}(\mathcal{R}_F)$, for all $t \in [0, \tau]$. Therefore, the augmented time-reverse arc $z(t) := (\tau - t, y(\tau - t))$ satisfies

$$\dot{z}(t) \in -\widetilde{F}(z(t)) \quad \text{a.e., } t \in [0, \tau],$$

with $z(t) \in \mathcal{G}(\mathcal{R}_F)$ for all $t \in [0, \tau]$, and $z(0) = (\tau, \alpha)$. The last argument guarantees the weak invariance of the pair $(\mathcal{G}(\mathcal{R}_F), -\widetilde{F})$, which in light of Theorem 2.1 yields to

$$\theta + H_F(x, \zeta) = -h_{-\widetilde{F}}(t, x, (\theta, \zeta)) \geq 0,$$

for all $(t, x) \in \mathcal{G}(\mathcal{R}_F)$ and all $(\theta, \zeta) \in N^P_{\mathcal{G}(\mathcal{R}_F)}(t, x)$. We now focus on property (b). Let $(\tau, \alpha) \in \mathcal{G}(\mathcal{R}_F)$ and $z(\cdot)$ be a trajectory of

$$\dot{w}(t) \in \widetilde{F}(w(t)) \quad \text{a.e., } t \in I,$$

with $z(0) = (\tau, \alpha)$. We claim that $z(t) \in \mathcal{G}(\mathcal{R}_F)$ for all $t \geq 0$. In fact, we first notice that necessarily $z(t) = (t + \tau, y(t + \tau))$, for some trajectory $y(\cdot)$ of (DI) defined on I. We distinguish two cases:

Case 1: If $\tau = 0$, the use of $z(\cdot)$ leads to $y(0) = \alpha$. Since $(0, \alpha) \in \mathcal{G}(\mathcal{R}_F)$, by definition of reachable set we must have $y(0) \in M$, which implies $z(t) = (t, y(t)) \in \mathcal{G}(\mathcal{R}_F)$, for $t \geq 0$.

Case 2: Let us assume $\tau > 0$. Using the definition of $\mathcal{G}(\mathcal{R}_F)$ we obtain a trajectory $x(\cdot)$ of (DI) defined on $[0, \tau]$, with $x(\tau) = \alpha$, and $x(0) \in M$. Then the piecewise arc

$$w(t) := \begin{cases} x(t), \ t \in [0, \tau), \\ y(t), \ t \in [\tau, \infty) \end{cases}$$

is a trajectory of (DI) satisfying $w(0) = x(0) \in M$. Again, from the definition of $\mathcal{G}(\mathcal{R}_F)$ we have $(t, w(t)) \in \mathcal{G}(\mathcal{R}_F)$, for all $t \geq 0$. In particular, $z(t) = (t + \tau, w(t + \tau)) \in \mathcal{G}(\mathcal{R}_F)$ for all $t \geq 0$.

The verified claim implies $(\mathcal{G}(\mathcal{R}_F), \widetilde{F})$ is strongly invariant. According to Remark 3.1, we have that $(\tau, y) \to_{(\theta, \zeta)} (t, x)$ if and only if $\tau \to_\theta t$ and $y \to_\zeta x$. Applying Proposition 2.1 we obtain

$$\theta + \limsup_{y \to_\zeta x} H_F(y, \zeta) = \limsup_{(\tau, y) \to_{(\theta, \zeta)} (t, x)} H_{\widetilde{F}}((\tau, y), (\theta, \zeta)) \leq 0,$$

for all $(\theta, \zeta) \in N^P_{\mathcal{G}(\mathcal{R}_F)}(t, x)$, and all $(t, x) \in \mathcal{G}(\mathcal{R}_F)$. The linear growth condition in (SP) implies boundedness on compact sets, and this is readily seen to be the key requirement for $\mathcal{G}(\mathcal{R}_F)$ to satisfy condition (c).

The proof of the uniqueness is based on the establishment of the double inclusion $S \subseteq \mathcal{G}(\mathcal{R}_F) \subseteq S$, for any closed $S \subseteq I \times \mathbb{R}^n$ for which (a), (b), and (c) hold. We remark that besides (SP), no additional assumption is required on F to prove the maximality of $\mathcal{G}(\mathcal{R}_F)$, since $S \subseteq \mathcal{G}(\mathcal{R}_F)$ follows, for instance, from the weak invariance Theorem 4.2.10 in Ref. 5, and the compactness of trajectories property given in Theorem 4.1.11 in Ref. 5. However, the complementary inclusion $\mathcal{G}(\mathcal{R}_F) \subseteq S$ is obtained by applying the replacement Hamiltonian condition for strong invariance given in Proposition 2.1 (instead of Theorem 4.3.8 in Ref. 5 for Lipschitz F), together with a slight modification of Theorem 4.3.11 of Ref. 5 (Lipschitz continuous dependence on initial conditions), which makes it count for US–DL dynamics. $\qquad\square$

Theorem 3.1 asserts that upper semicontinuity and the dissipative Lipschitz property are not in general sufficient ingredients to characterize $\mathcal{G}(\mathcal{R}_F)$ as a generalized solution to the exact HJ equation (1). However, the last property can be recovered by adding continuity to F, as the next statement confirms. Recall that F is continuous if $H_F(\cdot, \cdot)$ is continuous.

Corollary 3.1. *In addition to the assumptions given in Theorem 3.1, suppose F is continuous. Then the graph of the reachable set $\mathcal{G}(\mathcal{R}_F)$, is the unique closed subset S of $I \times \mathbb{R}^n$ satisfying the following for all $(\theta, \zeta) \in N_S^P(t, x)$, and all $(t, x) \in S$:*
(a) $\theta + H_F(x, \zeta) = 0$,
(b) $\lim_{T \downarrow 0} S_T = M$.

Proof. The result follows from Theorem 3.1(b), since in virtue of the continuity of H_F we have

$$\limsup_{y \to_\zeta x} H_F(y, \zeta) = H(x, \zeta). \qquad\square$$

References

1. J.-P. Aubin, *Viability Theory* (Birkhäuser, Boston, 1991).
2. F.H. Clarke, *Systems and control Letters* **27**, 195 (1996).
3. Clarke F., Ledyaev Yu., Radulescu M., *J. Dynam. Control Syst.* **3**, 493 (1997).
4. F.H. Clarke, Yu.S. Ledyaev, R.J. Stern, P.R. Wolenski, *J. Dynam. Control Systems* **1**, 1 (1995).
5. F. H. Clarke, Y. S. Ledyaev, R. J. Stern and P. R. Wolenski, *Nonsmooth analysis and control theory*, Graduate Texts in Mathematics, Vol. 178 (Springer-Verlag, New York, 1998).
6. K. Deimling, *Multivalued Differential Equations*, de Gruyter Series in Nonlinear Analysis and Applications, Vol. 1 (De Gruyter, Berlin, 1992).
7. T. Donchev, *Nonlinear Analysis* **16**, 543 (1991).

8. T.D. Donchev, V.R. Ríos., P.R. Wolenski, A characterization of strong invariance for perturbed dissipative systems, in *Optimal control, stabilization and nonsmooth analysis*, eds. M.S. de Queiroz, M. Malisoff and P. Wolenski, Lecture Notes in Control and Inform. Sci., Vol. 301 (Springer, Berlin, 2004), pp. 343–349.

9. T.D. Donchev, V.R. Ríos., P.R. Wolenski, *Nonlinear Anal.* **60**, 849 (2005).

10. Frankowska H. and Plaskacz S., *Nonlinear Analysis, Theory, Methods, and Applications* **26**, 565 (1996).

11. H. Frankowska, S. Plaskacz, T. Rzezuchowski, *Journal of Differential Equations* **116**, 265 (1995).

12. E. Hewitt, K. Stromberg, *Real and Abstract Analysis. A modern treatment of the theory of functions of a real variable* (Springer, New York, 1965).

13. M. Krastanov, Forward invariant sets, homogeneity and small-time local controllability, in *Geometry in Nonlinear Control and Differential Inclusions (Warsaw, 1993)*, eds. B. Jakubczyk, W. Respondek and T. Rzeżuchowski, Banach Center Publ. **32** (Polish Acad. Sci., Warsaw, 1995), pp. 287–300.

14. M. Kunze, *Non-smooth Dynamical Systems*, Lecture Notes in Mathematics, Vol. 1744 (Springer, Berlin, 2000).

15. A. Tolstonogov, *Differential Inclusions in a Banach Space*, Mathematics and its Applications, Vol. 524 (Kluwer Academic Publishers, Dordrecht, 2000). Translated from the 1986 Russian original and revised by the author.

16. V. M. Veliov, *Set-Valued Anal* **1**, 305 (1993).

17. P.R. Wolenski, Y. Zhuang, *SIAM J. Control Optim.* **36**, 1048 (1998).

LINEAR-CONVEX CONTROL AND DUALITY

R.T. ROCKAFELLAR[†] and R. GOEBEL[‡]

Department of Mathematics, University of Washington
Seattle, WA 98195-4350, USA
[†] *E-mail: rtr@math.washington.edu*
[‡] *E-mail: goebel@math.washington.edu*

An optimal control problem with linear dynamics and convex but not necessarily quadratic and possibly infinite-valued or nonsmooth costs can be analyzed in an appropriately formulated duality framework. The paper presents key elements of such a framework, including a construction of a dual optimal control problem, optimality conditions in open loop and feedback forms, and a relationship between primal and dual optimal values. Some results on differentiability and local Lipschitz continuity of the gradient of the optimal value function associated with a convex optimal control problem are included.

Keywords: Optimal control, generalized problem of Bolza, convex value function, duality, Hamilton-Jacobi theory

1. Introduction

The classical Linear Quadratic Regulator problem ($\mathcal{LQR}$) involves minimizing a quadratic cost subject to linear dynamics. See Ref. 1 for a detailed exposition. When the system is subject to constraints or actuator saturation, or when nonquadratic costs or barrier functions are involved, the linear and quadratic techniques applicable to $\mathcal{LQR}$ are no longer adequate. Often though, the optimal control problem one needs to solve is convex.

Much like what is appreciated in convex optimization, convex structure of an optimal control problem makes available techniques based on an appropriately formulated duality framework. Existence results, optimality conditions, and sensitivity to perturbations can be analyzed in such a framework even if constraints or nonsmooth costs are involved.

Following the contributions made over the years by the first author and his students and former students, we outline in this paper some key aspects of a duality framework for the study of general finite-horizon and infinite-horizon convex optimal control problems. For simplicity of presentation, but also with control engineering applications in mind, we specialize the key results to a Linear-Convex Regulator ($\mathcal{LCR}$): a problem in which the quadratic costs as in the $\mathcal{LQR}$ are replaced by general convex functions.

Besides expository purposes, the choice of the duality related results we state is motivated by the task of analyzing the regularity properties of optimal value functions associated with a control problem. The properties of our interest include finiteness, coercivity, differentiability, strict convexity, and finally, continuity of the gradient. In particular, based on some recent results of the authors on the duality between local strong convexity of a convex function and the local Lipschitz continuity of the gradient of its conjugate, we identify some mild assumptions on a convex optimal control problem that guarantee that both the primal value function and the dual value function have locally Lipschitz continuous gradients.

2. Duality in finite horizon optimal control

The full power of convex duality techniques was brought to the calculus of variations problems on finite time horizons by Rockafellar in Ref. 18, and led, in particular, to very general existence results in Ref. 20. We point the reader to Ref. 18 for an overview of other early developments relying, to an extent, on convex duality. Regarding linear-quadratic optimal control problems with constraints, a duality framework was suggested in Ref. 22 and led, among other things, to simple optimality conditions for problems where some state constraints are present; see Ref 23. The duality tools were specialized to the study of value functions, in the Hamilton-Jacobi framework, by Rockafellar and Wolenski in Refs. 25,26. Comparison to related Hamilton-Jacobi developments can be found in Refs. 9,25.

2.1. *The (finite horizon) Linear-Convex Regulator*

We will focus on the following optimal control problem:

$$\text{minimize} \quad \int_\tau^T \left\{ q(y(t)) + r(u(t)) \right\} dt + g(x(T)) \tag{1}$$

over all integrable control functions $u : [\tau, T] \to \mathbb{R}^k$, subject to linear dynamics

$$\dot{x}(t) = Ax(t) + Bu(t), \quad y(t) = Cx(t), \tag{2}$$

and the initial condition

$$x(\tau) = \xi. \tag{3}$$

The absolutely continuous arc $x : [\tau, T] \to \mathbb{R}^n$ denotes the state and $y : [\tau, T] \to \mathbb{R}^m$ denotes the output of the linear system (2). The matrices A, B, and C are any matrices of appropriate dimensions; we will explicitly make assumptions of observability or controllability when needed.

When the functions q, r, and g are quadratic and given by symmetric and positive definite matrices, the problem described above is the classical Linear-Quadratic Regulator; see Ref. 1 for background and detailed analysis. Here, we will allow q, r, and g to be more general convex functions. Consequently, we will call the problem defined by (1), (2), (3) a Linear-Convex Regulator, or just $\mathcal{LCR}$. When an explicit reference to initial conditions is needed, we will speak of the $\mathcal{LCR}(\tau, \xi)$ problem.

Below, $\overline{\mathbb{R}} = [-\infty, +\infty]$ and a function $f : \mathbb{R}^n \to \overline{\mathbb{R}}$ is called *coercive* if $\lim_{|x| \to \infty} f(x)/|x| = \infty$.

Standing Assumptions. *The functions $q : \mathbb{R}^m \to \overline{\mathbb{R}}$, $r : \mathbb{R}^k \to \overline{\mathbb{R}}$, and $g : \mathbb{R}^n \to \overline{\mathbb{R}}$ are elements of the class $\mathcal{C}$, where*

$$\mathcal{C} = \bigcup_{i=1}^{\infty} \left\{ f : \mathbb{R}^i \to \overline{\mathbb{R}} \;\middle|\; \begin{array}{l} f \text{ is convex, lower semicontinuous,} \\ f(0) = 0, f(x) \geq 0 \text{ for all } x \in \mathbb{R}^i \end{array} \right\}. \quad (4)$$

Furthermore, $q(y)$ is finite for all $y \in \mathbb{R}^m$ while r is coercive.

In what follows, we will say that $f : \mathbb{R}^n \to \overline{\mathbb{R}}$ is proper if it never equals $-\infty$ and is finite somewhere.

2.2. *Convex conjugate functions*

For any convex, lower semicontinuous (lsc.), and proper function $f : \mathbb{R}^n \to \overline{\mathbb{R}}$, its *conjugate function* $f^* : \mathbb{R}^n \to \overline{\mathbb{R}}$ is defined by

$$f^*(p) = \sup_{x \in \mathbb{R}^n} \{p \cdot x - f(x)\}.$$

The function f^* is itself a convex, lsc., and proper function, and the function conjugate to it is f. It is also immediate from the definition of f^* that $f \in \mathcal{C}$ if and only $f^* \in \mathcal{C}$; this will in particular imply that our assumptions on $\mathcal{LCR}$ and on the problem dual to $\mathcal{LCR}$ are symmetric.

A basic example of a pair of conjugate functions is

$$f(x) = \frac{1}{2} x \cdot Qx, \quad f^*(p) = \frac{1}{2} p \cdot Q^{-1} p, \quad (5)$$

where Q is a symmetric and positive definite matrix. Another is

$$f(x) = \begin{cases} \frac{1}{2} x \cdot Qx & \text{if } x \in X \\ +\infty & \text{if } x \notin X \end{cases}, \quad f^*(p) = \sup_{x \in X} \left\{ p \cdot x - \frac{1}{2} x \cdot Qx \right\} \quad (6)$$

for Q as above and a closed convex and nonempty set $X \subset \mathbb{R}^n$. Requiring that $f(0) = 0$ implies that $0 \in X$. If $0 \in \text{int} X$, then f^* is quadratic on some neighborhood of 0, and given there by $\frac{1}{2} p \cdot Q^{-1} p$. In general, this f^* is differentiable and ∇f^* is Lipschitz continuous; related properties of not necessarily quadratic convex functions will be discussed in Section 3.

A fundamental property of the pair f and f^*, that is the basis for Euler-Lagrange and Hamiltonian optimality conditions for convex optimal control problems, is that for any $x, p \in \mathbb{R}^n$,

$$f(x) + f^*(p) \geq x \cdot p, \tag{7}$$

while $f(x) + f^*(p) = x \cdot p$ if and only if $p \in \partial f(x)$ if and only if $p \in \partial f^*(x)$. Here, ∂f is the *subdifferential* of the convex function f:

$$\partial f(x) = \{p \in \mathbb{R}^n \mid f(x') \geq f(x) + p \cdot (x' - x) \text{ for all } x' \in \mathbb{R}^n\}.$$

A standard reference for the material just presented is Ref. 17.

2.3. *General duality framework*

Many of the results we will state or use were developed not in an optimal control setting, but rather, in the framework of duality for calculus of variations problems. We briefly outline this framework.

Given convex, lsc., and proper functions $L, l : \mathbb{R}^{2n} \to \overline{\mathbb{R}}$, consider

$$\widetilde{L}(p, w) = L^*(w, p), \qquad \widetilde{l}(p_\tau, p_T) = l^*(p_\tau, -p_T),$$

and two generalized problems of Bolza type: the primal problem $\mathcal{P}$ of minimizing, over all absolutely continuous $x : [\tau, T] \to \mathbb{R}^n$, the cost functional

$$\int_\tau^T L(x(t), \dot{x}(t)) \, dt + l(x(\tau), x(T)), \tag{8}$$

and the dual problem $\widetilde{\mathcal{P}}$ of minimizing, over all absolutely continuous $p : [\tau, T] \to \mathbb{R}^n$, the (dual) cost functional

$$\int_\tau^T \widetilde{L}(p(t), \dot{p}(t)) \, dt + \widetilde{l}(p(\tau), p(T)). \tag{9}$$

Note that the problem dual to $\widetilde{\mathcal{P}}$ is the original $\mathcal{P}$.

Directly from (7), it follows that it is always the case that $\inf(\mathcal{P}) \geq -\inf(\widetilde{\mathcal{P}})$, and that any absolutely continuous $p : [\tau, T] \to \mathbb{R}^n$ provides a lower bound on $\inf(\mathcal{P})$ through (9), while any absolutely continuous $x : [\tau, T] \to \mathbb{R}^n$ provides a lower bound on $\inf(\widetilde{\mathcal{P}})$ through (8). Under some mild assumptions, given in Refs. 18,20 for a more general time-dependent case and specialized in Ref. 25 to the autonomous case, the following holds:

$$-\infty < \inf(\mathcal{P}) = -\inf(\widetilde{\mathcal{P}}) < +\infty, \tag{10}$$

and moreover, optimal solutions for both the primal and the dual problem exist. Regarding L, Ref. 25 required that: (i) the set $F(x) = \{v \mid L(x,v) < \infty\}$ be nonempty for all x, and there exist a constant ρ such that $\mathrm{dist}(0, F(x)) \leq \rho(1 + |x|)$ for all x, and (ii) there exist constants α and β and a coercive, proper, nondecreasing function θ on $[0, \infty)$ such that $L(x,v) \geq \theta(\max\{0, |v| - \alpha|x|\}) - \beta|x|$ for all x and v. A particularly attractive feature of these assumptions, besides their generality, is that L satisfies them if and only if $\widetilde{L}$ does.

The Linear-Convex Regulator can be reformulated in the generalized Bolza framework. (For this, and other equivalences between optimal control problem formulations, see Chapter 1.3 in Ref 7.) To this end, for each fixed $\tau \leq T$, $\xi \in \mathbb{R}^n$, one considers

$$\begin{aligned}
L(x,v) &= q(Cx) + \min_{u} \{r(u) \mid Ax + Bu = v\}, \\
l(x_\tau, x_T) &= \delta_\xi(x_\tau) + g(x_T),
\end{aligned} \tag{11}$$

where δ_ξ is the indicator of ξ: $\delta_\xi(\xi) = 0$ while $\delta_\xi(x) = \infty$ if $x \neq \xi$. If a solution $x : [\tau, T] \to \mathbb{R}^n$ to the resulting Bolza problem $\mathcal{P}(\tau, \xi)$ is found, an optimal control $u : [\tau, T] \to \mathbb{R}^k$ for $\mathcal{LCR}$ can be then recovered, at almost every $t \in [\tau, T]$, as the minimizer of $r(u)$ over all u such that $Ax(t) + Bu = \dot{x}(t)$. Given such L and l, the dual Bolza problem $\widetilde{\mathcal{P}}(\tau, \xi)$ is defined by

$$\begin{aligned}
\widetilde{L}(p,w) &= r^*(B^*p) + \min_{z} \{q^*(z) \mid -A^*p + C^*z = w\}, \\
\widetilde{l}(p_\tau, p_T) &= \xi \cdot p_\tau + g^*(-p_T).
\end{aligned}$$

Our Standing Assumptions on q, r, and g guarantee that L and l as above (and equivalently, $\widetilde{L}$ and $\widetilde{l}$) meet the growth conditions of Ref. 25 and consequently, that (10) holds. We note that in most cases, and even if q and r are quadratic functions, L in (11) does not satisfy the classical coercivity conditions: $L(x,v)$ is not bounded below by a coercive function of v. This can easily be seen in the case of B being an identity: then $L(x,v) = q(Cx) + r(v - Ax)$.

2.4. *The primal and the dual value functions*

The *(primal) optimal value function* $V : (-\infty, T] \times \mathbb{R}^n \to \overline{\mathbb{R}}$ is defined, for each $(\tau, \xi) \in (-\infty, T] \times \mathbb{R}^n$, as the infimum in the problem $\mathcal{LCR}(\tau, \xi)$. For the quadratic case (of q, r, and g quadratic), $V(\tau, \cdot)$ is quadratic for each

$\tau \leq T$. In general, it is immediate that $V(\tau, \cdot)$ is a convex function. More precisely, Theorem 2.1 and Corollary 7.6. in Ref. 25 state the following:

Theorem 2.1 (value function – basic regularity).

- *For each $\tau \leq T$, $V(\tau, \cdot)$ is an element of C. If g is finite-valued, then so is $V(\tau, \cdot)$.*
- *$V(\tau, \cdot)$ depends epi-continuously on $\tau \in (-\infty, T]$.*

Epi-continuity above means that the epigraphs of $V(\tau, \cdot)$, i.e., the sets $\{(\xi, \alpha) \in \mathbb{R}^{n+1} \,|\, \alpha \geq V(\tau, \xi)\}$, depend continuously on τ. Such concept of continuity of V takes into account the fact that the effective domains of $V(\tau, \cdot)$ may depend on τ. (See Chapters 4 and 5 in Ref. 24 for details on set convergence and continuity of set-valued mappings.)

The value function V can be equivalently defined, at each (τ, ξ), as the infimum in the generalized problem of Bolza $\mathcal{P}(\tau, \xi)$ related to $\mathcal{LCR}(\tau, \xi)$ through (11). From (10) it follows that $V(\tau, \xi) = \inf(\mathcal{P}(\tau, \xi)) = -\inf(\widetilde{\mathcal{P}}(\tau, \xi))$. Since $\widetilde{l}(p_\tau, p_T) = \inf_{\pi \in \mathbb{R}^n}\{p_\tau \cdot \pi + \delta_{p_\tau}(\pi)\}$ and $\delta_{p_\tau}(\pi) = \delta_\pi(p_\tau)$, one obtains that $V(\tau, \xi)$ equals

$$-\inf\left\{\inf_{\pi \in \mathbb{R}^n}\left\{\xi \cdot \pi + \delta_{p(\tau)}(\pi)\right\} + g^*(-p(T)) + \int_\tau^T \widetilde{L}(p(t), \dot{p}(t))\, dt\right\}$$

$$= \sup_{\pi \in \mathbb{R}^n}\left\{-\xi \cdot \pi - \inf\left\{\delta_\pi(p(\tau)) + g^*(-p(T)) + \int_\tau^T \widetilde{L}(p(t), \dot{p}(t))\, dt\right\}\right\}.$$

The first and the last infimum above are taken over all arcs $p : [\tau, T] \to \mathbb{R}^n$. The presence of the term $\delta_\pi(p(\tau))$ ensures that the last infimum can be taken only over those arcs $p(\cdot)$ for which $p(\tau) = \pi$. This suggests that the last infimum defines a dual value function, parameterized by τ and π. We now make the formal definitions, following the ideas of Ref. 22 regarding the structure of a dual optimal control problem, and of Ref. 25 regarding the dual value function.

The *optimal control problem dual to $\mathcal{LCR}$* is as follows:

$$\text{minimize} \quad \int_\tau^T \left\{r^*(s(t)) + q^*(z(t))\right\} dt + g^*(-p(T)) \tag{12}$$

over all integrable control functions $z : [\tau, T] \to \mathbb{R}^m$, subject to linear dynamics

$$\dot{p}(t) = -A^* p(t) + C^* z(t), \quad s(t) = B^* p(t), \tag{13}$$

286

(here, A^* denotes the transpose of A, etc.) and the initial condition

$$p(\tau) = \pi. \tag{14}$$

The absolutely continuous arc $p : [\tau, T] \to \mathbb{R}^n$ denotes the state and $s : [\tau, T] \to \mathbb{R}^k$ denotes the output of the linear system (13). We will denote this problem by $\widetilde{\mathcal{LCR}}$, and when a direct reference to initial conditions is needed, by $\widetilde{\mathcal{LCR}}(\tau, \pi)$.

Note that the functions r^*, q^*, and $g^*(-\cdot)$ are in the class $\mathcal{C}$ (recall (4)). Furthermore, r^* is finite-valued (since r is coercive) and q^* is coercive (since q is finite-valued). Thus $\widetilde{\mathcal{LCR}}$ has the same growth properties that $\mathcal{LCR}$ has, based on the Standing Assumptions. Note also that if $\mathcal{LCR}$ is in fact a linear quadratic regulator, with q, r, and g quadratic, then so is $\widetilde{\mathcal{LCR}}$, thanks to (5). If $\mathcal{LCR}$ is a linear quadratic regulator with control constraints $u \in U$ for some closed convex set U (usually with $0 \in \mathrm{int}U$) then the dual problem is unconstrained, but the state cost r^* is not quadratic (recall (6)).

The *(dual) optimal value function* $\widetilde{V} : (-\infty, T] \times \mathbb{R}^n \to \overline{\mathbb{R}}$ is defined, at a given (τ, π), as the infimum in the problem $\widetilde{\mathcal{LCR}}(\tau, \pi)$. By the symmetry of the Standing Assumptions, the properties attributed to V in Theorem 2.1 are present also for $\widetilde{V}$.

The idea of the computation that followed Theorem 2.1 can be now summarized. For details, see Theorem 5.1 in Ref. 25.

Theorem 2.2 (value function duality). *For each $\tau \leq T$, the value functions $V(\tau, \cdot)$ and $\widetilde{V}(\tau, \cdot)$ are convex conjugates of each other, up to a minus sign. That is,*

$$\widetilde{V}(\tau, \pi) = \sup_{\xi} \left\{ -\pi \cdot \xi - V(\tau, \xi) \right\}, \qquad V(\tau, \xi) = \sup_{\pi} \left\{ -\xi \cdot \pi - \widetilde{V}(\tau, \pi) \right\}.$$

This result allows for studying properties of $V(\tau, \cdot)$ by analyzing corresponding properties of $\widetilde{V}(\tau, \cdot)$. For example, it is a general property of convex functions that a function is coercive if and only if its conjugate is finite-valued. Thus, if g is coercive, then g^* is finite-valued, then by Theorem 2.1 the function $\widetilde{V}(\tau, \cdot)$ is finite-valued for each τ, and so $V(\tau, \cdot)$ is coercive for each τ. Further correspondences, between differentiability and strict or strong convexity, will be explored in Section 3.

2.5. *The Hamiltonian and open-loop optimality conditions*

With the generalized problems of Bolza $\mathcal{P}$ as described in Subsection 2.3, one can associate the (maximized) *Hamiltonian* $H : \mathbb{R}^{2n} \to \overline{\mathbb{R}}$:

$$H(x,p) = \sup_{v} \{ p \cdot v - L(x,v) \} .$$

Convexity of L implies that $H(x,p)$ is convex in p for a fixed x and concave in x for a fixed p. In presence of the growth properties of L mentioned following (10), it is also finite-valued. (The said growth properties have equivalent characterizations in terms of H, see Theorem 2.3 in Ref. 25.) The finiteness of H in turn guarantees that

$$-H(x,p) = \sup_{w} \left\{ x \cdot w - \widetilde{L}(p,w) \right\} .$$

In other words, the Hamiltonian associated with the dual problem $\widetilde{\mathcal{P}}$ is exactly $(p,x) \mapsto -H(x,p)$. This further underlines the symmetry between $\mathcal{P}$ and $\widetilde{\mathcal{P}}$.

A *Hamiltonian trajectory* on $[\tau, T]$ is a pair of arcs $x, p : [\tau, T] \to \mathbb{R}^n$ such that

$$\dot{x}(t) \in \partial_p H(x(t), p(t)), \qquad -\dot{p}(t) \in \hat{\partial}_x H(x(t), p(t)) \tag{15}$$

for almost all $t \in [\tau, T]$. In (15), $\partial_p H(x,p)$ is the subdifferential, in the sense of convex analysis, of the convex function $H(x, \cdot)$, while $\hat{\partial}_x H(x,p) = -\partial_x (-H(x,p))$ where the second subdifferential is in the sense of convex analysis. Under suitable assumptions (which also guarantee (10)) the following are equivalent:

- $x(\cdot)$ solves $\mathcal{P}$, $p(\cdot)$ solves $\widetilde{\mathcal{P}}$;
- $x(\cdot), p(\cdot)$ form a Hamiltonian trajectory and

$$(p(\tau), -p(T)) \in \partial l(x(\tau), x(T)).$$

In the convex setting, the key behind this equivalence is the inequality (7) and the relationship between subdifferentials of L, $\widetilde{L}$ and H, rather than any regularity properties of these functions. For details, see Ref. 18.

When the problem $\mathcal{P}$ comes from reformulating $\mathcal{LCR}$ in the Bolza framework as done in (11), one has

$$H(x,p) = p \cdot Ax - q(Cx) + r^*(B^*p). \tag{16}$$

The Hamiltonian system (15) turns into

$$\begin{aligned}
\dot{x}(t) &= Ax(t) + Bu(t), & u(t) &\in \partial r^*(s(t)), \ s(t) = B^*p(t), \\
\dot{p}(t) &= -A^*p(t) + C^*z(t), & z(t) &\in \partial q(y(t)), \ y(t) = Cx(t),
\end{aligned} \tag{17}$$

if one additionally requires that $u(t)$ be a minimizer of $r(u)$ over all u such that $\dot{x}(t) = Ax(t) + Bu$ and that $z(t)$ be a minimizer of $q^*(z)$ over all z such that $\dot{p}(t) = -A^*p(t) + C^*z$. Details are given in Lemma 3.8 of Ref. 10; see also Theorem 4.7 in Ref. 23.

Theorem 2.3 (open-loop optimality). *For a control $u : [\tau, T] \to \mathbb{R}^k$, the following statements are equivalent:*

- *$u(\cdot)$ is an optimal control for $\mathcal{LCR}(\tau, \xi)$ and $-\pi \in \partial_\xi V(\tau, \xi)$;*
- *there exists an integrable control $z : [\tau, T] \to \mathbb{R}^m$ so that arcs $x, p : [\tau, T] \to \mathbb{R}^n$ corresponding to controls $u(\cdot), z(\cdot)$ through the initial condition $(x(\tau), p(\tau)) = (\xi, \pi)$ and (17) satisfy*

$$-p(T) \in \partial g(x(T)).$$

It can be added that if $z(\cdot)$ corresponds to an optimal $u(\cdot)$ as described in the theorem above, then $z(\cdot)$ is optimal for the problem defining $\widetilde{V}(\tau, \pi)$, and also $-\xi \in \partial_\pi \widetilde{V}(\tau, \pi)$.

2.6. *Hamilton-Jacobi results*

The Hamiltonian H and the associated Hamiltonian dynamical system (15) is involved in two different characterizations of the value function. One is the (generalized) Hamilton-Jacobi partial differential equation, as written in Theorem 2.5 of Ref. 25. It says that, for all $\tau < T$, we have $\sigma + H(x, p) = 0$ for each subgradient (σ, p) (in the sense of Definition 8.3 of Ref. 24) of $V(\tau, \xi)$. We stress that the subgradient is taken with respect to both time and space variables, and thus it can not be understood in the sense of convex analysis. Subject to a boundary condition, V is in fact the unique solution of this generalized Hamilton-Jacobi equation, as shown by Galbraith in Ref. 9.

The other characterization will involve the flow S_τ generated by the Hamiltonian dynamical system (15):

$$S_\tau(\xi_T, \pi_T) = \left\{ (\xi, \pi) \left| \begin{array}{l} \text{there exists a Hamiltonian traj. } (x(\cdot), p(\cdot)) \text{ on } [\tau, T] \\ \text{with } (x(\tau), p(\tau)) = (\xi, \pi), \ (x(T), p(T)) = (\xi_T, \pi_T) \end{array} \right. \right\}.$$

We have the following result, as stated in Theorem 2.4 of Ref. 25:

Theorem 2.4 (Hamiltonian flow of $\partial_\xi V(\tau, \cdot)$). *For any $\tau \leq T$,*

$$\mathrm{gph}\left(-\partial_\xi V(\tau, \cdot)\right) = S_\tau\left(\mathrm{gph}(-\partial g)\right).$$

In other words, the graph of the subdifferential of $V(\tau, \cdot)$ is essentially obtained from the graph of the subdifferential of g by flowing backwards along Hamiltonian trajectories. This result can also be thought of as saying that the (generalized) method of characteristics works globally in this convex setting. Thanks to Theorem 2.2 and the fact that subdifferential mappings of a pair of convex conjugate functions are inverses of one another, Theorem 2.4 also gives a description of $\partial_\pi \widetilde{V}(\tau, \cdot)$.

Since the subdifferential of a convex function is a monotone mapping, the graph of $-\partial g$ is an "anti-monotone set", that is, for any $(x_1, p_1), (x_2, p_2) \in \mathrm{gph}(-\partial g)$, we have $(x_1 - x_2) \cdot (p_1 - p_2) \leq 0$. Thanks to the concavity in x, convexity in p, of $H(x, p)$, the flow S_τ has a certain monotonicity, or rather anti-monotonicity preserving property:

$$(x_1(\tau) - x_2(\tau)) \cdot (p_1(\tau) - p_2(\tau)) \leq (x_1(T) - x_2(T)) \cdot (p_1(T) - p_2(T)) \quad (18)$$

for any pair of Hamiltonian trajectories $(x_i(\cdot), p_i(\cdot))$, $i = 1, 2$, on $[\tau, T]$. This was noted in Theorem 4 of Ref. 19. In what follows, we will need this result specified to the case of $\mathcal{LCR}$.

Lemma 2.1 (preservation of monotonicity). *Let* $(x_i(\cdot), p_i(\cdot))$, $i = 1, 2$, *be Hamiltonian trajectories on* $[\tau, T]$ *for the Hamiltonian (16). Then*

$$\frac{d}{dt}(x_1(t) - x_2(t)) \cdot (p_1(t) - p_2(t))$$

$$\in (\partial q(Cx_1(t)) - \partial q(Cx_2(t))) \cdot (Cx_1(t) - Cx_2(t))$$

$$+ (\partial r^*(B^* p_1(t)) - \partial r^*(B^* p_2(t))) \cdot (B^* p_1(t) - B^* p_2(t)) \geq 0$$

The inequality above can be easily shown directly from the definition of a Hamiltonian trajectory. It also obviously implies (18).

2.7. *Feedback optimality conditions*

Open-loop optimal solutions of optimal control problems are not well-suited to cope with uncertainty or disturbance. This, and the classical control engineering problem of feedback stabilization, motivate the pursuit of an optimal feedback: a mapping that at each state produces the set of optimal controls to be applied.

Theorem 2.3, Theorem 2.4, and the Hamiltonian system (15) suggest that optimal arcs $x(\cdot)$ satisfy the inclusion $\dot{x}(t) \in \partial_y H(x(t), -\partial_\xi V(t, x(t)))$. Indeed, if a Hamiltonian trajectory $(x(\cdot), p(\cdot))$ on $[\tau, T]$ is such that $p(T) \in -\partial g(x(T))$ then, by Theorem 2.4, $p(t) \in -\partial_\xi V(t, x(t))$ for almost all $t \in [\tau, T]$. A more difficult – and more important – issue is to what extent the stated inclusion is sufficient for optimality. To an extent, this was addressed in Ref. 10. We state the relevant result in Theorem 2.5.

The *optimal feedback* mapping $\Phi : (-\infty, T] \times \mathbb{R}^n$ for the problem $\mathcal{LCR}$ is defined, at each $t \le T$, $x \in \mathbb{R}^n$, by

$$\Phi(t, x) = \partial r^*(-B^* \partial_\xi V(t, x)) = \bigcup_{p \in \partial_\xi V(t,x)} \partial r^*(-B^* p).$$

Theorem 2.5 (optimal feedback). *Fix $(\tau, \xi) \in (-\infty, T] \times \mathbb{R}^n$. If $u : [\tau, T] \to \mathbb{R}^k$ is an optimal control for $\mathcal{LCR}(\tau, \xi)$ and $\partial_\xi V(\tau, \xi) \ne \emptyset$, then*

$$u(t) \in \Phi(t, x(t)) \text{ for almost all } t \in [\tau, T]. \tag{19}$$

On the other hand, if an integrable control $u : [\tau, T] \to \mathbb{R}^k$ satisfies (19), and the absolutely continuous $x : [\tau, T] \to \mathbb{R}^n$ satisfying (2) and $x(\tau) = \xi$ is such that $x(t) \in \mathrm{intdom} V(\tau, \cdot)$ for almost all $t \in [\tau, T]$, then $u(\cdot)$ is optimal for $\mathcal{LCR}(\tau, \xi)$.

The interior $\mathrm{intdom} V$ of the effective domain of V, i.e., of the set $\mathrm{dom} V$ of all points where V is finite, is nonempty if $\mathrm{intdom} g$ is nonempty, and $(\tau, \xi) \in \mathrm{intdom} V$ if and only if $\tau < T$ and $\xi \in \mathrm{intdom} V(\tau, \cdot)$; see Proposition 7.2 and Corollary 7.5 in Ref. 25. In general, the feedback mapping Φ it is outer semicontinuous and locally bounded on $\mathrm{intdom} V$, but need not have convex values; see Example 3.7 in Ref. 10.

3. Regularity of the value function

We now give some results on how regularity of the data of $\mathcal{LCR}$ may be reflected in the properties of V, and thus, of the optimal feedback mapping Φ. The central role played by the Hamiltonian dynamical system (15) in the analysis of convex optimal control problems is reflected below in the fact that the only tools from the previous chapter we use are Theorem 2.4, which describes how $\nabla_\xi V(\tau, \cdot)$ evolves in the Hamiltonian system, and Lemma 2.1, about the monotonicity-preserving properties of the Hamiltonian flow.

We note that similar regularity properties of the value function in a non-convex setting require stronger assumptions than the ones posed here; see for example Refs. 4,6. To an extent though, regularity properties (or singularities) of the value function are propagated via the Hamiltonian system; see Ref. 8.

3.1. *Convex-valued and single-valued optimal feedback*

A natural condition to guarantee that the optimal feedback Φ has convex values is that $\partial_\xi V(\tau, \xi)$ be single-valued, which is equivalent to essential differentiability of $V(\tau, \cdot)$. Theorem 3.1 will show that this property is in a sense inherited by V from g.

We need some preliminary material. Let $f : \mathbb{R}^n \to \overline{\mathbb{R}}$ be a proper, lsc., and convex function. The function f is *essentially differentiable* if $D = \operatorname{intdom} f$ is nonempty, f is differentiable on D, and $\lim_{i \to \infty} |\nabla f(x_i)| = \infty$ for any sequence $\{x_i\}_{i=1}^{\infty}$ converging to a boundary point of D. The function f is *essentially strictly convex* if f is convex on every convex subset of $\operatorname{dom} \partial f = \{x \in \mathbb{R}^n \,|\, \partial f(x) \neq \emptyset\}$; that is, if for each convex $S \subset \operatorname{dom} \partial f$, we have $(1 - \lambda)f(x_1) + \lambda f(x_2) > f((1 - \lambda)x_1 + \lambda x_2)$ for all $\lambda \in (0, 1)$, all $x_1, x_2 \in S$. The subdifferential mapping is a *monotone mapping*: for all $x_1, x_2 \in \operatorname{dom} \partial f$, all $y_1 \in \partial f(x_1)$, $y_2 \in \partial f(x_2)$, $(x_1 - x_2) \cdot (y_1 - y_2) \geq 0$. It is *strictly monotone* if $(x_1 - x_2) \cdot (y_1 - y_2) > 0$ as long as $x_1 \neq x_2$.

The following statements are equivalent: (i) f is essentially differentiable; (ii) ∂f is single-valued on $\operatorname{dom} \partial f$; (iii) ∂f^* is strictly monotone; and (iv) f^* is essentially strictly convex.

The result below is a special case of Theorem 4.4 in Ref. 10. Note that no regularity is assumed about r.

Theorem 3.1 (value function – essential differentiability).
Suppose that q and g are essentially differentiable. Then, for all $\tau \leq T$, $V(\tau, \cdot)$ is essentially differentiable.

Proof. By Theorem 2.4, any two points in $\operatorname{gph}(-\partial_\xi V(\tau, \cdot))$ can be expressed as $(x_i(\tau), p_i(\tau))$, $i = 1, 2$, where $(x_i(\cdot), p_i(\cdot))$ are Hamiltonian trajectories on $[\tau, T]$ with $p_i(T) = -\nabla g(x_i(T))$, $i = 1, 2$. If $x_1(\tau) = x_2(\tau)$, then by Lemma 2.1 one obtains

$$0 = (x_1(\tau) - x_2(\tau)) \cdot (p_1(\tau) - p_2(\tau)) \leq (x_1(T) - x_2(T)) \cdot (p_1(T) - p_2(T)) \leq 0,$$

with the last inequality above coming from monotonicity of $\nabla g(\cdot)$. Hence

$$(x_1(T) - x_2(T)) \cdot (p_1(T) - p_2(T)) = 0$$

which by essential strict convexity of g^* (a property dual to essential differentiability of g) implies that $p_1(T) = p_2(T)$. The fact that $(x_1(t) - x_2(t)) \cdot (p_1(t) - p_2(t))$ is constant on $[\tau, T]$ and Theorem 4 of Ref. 19 implies that $(x_1(\cdot), p_2(\cdot))$ is a Hamiltonian trajectory. In particular,

$$\begin{aligned}
\dot{p}_1(t) - \dot{p}_2(t) &= -A^*(p_1(t) - p_2(t)) + C^* \left(\nabla q(Cx_1(t)) - \nabla q(Cx_1(t))\right) \\
&= -A^*(p_1(t) - p_2(t)),
\end{aligned}$$

and since $p_1(T) = p_2(T)$, it must be the case that $p_1(\tau) = p_2(\tau)$.

We have shown that, for any two points $(x_i(\tau), p_i(\tau))$, $i = 1, 2$, in $\mathrm{gph}(-\partial_\xi V(\tau, \cdot))$, if $x_1(\tau) = x_2(\tau)$ then also $p_1(\tau) = p_2(\tau)$. This amounts to single-valuedness of $\partial_\xi V(\tau, \cdot)$ which in turn is equivalent to essential differentiability of $V(\tau, \cdot)$. $\qquad\square$

Under the assumptions of Theorem 3.1, the feedback Φ has convex values, since ∂r^* has convex values. If furthermore ∂r^* is single-valued, and so if r^* is differentiable (as r^* is finite, its essential differentiability is just differentiability), then Φ is single-valued and in fact continuous. Thus, if q and g are essentially differentiable and r is strictly convex, then the optimal feedback Φ is continuous on $\mathrm{intdom}V$. (C.f. Proposition 3.5 in Ref. 10.)

A different set of assumptions leading to essential differentiability of $V(\tau, \cdot)$ is strict convexity of both q and r^* and observability of (A, C) and $(-A^*, B^*)$. This combines arguments that apply to the case of strictly concave, strictly convex Hamiltonian, as in Ref. 21 and basic properties of observable linear systems (c.f. Theorem 3.1 in Ref. 13). We note though that strict convexity of r^* may be absent in simple cases, like when r represents a quadratic cost and a convex control constraint set (recall (6)).

The next natural property of the optimal feedback to pursue, with the uniqueness of solutions to the closed loop equation

$$\dot{x}(t) = Ax(t) + B\Phi(t, x(t)) \tag{20}$$

in mind, is the local Lipschitz continuity of $\Phi(t, \cdot)$. This motivates the developments in the next section.

3.2. *Convex functions with locally Lipschitz gradients*

Let $f : \mathbb{R}^n \to \overline{\mathbb{R}}$ be a proper, lsc., and convex function. The property of f being differentiable with ∇f Lipschitz continuous on $\mathbb{R}^n$ with constant $\rho > 0$ is equivalent to f^* being *strongly convex* with constant $\sigma = \rho^{-1}$:

$$(1-\lambda)f^*(p_1) + \lambda f^*(p_2) \geq f^*((1-\lambda)p_1 + \lambda p_2) + \frac{1}{2}\sigma\lambda(1-\lambda)|p_1 - p_2|^2 \tag{21}$$

for all $\lambda \in (0, 1)$, all $p_1, p_2 \in \mathbb{R}^n$. A manipulation of (21) shows that strong convexity of f^* with constant σ is equivalent to $f^* - \frac{1}{2}\sigma|\cdot|^2$ being convex. Furthermore, strong convexity of f^* is equivalent to its subdifferential being *strongly monotone* with constant σ: $(p_1 - p_2) \cdot (x_1 - x_2) \geq \sigma|p_1 - p_2|^2$ for all $(p_1, x_1), (p_2, x_2) \in \mathrm{gph}\partial f^*$.

We now report some recent results from Ref. 14 regarding localization of the properties just mentioned. In what follows, we will say that a proper, lsc., and convex function $g : \mathbb{R}^n \to \overline{\mathbb{R}}$ is *essentially locally strongly convex* if for any compact and convex $K \subset \mathrm{dom}\partial g$, g is strongly convex on K, in the sense that there exists $\sigma > 0$ such that (21) is satisfied for all $p_1, p_2 \in K$ and all $\lambda \in (0, 1)$.

Theorem 3.2 (strong convexity and gradient continuity). *For a proper, lsc., and convex function $f : \mathbb{R}^n \to \overline{\mathbb{R}}$, the following conditions are equivalent:*

(a) *f is essentially differentiable and ∇f is locally Lipschitz continuous on* $\mathrm{intdom}f$;

(b) *the mapping ∂f is single-valued on $\mathrm{dom}\partial f$ and locally Lipschitz continuous relative to $\mathrm{gph}\partial f$ in the sense that for each compact subset $S \subset \mathrm{gph}\partial f$ there exists $\rho > 0$ such that for any $(x_1, y_1), (x_2, y_2) \in S$ one has*

$$|y_1 - y_2| \leq \rho|x_1 - x_2|;$$

(c) *the mapping ∂f^* is locally strongly monotone relative to $\mathrm{gph}\partial f^*$ in the sense that for each compact subset $S \subset \mathrm{gph}\partial f^*$ there exists $\sigma > 0$ such that for any $(y_1, x_1), (y_2, x_2) \in S$ one has*

$$(y_1 - y_2) \cdot (x_1 - x_2) \geq \sigma|y_1 - y_2|^2.$$

If f is such that $\mathrm{dom}\partial f^*$ is convex and open, then each of the conditions above is equivalent to:

(d) *f^* is essentially locally strongly convex.*

This in particular says that for a coercive (and proper and lsc) convex function f, essential differentiability of f and local Lipschitz continuity of ∇f is equivalent to (naturally understood) local strong convexity of f^*.

For convenience, we define the following classes of convex functions:

$$\mathcal{C}^+ = \left\{ f \in \mathcal{C} \ \middle| \ \begin{array}{c} f \text{ is essentially differentiable and} \\ \nabla f \text{ is locally Lipschitz continuous on } \mathrm{intdom}f \end{array} \right\},$$

$$\mathcal{C}^{++} = \{ f \in \mathcal{C}^+ \mid f \text{ is essentially locally strongly convex } \}.$$

Since for any essentially differentiable f, $\mathrm{dom}\partial f = \mathrm{intdom}f$ is open and convex, for such f the condition that f is essentially locally strongly convex is equivalent, by Theorem 3.2, to $f^* \in \mathcal{C}^+$. Thus, $f \in \mathcal{C}^{++}$ if and only if $f, f^* \in \mathcal{C}^+$. This immediately implies that $f \in \mathcal{C}^{++}$ if and only if $f^* \in \mathcal{C}^{++}$.

3.3. *Locally Lipschitz continuous optimal feedback*

The result below is a local version of Theorem 3.6 of Ref. 11, where global Lipschitz continuity of ∇H and ∇g was assumed. Here though, we make stronger symmetric assumptions on g. This simplifies the proof.

Theorem 3.3 (value function – strong local regularity).
Suppose that $q, r^* \in \mathcal{C}^+$ while $g \in \mathcal{C}^{++}$. Then, for all $\tau \leq T$, $V(\tau, \cdot)$ is an element of $\mathcal{C}^{++}$.

Proof. By Theorem 2.4, any point in $\text{gph}\partial_\xi V(\tau, \cdot)$ can be expressed as $(x(\tau), p(\tau))$ where $(x(\cdot), p(\cdot))$ is a Hamiltonian trajectories on $[\tau, T]$ with $p(T) = -\nabla g(x(T))$. By local Lipschitz continuity of $\nabla H(\cdot, \cdot)$, one can find a compact neighborhood $X_\tau \times P_\tau$ of $(x(\tau), p(\tau))$ and a compact neighborhood $X_T \times P_T$ of $(x(T), p(T))$ such that:

(i) $X_\tau \times P_\tau \subset S_\tau(X_T \times P_T)$;

(ii) no Hamiltonian trajectory from $X_\tau \times P_\tau$ blows up in time less or equal to $T - \tau$ (and so in particular, there exists a compact set $K \subset \mathbb{R}^{2n}$ so that all Hamiltonian trajectories on $[\tau, T]$ from $X_\tau \times P_\tau$ are in K);

(iii) ∇H is Lipschitz continuous with constant M_1 on K;

(iv) $X_T \subset \text{intdom}g$, $P_T \subset \text{intdom}g^*$;

(v) $\nabla g(\cdot)$ is Lipschitz continuous with constant M_2 and strongly monotone with constant $1/M_2$ relative to $X_T \times P_T$.

Now, pick any two different points $(x_i(\tau), p_i(\tau))$, $i = 1, 2$, in $\text{gph}\partial_\xi V(\tau, \cdot)$ with $x_i(\tau) \in X_\tau$, $p_i(\tau) \in P_\tau$, and let $(x_i(\cdot), p_i(\cdot))$, $i = 1, 2$, to be the associated Hamiltonian trajectories on $[\tau, T]$ with $p_i(T) = -\nabla g(x_i(T))$. Note that by Theorem 3.1 $x_1(\tau) \neq x_2(\tau)$ and $p_1(\tau) \neq p_2(\tau)$. To shorten the notation, we will write $\alpha(t) = x_1(t) - x_2(t)$, $\beta(t) = p_1(t) - p_2(t)$. We have

$$-\|\alpha(\tau)\|\|\beta(\tau)\| \leq \alpha(\tau) \cdot \beta(\tau) \leq \alpha(T) \cdot \beta(T) \leq -\frac{1}{M_2}\|\beta(T)\|^2$$

where the second inequality comes from Lemma 2.1, and the last inequality comes the strong monotonicity of ∇g (as in (v) above). In particular,

$$\frac{1}{\|\alpha(\tau)\|\|\beta(\tau)\|} \leq \frac{M_2}{\|\beta(T)\|^2}.$$

Lipschitz continuity of ∇H yields *

$$\|\beta(\tau)\| \leq \|\beta(T)\| + \left(e^{M_1(T-\tau)} - 1\right)\|(\alpha(T), \beta(T)\|.$$

---

*This bound comes from reversing time, i.e., considering the backward Hamiltonian flow, with time 0 corresponding to T and $T - \tau$ corresponding to τ. By Lipschitz continuity

Squaring both sides and multiplying by the previous displayed inequality yields

$$\frac{\|\beta(\tau)\|}{\|\alpha(\tau)\|} \leq M_2 \left(1 + \left(e^{M_1(T-\tau)} - 1 \right) \sqrt{1 + \frac{\|\alpha(T)\|^2}{\|\beta(T)\|^2}} \right)^2$$

$$\leq M_2 \left(1 + \left(e^{M_1(T-\tau)} - 1 \right) \sqrt{1 + M_2^2} \right)^2 .$$

This shows that $\nabla_\xi V(\tau, \cdot)$ is Lipschitz continuous relative to $X_\tau \times P_\tau$. Strong monotonicity of $\nabla_\xi V(\tau, \cdot)$ can be shown similarly, by using a bound on $\|\alpha(\tau)\|$ in place of the bound on $\|\beta(\tau)\|$ in the arguments above. $\qquad\square$

Minor modifications of the proof above do show that the Lipschitz constant for $\nabla_\xi V(\tau, \cdot)$ is locally bounded in both τ and ξ. This, and local Lipschitz continuity of ∇r^*, implies that the solutions to (20) are unique.

4. Infinite horizon problems

The original control engineering motivation for considering convex optimal control problems – construction of stabilizing feedbacks – calls for posing problems on an infinite time horizon. (Finite time horizon problems can then be used to approximate the infinite horizon ones, or, as seen in Receding Horizon control (also referred to as Model Predictive Control), to obtain stabilizing feedbacks without necessarily having them optimal; see Ref. 16 for a survey.) Besides control engineering, infinite-horizon problems have seen treatment in theoretical economics; see the book by Carslon, Haurie, and Leizarowitz, Ref. 5, for an overview, Ref. 2 for an example of duality techniques, and Ref. 21 for an illustration of the role played by the Hamiltonian dynamical system. For an infinite-horizon $\mathcal{LCR}$, some convexity tools were used in Ref. 3. A detailed, and largely self-contained analysis of a Linear-Quadratic Regulator with control constraints was carried out in Ref 15. General calculus of variations problems on the infinite-time horizon were

---

of ∇H with constant M_1 one gets

$$\left\| \frac{d}{dt}(\alpha(t), \beta(t)) \right\| \leq M_1 \|(\alpha(t), \beta(t))\|$$

and thus $\|(\alpha(t), \beta(t))\| \leq \|(\alpha(0), \beta(0))\| e^{M_1 t}$. One also gets

$$\|\dot{\beta}(t)\| \leq M_1 \|(\alpha(t), \beta(t))\| \leq M_1 \|(\alpha(0), \beta(0))\| e^{M_1 t}$$

which then yields $\|\beta(t)\| \leq \|\beta(0)\| + \|(\alpha(0), \beta(0))\| (e^{M_1 t} - 1)$.

treated in Ref. 12, based to an extent on the duality and Hamilton-Jacobi developments of Ref. 25.

The (primal) infinite horizon optimal value function $W : \mathbb{R}^n \to \overline{\mathbb{R}}$ is defined as the infimum of

$$\int_0^\infty \left\{ q(y(t)) + r(u(t)) \right\} dt,$$

taken over all locally integrable controls $u : [0,\infty) \to \mathbb{R}^k$, subject to dynamics (2) and the initial condition

$$x(0) = \xi.$$

The (dual) infinite horizon optimal value function $\widetilde{W} : \mathbb{R}^n \to \overline{\mathbb{R}}$ is defined as the infimum of

$$\int_0^\infty \left\{ r^*(s(t)) + q^*(z(t)) \right\} dt,$$

taken over all locally integrable controls $z : [0,\infty) \to \mathbb{R}^m$, subject to dynamics (13) and the initial condition

$$p(0) = \pi.$$

Addition to the Standing Assumptions. *The functions q, q^*, r, r^* are positive definite (0 at 0, positive elsewhere). The pairs of matrices (A, C) and $(-A^T, B^T)$ are observable.*

Of course, the value functions W and $\widetilde{W}$ are convex. Results regarding their lower semicontinuity and the existence of optimal arcs in the problems defining them are in Ref. 12. The functions W and $\widetilde{W}$ are both positive definite, and the optimal arcs, for both W and $\widetilde{W}$, tend to 0 as $t \to \infty$. Indeed, for W, note that $W(\xi) = 0$ only if the optimal arc x satisfies $Cx(t) = 0$ while the optimal control is $u(t) = 0$ for all $t \geq 0$. But then $\dot{x}(t) = Ax(t)$, and observability of (A, C) gives $x(t) = 0$ for all $t \geq 0$, in particular $\xi = x(0) = 0$. Given an optimal arc $x(\cdot)$ with $W(x(0)) < \infty$, we have $W(x(T)) = \int_T^\infty \{ q(Cx(t)) + r(u(t)) \} dt$ and by finiteness of $W(\xi)$, the integral tends to 0 as $T \to \infty$. Positive definiteness of $W(x(T))$ now implies that $x(T) \to 0$. Arguments for $\widetilde{W}$ are symmetric.

Now, [12, Corollary 3.5] says that:

- $W(\xi) = \widetilde{W}^*(-\xi)$, equivalently, $\widetilde{W}(\pi) = W^*(-\pi)$;
- $f = W$ is the unique $f \in \mathcal{C}$ such that $H(x, -\partial f(x)) = 0$ for all $x \in \mathbb{R}^n$;
- $f = \widetilde{W}$ is the unique $f \in \mathcal{C}$ such that $H(-\partial f(p), p) = 0$ for all $p \in \mathbb{R}^n$.

In other words, the infinite horizon value functions are conjugate to each other, and are the unique solutions to the generalized stationary Hamilton-Jacobi equations. (The equation $H(x, -\partial f(x)) = 0$ is to be understood as $H(x, -p) = 0$ for all $p \in \partial f(x)$.) It is interesting to note that with such a concept of a solution, it is no longer true that under standard assumptions of (A, B) being stabilizable and (A, C) being detectable, and with q and r quadratic, the (quadratic) value function W is the unique solution to the Hamilton-Jacobi equation. (This occurs even though the matrix defining W is the unique positive definite solution to the matrix Riccati equation, to which the Hamilton-Jacobi equation simplifies if one only looks for quadratic solutions.) See Section 4.1 in Ref. 12 for details.

Another consequence of the fact that optimal arcs for both W and $\widetilde{W}$ converge to 0 is a result relating these two value functions to the finite horizon value functions discussed earlier. It turns out that both W and $\widetilde{W}$ can be simultaneously approximated by a pair of conjugate finite horizon value functions $V(\tau, \cdot)$ and $\widetilde{V}(\tau, \cdot)$. Indeed, as the proof of Theorem 3.4 in Ref. 12 suggests, we have the following:

Theorem 4.1 (finite-horizon approximation). *Consider the functions* $W^T, \widetilde{W}^T : \mathbb{R}^n \to \overline{\mathbb{R}}$ *defined at points* $\xi, \pi \in \mathbb{R}^n$ *as the optimal values in* $\mathcal{LCR}(0, \xi)$, $\widetilde{\mathcal{LCR}}(0, \pi)$. *Suppose that g and g^* are finite-valued on some neighborhood of* 0. *Then W^T converges epi-graphically to W while $\widetilde{W}^T$ converges epi-graphically to $\widetilde{W}$ as $T \to \infty$.*

For the infinite horizon case, the Hamiltonian dynamical system (15) again yields a description of the subdifferential of W (and of $\widetilde{W}$). Indeed, $\text{gph} - \partial W$ consists of all points (ξ, π) for which there exists a Hamiltonian trajectory x, p on $[0, \infty)$ with $x(0) = \xi$, $p(0) = \pi$, and $x(t) \to 0$, $p(t) \to 0$ as $t \to \infty$. See Proposition 3.7 in Ref. 12 which generalizes the results given for strictly concave / strictly convex Hamiltonians in Ref. 21. We add that for such a Hamiltonian trajectory, x is optimal for $W(\xi)$ while p is optimal for $\widetilde{W}(\pi)$.

The said description of ∂W leads to the following result.

Proposition 4.1. *Suppose that there exists a neighborhood of* 0 *on which both q and r^* are strictly convex. Then W is both essentially differentiable and essentially strictly convex.*

The assumption is equivalent to strict convexity of q and essential differentiability of r. By duality, the same assumptions imply that $\widetilde{W}$ is also essentially differentiable and essentially strictly convex. The result can be shown via arguments combining Lemma 2.1 and controllability assumptions we have. (Similar results, shown via slightly different methods, are in Theorem 4.1 of Ref. 21 and Theorem 3.1 of Ref. 13.)

An interesting application of Proposition 4.1 lies in showing that for a wide class of linear systems with saturation nonlinearities there does exist a continuous and stabilizing feedback. (Such an issue was partially addressed in Refs. 27,28 via tools not related to optimality.) Consider a system

$$\dot{x}(t) = Ax(t) + B\sigma(u(t)) \tag{22}$$

where $\sigma : \mathbb{R}^k \to \mathbb{R}^k$ is a saturation nonlinearity. Often, for single input systems, $\sigma(u) = \arctan(u)$ or $\sigma(u) = u$ for $u \in [-1,1]$, $\sigma(u) = -1$ for $u < -1$, $\sigma(u) = 1$ for $u > 1$. For these, and for many other commonly encountered in the control literature saturation functions, $\sigma = \nabla s$ for a convex function $s \in \mathcal{C}$ which is strictly convex near 0. Taking $r = s^*$ and, for simplicity, any positive definite quadratic q, leads to an essentially differentiable W. The optimal feedback $\Phi(x) = \nabla s(-B^*\nabla W(x))$ leads to asymptotic stability for the closed loop system $\dot{x}(t) = Ax(t) + B\Phi(x(t))$. This asymptotic stability can in fact be verified with W as a Lyapunov function. But since $\nabla s = \sigma$, this means that choosing $u(x) = -B^*\nabla W(x)$ in (22) also leads to asymptotic stability. As ∇W is continuous, there does exist a continuous stabilizing feedback for the nonlinear system (22). For details, see Ref. 13.

We conclude by stating a result on local Lipschitz continuity of ∇W. Since optimal trajectories for problems defining both W and $\widetilde{W}$ converge to 0, local properties around 0 of these functions to an extent determine global properties. Another way to say this is that, on each compact subset of dom$W$, W equals to a finite horizon value function defined by $\mathcal{LCR}$ with $g = W$ in a neighborhood of 0, as long as T is large enough. This and Theorem 3.3 lead to the following.

Proposition 4.2. *Suppose that there exists a neighborhood of 0 on which W is differentiable and strongly convex and ∇W is Lipschitz continuous. Then $W \in \mathcal{C}^{++}$.*

An important example of problems where W is differentiable and strongly convex around 0 is provided by a $\mathcal{LQR}$ problem with control constraints: u is restricted to some closed convex set U and $0 \in \text{int}U$. Indeed, then on some neighborhood of 0, W is quadratic and equals to the value function of the unconstrained $\mathcal{LQR}$ problem.

Acknowledgments

Research by R.T. Rockafellar was partially supported by the NSF Grant DMS-0104055.

References

1. B. Anderson and J. Moore, *Optimal Control: Linear Quadratic Methods* (Prentice-Hall, Upper Saddle River, NJ, 1990).
2. L. Benveniste and J. Scheinkman, *J. Economic Theory* **27**, 1 (1982).
3. G. D. Blasio, *SIAM J. Control Optim.* **29**, 909 (1991).
4. C. Byrnes and H. Frankowska, *C.R. Acad. Sci. Paris* **315**, 427 (1992).
5. D. Carlson, A. Haurie and A. Leizarowitz, *Infinite Horizon Optimal Control: Deterministic and Stochastic Systems*, 2nd edition (Springer-Verlag, New York, 1991).
6. N. Caroff and H. Frankowska, *Trans. Amer. Math. Soc.* **348**, 3133 (1996).
7. F. Clarke, *Optimization and Nonsmooth Analysis*, Canadian Math. Soc. Series of Monogr. and Advanc. Texts (Wiley & Sons Inc., New York, 1983).
8. H. Frankowska and A. Ochal, *J. Math. Anal. Appl.* **306**, 714 (2005).
9. G. Galbraith, *SIAM J. Control Optim.* **39**, 281 (2000).
10. R. Goebel, *Set-Valued Analysis* **12**, 127 (2004).
11. R. Goebel, *SIAM J. Control Optim.* **43**, 1781 (2005).
12. R. Goebel, *Trans. Amer. Math. Soc.* **357**, 2187 (2005).
13. R. Goebel, *IEEE Trans. Automat. Contr.* **50**, 650 (2005).
14. R. Goebel and R. Rockafellar, Local strong convexity and local Lipschitz continuity of the gradient of convex functions, *J. Convex Anal.* (to appear).
15. R. Goebel and M. Subbotin, *IEEE Trans. Automat. Contr.* **52**, 886 (2007).
16. D. Mayne, J. Rawlings, C. Rao and P. Scokaert, *Automatica* **36**, 789 (2000).
17. R. Rockafellar, *Convex Analysis*, Princeton Mathematical Series, No. 28 (Princeton University Press, Princeton, N.J, 1970).
18. R. Rockafellar, *J. Math. Anal. Appl.* **32**, 174 (1970).
19. R. Rockafellar, *Pacific J. Math.* **33**, 411 (1970).
20. R. Rockafellar, *Trans. Amer. Math. Soc.* **159**, 1 (1971).
21. R. Rockafellar, *J. Optim. Theory Appl.* **12** (1973).
22. R. Rockafellar, *SIAM J. Control Optim.* **25**, 781 (1987).
23. R. Rockafellar, *SIAM J. Control Optim.* **27**, 1007 (1989).
24. R. Rockafellar and R. J.-B. Wets, *Variational Analysis*, Grundl. der Mathemat. Wissensch. **317** (Springer, Berlin, 1998).
25. R. Rockafellar and P. Wolenski, *SIAM J. Control Optim.* **39**, 1323 (2000).
26. R. Rockafellar and P. Wolenski, *SIAM J. Control Optim.* **39**, 1351 (2000).
27. E. Sontag and H. Sussmann, Nonlinear output feedback design for linear systems with saturating controls, in *Proc. 29 IEEE Conf. Decision and Control (Honolulu, Hawaii, 1990)*: 3414-3416.
28. E. Sontag, H. Sussmann and Y. Yang, *IEEE Trans. Automat. Control* **39**, 2411 (1994).

STRONG OPTIMALITY OF SINGULAR TRAJECTORIES

G. STEFANI

Dipartimento di Matematica Applicata, Università di Firenze,
via di S.Marta 3, 50139 Firenze, Italia
E-mail: gianna.stefani@unifi.it

The paper proves second order sufficient conditions for the strong optimality of singular extremals of the first kind. The conditions are given both in Hamiltonian formulation and by means of the coercivity of a suitable coordinate–free second variation. The conditions are *close* to the necessary ones in the usual sense, namely we require strict inequalities where the necessary conditions have mild inequalities.

Keywords: sufficient conditions, singular control, second variation, Hamiltonian methods.

1. Introduction

In this paper we use a Hamiltonian approach to prove sufficient second order conditions for the strong optimality of a singular extremal of the following Mayer problem on the given interval $[0, T]$:

$$\text{minimize} \quad c_0(\xi(0)) + c_T(\xi(T))$$

subject to

$$\dot{\xi}(t) = f_0(\xi(t)) + u(t)f_1(\xi(t)) \tag{1}$$

$$\xi(0) \in N_0, \quad \xi(T) \in N_T \tag{2}$$

$$u(t) \in [u_0, u_1]. \tag{3}$$

The state space is a smooth manifold M, N_0 and N_T are smooth submanifolds, possibly reduced to a point, c_0 and c_1 are smooth functions, and $f_0, f_1 \colon M \to TM$ are smooth vector fields. By smooth we mean $\mathcal{C}^\infty$, although the proof of the results requires only $\mathcal{C}^2$ data.

Since the dynamics is linear with respect to the control, by Pontryagin Maximum Principle, generically optimal controls take values on the extremal points of the interval, nevertheless Filippov's theorem implies that interior points cannot be avoided. Here we study the strong local optimality of a reference trajectory relative to a singular control, i.e. to a control which takes values in the interior of the interval. Namely we study the strong local optimality of a reference couple $(\widehat{\xi}, \widehat{u})$ satisfying (1), (2) and such that

$$\widehat{u}(t) \in (u_0, u_1), \quad \forall t \in [0, T].$$

The trajectory $\widehat{\xi}$ is *a strong local minimizer* if it is a local minimizer in the C^0 topology of the trajectories on $[0, T]$, independently on the values of the associated controls.

Since we deal with a minimizer corresponding to an interior control we require that the reference couple satisfies the first order optimality conditions in the form of Lagrange multipliers rule (see Lemma 2.1), which defines a lift of $\widehat{\xi}$ to the cotangent bundle, say $\widehat{\lambda} : [0, T] \to T^*M$, called *adjoint covector*. Remark that the adjoint covector given by Pontryagin Maximum Principle is opposite to the one we consider, which instead is linked to a *minimization principle*.

A trajectory $\widehat{\xi}$ of the control system (1) associated to a singular control, satisfying first order conditions is called *singular state-extremal*.

The Hamiltonian approach to sufficient conditions in optimal control corresponds to the classical construction in Calculus of Variations of a field of state extremals covering a neighborhood of the reference trajectory, see for example Ref. 10. The idea is to lift admissible trajectories to the cotangent bundle and to compare their costs, independently on the associated controls.

When the minimized Hamiltonian H^{min} is sufficiently smooth the field of extremals can be obtained by projecting on the state manifold its flow emanating from a suitable Lagrangian sub-manifold and the coercivity of the second variation allows to invert the projection and lift the admissible trajectories, see Refs. 4,5 and the references therein. An important feature is that the comparison of the costs can be also obtained with a Hamiltonian which is smaller than or equal to H^{min}.

In the present case we have to face two difficulties: from one hand H^{min} is not sufficiently smooth near the adjoint covector, on the other hand the classical second variation is completely degenerate. To overcome the problem we add a negative term to the minimized Hamiltonian which keeps the trajectories of the associated Hamiltonian system on the hyper-surface where H^{min} is smooth, for more details see Section 3. It turns out that this new Hamiltonian is naturally linked to the *extended second variation*, defined in Section 5. The construction is possible by assuming *the strengthened generalized Legendre condition* (9).

For references on singular trajectories see Refs. 2,7–9,11,21, and the references therein. We underline that we study sufficient conditions for strong local optimality which to our knowledge do not appear in the literature on singular trajectories. We consider also the *abnormal case*, that is when the multiplier associated to the cost is zero.

As the second variation is concerned, we start from the coordinate-free second variation defined in Ref. 3, we apply the Goh transformation and we end up with an extended second variation which is coordinate-free too and which is proved to be the standard second variation of a new control system where the control appears up to the second order. In Section 5 we give both the definition of the extended second variation and the necessary and sufficient conditions for its coercivity in terms of a suitable Hamiltonian system. In the particular case when the singular state extremals are integral curves of the same vector field f_S, the conditions are given through the behavior of f_S and the controlled vector field f_1, see Subsection 6.2.

The main theorem of the paper is the following:

Theorem 1.1 (Main Theorem). *Suppose that $\widehat{\xi}$ is a singular state-extremal and that the associated extended second variation J_E'', as defined in Section 5, is coercive, then in the normal case ($p_0 = 1$), $\widehat{\xi}$ is a strict strong local minimizer for the original problem, in the abnormal case ($p_0 = 0$), $\widehat{\xi}$ is locally isolated between the solutions of (1) and (2)*

We remark that the coercivity of J_E'' implies the strengthened generalized Legendre condition, moreover the above stated sufficient conditions are *close* to the necessary ones in the usual sense, namely we require strict inequalities where the necessary conditions have mild inequalities, see Theorem 5.1.

The plan of the paper is as follows.

In Section 2 we give the used notations and preliminary results including the first order conditions and the description of the geometrical picture of the cotangent space near the adjoint covector.

In Section 3 we describe the Hamiltonian approach to strong optimality and we define the Hamiltonian χ which allows the use of the method in the present case.

In Section 4 we prove sufficient conditions from a Hamiltonian viewpoint (Theorem 4.1), describing how to compare the costs starting from the construction of a field of non intersecting almost extremals.

In Section 5 we define the coordinate-free extended second variation J_E''

and we prove that its coercivity can be tested via a non singular problem.

In Section 6 we describe the links between the extended second variation and the modified minimizing Hamiltonian and we prove the main theorem. In particular Lemma 6.3 gives necessary and sufficient conditions for the coercivity of J_E'' in terms of almost extremals.

In Section 7 we give some summarizing remarks on the paper and we describe future research perspectives.

2. Notations and preliminary results

2.1. *Notations*

In this paper we use some basic element of the theory of symplectic manifolds referred to the cotangent bundle T^*M. For a general introduction see Ref. 6, for specific application to Control Theory we refer to Ref. 1 and Ref. 2.

Let us recall some basic facts and let us introduce some specific notations. We identify any bilinear form Q on a vector space V with a linear form $Q : V \to V^*$ and we denote the associated quadratic form as

$$Q(v, v) = Q[v]^2.$$

The kernel of Q as a quadratic form is the kernel of Q as a linear operator. We call nullity the dimension of $\ker Q$ and index the dimension of the maximal subspace on which Q is negative.

For any subspace $V \subset T_x M$, $V^\perp \subset T_x^* M$ denotes its annihilator. If N is a sub-manifold of M, we write $T_x^\perp N$ for $(T_x N)^\perp$.

We use two equivalent notations for the Lie derivative of a function ϕ in the direction of the vector field f, namely we write

$$L_f \phi := f \cdot \phi : x \mapsto \langle d\phi(x), f(x) \rangle,$$

depending on typographic convenience.

Denote by $\pi : T^*M \to M$ the canonical projection, the space $T_{\pi\ell}^* M$ is canonically embedded in $T_\ell T^*M$ as the space of tangent vectors to the fibers.

The canonical Liouville one-form s on T^*M and the associated canonical symplectic two-form $\sigma = ds$ allow to associate to any, possibly time-dependent, smooth Hamiltonian $H_t : T^*M \to \mathbb{R}$, a Hamiltonian vector field $\vec{H}_t$, by

$$\sigma(v, \vec{H}_t(\ell)) = \langle dH_t(\ell), v \rangle, \quad \forall v \in T_\ell T^*M$$

and we denote its flow from time 0 to time t by

$$\mathcal{H} : (t, \ell) \mapsto \mathcal{H}(t, \ell) = \mathcal{H}_t(\ell).$$

We keep these notation throughout the paper, namely the overhead arrow denotes the vector field associated to a Hamiltonian and the script letter denotes its flow from time 0.

Finally recall that any vector field f on the manifold M defines, by lifting to the cotangent bundle, a Hamiltonian

$$\ell \in T^*M \mapsto \langle \ell, f(\pi\ell) \rangle \in \mathbb{R}.$$

Coming back to our problem, we define the time-dependent reference vector field by

$$\widehat{f}_t : M \to TM, \quad x \mapsto f_0(x) + \widehat{u}(t) f_1(x).$$

The flow, from time 0 to time t, of $\widehat{f}_t$ is defined for all $t \in [0, T]$ in a neighborhood $\mathcal{O}$ of

$$\hat{x}_0 := \widehat{\xi}(0),$$

and it is denoted by $\widehat{S}_t : \mathcal{O} \to M$, $t \in [0, T]$, while the time-dependent reference Hamiltonian defined by the vector field $\widehat{f}_t$ is denoted by $\widehat{H}_t$.

Following the notation of Ref. 2, we denote by H_0, H_1 the Hamiltonians associated to f_0, f_1 respectively and by

$$H_{i_1 i_2 \ldots i_k} := \{H_{i_1}, \{\ldots \{H_{i_{k-1}}, H_{i_k}\} \ldots\}$$

the Hamiltonian associated to $f_{i_1 i_2 \ldots i_k} := [f_{i_1}, [\ldots [f_{i_{k-1}}, f_{i_k}] \ldots]]$, where $\{,\}$ denotes the Poisson parentheses between Hamiltonians and $[,]$ denotes the Lie brackets between vector fields.

2.2. The Weak Maximum Principle

As we pointed out in Section 1 we assume the reference couple to satisfy the first order optimality conditions in the form of Lagrange multipliers rule, usually called the weak maximum principle (WMP). Recall that the adjoint covector given by Pontryagin Maximum Principle is the opposite to the one we consider, which instead is linked to the "minimization principle".

Lemma 2.1 (WMP). *If $(\widehat{\xi}, \widehat{u})$ is a minimizing couple with $\widehat{u}(t) \in (u_0, u_1)$, a.e. $t \in [0, T]$, then there is $p_0 \in \{0, 1\}$ and an absolutely con-*

tinuous solution $\widehat{\lambda}$ of the Hamiltonian system with transversality conditions

$$\dot{\lambda}(t) = \vec{\widehat{H}}_t(\lambda(t))$$

$$\lambda(0) = -p_0 \, dc_0(\hat{x}_0) \quad on \ T_{\hat{x}_0} N_0 \tag{4}$$

$$\lambda(T) = p_0 \, dc_T(\widehat{\xi}(T)) \quad on \ T_{\widehat{\xi}(T)} N_T \tag{5}$$

which satisfies $\pi \circ \widehat{\lambda} = \widehat{\xi}$, $p_0 + \|\hat{\lambda}(0)\| > 0$ *and*

$$\partial_u \langle \ell, f_0(\pi\ell) + u \, f_1(\pi\ell) \rangle_{|(\widehat{\lambda}(t), \widehat{u}(t))} = 0 \,, \quad a.e. \ t \in [0, T].$$

In our case WMP implies

$$H_1\left(\widehat{\lambda}(t)\right) = \langle \widehat{\lambda}(t), f_1(\hat{\xi}(t)) \rangle = 0 \,, \ \forall \ t \in [0, T] \tag{6}$$

and, by derivation, equation (6) leads to

$$H_{01}\left(\widehat{\lambda}(t)\right) = \langle \widehat{\lambda}(t), [f_0, f_1](\widehat{\zeta}(t)) \rangle = 0 \tag{7}$$

$$(H_{001} + \widehat{u}(t) H_{101})\left(\widehat{\lambda}(t)\right) = 0 \,, \quad a.e. \ t \in [0, T] \,. \tag{8}$$

In what follows we assume *the strengthened generalized Legendre condition* (SGLC), i.e.

$$H_{110}(\widehat{\lambda}(t)) = \langle \widehat{\lambda}(t), [f_1, [f_1, f_0]](\widehat{\xi}(t)) \rangle > 0, \quad t \in [0, T] \,, \tag{9}$$

so that we obtain

$$\widehat{u}(t) = -\frac{H_{001}}{H_{101}}(\widehat{\lambda}(t)) = \frac{H_{001}}{H_{110}}(\widehat{\lambda}(t))$$

and $\widehat{\lambda}$ is called a singular extremal of the first kind, see for example Ref. 21.

Definition 2.1. Let $u \in \mathbf{L}^\infty([0, T], \mathbb{R})$ be such that $u(t) \in [u_0, u_1]$, $a.e. \, t \in [0, T]$. We call any non trivial trajectory $\lambda : [0, T] \to T^* M$ of the Hamiltonian vector field $\vec{H}_0 + u(t)\vec{H}_1$, which satisfies (6), *extremal* of the control system, while we call its projection $\xi : [0, T] \to M$ *state extremal*.

2.3. *Geometry near singular extremals of the first kind*

By (6), (7) and (8), any singular extremal of the first kind belongs to the $(2n - 2)$–dimensional symplectic manifold

$$\mathcal{S} = \{\ell : H_1(\ell) = H_{01}(\ell) = 0, H_{110}(\ell) > 0, \frac{H_{001}}{H_{110}}(\ell) \in (u_0, u_1)\}$$

which is contained in the hyper-surface

$$\Sigma := \{\ell \in T^* M : H_1(\ell) = 0\},$$

where the minimized Hamiltonian

$$H^{min} : \ell \mapsto \min\{H_0(\ell) + uH_1(\ell) : u \in [u_0, u_1]\}$$

coincides with H_0. By SGLC it is easy to prove the following result.

Lemma 2.2. *In a neighborhood $\mathcal{U}$ of $\mathcal{S}$ in T^*M the following statements hold true.*

(1) Σ is an hyper-surface containing the symplectic manifold $\mathcal{S}$
(2) $\vec{H}_1$ is tangent to Σ
(3) $\vec{H}_1$ is transversal to $\mathcal{S}$
(4) $\vec{H}_{01}$ is transversal to Σ
(5) The maps $(s, \ell) \mapsto \exp s\vec{H}_1(\ell)$ and $(\tau, s, \ell) \mapsto \exp \tau \vec{H}_{01} \circ \exp s\vec{H}_1(\ell)$ are local diffeomorphisms from $\mathbb{R} \times \mathcal{S}$ to Σ and from $\mathbb{R} \times \mathbb{R} \times \mathcal{S}$ to T^*M respectively.
(6) Σ separates in $\mathcal{U}$ the regions defined by

$$H^{min} = H_0 + H_1, \quad H^{min} = H_0 - H_1.$$

Properties *(5)* in Lemma 2.2 yields to the possibility of defining the smooth function $v : \mathcal{U} \to \mathbb{R}$ by

$$v := H_{001}/H_{110} \quad \text{on} \quad \mathcal{S} \tag{10}$$

and extend it constant first on the integral lines $\vec{H}_1$ and then on the ones of $\vec{H}_{01}$. In this way we get *the Hamiltonian of singular extremal of the first kind* by defining

$$H_S := H_0 + v\,H_1.$$

Indeed $\vec{H}_S$ is tangent to $\mathcal{S}$ and any singular extremal of the first kind of our problem is an integral curve of $\vec{H}_S$ contained in $\mathcal{S}$: as a consequence $\widehat{\lambda}$ is $\mathcal{C}^\infty$ and the same is $\widehat{u} = v \circ \widehat{\lambda}$.

Remark 2.1. A particular case occurs when the function v depends only on $\pi\ell \in M$, indeed in this case the Hamiltonian H_S is the lift of a vector field $f_S = f_0 + vf_1$ and any singular state extremal associated to the dynamics (1) is an integral line of f_S. The vector field f_S, after a reparametrization of the control, can be thought as the drift vector field f_0 and we can also think that the equality $H_{001|\mathcal{S}} = 0$ holds.

Remark that the above described case can be considered generic in dimensions 2 and 3, as it can be easily seen in any chart.

3. Hamiltonian approach to strong local optimality

The Hamiltonian approach to prove sufficient conditions to strong optimality consists in constructing a field of non-intersecting *state almost-extremals* covering a neighborhood of the given trajectory. We call almost extremal a solution of the Hamiltonian system associated to a Hamiltonian H_t (possibly time-dependent) with the following properties

$$H_t \leq H^{min}, \quad H_t \circ \widehat{\lambda} = \widehat{H}_t \circ \widehat{\lambda}, \quad \dot{\widehat{\lambda}} = \overrightarrow{H}_t \circ \widehat{\lambda}. \tag{11}$$

The field of state almost-extremals is obtained by projecting on the state manifold the flow associated to H_t emanating from a suitable horizontal Lagrangian sub-manifold Λ. If the map

$$id \times \pi\mathcal{H} : [0, \widehat{T}] \times \Lambda \to [0, \widehat{T}] \times M, \quad (t, \ell) \mapsto (t, \pi\mathcal{H}_t(\ell))$$

is a diffeomorphism, then we can use symplectic arguments to compare the costs by lifting admissible trajectories to the cotangent bundle, independently on the associated controls.

Remark that the third property in (11) is a consequence of the first two if H_t is differentiable along the adjoint covector and that the minimizing Hamiltonian H^{min} satisfies the properties if it is sufficiently smooth.

When the trajectory is singular we cannot satisfy (11) with H^{min} unless $\widehat{u} \equiv 0$ and in any case H^{min} does not guarantees that we can find the Lagrangian manifold Λ with the required properties, since generically the associated Hamiltonian system does not have the continuity property with respect to the initial conditions near the adjoint covector.

Our idea for overcoming the problem is to use a Hamiltonian having the properties (11) and such that the associated vector field is tangent to Σ, so that $H^{min} = H_0$ along the chosen almost extremals, as explained in the following subsection.

3.1. *The Hamiltonian χ*

For the singular trajectories of the first kind, we pursue the above described strategy by adding to H_0 a negative Hamiltonian χ. This possibility is given by the following Lemma 3.1. Remark that the result has been proved in a slight different form in Ref. 18.

Lemma 3.1. *By possibly restricting $\mathcal{U}$, it is possible to define a smooth function $\rho : \mathcal{U} \subset T^*M \to \mathbb{R}$, with the following properties*

(1) The Hamiltonian $\chi = \frac{\rho}{2} H_{01}^2$ is such that the Hamiltonian vector field $\overrightarrow{H}_0 + \overrightarrow{\chi}$ is tangent to Σ.

(2) $\rho(\ell) = \frac{1}{H_{101}(\ell)}$, *for all* $\ell \in \mathcal{S}$, *hence, without loss of generality, we can suppose* $\rho < 0$

(3) ρ *can be chosen so that*

$$(\vec{H}_0 + \vec{\chi})(\ell) = (\exp(-\vartheta(\ell))\vec{H}_1)_* \vec{H}_0 \circ (\exp \vartheta(\ell)\vec{H}_1)(\ell)\,, \quad \forall \ell \in \Sigma$$

where $\vartheta(\ell)$ *is defined by*

$$H_{01}(\exp \vartheta(\ell)\vec{H}_1(\ell)) = 0 \quad and \quad \vartheta(\ell) = 0\,, \ \forall \ell \in \mathcal{S}.$$

Proof. $\vec{H}_0 + \vec{\chi}$ is tangent to Σ if and only if on Σ we have

$$\langle dH_1, \vec{H}_0 + \vec{\chi}\rangle = H_{01}(1 - \frac{1}{2}H_{01}\langle d\rho, \vec{H}_1\rangle + \rho H_{110}) = 0$$

and hence

$$1 - \frac{1}{2}H_{01}\langle d\rho, \vec{H}_1\rangle + \rho H_{110} = 0. \tag{12}$$

Since $\vec{H}_1$ is tangent to Σ, then we need to define ρ only on Σ. Any smooth extension of such a ρ has property *(1)*. The previous equation reads $\rho = 1/H_{101}$ on $\mathcal{S}$, hence we can solve (12) by the characteristic method, obtaining for $\ell \in \mathcal{S}$, and $|t|$ sufficiently small

$$\rho(\exp t\vec{H}_1(\ell)) =$$
$$\frac{-2}{\left(H_{01}(\exp t\vec{H}_1(\ell))\right)^2} \int_0^t \int_0^\sigma H_{110}(\exp s\vec{H}_1(\ell))\, ds\, d\sigma =$$
$$\frac{-1}{H_{110}(\ell)} + o(1).$$

This proves *(1)* and *(2)*.

From the above expression for ρ, taking into account that:
$\frac{\partial}{\partial t}H_{10} \circ \exp t\vec{H}_1 = H_{110} \circ \exp t\vec{H}_1$ and $\frac{\partial}{\partial t}H_0 \circ \exp t\vec{H}_1 = H_{10} \circ \exp t\vec{H}_1$
we obtain in $\ell_\Sigma = \exp t\vec{H}_1(\ell_\mathcal{S}) \in \Sigma$

$$\chi(\ell_\Sigma) = H_0(\ell_\mathcal{S}) - H_0(\ell_\Sigma)\,. \tag{13}$$

For $\ell \in \Sigma$ sufficiently close to $\mathcal{S}$, it is possible to define $\vartheta(\ell)$ by

$$H_{01}(\exp \vartheta(\ell)\vec{H}_1(\ell)) = 0 \quad and \quad \vartheta(\ell) = 0\,, \ \forall \ell \in \mathcal{S}.$$

With this notation, if we choose to extend χ constant along the integral lines of $\vec{H}_{01}$, equation (13) reads

$$d\chi = dH_0 \circ (\exp \vartheta\vec{H}_1)_* - dH_0$$

and *(3)* is proved. $\qquad\square$

Remark 3.1. For each smooth function

$$\nu : \mathbb{R} \times T^*M \to \mathbb{R}, \quad (t,\ell) \mapsto \nu(t,\ell) = \nu_t(\ell)$$

consider the Hamiltonian $H_t^\nu := H_0 + \nu_t H_1 + \chi$. We have $H_t^\nu \leq H_0 = H^{min}$ on Σ and the associated vector field $\vec{H}_t^\nu$ is tangent to Σ. Moreover if $\nu(t,\widehat{\lambda}(t)) = \widehat{u}(t)$, then $dH_t^\nu(\widehat{\lambda}(t)) = d\widehat{H}_t(\widehat{\lambda}(t))$ therefore, in this case, $\widehat{\lambda}$ is a trajectory of $\vec{H}_t^\nu$. In particular

$$\vec{\widehat{H}}_t := \vec{\widehat{H}}_t + \vec{\chi} \quad \text{and} \quad \vec{\widehat{H}}_S := \vec{H}_S + \vec{\chi}$$

are tangent to Σ and $\widehat{\lambda}$ is a trajectory of both of them.

Lemma 3.2. *If the function ν is such that*

$$\nu(t,\widehat{\lambda}(t)) = \widehat{u}(t) \quad \text{and} \quad \langle d\nu_t \vec{H}_1\rangle(\widehat{\lambda}(t)) = 0, \tag{14}$$

then $\vec{H}_1$ and $\vec{H}_0$ are invariant with respect to the flow of $\mathcal{H}_t^\nu$ along the adjoint covector, i.e.

$$\vec{H}_1(\widehat{\lambda}(t)) = (\mathcal{H}_t^\nu)_* \vec{H}_1(\widehat{\lambda}(0)) \quad \text{and} \quad \vec{H}_0(\widehat{\lambda}(t)) = (\mathcal{H}_t^\nu)_* \vec{H}_0(\widehat{\lambda}(0)).$$

Proof.

$$\partial_t \left((\mathcal{H}_t^\nu)_*^{-1} \vec{H}_1 \circ \mathcal{H}_t^\nu \right)(\widehat{\lambda}(0)) = (\mathcal{H}_t^\nu)_*^{-1} \left[\vec{H}_t^\nu, \vec{H}_1 \right](\widehat{\lambda}(t)) =$$

$$= (\mathcal{H}_t^\nu)_*^{-1} \left(\vec{H}_{01} + \left[\nu_t \vec{H}_1 + H_1 \vec{\nu}_t + H_{01}^2 \vec{\rho}/2 + H_{01}\rho\vec{H}_{01}, \vec{H}_1 \right] \right)(\widehat{\lambda}(t)) =$$

$$= (\mathcal{H}_t^\nu)_*^{-1} \left(1 - \rho\langle dH_{01}, \vec{H}_1\rangle \right) \vec{H}_{01}(\widehat{\lambda}(t)) \equiv 0,$$

the last equality follows by (12). In an analogous way we get

$$\partial_t \left((\mathcal{H}_t^\nu)_*^{-1} \vec{H}_0 \circ \mathcal{H}_t^\nu \right)(\widehat{\lambda}(0)) = (\mathcal{H}_t^\nu)_*^{-1} \left(\widehat{u}(t) + \rho\langle dH_{01}, \vec{H}_0\rangle \right) \vec{H}_{10}(\widehat{\lambda}(t)),$$

which is identically zero by (10), since $\rho(\widehat{\lambda}(t)) = 1/H_{101}(\widehat{\lambda}(t))$. $\square$

4. Sufficient conditions from Hamiltonian viewpoint

To apply the Hamiltonian approach to prove sufficient conditions to strong optimality of singular extremals as explained in the previous Section 3, first we extend in a suitable way the cost functions, namely the smooth functions $p_0\, c_0$ and $p_0\, c_T$ are defined on N_0 and N_T respectively, but they can be extended to the whole manifold M in such a way that the transversality conditions (4) and (5) hold on the whole tangent space. Therefore we denote by $\alpha, \beta : M \to \mathbb{R}$ functions such that

$$\alpha = -p_0\, c_0 \quad \text{on } N_0\, , \quad \beta = p_0\, c_T \quad \text{on } N_T \tag{15}$$

$$\widehat{\lambda}(0) = -d\alpha(\widehat{x}_0) \quad \text{on } T_{\widehat{x}_0}M\, , \quad \widehat{\lambda}(T) = d\beta(\widehat{x}_T) \quad \text{on } T_{\widehat{x}_T}M. \tag{16}$$

In the normal case ($p_0 = 1$) α, β are cost functions equivalent to the original ones while in the abnormal case ($p_0 = 0$) they are extensions of the zero function. When $p_0 = 0$ all the costs disappear, we study a problem with a zero cost and indeed we are studying the constraints. Proving that $\hat{\xi}$ is a *strict* strong minimizer with $p_0 = 0$ implies that it is isolated among the admissible trajectories. The sufficient conditions will be derived studying the following optimal control problem

$$\textbf{Minimize} \quad J(\xi) := \alpha(\xi(0)) + \beta(\xi(T)) \tag{17}$$

subject to (1), (2) and (3).

For any α with the properties (15), (16) and

$$L_{f_1} L_{f_1} \alpha(\hat{x}_0) > 0, \tag{18}$$

the set $\tilde{\mathcal{O}} = \{x \in \mathcal{O} : L_{f_1}\alpha(x) = 0\}$ is a hyper-surface of $\mathcal{O}$ and, by possibly restricting $\mathcal{O}$, for $\bar{\epsilon}$ sufficiently small, we can state

$$\mathcal{O} = \{\exp s f_1(x) : x \in \tilde{\mathcal{O}}, \ |s| < \bar{\epsilon}\}.$$

We define $\tilde{\alpha}$ on $\mathcal{O}$ by setting

$$\tilde{\alpha}(\exp s f_1(x)) = \alpha(x), \ x \in \tilde{\mathcal{O}}, \ |s| < \bar{\epsilon},$$

$\tilde{\alpha}$ is constant along the integral lines of f_1, therefore the Lagrangian sub-manifold

$$\Lambda_0^\alpha := \{\ell = -d\tilde{\alpha}(x), \ x \in \mathcal{O}\}$$

is contained in Σ, so that $\vec{H}_1 = d\tilde{\alpha}_* f_1$ is tangent to Λ_0^α. Remark that Λ_0^α can be also obtained as

$$\Lambda_0^\alpha = \left\{\exp s\vec{H}_1(\ell) : \ell = -d\alpha(x), \ x \in \tilde{\mathcal{O}}, \ |s| < \bar{\epsilon}\right\}.$$

Choose a function ν with the properties (14) and consider the Hamiltonian H_t^ν as defined in Remark 3.1, with associated vector field $\vec{H}_t^\nu$ and flow $\mathcal{H}_t^\nu$.

Standard symplectic arguments show that the one-form $\boldsymbol{\omega}$ defined by

$$\boldsymbol{\omega} = (id \times \mathcal{H}^\nu)^*(\boldsymbol{s} - H_t^\nu \, dt)$$

is exact on $[0,T] \times \Lambda_0^\alpha$, see Ref. 6. Setting $\boldsymbol{\omega} = d\theta$ we can choose θ so that $\theta(0,\ell) = -\tilde{\alpha} \circ \pi(\ell)$, $\ell \in \Lambda_0^\alpha$, moreover whenever $\pi \circ \mathcal{H}_t^\nu$ is a local diffeomorphism from Λ_0^α onto M we get

$$d\left(\theta_t \circ (\pi \circ \mathcal{H}_t^\nu)^{-1}\right) = (\mathcal{H}^{\nu *} \boldsymbol{s})\,(\pi \circ \mathcal{H}_t^\nu)_*^{-1} = \mathcal{H}_t^\nu \circ (\pi \circ \mathcal{H}_t^\nu)^{-1}.$$

Theorem 4.1. *Suppose that $\widehat{\lambda}(t) \in \mathcal{S}$ for all $t \in [0,T]$. If there are functions α, β, ν with the properties (15), (16), (18), (14) and such that:*

(1) $\pi_* \mathcal{H}^\nu_{t*} : T_{\widehat{\lambda}(0)} \Lambda_0^\alpha \to T_{\widehat{\xi}(t)} M$ is an isomorphism for all $t \in [0, T]$,
(2) the quadratic map

$$\delta x \mapsto \boldsymbol{\sigma} \left(d\beta_* \delta x, \mathcal{H}^\nu_{T*} (\pi \mathcal{H}^\nu_T)^{-1}_* \delta x \right)$$

is positive definite on $T_{\widehat{\xi}(T)} N_T$,

then $\widehat{\xi}$ is a strict strong local minimizer for the problem (17), therefore in the normal case ($p_0 = 1$) $\widehat{\xi}$ is a strict strong local minimizer for the original problem, in the abnormal case ($p_0 = 0$) $\widehat{\xi}$ is locally isolated between the solutions of (1) and (2).

Proof. We consider the map

$$id \times \mathcal{H}^\nu : [0, T] \times \Lambda_0 \to [0, T] \times \Sigma$$
$$(t, \ell) \mapsto (t, \mathcal{H}^\nu_t(\ell)).$$

If assumption *(1)* is satisfied, then

$$\pi \circ \mathcal{H}^\nu_t : \Lambda_0^\alpha \to M$$

is a local diffeomorphism covering a neighborhood of $\widehat{\xi}(t)$, therefore, by possibly restricting $\mathcal{O}$, there is a neighborhood $\mathcal{V}$ of the range of $(id \times \widehat{\xi})$ in $[0, T] \times M$ such that the map

$$id \times \pi \circ \mathcal{H}^\nu : \Lambda_0^\alpha \times [0, T] \to \mathcal{V}$$

is a diffeomorphism.

Consider an admissible couple (u, ξ) of the problem such that

$$\{(t, \xi(t)) : t \in [0, T]\} \subset \mathcal{V}$$

and denote

$$\lambda(t) = (\pi \circ \mathcal{H}^\nu_t)^{-1}(\xi(t))$$
$$\mu(t) = (t, \lambda(t)), \quad \widehat{\mu}(t) = (t, \widehat{\lambda}(0)).$$

Since $\lambda(t) \subset \Sigma$, then $\mathcal{H}^\nu_t(\lambda(t)) \subset \Sigma$, therefore using that $H^\nu_t \leq H_0$ and $H_1 = 0$ along $\mathcal{H}^\nu_t(\lambda(t))$, we compute

$$0 = \int_\mu \omega - \int_{\widehat{\mu}} \omega + \theta(T, \widehat{\ell}_0) - \theta(T, \lambda(T)) + \theta(0, \lambda(0)) - \theta(0, \widehat{\ell}_0)$$

$$= \int_\mu \omega + \theta(T, \widehat{\ell}_0) - \theta(T, \lambda(T)) - \tilde{\alpha}(x) + \alpha(\hat{x}_0)$$

$$\geq \theta(T, \widehat{\ell}_0) - \theta(T, \lambda(T)) - \tilde{\alpha}(\xi(0)) + \alpha(\hat{x}_0)$$

312

Hence

$$\tilde{\alpha}(\xi(0)) - \alpha(\hat{x}_0) + \beta(\xi(T)) - \beta(\widehat{\xi}(T))$$

$$\geq \beta(\xi(T)) - \theta(T, \lambda(T)) - \left(\beta(\widehat{\xi}(T)) - \theta(T, \widehat{\ell}_0)\right)$$

Setting $\varphi := \beta - \theta_T \circ (\pi \mathcal{H}_T^\nu)^{-1}$, we have

$$d\varphi = d\beta - \mathcal{H}_T^\nu \circ (\pi \mathcal{H}_T^\nu)^{-1}$$
$$d\varphi(\widehat{\xi}(T)) = d\beta(\widehat{\xi}(T)) - \widehat{\lambda}(T) = 0$$
$$D^2\varphi(\widehat{\xi}(T))(\cdot, \cdot) = \boldsymbol{\sigma}\left(d\beta_*(\cdot), \mathcal{H}_{T*}^\nu \circ (\pi \mathcal{H}_T^\nu)_*^{-1}(\cdot)\right),$$

therefore $\widehat{\xi}(T)$ is a local strict minimizer for ϕ on N_T. By possibly restricting $\mathcal{O}$ we have $\alpha \geq \tilde{\alpha}$ by (18), so that *(2)* implies

$$J(u, \xi) - J(\widehat{u}, \widehat{\xi}) \geq 0.$$

To prove that the minimum is strict, remark that $J(u, \xi) - J(\widehat{u}, \widehat{\xi}) = 0$ implies $L_{f_1}\alpha(\xi(0)) = 0$, $\xi(T) = \widehat{\xi}(T)$, $\int_\mu \boldsymbol{\omega} = 0$. In particular the last inequality implies

$$H_{01} \circ \mathcal{H}_t^\nu(\lambda(t)) \equiv 0$$

so that $\mathcal{H}_t^\nu \circ \lambda(t) \in \mathcal{S}$. From Lemma 3.2 we obtain $(\pi\mathcal{H}_t^\nu)_* \vec{H}_1(\lambda(t)) = f_1(\xi(t))$ and hence $\mathcal{H}_{t*}^\nu(\pi\mathcal{H}_t^\nu)_*^{-1} f_1(\xi(t)) = \vec{H}_1 \circ \mathcal{H}_t^\nu(\lambda(t))$ for all $t \in [0, T]$. Finally we compute

$$\frac{d}{dt}\left(\mathcal{H}_t^\nu \circ \lambda(t)\right) = \vec{H}_t^\nu \circ \mathcal{H}_t^\nu(\lambda(t)) + (u(t) - \nu_t \circ \mathcal{H}_t^\nu(\lambda(t)))\, \mathcal{H}_{t*}^\nu(\pi\mathcal{H}_t^\nu)_*^{-1} f_1(\xi(t))$$

$$= \left(\vec{H}_0 + u(t)\vec{H}_1\right) \circ \mathcal{H}_t^\nu(\lambda(t)).$$

In other words $\mathcal{H}_t^\nu \circ \lambda$ is a singular extremal of the first kind, hence it is the integral curve of $\vec{H}_S$ passing through $\widehat{\lambda}(T)$ so that $\xi = \widehat{\xi}$. □

4.1. *Remarks on Theorem 4.1*

If $f_1(\hat{x}_0)$ belongs to $T_{\hat{x}_0} N_0$, then equation (18) is an assumption to be fulfilled by the function α. If $f_1(\hat{x}_0)$ does not belong to $T_{\hat{x}_0} N_0$, then α with the properties (15), (16) can be chosen so that (18) is fulfilled and $\tilde{\mathcal{O}} = N_0 \cap \mathcal{O}$, hence $\alpha = \tilde{\alpha}$ on N_0. In any case α and $\tilde{\alpha}$ are determined on N_0 by the data, while we are free to chose the second order terms on any complement of N_0.

Remark that we can choose both

$$H_t^\nu = \tilde{H}_S = H_S + \chi \quad \text{and} \quad H_t^\nu = \widehat{H}_t + \chi$$

in Theorem 4.1. The main theorem will be proved by using Theorem 4.1 with the Hamiltonian

$$H_t := \widehat{H}_t + \chi. \tag{19}$$

When the initial point of the original problem is free, we have the following sufficient conditions in terms of the data.

Corollary 4.1. *Suppose that $N_0 = M$ and $\widehat{\lambda}(t) \in \mathcal{S}$ for all $t \in [0, T]$. Let H_t be the Hamiltonian defined in (19) and suppose*

(1) $L_{f_1} L_{f_1} c_0(\hat{x}_0) > 0$
(2) $\pi_* \mathcal{H}_{t*} : T_{\widehat{\lambda}(0)} \Lambda_0^{co} \to T_{\widehat{\xi}(t)} M$ is an isomorphism for all $t \in [0, T]$.
(3) The quadratic form

$$\delta x \mapsto \sigma \left(dc_{T*} \delta x, \mathcal{H}_{T*}^{\nu} \pi \mathcal{H}_T^{\nu})_*^{-1} \delta x \right)$$

is positive definite on $T_{\widehat{\xi}(T)} N_T$

then $\widehat{\xi}$ is a strict strong local minimizer for the original problem. Recall that in this case the extremal is normal, hence $\alpha = c_0$.

Remark 4.1. We have chosen to apply Hamiltonian methods starting from an initial Lagrangian sub-manifold and considering the forward flow of $\vec{H}_t^{\nu}$. We could have chosen to start at any time, in particular some authors prefer to consider a final Lagrangian sub-manifold and integrate the Hamiltonian vector field backwards. This last strategy is obviously preferable when the final point is free. It is not difficult to reformulate Theorem 4.1 and Corollary 4.1 starting from the final time.

5. The extended second variation

In this Section we use the second variation defined in Ref. 3, hence we need to define the *pull back system*. In the neighborhood $\mathcal{O}$ of $\hat{x}_0$ set

$$g_t = \widehat{S}_{t*}^{-1} f_1 \circ \widehat{S}_t : \mathcal{O} \to TM$$
$$\widehat{\beta} = \beta \circ \widehat{S}_T : \mathcal{O} \to \mathbb{R}.$$

Choosing α and β with the properties (15), (16), we have

$$d(\alpha + \widehat{\beta})(\hat{x}_0) = -\widehat{\lambda}(0) + \widehat{\lambda}(T)\widehat{S}_{T*} = 0,$$

so that $\gamma'' = D^2(\alpha + \widehat{\beta})(\hat{x}_0)$ is a well defined bilinear form on $T_{\hat{x}_0} M$. The second variation, as defined in Ref. 3, is a linear quadratic form on

$T_{\hat{x}_0}N_0 \times L^2([0,T],\mathbb{R})$ realized as a linear–quadratic (LQ) problem on the vector space $T_{\hat{x}_0}M$. Set

$$\gamma'' = D^2(\alpha + \widehat{\beta})(\hat{x}_0)$$
$$N_T'' = S_{T*}^{-1}T_{\hat{x}_T}N_T$$
$$\delta e = (\delta x, v) \in T_{\hat{x}_0}N_0 \times L^2([0,T],\mathbb{R}),$$

the second order approximation of the cost is given by

$$J''[\delta e]^2 = \frac{1}{2}\gamma''[\delta x]^2 + \int_0^T v(t)\, L_{\eta(t)}\, L_{g_t}\, \widehat{\beta}(\hat{x}_0)dt$$

where η satisfies

$$\dot{\eta}(t) = v(t)g_t(\hat{x}_0)$$
$$\eta(0) = \delta x \in T_{\hat{x}_0}N_0, \quad \eta(T) \in N_T''.$$

Remark 5.1. The second variation is independent on the choice of the functions α, β with the properties (15) and (16), see Ref. 3.

Introducing

$$w(t) = \int_t^T v(s)ds, \quad w_0 = w(0), \quad \dot{g}_t(x) = \partial_t g(t,x)$$

and integrating by parts we obtain

$$\eta(T) = \delta x + \int_0^T v(s)g_s(\hat{x}_0)ds = \delta x + w_0 f_1(\hat{x}_0) + \int_0^T w(s)\dot{g}_s(\hat{x}_0)ds$$

and

$$J''\,[\delta e]^2 = \frac{1}{2}\gamma''[\delta x]^2 + \int_0^T v(t)L_{\delta x}L_{g_t}\widehat{\beta}(\hat{x}_0)dt + \int_0^T v(t)L_{\left(\int_0^t v(s)g_s\,ds\right)}L_{g_t}\widehat{\beta}(\hat{x}_0)dt$$

$$= \frac{1}{2}\gamma''[\delta x]^2 + w_0 L_{\delta x}L_{f_1}\widehat{\beta}(\hat{x}_0) + \int_0^T w(t)L_{\delta x}L_{\dot{g}_t}\widehat{\beta}(\hat{x}_0)\, dt + \frac{1}{2}L^2_{w_0 f_1 + \int_0^T w(t)\dot{g}_t\,dt}\widehat{\beta}(\hat{x}$$

$$- \frac{1}{2}\int_0^T w_0 w(t)L_{[\dot{g}_t\,,\,f_1]}\widehat{\beta}(\hat{x}_0)\, dt + \frac{1}{2}\int_0^T L_{\left[w(t)\dot{g}_t\,,\,w(t)g_t - \int_0^t w(s)\dot{g}_s\,ds\right]}\widehat{\beta}(\hat{x}_0)\, dt$$

$$= \frac{1}{2}\gamma''[\delta x]^2 + \frac{1}{2}w_0^2 L^2_{f_1}\widehat{\beta}(\hat{x}_0) + w_0\, L_{\delta x}L_{f_1}\widehat{\beta}(\hat{x}_0) + \frac{1}{2}\int_0^T w(t)^2 L_{[\dot{g}_t\,,\,g_t]}\widehat{\beta}(\hat{x}_0)\, dt +$$

$$+ \int_0^T L_{\left(\delta x + w_0 f_1 + \int_0^t w(s)\dot{g}_s\,ds\right)}L_{w(t)\dot{g}_t}\widehat{\beta}(\hat{x}_0)\, dt.$$

Notice that

$$\dot{g}_t = \widehat{S}_{t*}^{-1}[f_0, f_1] \circ \widehat{S}_t$$

$$[\dot{g}_t, g_t] = \widehat{S}_{t*}^{-1}[[f_0, f_1], f_1] \circ \widehat{S}_t$$

$$[\dot{g}_t, g_t] \cdot \widehat{\beta}(\hat{x}_0) = \langle \widehat{\lambda}(0), \widehat{S}_{t*}^{-1}[[f_0, f_1], f_1] \circ \widehat{S}_t(\hat{x}_0)\rangle = H_{110} \circ \widehat{\lambda}(t)$$

To simplify the notation set

$$\Gamma_0[\delta x, w_0]^2 = \frac{1}{2}\gamma''[\delta x]^2 + \frac{1}{2}w_0^2 L_{f_1}^2 \widehat{\beta}(\hat{x}_0) + w_0 \, L_{\delta x} L_{f_1}\widehat{\beta}(\hat{x}_0) \qquad (20)$$

$$R(t) = L_{[\dot{g}_t, g_t]}\widehat{\beta}(\hat{x}_0) \quad \text{and} \quad Q(t) = L_{(\cdot)} \, L_{\dot{g}_t} \, \widehat{\beta}(\hat{x}_0). \qquad (21)$$

Simple computations show that $\Gamma_0[\delta x, w_0]^2$ can be equivalently written as

$$\frac{1}{2}\gamma''[\delta x + w_0 f_1(\hat{x}_0)]^2 - \frac{1}{2}w_0^2 \, L_{f_1}^2 \alpha\,(\hat{x}_0) - w_0 \, L_{\delta x} L_{f_1}\alpha\,(\hat{x}_0). \qquad (22)$$

Since the map

$$v \in L^2([0, T], \mathbb{R}) \mapsto (w_0, w) \in \mathbb{R} \times L^2([0, T], \mathbb{R}) \qquad (23)$$

is continuous and has a dense image, we extend J'' to $T_{\hat{x}_0}N_0 \times \mathbb{R} \times L^2([0, T], \mathbb{R})$ by continuity and we call this extension the *extended second variation*. For all $\delta e = (\delta x, w_0, w)$ we set

$$J_E''[\delta e]^2 = \Gamma_0[\delta x, w_0]^2 + \frac{1}{2}\int_0^T w(t)^2 R(t) + 2w(t)\langle Q(t), \zeta(t)\rangle \, dt \qquad (24)$$

where $\zeta(t)$ is a solution of

$$\dot{\zeta}(t) = w(t)\dot{g}_t(\hat{x}_0) \qquad (25)$$

$$\zeta(0) = \delta x + w_0 f_1(\hat{x}_0), \quad \zeta(T) \in N_T''$$

J_E'' is a coordinate free version of a classical one obtained by using the so called Goh transformation, Ref. 11, see also Ref. 2 where the same second variation is given in Hamiltonian form for a problem with fixed end–points.

Theorem 5.1 (Necessary Conditions). *If $(\hat{u}, \hat{\xi})$ is a singular optimizer for the problem (17), then*

(1) The generalized Legendre condition (GLC) holds true

$$R(t) = H_{110}\left(\widehat{\lambda}(t)\right) = \langle \widehat{\lambda}(t), [f_1, [f_1, f_0]](\hat{\xi}(t))\rangle \geq 0.$$

(2) If the adjoint covector is unique up to a positive constant, then

$$J'' \geq 0 \quad \text{so that} \quad J_E'' \geq 0$$

by the properties of the embedding (23).

Proof. The proof from a geometric viewpoint can be found in Ref. 2: Corollary 20.18 pag.318 proves (1), while Proposition 21.5 proves (2). Remark that in Ref. 2 the sign of the adjoint covector is opposite to the one in this paper, hence all the inequalities are reversed.

However both conditions are classical results: see for example the survey paper [9, eq. (47), p. 141] for (1), and [13, Theorem 9.1, p. 285] for (2). $\square$

Now we analyze the extended second variation more deeply.

The case $f_1(\hat{x}_0) \notin T_{\hat{x}_0} N_0$.
In this case J_E'' is realized as a LQ control problem on $T_{\hat{x}_0} M$ with initial constraint given by $N_0'' = T_{\hat{x}_0} N_0 \oplus \mathbb{R} f_1(\hat{x}_0)$.

The case $f_1(\hat{x}_0) \in T_{\hat{x}_0} N_0$ and $L_{f_1}^2 \alpha(\hat{x}_0) = 0$.
In this case, which includes the case $f_1(\hat{x}_0) = 0$, we have two possibilities: either $L_{\delta x} L_{f_1} \alpha (\hat{x}_0) = 0$ for all $\delta e \in T_{\hat{x}_0} N_0 \times \mathbb{R} \times L^2([0,T],\mathbb{R})$ such that $\zeta(T) \in N_T''$ or there is δe such that $L_{\delta x} L_{f_1} \alpha (\hat{x}_0) \neq 0$ and $\zeta(T) \in N_T''$. In the first case, by (22), $\delta e = (f_1, -1, w \equiv 0)$ belongs to the kernel of J_E'', so that J_E'' cannot be coercive and its index and nullity can be checked on the quadratic form obtained by setting $w_0 = 0$. In the second one the index of J_E'' is infinite and $(\widehat{\xi}, \widehat{u})$ cannot be a minimizer, see for example Ref. 2.

The case $f_1(\hat{x}_0) \in T_{\hat{x}_0} N_0$ and $L_{f_1}^2 \alpha(\hat{x}_0) \neq 0$.
In this case

$$N_\Sigma'' = \{\delta x \in T_{\hat{x}_0} N_0 : L_{\delta x} L_{f_1} \alpha(\hat{x}_0) = 0\} \tag{26}$$

is a linear subspace and we get

$$T_{\hat{x}_0} N_0 = N_\Sigma'' \oplus \mathbb{R} f_1(\hat{x}_0).$$

Substituting $\delta x = \delta y + w_1 f_1(\hat{x}_0)$ in (22), with $\delta y \in N_\Sigma''$ and performing the substitution $(w_0 + w_1, w_1) \mapsto (v_0, v_1)$ we obtain

$$J_E''[\delta y, v_1, v_0, w]^2 = v_1^2 L_{f_1}^2 \alpha (\hat{x}_0) + \Gamma_0[\delta y, v_0]^2 +$$

$$+ \frac{1}{2} \int_0^T w(t)^2 R(t) + 2w(t)\langle Q(t), \zeta(t)\rangle dt \tag{27}$$

where

$$\dot{\zeta}(t) = w(t)\dot{g}_t(\hat{x}_0)$$

$$\zeta(0) = \delta y + v_0 f_1(\hat{x}_0), \quad \zeta(T) \in N_T''.$$

We underline that J_E'' is a direct sum of an LQ form on $\mathbb{R}$ and of an LQ form realized as a control problem on $T_{\hat{x}_0} M$.

Remark 5.2. The coercivity of J_E'' implies that SGLC must be satisfied. Moreover the sufficient condition implies $f_1(\hat{x}_0) \neq 0$, and if $f_1(\hat{x}_0) \in T_{\hat{x}_0} N_0$, then it splits in two conditions: $L_{f_1} L_{f_1} \alpha(\hat{x}_0) > 0$ and J_E'' coercive on $N_\Sigma'' \times \mathbb{R} \times L^2([0,T], \mathbb{R})$. When $f_1(\hat{x}_0) \notin T_{\hat{x}_0} N_0$, without loss of generality, we can assume $L_{f_1}^2 \alpha(\hat{x}_0) > 0$ and $L_{\delta x} L_{f_1} \alpha(\hat{x}_0) = 0$ for all $\delta x \in T_{\hat{x}_0} N_0$, see Remark 5.1.

5.1. *Reduction to a non-singular problem*

In this Subsection we suppose that either $f_1(\hat{x}_0) \notin T_{\hat{x}_0} N_0$, or that $L_{f_1}^2 \alpha(\hat{x}_0) \neq 0$ and we show that index and nullity of the extended second variation can be calculated via the standard second variation associated to a suitable non necessarily singular problem.

Before addressing the proof of the above statement we unify the notations denoting by $J_M'' : N_0'' \times L^2([0,T], \mathbb{R}) \to \mathbb{R}$ the following quadratic forms realized as an LQ-problem on $T_{\hat{x}_0} M$.

(1) If $f_1(\hat{x}_0) \notin T_{\hat{x}_0} N_0$, then we set

$$N_0'' = T_{\hat{x}_0} N_0 \oplus \mathbb{R} f_1(\hat{x}_0), \quad N_\Sigma = N_0$$

$$J_M''[\delta x + v_0 f_1(\hat{x}_0), w]^2 = J_E''[\delta x, v_0, w]^2$$

$$\tilde{N}_0 = \{\exp s f_1(y) : |s| < \bar{s}, \ y \in N_0\} \quad \text{and} \quad \tilde{\alpha}(\exp s f_1(y)) = \alpha(y).$$

(2) If $f_1(\hat{x}_0) \in T_{\hat{x}_0} N_0$ and $L_{f_1}^2 \alpha(\hat{x}_0) \neq 0$, then we set

$$N_0'' = N_\Sigma'' \oplus \mathbb{R} f_1(\hat{x}_0) \quad \text{and} \quad N_\Sigma = \{y \in N_0 : L_{f_1} \alpha(y) = 0\}$$

$$J_M''[\delta y + v_0 f_1(\hat{x}_0), w]^2 = J_E''[\delta y, v_0, w]^2$$

$$\tilde{N}_0 = \{\exp s f_1(y) : |s| < \bar{s}, \ y \in N_\Sigma\} \quad \text{and} \quad \tilde{\alpha}(\exp s f_1(y)) = \alpha(y).$$

With the above notations consider the non-singular problem

$$\text{minimize} \quad \tilde{\alpha}(\xi(0)) + \beta(\xi(T)) \quad \text{subject to} \tag{28}$$

$$\dot{\xi}(t) = \widehat{f_t}(\xi(t)) + w(t)[f_0, f_1](\xi(t)) + \tfrac{1}{2} w(t)^2 [f_1, [f_1, f_0]](\xi(t)) \tag{29}$$

$$\xi(0) \in \tilde{N}_0, \quad \xi(T) \in N_T. \tag{30}$$

Lemma 5.1. *J_M'' is the second variation of the above problem* (28) *subject to* (29) *and* (30), *relative to the reference couple* $(\widehat{\xi}, \widehat{w}(t) \equiv 0)$ *with associated adjoint covector* $\widehat{\lambda}$.

Proof. It is easy to see that the couple $(\widehat{\lambda}, \widehat{w}(t) \equiv 0)$ is an extremal. Since $\langle d\alpha(\hat{x}_0), f_1(\hat{x}_0) \rangle = 0$, then $d\tilde{\alpha}(\hat{x}_0) = d\alpha(\hat{x}_0)$, hence, denoting $\tilde{\gamma} = \tilde{\alpha} + \widehat{\beta}$,

$d\tilde{\gamma}(\hat{x}_0) = 0$. Set $\tilde{\gamma}'' = D^2\tilde{\gamma}(\hat{x}_0)$, the second variation as defined in Ref. 3 is given by

$$J''[\delta e]^2 = \frac{1}{2}\tilde{\gamma}''[\delta x]^2 + \frac{1}{2}\int_0^T w(t)^2 R(t) + 2w(t)\,Q(t)\zeta(t)\,dt \qquad (31)$$

where

$$\delta e = (\delta x, w) \in N_0'' \times L^2([0,T], \mathbb{R})$$
$$\dot{\zeta}(t) = w(t)\dot{g}_t(\hat{x}_0)$$
$$\zeta(0) \in N_0'', \quad \zeta(T) \in N_T''.$$

We can write $\delta x = \delta y + v_0 f_1(\hat{x}_0)$ with $L_{\delta y}L_{f_1}\alpha(\hat{x}_0) = 0$, so that

$$\tilde{\gamma}''[\delta x]^2 = \tilde{\gamma}''[\delta y]^2 + v_0^2\tilde{\gamma}''[f_1(\hat{x}_0)]^2 + 2v_0\tilde{\gamma}''(\delta y, f_1(\hat{x}_0))^2.$$

Since $d\tilde{\gamma}(\hat{x}_0) = 0$, then we have

$$\tilde{\gamma}''[f_1(\hat{x}_0)]^2 = L_{f_1}L_{f_1}(\tilde{\alpha} + \widehat{\beta})(\hat{x}_0)$$
$$\tilde{\gamma}''(f_1(\hat{x}_0), \delta y) = L_{\delta y}L_{f_1}(\tilde{\alpha} + \widehat{\beta})(\hat{x}_0)$$

and we obtain the result noting that $L_{f_1}\tilde{\alpha}(x) = 0$ for all $x \in \tilde{N}_0$. $\qquad\square$

Remark 5.3. Remark that SGLC is the strong Legendre condition for the above non-singular problem. Moreover we can always suppose that $L_{f_1}^2\alpha(\hat{x}_0) > 0$ and that $N_\Sigma = \{y \in \tilde{N}_0 : L_{f_1}\alpha(y) = 0\}$.

Remark 5.4. If the initial point is free and the second variation is coercive, then α and $\tilde{\alpha}$ are the same defined in Section 4.

Remark 5.5. The extended second variation is a quadratic form on a Hilbert space, hence its index and nullity can be studied using the large literature on the subject. In particular if SGLC is satisfied, then we deal with a non singular LQ problem on a vector space so that it is possible to calculate the index and the nullity of J_M'' through the idea of conjugate and semi-conjugate points, or by Riccati equation, see Ref. 20 and the references therein. In particular index and nullity can be calculated via the properties of the linear Hamiltonian flow $\mathcal{H}_t''$ associated to the quadratic Hamiltonian $H_t'' : T_{\hat{x}_0}^* M \times T_{\hat{x}_0} M \to \mathbb{R}$ defined by

$$H_t'' : (p, \delta x) \mapsto \frac{-1}{2R(t)}\left(\langle p, \dot{g}_t(\hat{x}_0)\rangle + L_{\delta x}L_{\dot{g}_t}\widehat{\beta}(\hat{x}_0)\right)^2 \qquad (32)$$

and starting from the Lagrangian subspace

$$L_0'' = \{(-\tilde{\gamma}(\delta x, \cdot), \delta x) : \delta x \in N_0''\} \oplus N_0''^{\perp}. \qquad (33)$$

6. Coercivity of J_M''

In this section we study the coercivity of J_M'' from a Hamiltonian point of view, in particular we apply the results of Ref. 3 to problem (28) and we transfer the results to the Hamiltonian H_t defined in (19) by linking the associated Hamiltonian flows.

The coercivity of J_M'' implies SGLC, then we can define the "minimizing Hamiltonian" associated to the non-singular problem (28) and defined in a neighborhood of $\mathcal{S}$, given by

$$\tilde{H}_t := \widehat{H}_t + \frac{H_{01}^2}{2\,H_{101}}. \tag{34}$$

$\tilde{H}_t$ and H_t coincide on $\mathcal{S}$ up to the third order, but the Hamiltonian vector field $\vec{\tilde{H}_t}$ is not tangent to Σ out from $\mathcal{S}$, nevertheless the associated flows $\tilde{\mathcal{H}}_t$ and $\mathcal{H}_t$ have the same tangent map at $\widehat{\ell}_0$, i.e. we can prove the following lemma.

Lemma 6.1. *Let $\tilde{\mathcal{H}}_t$ and $\mathcal{H}_t$ be respectively the flow of $\tilde{H}_t$ and H_t, then their tangent maps $\tilde{\mathcal{H}}_{t*}$ and $\mathcal{H}_{t*}$ coincide on $T_{\widehat{\lambda}(0)}T^*M$.*

Proof. The map $\tilde{\mathcal{H}}_t^{-1}\circ\mathcal{H}_t$ is the Hamiltonian flow associated to the Hamiltonian

$$\Phi_t = (\tilde{H}_t - H_t)\circ\tilde{\mathcal{H}}_t = \frac{H_{01}^2}{2}\left(\rho - \frac{1}{H_{101}}\right)\circ\tilde{\mathcal{H}}_t.$$

Since $d\Phi_t(\widehat{\lambda}(0)) = 0$, then it is easy to verify (see Ref. 14) that

$$(\tilde{\mathcal{H}}_t^{-1}\circ\mathcal{H}_t)_* = \tilde{\mathcal{H}}_{t*}\mathcal{H}_{t*} : T_{\widehat{\lambda}(0)}T^*M \to T_{\widehat{\lambda}(0)}T^*M$$

is a linear Hamiltonian flow associated to $D^2\Phi_t(\widehat{\lambda}(0))$, which is easily seen to be equal to zero. $\square$

The links between the quadratic Hamiltonian H_t'' associated to J_M'' and defined in (32) and the Hamiltonian H_t can be described via *the symplectic isomorphism* described in Ref. 3

$$\imath : T_{\widehat{x}_0}^*M \times T_{\widehat{x}_0}M \to T_{\widehat{\lambda}(0)}T^*M$$

$$(\delta p, \delta x) \mapsto \delta p + d(\widehat{\beta})_*\delta x \quad . \tag{35}$$

Lemma 6.2. *Let $\mathcal{G}_t$ be the* pull-back Hamiltonian flow *given by $\widehat{\mathcal{H}}_t^{-1}\circ\mathcal{H}_t$, then the equality $\mathcal{H}_t'' = \imath^{-1}\mathcal{G}_{t*}\imath$ holds through.*

Proof. Consider the pull-back Hamiltonian flow $\tilde{\mathcal{G}}_t = \widehat{\mathcal{H}}_t^{-1} \circ \tilde{\mathcal{H}}_t$, Corollary 3, pag. 702 in Ref. 3 reads in our case $\mathcal{H}_t'' = \imath^{-1}\tilde{\mathcal{G}}_{t*}\imath$, hence we get the statement by Lemma 6.1. $\qquad\qquad\square$

Using the symplectic isomorphism defined in (35) and the results in Ref. 3 we state the conditions for J_M'' to be coercive in terms of properties of the flow $\mathcal{H}_{t*}$ starting from the subspace L_0 of $T_{\hat{\ell}_0}T^*M$ defined by

$$L_0 = \imath L_0'' = \left\{-d\tilde{\alpha}_*\delta x : \delta x \in T_{\hat{x}_0}\tilde{N}_0\right\} \oplus T_{\hat{x}_0}^{\perp}\tilde{N}.$$

Using the results in Ref. 20 and the techniques of Ref. 3 we can prove the following result.

Lemma 6.3. J_M'' *is coercive if and only if the following properties hold true*

(1) If $\pi_*\mathcal{H}_{t*}\delta\ell = 0$, $\delta\ell \in L_0$, then $\pi_*\mathcal{H}_{s*}\delta\ell = 0$ and $\mathcal{H}_{s*}\delta\ell = \widehat{\mathcal{H}}_{s*}\delta\ell$ for all $s \in [0, t]$.

(2) $\sigma\left(d\beta_*\pi_*\mathcal{H}_{T*}\delta\ell, \mathcal{H}_{T*}\delta\ell\right) > 0$ for all $\delta\ell \in L_0$ such that $\pi_*\mathcal{H}_{T*}\delta\ell$ is a nonzero element of $T_{\hat{x}_T}N_T$.

Proof. Denote by $\mathcal{V}$ the space of the first order variations with fixed final point, i.e.

$$\mathcal{V} = \{\delta e = (\delta x, w) \in N_0'' \times L^2([0, T], \mathbb{R}) : \zeta(T) = 0\}.$$

J_M'' is coercive if and only if it is coercive on $\mathcal{V}$ and on the space $\mathcal{V}^{\perp}$ orthogonal to $\mathcal{V}$ with respect to J_M''. Recall that $\mathcal{V}^{\perp}$ is finite dimensional. Applying Corollary 5 in Ref. 3 to a problem with fixed final point we get that $J_{M|\mathcal{V}}''$ is coercive if and only if

$$\pi_*\mathcal{H}_t''\ell'' = 0, \ \ell'' \in L_0'' \ \Rightarrow \ \mathcal{H}_s''\ell'' = \ell'', \ \forall s \in [0, t]. \tag{36}$$

Noting that $\widehat{\mathcal{H}}_{t*}\delta\ell$ is vertical if and only if $\delta\ell$ is and using Lemma 6.1, we get that $J_{M|\mathcal{V}}''$ is coercive if and only if the kernel of $\pi_*\tilde{\mathcal{H}}_{t*}$ is given by

$$\{\delta\ell \in L_0 : \pi_*\tilde{\mathcal{H}}_{s*}\delta\ell = 0, \ \tilde{\mathcal{H}}_{s*}\delta\ell = \widehat{\mathcal{H}}_{s*}\delta\ell, \ \forall s \in [0, t]\},$$

hence $J_{M|\mathcal{V}}''$ is coercive if and only if *(1)* holds true.

To prove *(2)* consider that $\delta e = (\delta x, w)$ belongs to $\mathcal{V}^{\perp}$ if and only if there are $p_0 \in N_0''^{\perp}$ and $p_T \in T_{\hat{x}_0}^*M$ such that

$$J_M''(\delta e, \bar{\delta} e) = \frac{1}{2}\langle p_0, \bar{\delta} x\rangle + \frac{1}{2}\langle p_T, \bar{\zeta}(T)\rangle, \quad \forall \bar{\delta} e = (\bar{\delta} x, \bar{w}) \in T_{\hat{x}_0}M \times L^2([0, T], \mathbb{R}).$$

Denote $\ell'' = (\delta p, \delta x) = (-\tilde{\gamma}(\delta x, \cdot) + p_0, \delta x) \in L_0''$, $\zeta(t)$ the solution of (25) with initial condition $\zeta(0) = \delta x$ and $\mu(t)$ the solution of

$$\dot{\mu}(t) = -w(t)Q(t), \quad \mu(0) = \delta p.$$

Integrating by parts we obtain that $\delta e \in \mathcal{V}^\perp$ if and only if for all $\bar{\delta e}$

$$\tfrac{1}{2}\langle p_T \bar{\zeta}(T)\rangle = -\tfrac{1}{2}\langle \mu(T)\bar{\zeta}(T)\rangle +$$
$$+\tfrac{1}{2}\int_0^T \bar{w}(t)\left(w(t)R(t) + Q(t)\zeta(t) + \mu(t)\dot{g}_t(\hat{x}_0)\right) dt,$$

that is if and only if

$$\mu(T) = -p_T, \quad w(t) = -\frac{Q(t)\zeta(t) + \langle \mu(t), \dot{g}_t(\hat{x}_0)\rangle}{R(t)},$$

so that $(\mu(t), \zeta(t)) = \mathcal{H}_t'' \ell''$.

Reasoning as in the proof of Lemmas 4 and 5 in Ref. 3 we can prove that $\delta e \in \mathcal{V}^\perp$ if and only if there is $\ell'' \in L_0''$ such that

$$\pi_* \ell'' = \delta x, \quad \text{and} \quad \pi_* \mathcal{H}_T'' \ell'' \in N_T'',$$

moreover in this case

$$J_M''[\delta e]^2 = -\frac{1}{2}\langle \mu(T)\zeta(T)\rangle = -\frac{1}{2}\,\sigma\left(\mathcal{H}_T''\ell'', (0, \pi\mathcal{H}_T''\ell'')\right).$$

On the other hand if $\zeta(T) = \pi_* \mathcal{H}_T'' \ell'' = 0$, then by (36) we get $\zeta(t) \equiv 0$, so that $\delta x = 0$ and $w(t)\dot{g}_t(\hat{x}_0) = 0$, $a.e.\, t \in [0, T]$, hence w is the zero element of $L^2([0, T], \mathbb{R})$, since $\ddot{g}_t$ cannot be zero in a subinterval of $[0, T]$.

The proof of *(2)* follows easily applying the symplectic isomorphism, the invariance of σ with respect to $\widehat{\mathcal{H}}_{t*}$ and Lemma 6.1. $\qquad\square$

For the free initial point case, the above conditions *(1)* and *(2)* coincide with conditions of Theorem 4.1, namely we get:

Corollary 6.1. *If $N_0 = M$, then J_M'' is coercive if and only if*

(1) $\pi_* \mathcal{H}_{t*} : T_{\widehat{\lambda}(0)}\Lambda_0^\alpha \to T_{\widehat{\xi}(t)}M$ *is an isomorphism for all $t \in [0, T]$.*
(2) *The quadratic map*

$$\delta x \mapsto \sigma\left(d\beta_* \delta x, \mathcal{H}_{T*}(\pi_* \mathcal{H}_{T*})^{-1}\delta x\right)$$

is positive definite on $T_{\widehat{\xi}(T)} N_T$

Proof. *(1)* follows easily noting that $L_0 = T_{\widehat{\lambda}(0)}\Lambda_0^\alpha$. To obtain *(2)* it is sufficient to substitute $\delta x = \pi_* \mathcal{H}_{T*}\delta\ell$. $\qquad\square$

6.1. *Proof of the Main Theorem*

In this section we assume that the extended second variation is coercive, therefore we assume SGLC, J_M'' coercive and $L_{f_1} L_{f_1} \alpha(\hat{x}_0) > 0$, see Remark 5.2. We show that under the above stated assumptions we can find a function α_s which coincides with α on N_0 and satisfies the assumptions of Theorem 4.1 with the Hamiltonian H_t defined in equation (19).

If $N_0 = M$ we have nothing to do, otherwise since $f_1(\hat{x}_0) \neq 0$, then we can choose coordinates $(x_1, \cdots, x_n)$ at $\hat{x}_0$ such that

$$f_1 = \frac{\partial}{\partial x_1} \text{ and } \tilde{N}_0 \text{ is defined by } x_{r+1} = \cdots = x_n = 0. \tag{37}$$

Since the second variation does not depend on α, β with the properties (15), (16), then without loss of generality we can assume

$$\tilde{\alpha}(x) = \alpha(0, x_2, \cdots, x_r, 0, \cdots, 0) + \sum_{i=r+1}^{n} \partial_i \alpha(\hat{x}_0) x_i. \tag{38}$$

In these coordinates choose the non-negative quadratic form on $T_{\hat{x}_0} M$

$$\Gamma = \frac{1}{2} \sum_{i=r+1}^{n} dx_i \otimes dx_i$$

and extend it to $T_{\hat{x}_0} M \times L^2([0, T], \mathbb{R})$ by $\Gamma[\delta e]^2 = \Gamma[\delta x]^2$.

On the other hand we consider J_M'' as defined on the whole space $T_{\hat{x}_0} M \times L^2([0, T], \mathbb{R})$, and, since J_M'' is coercive on the kernel of Γ, we can apply [12, Theorem 13.2] and conclude that there is a positive constant s such that

$$J_M'' + s\Gamma \text{ is coercive on } \mathcal{W} = \{\delta e \in T_{\hat{x}_0} M \times L^2([0, T], \mathbb{R}) : \zeta(T) \in N_T''\}.$$

Defining

$$\alpha_s(x) = \alpha(x_1, \cdots, x_r, 0, \cdots, 0) + \sum_{i=r+1}^{n} \left(\partial_i \tilde{\alpha}(\hat{x}_0) x_i + \frac{1}{2} s\, x_i^2 \right),$$

we get

(1) $L_{f_1}^2 \alpha_s(\hat{x}_0) = L_{f_1}^2 \alpha(\hat{x}_0) > 0$
(2) $d\alpha_s(\hat{x}_0) = -\widehat{\lambda}(0)$
(3) $\tilde{\alpha}_s = \tilde{\alpha}$ on $\tilde{N}_0$
(4) $D^2(\tilde{\alpha}_s + \widehat{\beta})(\hat{x}_0) = \tilde{\gamma}'' + s\Gamma.$

Corollary 6.1 proves that the assumptions of Theorem 4.1 are satisfied and hence the main theorem is proved.

6.2. *A particular case*

A particular case occurs when the Hamiltonian of singular extremals H_S as defined in Subsection 2.3 is the lifting of a vector field f_S, see Remark 2.1. In this case we can give conditions for the extended second variation to be coercive in terms of the vector fields f_S and f_1. In particular the following Lemma gives necessary and sufficient conditions for the first property of Lemma 6.3 to be verified and describes $\pi_* \mathcal{H}_{T*} L_0$, so that the second property of the same lemma can be verified in any chart at $\widehat{\xi}(T)$.

Lemma 6.4. *Let the Hamiltonian H_S be the lift of the vector field f_S.*

(1) If $[f_S, f_1](\hat{x}_0) \in N_0''$, then $J_{M|\mathcal{V}}''$ is coercive if and only if for all $t \in (0, T]$
$$f_1(\widehat{\xi}(t)) \notin (\exp t f_S)_* V \quad where$$

$$V = \{v \in N_0'' : L_v L_{[f_S, f_1]} \tilde{\alpha}(\hat{x}_0) = L_v L_{f_{01}} \tilde{\alpha}(\hat{x}_0) = 0\}.$$

Moreover $\pi_* \mathcal{H}_{T*} L_0$ is the linear space spanned by $f_1(\widehat{\xi}(T))$ and $(\exp T f_S)_* V$.

(2) $[f_S, f_1](\hat{x}_0) \notin N_0''$, then $J_{M|\mathcal{V}}''$ is coercive if and only if for all $t \in (0, T]$

$$f_1(\widehat{\xi}(t)) \notin (\exp t f_S)_* N_0''.$$

Moreover $\pi_* \mathcal{H}_{T*} L_0$ is the linear space spanned by $f_1(\widehat{\xi}(T))$ and $(\exp T f_S)_* N_0''$.

Proof. Without loss of generality we can assume that $f_S = f_0$, see Remark 2.1, so that $\vec{H}_t = \vec{H}_0 + \chi$ and $\vec{H}_{t|S} = \vec{H}_{0|S}$ is tangent to $\mathcal{S}$.

Consider the space $V = \{v \in N_0'' : L_v L_{f_{01}} \tilde{\alpha}(\hat{x}_0) = 0\}$ and observe that, $N_0'' = \mathbb{R} f_1(\hat{x}_0) \oplus V$, therefore we can choose coordinates $(x_1, \cdots, x_n)$ at $\hat{x}_0$, with the properties (37), (38) and such that V is spanned by ∂_i, $i = 2, .., r$. In these coordinates $\vec{H}_1(\widehat{\lambda}(0)) = d\tilde{\alpha}_* \partial_1$, $d\tilde{\alpha}_* \partial_i \in \mathcal{S}$, $i = 2, .., r$, and

$$L_0 = span\{\vec{H}_1, d\tilde{\alpha}_* \partial_i, dx_j : i = 1, .., r, \ j = r+1, .., n\}.$$

(1) If $f_{01}(\hat{x}_0) \in N_0''$, then $dx_j \in \mathcal{S}$, $j = r+1, .., n$, so that $\pi_* \mathcal{H}_{t*} dx_j = 0$ and

$$\pi_* \mathcal{H}_{t*} L_0 = \mathbb{R} f_1(\widehat{\xi}(t)) + (\exp t f_0)_* V.$$

Therefore Lemma 6.3 proves the statements for the first case since if $\delta\ell \in span\{d\tilde{\alpha}_* \partial_i : i = 1, .., r\}$ belongs to $\ker \pi_* \mathcal{H}_{t*}$ and $\pi_* \delta\ell = 0$, then $\delta\ell = 0$.

(2) If $f_{01}(\hat{x}_0) \notin N_0''$, then we can suppose without loss of generality, $f_{01}(\hat{x}_0) = \partial_{r+1}$ so that $dx_j \in \mathcal{S}$ for all $j = r + 2, .., n$ and $dx_{r+1} = \delta\ell + \gamma\, d\tilde{\alpha}_* \partial_1$, with $\delta\ell \in \mathcal{S}$ and $\gamma = 1/H_{101}(\widehat{\lambda}(0))$. In this case $\pi_* \mathcal{H}_{t*} dx_j = 0$, for all $j = r + 2, .., n$, $\pi_* \mathcal{H}_{t*} dx_{r+1} = -\gamma(\exp t f_0)_* \partial_1$ and

$$\pi_* \mathcal{H}_{t*} L_0 = \mathbb{R} f_1(\widehat{\xi}(t)) + (\exp t f_0)_* N_0''.$$

Therefore Lemma 6.3 proves the statements for the second case since if $\delta\ell \in span\{d\tilde{\alpha}_* \partial_i, dx_{r+1} : i = 1, .., r\}$ belongs to $\ker \pi_* \mathcal{H}_{t*}$ and $\pi_* \delta\ell = 0$, then $\delta\ell = 0$. $\qquad\square$

7. Final remarks

The paper proves second order sufficient optimality conditions for the strong local optimality of a singular trajectory $\widehat{\xi}$, that is the optimality is with respect to the admissible trajectories belonging to a neighborhood of $\widehat{\xi}$ in $\mathcal{C}^0([0, T], M)$, independently on the values of the control. Remark that as a consequence we obtain that the trajectory is strong locally optimal even if the controls are unbounded. The proof is obtained in the framework of Hamiltonian approach to optimal control and points out the connections between the coordinate-free extended second variation and the modified minimizing Hamiltonian.

The ideas of the paper have been applied to prove strong local minimum time optimality of a singular trajectory,[19] and of a bang-singular trajectory,[16][17] Strong local minimum time optimality may be of two kinds: either local with respect to both state and time (i.e. with respect to a neighborhood of the graph of the reference trajectory in $\mathbb{R} \times M$), or local only with respect to the state (i.e. with respect to a neighborhood of the range of the reference trajectory in M).

The optimality considered in Refs. 16,17,19 is local with respect to both state and time, moreover the conditions for the bang-singular trajectories are not completely satisfactory. The work is in progress both to extend the conditions to the state local optimality and to improve the conditions for a reference extremal which contains bang arcs and singular arcs.

A further research direction is to consider singular (or partially singular) trajectories of multi-input systems. The geometric picture is clear when only one control is singular and the others are bang: the result will appear elsewhere. It is the opinion of the author that the theory can be extended to totally singular control if the controlled vector fields commute, while the general case seems to be much more difficult.

Finally it is the opinion of the author that the theory can give a deeper insight to understand the possible sufficient conditions for the optimality of singular extremals of the second kind.

References

1. A. A. Agrachev and R. V. Gamkrelidze, Symplectic methods for optimization and control, in *Geometry of Feedback and Optimal Control*, eds. B. Jacubczyk and W. Respondek, Monogr. Textbooks Pure Appl. Math. Vol. 207 (Dekker, New York, 1998), pp. 19–77.
2. A. A. Agrachev and Y. L. Sachkov, *Control theory from the geometric viewpoint*, Encyclopaedia of Mathematical Sciences, Vol. 87. Control Theory and Optimization, II. (Springer-Verlag, Berlin, 2004).
3. A. A. Agrachev, G. Stefani, and P. Zezza, *Internat. J. Control* **71**, 689 (1998).
4. A. A. Agrachev, G. Stefani, and P. Zezza, *Proceedings of the Steklov Mathematical Institute* **220**, 4 (1998).
5. A. A. Agrachev, G. Stefani, and P. Zezza, *SIAM J. Control Optimization* **41**, 991 (2002).
6. V. I. Arnold, *Mathematical methods of classical mechanics*. Translated from the Russian by K. Vogtmann and A. Weinstein. Graduate Texts in Mathematics, Vol. 60, second edition (Springer-Verlag, New York, 1989).
7. A.V. Dmitruk, English translation: *Soviet Math. Dokl.* **18**, (1977).
8. A.V. Dmitruk, *Dokl. Akad. Nauk SSSR* **272**, 285 (1983) [in Russian]. English translation: *Soviet Math. Dokl.* **28**, 275 (1983).
9. V. Gabasov, F.M. Kirillova, *SIAM J. Control Optimization* **10**, 127 (1972).
10. M. Giaquinta and S. Hildebrandt, *Calculus of variations, I and II*, Grundl. der Mathemat. Wissensch. **310**, **311** (Springer, Berlin, 1996).
11. B.S. Goh, *SIAM J. Control Optimization* **4**, 309 (1966).
12. M. R. Hestenes, *Pac. J. Math.* **1**, 525 (1951).
13. M. R. Hestenes, *Calculus of variations and optimal control theory* (John Wiley & Sons, New York, 1966).
14. J.E. Marsden and T.S. Ratiu, *Introduction to Mechanics and Symmetry*, A basic exposition of classical mechanical systems. Texts in Applied Mathematics, Vol. 17, (Springer-Verlag, New York, 1994).
15. Z. Páles and V. Zeidan, *SIAM J. Control Optimization* **32**, 1476 (1994).
16. L. Poggiolini and G. Stefani, Minimum time optimality for a bang-singular arc: second order sufficient conditions, in *Proc. 44th IEEE Conf. Decision and Control, and the European Control Conference ECC 2005 (Seville, Spain, 2005)*.
17. L. Poggiolini and G. Stefani, Minimum time optimality of a partially singular arc: second order sufficient conditions, in *Lagrangian and Hamiltonian Methods for Nonlinear Control 2006*, eds. F. Bullo and K. Fujimoto, Lecture Notes in Control and Inform. Sci., Vol. 366 (Springer, Berlin, 2006), pp. 281–291.

18. G. Stefani, On sufficient optimality conditions for a singular extremal, in *Proc. 42nd IEEE Conf. Decision and Control (Maui, Hawaii, 2003)*: 2746-2749.

19. G. Stefani, Minimum-time optimality of a singular arc: second order sufficient conditions, in *Proc. 43rd IEEE Conf. Decision and Control (Atlantis, Paradise Island, Bahamas, 2004)*: 450-454.

20. G. Stefani and P. Zezza, *SIAM J. Control Optim.* **35**, 876 (1997).

21. M. I. Zelikin and V. F. Borizov, *Theory of Chattering Control*, With applications to astronautics, robotics, economics, and engineering. Systems & Control: Foundations & Applications (Birkhäuser Boston Inc., Boston, MA, 1994).

HIGH-ORDER POINT VARIATIONS
AND GENERALIZED DIFFERENTIALS

H. J. SUSSMANN*

*Department of Mathematics
Rutgers, the State University of New Jersey
Hill Center—Busch Campus
110 Frelinghuysen Road
Piscataway, NJ 08854-8019, USA
E-mail: sussmann@math.rutgers.edu
http://www.math.rutgers.edu*

In a series of nonsmooth versions of the Pontryagin Maximum Principle, we used generalized differentials of set-valued maps, flows, and abstract variations. Bianchini and Stefani have introduced a notion of possibly high-order variational vector that has the summability property. We consider a slightly more general class of variational vectors than that defined by Bianchini and Stefani, and prove that the convex combinations of these vectors arise as "differentials" of variations that are differentiable in the sense of one of our generalized differentiation theories, namely, that of "approximate generalized differential quotients" (AGDQs).

1. Introduction

In a series of papers (cf. Refs. 5–7,9,10), we showed how to derive general, nonsmooth versions of the Pontryagin Maximum Principle using generalized differentials of set-valued maps, flows, and abstract point variations. The use of general variations rather than the nedle variations used to prove the ordinary maximum principle makes it possible to obtain high-order versions of the maximum principle. The main technical difficulty with these general abstract variations is that they need not have the summability property, which is absolutely essential in order to derive the necessary conditions for optimality.

R. M. Bianchini and G. Stefani (cf. Refs. 1–4) proposed a concept of high-order point variation that has good summability properties. The goal

---

*Supported in part by NSF Grant DMS-05-09930

of this note is to relate this concept to a theory of generalized differentials, by describing a slightly more general version of the Bianchini-Stefani variations, and showing that they are differentiable in the precise sense of the theory of "Approximate Generalized Differential Quotients" (AGDQs). This makes it possible to use these variations in order to get additional necessary conditions for an optimum in situations such as the very general one described in Ref. 9, where the differentials involved are generalized differential quotients, and *a fortiori* AGDQs.

1.1. *Preliminary remarks on notation*

We will use the notations and abbreviations of Ref. 9. In particular, "FDRLS" stands for "finite-dimensional real linear space," "FDNRLS" for "normed FDRLS," and "SVM" for "set-valued map." If f is a SVM, then $\mathrm{So}(f)$, $\mathrm{Ta}(f)$, $\mathrm{Gr}(f)$, $\mathrm{Do}(f)$, $\mathrm{Im}(f)$ are, respectively, the source, target, graph, domain and image of a SVM f. (We recall that a **SVM** is a triple (A, B, G) such that A, B are sets and G is a subset of $A \times B$, in which case we say that f is a **SVM from A to B**, and define $G^{-1} \overset{\mathrm{def}}{=} \{(x, y) : (y, x) \in G\}$, $f^{-1} \overset{\mathrm{def}}{=} (B, A, G^{-1})$, $\mathrm{So}(f) \overset{\mathrm{def}}{=} A$, $\mathrm{Ta}(f) \overset{\mathrm{def}}{=} B = \mathrm{So}(f^{-1})$, $\mathrm{Gr}(f) = G$, $f(x) = \{y : (x, y) \in \mathrm{Gr}(f)\}$, $\mathrm{Do}(f) = \{x : f(x) \neq \emptyset\}$, $\mathrm{Im}(f) = \mathrm{Do}(f^{-1})$.) We use $SVM(A, B)$ to denote the set of all set-valued maps from A to B. The notation "$f : A \mapsto B$" means "f is a set-valued map from A to B." If $f \in SVM(A, B)$ then f is (i) **single-valued** if the set $f(x)$ consists of a single member for every $x \in \mathrm{Do}(f)$, (ii) **one-to-one** if f^{-1} is single-valued, (iii) **surjective** if $\mathrm{Im}(f) = \mathrm{Ta}(f)$, (iv) **everywhere defined** if $\mathrm{Do}(f) = \mathrm{So}(f)$, i.e., if f^{-1} is surjective, (v) **a ppd map** (where "ppd" stands for "possibly partially defined") if it is single-valued, and (vi) **an ordinary map** if it is an everywhere defined ppd map. The notation "$f : A \hookrightarrow B$" means "f is a ppd map from A to B."

If S is a set, then $\mathbb{I}_S$ is the identity map of S, i.e., the triple (S, S, Δ_S), where $\Delta_S = \{(x, x) : x \in S\}$.

The abbreviation "CCA" stands for for "Cellina continuously approximable." (We recall that a **CCA map** from a metric space X to a metric space Y is a set-valued map $F : X \mapsto Y$ such that, for every compact subset K of X, (i) the set $(K \times Y) \cap \mathrm{Gr}(F)$ is compact, and (ii) there exists a sequence $\{F_j\}_{j=1}^{\infty}$ of single-valued continuous maps from K to Y such that the graphs $\mathrm{Gr}(F_j)$ converge to $\mathrm{Gr}(F)$, in the sense that

$$\lim_{j \to \infty} \sup\{\mathrm{dist}_{X \times Y}(q, \mathrm{Gr}(F)) : q \in \mathrm{Gr}(F_j)\} = 0.$$

(A detailed study of CCA maps appears in Ref. 9.) We use $CCA(X,Y)$ to denote the set of all CCA maps from X to Y.

If I is a totally ordered set, then we use $\leq_I$ to denote the order relation on I, and simply write $\leq$ when the context makes I unambiguous. Also, "$a <_I b$"—or, simply, "$a < b$"—means "$a \leq_I b$ and $a \neq b$." A **subinterval** of I is a subset J of I such that $c \in J$ whenever $a \in J$, $b \in J$, $c \in I$, and $a \leq c \leq b$. We use square bracket notation for subintervals of I that have an infimum and a supremum in I. (That is, if $a, b \in I$ and $a \leq b$, we write $]a, b[_I \overset{\text{def}}{=} \{t \in I : a < t < b\}$, $[a, b[_I \overset{\text{def}}{=} \{a\} \cup]a, b[_I$, $]a, b]_I \overset{\text{def}}{=}]a, b[_I \cup \{b\}$, and $[a, b]_I \overset{\text{def}}{=} \{a, b\} \cup]a, b[_I$, and we omit the subscript when I is uniquely determined by the context.). Then every subinterval of I that has an infimum and a supremum in I is of one of the forms $]a, b[$, $[a, b]$, $[a, b[$, $]a, b]$). When the totally ordered set is not specified, it is understood that it is the extended real line $\bar{\mathbb{R}} \overset{\text{def}}{=} \mathbb{R} \cup \{-\infty, +\infty\}$. We define $\mathbb{R}_+ \overset{\text{def}}{=} \{x \in \mathbb{R} : x \geq 0\}$ and $\mathbb{R}_{+,>} \overset{\text{def}}{=} \{x \in \mathbb{R} : x > 0\}$, and let $\bar{\mathbb{R}}_+ \overset{\text{def}}{=} \mathbb{R}_+ \cup \{+\infty\}$.

We use Θ to denote the class of all functions $\theta : \mathbb{R}_{+,>} \mapsto \bar{\mathbb{R}}_+$ such that

- θ is monotonically nondecreasing (that is, $\theta(s) \leq \theta(t)$ whenever $s, t \in \mathbb{R}$ are such that $0 \leq s \leq t < +\infty$);
- $\lim_{s \downarrow 0} \theta(s) = 0$.

If X is a FDNRLS, $x_* \in X$, $r > 0$, then $\mathbb{B}_X(x_*, r)$, $\bar{\mathbb{B}}_X(x_*, r)$ are, respectively, the open ball $\{x \in X : \|x - x_*\| < r\}$ and the closed ball $\{x \in X : \|x - x_*\| \leq r\}$. If X, Y are FDRLSs, then $Lin(X, Y)$, $Aff(X, Y)$ will denote, respectively, the set of all linear maps and the set of all affine maps from X to Y. By definition, the members of $Aff(X, Y)$ are the maps $affm_{L,h}$, for $L \in Lin(X, Y)$, $h \in Y$, where $affm_{L,h}$ denotes the **affine map with linear part L and constant part h**, defined by $affm_{L,h}(x) \overset{\text{def}}{=} L \cdot x + h$. We identify $Aff(X, Y)$ with $Lin(X, Y) \times Y$ by identifying each map $affm_{L,h} \in Aff(X, Y)$ with the pair $(L, h) \in Lin(X, Y) \times Y$.

If X and Y are FDNRLSs, then we endow $Lin(X, Y)$ with the operator norm $\|\cdot\|_{op}$ given by $\|L\|_{op} = \sup\{\|Lx\| : x \in X, \|x\| \leq 1\}$, so $Lin(X, Y)$ is a FDNRLS as well. Also, we endow the linear space $Aff(X, Y)$ with the norm given by $\|affm_{L,h}\| = \|L\| + \|h\|$.

If Λ is subset of $Lin(X, Y)$, and $\delta \in \mathbb{R}_{+,>}$, we define

$$\Lambda^\delta = \{L \in Lin(X, Y) : \text{dist}(L, \Lambda) \leq \delta\},$$

where $\text{dist}(L, \Lambda) = \inf\{\|L - L'\|_{op} : L' \in \Lambda\}$. Also, if $\delta, \varepsilon \in \mathbb{R}_{+,>}$, and we still assume that $\Lambda \subseteq Lin(X,Y)$, we let

$$\Lambda^{(\delta,\varepsilon)} = \{ \mathit{affm}_{L,h} : L \in Lin(X,Y), \, \mathrm{dist}(L,\Lambda) \leq \delta, \, h \in Y, \, \|h\| \leq \delta\varepsilon \} \, ,$$

Notice that if $L \in Lin(X,Y)$, then $\mathrm{dist}(L,\emptyset) = +\infty$. In particular, if $\Lambda = \emptyset$ then $\Lambda^{\delta} = \emptyset$. and $\Lambda^{(\delta,\varepsilon)} = \emptyset$. Notice also that if Λ is compact (resp. convex) then Λ^{δ} and $\Lambda^{(\delta,\varepsilon)}$ are compact (resp. convex).

If X is a FDRLSs, then we use $X^{\dagger}$ to denote the dual space of X, i.e., the space $Lin(X,\mathbb{R})$.

The word "manifold" will mean "finite-dimensional paracompact differentiable manifold without boundary." If M is a manifold of class C^1, and $x \in M$, then $T_x M$, $T_x^* M$ denote, respectively, the tangent and cotangent space of M at x.

1.2. *Approximate Generalized Differential Quotients*

Definition 1.1. Assume that X, Y are FDNRLSs, $F : X \longmapsto Y$ is a set-valued map, Λ is a compact subset of $Lin(X,Y)$, $\bar{x}_* \in X$, $\bar{y}_* \in Y$, and $S \subseteq X$. We say that Λ is an **approximate generalized differential quotient of** F **at** $(\bar{x}_*, \bar{y}_*)$ **in the direction of** S—and write $\Lambda \in AGDQ(F, \bar{x}_*, \bar{y}_*, S)$—if there exists a function $\theta \in \Theta$—called an **AGDQ modulus for** $(\Lambda, F, \bar{x}_*, \bar{y}_*, S)$—having the property that

(*) *for every $\varepsilon \in \mathbb{R}_{+,>}$ such that $\theta(\varepsilon) < \infty$ there exists a set-valued map $A^{\varepsilon} \in CCA(\bar{\mathbb{B}}_X(\bar{x}_*,\varepsilon) \cap S, \mathit{Aff}(X,Y))$, with values in $\Lambda^{(\theta(\varepsilon),\varepsilon)}$, such that $\bar{y}_* + A(x - \bar{x}_*) \in F(x)$ whenever $x \in \bar{\mathbb{B}}_X(\bar{x}_*,\varepsilon) \cap S$ and $A \in A^{\varepsilon}(x)$.* □

1.2.1. *Properties of AGDQs*

If A, B, C are sets, and Ξ, Z are sets of maps from A to B and from B to C, respectively, then the **composite** $Z \circ \Xi$ is the set of maps from A to C given by $Z \circ \Xi = \{ \zeta \circ \xi : \zeta \in Z, \, \xi \in \Xi \}$.

The following statement, proved in Ref. 9, is the **chain rule** for AGDQs.

Theorem 1.1. *For $i = 1, 2, 3$, let X_i be a FDNRLS, and let $\bar{x}_{*,i}$ be a point of X_i. Assume that, for $i = 1, 2$, (i) $F_i : X_i \longmapsto X_{i+1}$ is a set-valued map, (ii) S_i is a subset of X_i, and (iii) $\Lambda_i \in AGDQ(F_i, \bar{x}_{*,i}, \bar{x}_{*,i+1}, S_i)$. Assume, in addition, that (iv) $F_1(S_1) \subseteq S_2$, and either (v) S_2 is a local quasiretract (cf. Remark 1.1) of X_2 at $\bar{x}_{*,2}$ or (v') there exists a neighborhood U of $\bar{x}_{*,1}$ in X_1 such that the restriction $F_1 \lceil (U \cap S_1)$ of F_1 to $U \cap S_1$ is single-valued. Then $\Lambda_2 \circ \Lambda_1 \in AGDQ(F_2 \circ F_1, \bar{x}_{*,1}, \bar{x}_{*,3}, S_1)$.* □

Remark 1.1. The notion of a "local quasiretract" is defined in Ref. 9. The precise definition is as follows. First, if T is a topological space and $S \subseteq T$, we say that S is a **quasiretract of** T if for every compact subset K of S there exist a neighborhood U of K and a continuous map $\rho : U \mapsto S$ such that $\rho(s) = s$ for every $s \in K$. Then, if $S \subseteq T$ and $\bar{s} \in S$, we say that S is a **local quasiretract of** T **at** $\bar{s}$ if there exists a neighborhood U of $\bar{s}$ such that $S \cap U$ is a quasiretract of U.

An important example of a local quasiretract of a manifold M at a point $s \in M$ is a subset S of M such that, for some open neighborhood U of s, the set $S \cap U$ is the image of a convex subset of an open neighborhood V of 0 in $\mathbb{R}^{\dim M}$ under a diffeomorphism Φ of class C^1 from V onto U such that $\Phi(0) = s$. In particular, any set whose germ at s is, relative to some coordinate chart near s, the germ at s of a convex subset of $\mathbb{R}^{\dim M}$, is a local quasiretract of M at s. $\qquad\square$

If M and N are manifolds of class C^1, $\bar{x}_* \in M$, $\bar{y}_* \in N$, $S \subseteq M$, and $F : M \mapsto N$, then it is possible to define a set $AGDQ(F, \bar{x}_*, \bar{y}_*, S)$ of compact subsets of the space $Lin(T_{\bar{x}_*} M, T_{\bar{y}_*} N)$ of linear maps from $T_{\bar{x}_*} M$ to $T_{\bar{y}_*} N$ as follows. We let $m = \dim M$, $n = \dim N$, and pick coordinate charts $\xi : M \hookrightarrow \mathbb{R}^m$, $\eta : N \hookrightarrow \mathbb{R}^n$, defined near $\bar{x}_*$, $\bar{y}_*$ and such that $\xi(\bar{x}_*) = 0$ and $\eta(\bar{y}_*) = 0$, and declare that a subset Λ of $Lin(T_{\bar{x}_*} M, T_{\bar{y}_*} N)$ belongs to $AGDQ(F, \bar{x}_*, \bar{y}_*, S)$ if the composite set of maps $D\eta(\bar{y}_*) \circ \Lambda \circ D\xi(\bar{x}_*)^{-1}$ is in $AGDQ(\eta \circ F \circ \xi^{-1}, 0, 0, \xi(S)))$. It then follows easily from the chain rule that, with this definition, **the set** $AGDQ(F, \bar{x}_*, \bar{y}_*, S)$ **does not depend on the choice of the charts** ξ, η. In other words, **the notion of an AGDQ is invariant under** C^1 **diffeomorphisms and therefore makes sense intrinsically on manifolds of class** C^1.

Then the chain rule also holds on manifolds, as pointed out in Ref. 9.

Proposition 1.1. *Assume that*

    (I) for $i = 1, 2, 3$, M_i is a manifold of class C^1 and $\bar{x}_{*,i} \in M_i$,

    (II) $S_i \subseteq M_i$, $F_i : M_i \mapsto M_{i+1}$, and $\Lambda_i \in AGDQ(F_i, \bar{x}_{*,i}, \bar{x}_{*,i+1}, S_i)$ for $i = 1, 2$,

    (III) either S_2 is a local quasiretract of M_2 or F_1 is single-valued on $U \cap S_1$ for some neighborhood U of $\bar{x}_{*,1}$.

Then the composite $\Lambda_2 \circ \Lambda_1$ belongs to $AGDQ(F_2 \circ F_1, \bar{x}_{*,1}, \bar{x}_{*,3}, S_1)$. $\qquad\square$

Furthermore, AGDQs have several natural properties. First, the following statement, proved in Ref. 9, says that classical differentials at one point of continuous maps and Clarke generalized Jacobians of Lipschitz maps are AGDQs.

Proposition 1.2. *If M, N are manifolds of class C^1, $S \subseteq M$, $\bar{x}_* \in M$, $\bar{y}_* \in N$, $F : M \mapsto N$, U is an open neighborhood of $\bar{x}_*$ in M, and $F(\bar{x}_*) = \{\bar{y}_*\}$, then*

(1) If (i) the restriction $F \lceil (U \cap S)$ is a continuous everywhere defined map, (ii) L is a differential of F at $\bar{x}_*$ in the direction of S (that is, $L \in Lin(T_{\bar{x}_*} M, T_{\bar{y}_*} N)$ and $\|F(x) - F(\bar{x}_*) - L \cdot (x - \bar{x}_*)\| = o(\|x - \bar{x}_*\|)$ as $x \to \bar{x}_*$ via values in S, relative to some choice of coordinate charts about $\bar{x}_*$ and $\bar{y}_*$), then $\{L\}$ belongs to $AGDQ(F, \bar{x}_*, \bar{y}_*, S)$.

(2) If (i) the restriction $F \lceil U$ is a locally Lipschitz everywhere defined map, and (ii) Λ is the Clarke generalized Jacobian of F at $\bar{x}_*$, then Λ belongs to $AGDQ(F, \bar{x}_*, \bar{y}_*, M)$. $\square$

The following two propositions, also proved in Ref. 9, are the Cartesian product rule and the assertion that AGDQs are local, in the sense that the set $AGDQ(F, \bar{x}_*, \bar{y}_*, S)$ is completely determined by the germ of the set S at $\bar{x}_*$ and the germ of the graph of F at $(\bar{x}_*, \bar{y}_*)$. In Proposition 1.3, if A, B, C, D are sets and $\mu : A \mapsto C$, $\nu : B \mapsto D$, then $\mu \times \nu$ is the set-valued map from $A \times B$ to $C \times D$ that sends each point $(a, b) \in A \times B$ to the subset $\mu(a) \times \nu(b)$ of $C \times D$. (In particular, if μ and ν are ordinary single-valued maps, then $\mu \times \nu$ is an ordinary single-valued map, given by $(\mu \times \nu)(a, b) = (\mu(a), \nu(b))$ for $a \in A$, $b \in B$.) If $\mathcal{M}$, $\mathcal{N}$ are sets of SVMs from A to C and from B to D, respectively, then $\mathcal{M} \times \mathcal{N}$ is the set of all SVMs $\mu \times \nu$, $\mu \in \mathcal{M}$, $\nu \in \mathcal{N}$. The spaces $T_{\bar{x}_*,1} M_1 \times T_{\bar{x}_*,2} M_2$ $T_{\bar{y}_*,1} N_1 \times T_{\bar{y}_*,2} N_2$ are identified with $T_{(\bar{x}_*,1,\bar{x}_*,2)}(M_1 \times M_2)$ and $T_{(\bar{y}_*,1,\bar{y}_*,2)}(N_1 \times N_2)$, respectively.

Proposition 1.3. (The product rule.) *Assume that*

(1) for $i = 1, 2$, M_i and N_i are manifolds of class C^1, $S_i \subseteq M_i$, $\bar{x}_{*,i} \in M_i$, $\bar{y}_{*,i} \in N_i$, $F_i : M_i \mapsto N_i$, and $\Lambda_i \in AGDQ(F_i, \bar{x}_{*,i}, \bar{y}_{*,i}, S_i)$;

(2) $\bar{x}_* = (\bar{x}_{*,1}, \bar{x}_{*,2})$, $\bar{y}_* = (\bar{y}_{*,1}, \bar{y}_{*,2})$, $S = S_1 \times S_2$, and $F = F_1 \times F_2$.

Then $\Lambda_1 \times \Lambda_2 \in AGDQ(F, \bar{x}_*, \bar{y}_*, S)$. $\square$

Proposition 1.4. (Locality.) *Assume that (1) M, N, are manifolds of class C^1, (2) $\bar{x}_* \in M$, (3) $\bar{y}_* \in N$, (4) $S_i \subseteq M$ and $F_i : M \mapsto N$ for $i = 1, 2$, and (5) the sets S_1 and S_2 have the same germ at $\bar{x}_*$, and the graphs $\mathrm{Gr}(F_1)$, $\mathrm{Gr}(F_2)$, have the same germ at $(\bar{x}_*, \bar{y}_*)$ (that is, there exist neighborhoods U, V of $\bar{x}_*$, $\bar{y}_*$, in M, N, respectively, such that $U \cap S_1 = U \cap S_2$ and $(U \times V) \cap \mathrm{Gr}(F_1) = (U \times V) \cap \mathrm{Gr}(F_2)$). Then $AGDQ(F_1, \bar{x}_*, \bar{y}_*, S_1) = AGDQ(F_2, \bar{x}_*, \bar{y}_*, S_2)$.* $\square$

1.2.2. *Uniform AGDQs*

Assume that X and Y are FDNRLSs, and we are given a family $\{(F_\alpha, x_\alpha, y_\alpha, S_\alpha)\}_{\alpha \in A}$ of 4-tuples, such that each F_α is a set-valued map from X to Y, each x_α is a point of X, each y_α is a point of Y, and each S_α is a subset of X. We say that a family $\{\Lambda_\alpha\}_{\alpha \in A}$ of compact subsets of $Lin(X, Y)$ is a **uniform AGDQ of the maps F_α at the points (x_α, y_α) in the direction of the S_α** if there exists a function $\theta \in \Theta$ which is an AGDQ modulus for $(\Lambda_\alpha, F_\alpha, x_\alpha, y_\alpha, S_\alpha)$ for each $\alpha \in A$.

The concept of a uniform AGDQ makes sense as well, in an intrinsic way, when X and Y are manifolds, provided that the family $\{(F_\alpha, x_\alpha, y_\alpha, S_\alpha)\}_{\alpha \in A}$ is such that the set $Q = \{(x_\alpha, y_\alpha) : \alpha \in A\}$ is precompact in $X \times Y$. Indeed, let d_X, d_Y be the dimensions of X and Y. If Q is precompact in $X \times Y$, then we can find a finite family $\Sigma = \{(\xi_j, U_j, \eta_j, V_j, K_j, L_j)\}_{1 \leq j \leq m}$ such that

(1) for each j, (i) ξ_j is a coordinate chart of X with domain U_j, (ii) η_j is a coordinate chart of Y with domain V_j, (iii) K_j is a compact subset of U_j, and (iv) L_j is a compact subset of V_j;

(2) $Q \subseteq \bigcup_{j=1}^m (K_j \times L_j)$.

Then, if we let $A_j = \{\alpha \in A : (x_\alpha, y_\alpha) \in K_j \times L_j\}$, it is clear that $A = \bigcup_{j=1}^m A_j$, and we can consider, for each j, the family $\Phi_j = \{(\tilde{F}_{j,\alpha}, x_\alpha, y_\alpha, \tilde{S}_{j,\alpha})\}_{\alpha \in A_j}$, where $\tilde{F}_{j,\alpha}$ is the set-valued map from U_j to V_j whose graph is $\mathrm{Gr}(F_\alpha) \cap (U_j \times V_j)$, and $\tilde{S}_{j,\alpha} = S_\alpha \cap U_j$. If we identify U_j, V_j with open subsets $\tilde{U}_j$, $\tilde{V}_j$ of $\mathbb{R}^{d_X}$, $\mathbb{R}^{d_Y}$, then $\{\tilde{F}_{j,\alpha}\}_{\alpha \in A_j}$ is a family of set-valued maps from $\mathbb{R}^{d_X}$ to $\mathbb{R}^{d_Y}$, the x_α belong to $\mathbb{R}^{d_X}$, the y_α belong to $\mathbb{R}^{d_Y}$, and $\tilde{S}_\alpha$ is a subset of $\mathbb{R}^{d_X}$, so we are in the situation of the previous paragraph, and it makes sense to talk about a "uniform AGDQ" $\{\Lambda_\alpha\}_{\alpha \in A_j}$ of the family Φ_j. We then say that a family $\{\Lambda_\alpha\}_{\alpha \in A}$ is a **uniform AGDQ of the family** $\{(F_\alpha, x_\alpha, y_\alpha, S_\alpha)\}_{\alpha \in A}$ if, for some choice of m and the family $\Sigma = \{(\xi_j, U_j, \eta_j, V_j, K_j, L_j)\}_{1 \leq j \leq m}$ as above, it turns out that $\{\Lambda_\alpha\}_{\alpha \in A_j}$ is a uniform AGDQ of Φ_j for each j. (It is easily seen that if this condition holds for one choice of m and Σ, then it holds for all such choices.)

1.2.3. *AGDQ approximating multicones.*

A **cone** in a FDRLS X is a nonempty set C which is closed under multiplication by nonnegative real numbers, i.e., such that $rc \in C$ whenever $c \in C$ and $r \geq 0$. The **polar** of a cone C in X is the subset $C^\dagger$ of $X^\dagger$ defined by $C^\dagger = \{\mu \in X^\dagger : \mu(c) \leq 0 \text{ whenever } c \in C\}$. Clearly, $C^\dagger$ is always a closed convex cone. If we identify $X^{\dagger\dagger}$ with X in the usual way, then $C \subseteq C^{\dagger\dagger}$, and $C = C^{\dagger\dagger}$ if and only if C is closed and convex.

A *multicone* is a nonempty set of cones. A multicone $\mathcal{M}$ is *convex* if all the members of $\mathcal{M}$ are convex cones. The *polar* $\mathcal{M}^{\dagger}$ of a multicone $\mathcal{M}$ is the closure of the union of the polars $M^{\dagger}$, $M \in \mathcal{M}$. Therefore $\mathcal{M}^{\dagger}$ is a always a closed cone in $X^{\dagger}$. Naturally, $\mathcal{M}^{\dagger}$ need not be convex in general.

Definition 1.2. Assume that M is a manifold of class C^1, S is a subset of M, and $\bar{x}_* \in S$. An *AGDQ approximating multicone to S at $\bar{x}_*$* is a convex multicone $\mathcal{C}$ in $T_{\bar{x}_*}M$ such that there exist a nonnegative integer m, a set-valued map $F : \mathbb{R}^m \mapsto M$, a convex cone D in $\mathbb{R}^m$, and a $\Lambda \in AGDQ(F, 0, \bar{x}_*, D)$, such that $F(D) \subseteq S$ and $\mathcal{C} = \{LD : L \in \Lambda\}$. $\qquad\square$

1.3. *Transversality of cones and multicones.*

If S_1, S_2 are subsets of a linear space X, we define the *sum* $S_1 + S_2$ and the *difference* $S_1 - S_2$ by letting $S_1 + S_2 = \{s_1 + s_2 : s_1 \in S_1, s_2 \in S_2\}$, $S_1 - S_2 = \{s_1 - s_2 : s_1 \in S_1, s_2 \in S_2\}$.

Recall that if S_1, S_2 are linear subspaces of a FDRLS X, then S_1 and S_2 are *transversal* if $S_1 + S_2 = X$, or, equivalently, if $S_1 - S_2 = X$. If two submanifolds M_1, M_2 of class C^1 intersect at a point x_*, and S_1, S_2 are their tangent spaces at x_*, then it is well known that if S_1 and S_2 are transversal then $M_1 \cap M_2$ looks, near x_*, like $S_1 \cap S_2$. In particular, if $S_1 \cap S_2 \neq \{0\}$ (i.e., if $\dim S_1 \cap S_2 \geq 1$), then $M_1 \cap M_2$ contains a nontrivial curve going through x_*. The following definitions generalize the concept of transversality and that of "transversality with a nontrivial intersection," first to cones and then to multicones.

Definition 1.3. Let X be a FDRLS, and let C_1, C_2 be two convex cones in X. We say that C_1 and C_2 are *transversal*, and write $C_1 \pitchfork C_2$, if $C_1 - C_2 = X$. We say that C_1 and C_2 are *strongly transversal*, and write $C_1 \pitchfork C_2$, if $C_1 \pitchfork C_2$ and in addition $C_1 \cap C_2 \neq \{0\}$. $\qquad\square$

In order to extend Definition 1.3 to multicones, it is convenient to start by reformulating the concept of strong transversality of cones, by making the trivial observation that $C_1 \pitchfork C_2$ if and only if the following two conditions hold: (i) $C_1 \pitchfork C_2$ and (ii) there exists a linear functional $\mu \in X^{\dagger}$ such that $\mu(v) > 0$ for some $v \in C_1 \cap C_2$.

In view of the above reformulation, we define a linear functional $\mu : X \mapsto \mathbb{R}$ to be *intersection-positive* on a pair $(\mathcal{C}_1, \mathcal{C}_2)$ of multicones, if the set $\{c \in C_1 \cap C_2 : \mu(c) > 0\}$ is nonempty for every $C_1 \in \mathcal{C}_1$ and every $C_2 \in \mathcal{C}_2$. Using this concept, the definitions of "transversality" and "strong transversality" of convex multicones are nearly identical to the definitions for cones.

Definition 1.4. Let X be a FDRLS. We say that two convex multicones $\mathcal{C}_1$ and $\mathcal{C}_2$ in X are **transversal**, and write $\mathcal{C}_1 \overline{\pitchfork} \mathcal{C}_2$, if $C_1 \overline{\pitchfork} C_2$ for all $C_1 \in \mathcal{C}_1$, $C_2 \in \mathcal{C}_2$. We say that $\mathcal{C}_1$ and $\mathcal{C}_2$ are **strongly transversal**, and write $\mathcal{C}_1 \overline{\pitchfork} \mathcal{C}_2$, if (i) $\mathcal{C}_1 \overline{\pitchfork} \mathcal{C}_2$, and (ii) there exists a linear functional $\mu \in X^\dagger$ which is intersection-positive on $(\mathcal{C}_1, \mathcal{C}_2)$. $\qquad\square$

Two convex cones C_1, C_2 in a FDRLS X are **linearly separated** if there exists a nontrivial linear functional $\lambda \in X^\dagger$ such that $\lambda(c) \leq 0$ whenever $c \in C_1$, and $\lambda(c) \geq 0$ whenever $c \in C_2$. (Equivalently, C_1 and C_2 are linearly separated if and only if $C_1^\dagger \cap (-C_2)^\dagger \neq \{0\}$.) It is easy to see that C_1 and C_2 are linearly separated if and only if they are not transversal. In view of this, we will call two convex multicones $\mathcal{C}_1$, $\mathcal{C}_2$, **linearly separated** if they are not transversal. Since strong transversality is a stronger property than transversality, its negation is weaker than the negation of transversality, i.e., than linear separation. So we will say that two convex multicones $\mathcal{C}_1$, $\mathcal{C}_2$, are **weakly linearly separated** if they are not strongly transversal.

The following characterization of weak linear separation is proved in Ref. 10.

Proposition 1.5. *Let $\mathcal{C}_1$, $\mathcal{C}_2$ be convex multicones in a FDRLS X. Then the following conditions are equivalent:*

1. *$\mathcal{C}_1$ and $\mathcal{C}_2$ are weakly linearly separated;*
2. *for every $\mu \in X^\dagger \backslash \{0\}$ there exist $\pi_0, \pi_1, \pi_2, C_1, C_2$ such that*

    i. $\pi_0 \in \mathbb{R}$ and $\pi_0 \geq 0$,
    ii. $C_1 \in \mathcal{C}_1$ and $C_2 \in \mathcal{C}_2$,
    iii. $\pi_1 \in C_1^\dagger$ and $\pi_2 \in C_2^\dagger$,
    iv. $\pi_0 \mu = \pi_1 + \pi_2$,
    v. $(\pi_0, \pi_1, \pi_2) \neq (0, 0, 0)$. $\qquad\square$

1.4. *The nonseparation theorem.*

The crucial fact about AGDQs that leads to the maximum principle is the transversal intersection property, that we now state (cf. Ref. 9 for the proof).

Theorem 1.2. *Let M be a manifold of class C^1, let S_1, S_2 be subsets of M, and let $\bar{s}_* \in S_1 \cap S_2$. Let C_1, C_2 be AGDQ-approximating multicones to S_1, S_2 at $\bar{s}_*$ such that $C_1 \pitchfork C_2$. Then S_1 and S_2 are not locally separated at $\bar{s}_*$. (That is, the set $S_1 \cap S_2$ contains a sequence of points s_j converging to $\bar{s}_*$ but not equal to $\bar{s}_*$.)* $\qquad\square$

Theorem 1.2 and Proposition 1.5 trivially imply the following result.

Corollary 1.1. *Let M be a manifold of class C^1, let S_1, S_2 be subsets of M, and let $\bar{s}_* \in S_1 \cap S_2$. Let C_1, C_2 be AGDQ-approximating multicones to S_1, S_2 at $\bar{s}_*$. Assume that S_1 and S_2 are locally separated at $\bar{s}_*$. (That is, there exists a neighborhood U of $\bar{s}_*$ such that $S_1 \cap S_2 \cap U = \{\bar{s}_*\}$). Then Condition 2 of the statement of Proposition 1.5 holds.* $\qquad\square$

The more familiar forms of the maximum principle for optimal control follow by applying Corollary 1.1 to suitable choices of M, S_1, S_2, C_1, C_2, $\bar{s}_*$, and using the conclusion of the corollary with a suitable μ. For example, consider a fixed time interval optimal control problem $\mathcal{P}$ whose data 9-tuple $D = (M_0, U, a, b, \mathcal{U}, f, L, \bar{x}_*, S)$ satisfies (D1) the state space M_0 is a smooth manifold, (D2) U is a set, (D3) $a, b \in \mathbb{R}$ and $a < b$, (D4) $\mathcal{U}$ (the class of "admissible controls") is a set of U-valued functions on $[a, b]$, (D5) $(L(x, u, t,), f(x, u, t)) \in \mathbb{R} \times T_x M_0$ for each $(x, u, t) \in M_0 \times U \times [a, b]$, (D6) $\bar{x}_* \in M_0$, and (D7) $S \subseteq M_0$. Suppose that the objective of $\mathcal{P}$ is to minimize the integral $\int_a^b L(\xi(t), \eta(t), t)\, dt$, subject to the following conditions: (C1) $\xi : [a, b] \mapsto M_0$ is absolutely continuous, (C2) $\eta \in \mathcal{U}$, (C3) $\dot{\xi}(t) = f(\xi(t), \eta(t), t)$ for almost all $t \in [a, b]$, (C4) $\xi(a) = \bar{x}_*$, and (C5) $\xi(b) \in S$. We then take $M = \mathbb{R} \times M_0$. If a trajectory-control pair (ξ_*, η_*) is a solution of $\mathcal{P}$, we take S_1 to be the set of all points $(r, x) \in M$ such that x is reachable from $\bar{x}_*$ over $[a, b]$ with cost r, and we take S_2 to be the set $(\,]-\infty, r_*[\times S) \cup \{q_*\}$, where $q_* = (r_*, \xi_*(b))$, and r_* is the cost of (ξ_*, η_*). Then the optimality of (ξ_*, η_*) implies that S_1 and S_2 are locally separated at q_*. We then take C_1 to be an AGDQ-approximating multicone to S_1 at q_* obtained by constructing variations and propagating their effects to the terminal point of ξ_*, and take $C_2 = \{\,]-\infty, 0] \times C : C \in \mathcal{C}\}$, where $\mathcal{C}$ is an AGDQ-approximating multicone to S at $\xi_*(b)$. We choose μ to be the linear functional on $T_{q_*} M \sim \mathbb{R} \times T_{\xi_*(b)} M_0$ given by $\mu(r, v) = -r$, so $-\mu \in C_2^\dagger$ for every $C_2 \in \mathcal{C}_2$. Corollary 1.1 then yields a decomposition $\pi_0 \mu = \pi_1 + \pi_2$, where $\pi_1 \in C_1^\dagger$, $\pi_2 \in C_2^\dagger$, $\pi_0 \geq 0$, $(\pi_0, \pi_1, \pi_2) \neq (0, 0, 0)$, $C_1 \in \mathcal{C}_1$, and $C_2 \in \mathcal{C}_2$. Then $-\pi_1 = \pi_2 - \pi_0 \mu$. Since $\pi_2 \in C_2^\dagger$, $\pi_0 \geq 0$, and $-\mu \in C_2^\dagger$, it follows that $-\pi_1 \in C_2^\dagger$. If we write $C_2 = \,]-\infty, 0] \times C$, $C \in \mathcal{C}$, and let $\pi_1 = (-\rho, \bar{\pi})$, then the fact that $-\pi_1 \in C_2^\dagger$ implies that $\rho \geq 0$ and $-\bar{\pi} \in C^\dagger$. Then $\bar{\pi}$ and ρ are, respectively, the terminal adjoint vector (often called $\psi(b)$ or $\lambda(b)$ in the literature) and the additional multiplier (often called ψ_0 or λ_0) conjugate to the cost r, and the familiar conclusions of the maximum principle follow.

2. Flows, trajectories, and generalized differentials of flows.

2.1. *State space bundles and their sections*

A *time set* is a nonempty totally ordered set. If I is a time set, we define $I^{2,\geq} = \{(t,s) \in I \times I : t \geq s\}$, and $I^{3,\geq} = \{(t,s,r) \in I \times I \times I : t \geq s \geq r\}$. A *state-space bundle* (abbr. SSB) *over* I is an indexed family $\mathbf{X} = \{X_t\}_{t \in I}$ of sets. A *state-space bundle* is a pair $\mathcal{X} = (\mathbf{X}, I)$ such that I is a nonempty totally ordered set and $\mathbf{X}$ is an SSB over I. The set I is the *time set* of the SSB $\mathcal{X}$.

Remark 2.1. There are several reasons for using general totally ordered sets, rather than real intervals, as time sets for control systems. For a simple example, cf. Ref. 8, where an example is given of a problem for which the natural time set consists of a compact interval minus one interior point. $\square$

If $\mathcal{C}$ is a category whose objects are sets with some additional structure (for example, topological spaces, metric spaces, manifolds of class C^k, linear spaces, FDRLSs), then an SSB $(\mathbf{X}, I) = (\{X_t\}_{t \in I}, I)$ is a *bundle of $\mathcal{C}$-objects* if each X_t is a member of $\mathcal{C}$. In particular, if k is a nonnegative integer, a C^k *SSB* is an SSB of manifolds of class C^k. Also, an *FDRLS SSB* is an SSB of finite-dimensional real linear spaces.

Definition 2.1. Assume that $\mathcal{X} = (\mathbf{X}, I) = (\{X_t\}_{t \in I}, I)$ is an SSB. A *section of $\mathcal{X}$* is a single-valued everywhere defined map ξ on I such that $\xi(t) \in X_t$ for every $t \in I$. We use $Sec(\mathcal{X})$ to denote the set of all sections of $\mathcal{X}$. $\square$

Definition 2.2. Let $\mathcal{X} = (\mathbf{X}, I) = (\{X_t\}_{t \in I}, I)$ be a C^1 state-space bundle, and assume that $\xi \in Sec(\mathcal{X})$. The family $\mathbf{T}_\xi \mathcal{X} = \{T_{\xi(t)} X_t\}_{t \in I}$ is the *tangent bundle* of $\mathcal{X}$ along ξ. $\square$

Clearly, the tangent bundle $\mathbf{T}_\xi \mathcal{X}$ of a C^1 SSB $\mathcal{X}$ along a section $\xi \in Sec(\mathcal{X})$ is an FDRLS SSB.

2.2. *Flows and trajectories*

Definition 2.3. Assume that $\mathcal{C}$ is a category whose objects are sets with some additional structure, and $\mathcal{X} = (\mathbf{X}, I) = (\{X_t\}_{t \in I}, I)$ is an SSB of $\mathcal{C}$-objects. A $\mathcal{C}$-*flow* on $\mathcal{X}$ is an indexed family $\mathbf{f} = \{f_{t,s}\}_{(t,s) \in I^{2, \geq}}$ such that

(1) $f_{t,s}$ is a $\mathcal{C}$-morphism from X_s to X_t whenever $(t, s) \in I^{2, \geq}$;
(2) $f_{t,t}$ is the identity morphism of X_t whenever $t \in I$;
(3) $f_{t,s} \circ f_{s,r} = f_{t,r}$ whenever $(t, s, r) \in I^{3, \geq}$.

A $\mathcal{C}$-*flow* is a pair $\mathcal{F} = (\mathcal{X}, \mathbf{f})$ such that $\mathcal{X}$ is an SSB of $\mathcal{C}$-objects and $\mathbf{f}$ is a $\mathcal{C}$-flow on $\mathcal{X}$. $\qquad\square$

Example 2.1. If $\mathcal{C}$ is the category whose objects are all the sets, and whose morphisms are the set-valued maps, then a $\mathcal{C}$-flow on an SSB $\mathcal{X}$ will just be called a *flow* on $\mathcal{X}$. $\qquad\square$

Example 2.2. If $\mathcal{C}$ is the category whose objects are all FDRLSs, and whose morphisms are the linear maps, then a $\mathcal{C}$-flow on an FDRLS SSB $\mathcal{X}$ will be called a *linear FD flow*. $\qquad\square$

Example 2.3. We use $FDCLin$ to denote the category whose objects are all FDRLSs, and whose morphisms are defined as follows: if X, Y are FDRLSs, then the set of morphisms from X to Y is the set $CLin(X, Y)$ of all nonempty compact subsets of $Lin(X, Y)$. (Composition of morphisms is defined in the obvious way: if $\Lambda_1 \in CLin(X, Y)$ and $\Lambda_2 \in CLin(Y, Z)$, then $\Lambda_2 \circ \Lambda_1 \overset{\text{def}}{=} \{L_2 \circ L_1 : L_2 \in \Lambda_2, L_1 \in \Lambda_1\}$.)
An $FDCLin$-flow is a *linear FD multiflow*. $\qquad\square$

Remark 2.2. It is well known that every time set I can be regarded as a category $cat(I)$, by taking the objects of $cat(I)$ to be the members of I, and the set $Hom_{cat(I)}(a, b)$ of morphisms from $a \in I$ to $b \in I$ to consist of a single object if $a \leq_I b$, and to be empty if $b <_I a$. In terms of this identification, a $\mathcal{C}$-flow with time set I is exactly the same as a functor from $cat(I)$ to $\mathcal{C}$. $\qquad\square$

2.2.1. *Comparison of maps and flows*

If f, f' are SVMs, we write $f \preceq f'$ if $\mathrm{So}(f) = \mathrm{So}(f')$, $\mathrm{Ta}(f) = \mathrm{Ta}(f')$, and $\mathrm{Gr}(f) \subseteq \mathrm{Gr}(f')$. If, for $i = 1, 2$, $\mathcal{F}^i = (\mathcal{X}, \mathbf{f}^i)$ are flows on the same SSB $\mathcal{X}$, and $\mathbf{f}^i = \{f^i_{t,s}\}_{(t,s) \in I^{2, \geq}}$, we say that $\mathcal{F}^1$ is a *subflow* of $\mathcal{F}^2$, or $\mathcal{F}^2$ is a *superflow* of $\mathcal{F}^1$, and write $\mathcal{F}^1 \preceq \mathcal{F}^2$, if $f^1_{t,s} \preceq f^2_{t,s}$ for all $(t, s) \in I^{2, \geq}$.

2.2.2. *Trajectories*

Definition 2.4. Assume that $\mathcal{X} = (\mathbf{X}, I)$ is a state-space bundle, $\mathcal{F} = (\mathcal{X}, \mathbf{f})$ is a flow, $\mathbf{X} = (\{X_t\}_{t \in I}, I)$, and $\mathbf{f} = \{f_{t,s}\}_{(t,s) \in I^{2,\geq}}$. A **trajectory** of $\mathcal{F}$ is a section ξ of $\mathcal{X}$ such that $\xi(t)$ belongs to $f_{t,s}(\xi(s))$ whenever $(t,s) \in I^{2,\geq}$.

We use $Traj(\mathcal{F})$ to denote the set of all trajectories of the flow $\mathcal{F}$. $\quad\square$

2.3. *AGDQs of flows along trajectories*

Definition 2.5. Assume that $\mathcal{X} = (\mathbf{X}, I) = (\{X_t\}_{t \in I}, I)$ is a C^1 SSB, $\mathcal{F} = (\mathcal{X}, \mathbf{f})$ is a flow, $\mathbf{f} = \{f_{t,s}\}_{(t,s) \in I^{2,\geq}}$, and $\xi \in Traj(\mathcal{F})$. An **AGDQ** of $\mathcal{F}$ along ξ is a linear FD multiflow $\mathbf{g} = \{g_{t,s}\}_{(t,s) \in I^{2,\geq}}$ on the tangent bundle $T_\xi \mathcal{X}$ such that $g_{t,s} \in AGDQ(f_{t,s}; \xi(s), \xi(t); X_s)$ whenever $(t,s) \in I^{2,\geq}$.

Remark 2.3. In view of our definitions, the condition that $\mathbf{g}$ is a linear FD multiflow on $T_\xi \mathcal{X}$ means that

(1) if $(t,s) \in I^{2,\geq}$, then $g_{t,s}$ is a nonempty compact set of linear maps from $T_{\xi(s)} X_s$ to $T_{\xi(t)} X_t$;
(2) $g_{t,t} = \{\mathbb{I}_{T_{\xi(t)} X_t}\}$ whenever $t \in I$;
(3) $g_{t,s} \circ g_{s,r} = g_{t,r}$ whenever $(r,s,t) \in I^{3,\geq}$. $\quad\square$

2.3.1. *Compatible selections*

Definition 2.6. Assume that $\mathbf{g} = \{g_{t,s}\}_{(t,s) \in I^{2,\geq}}$ is a linear FD multiflow on an FDRLS SSB $(\mathcal{Y}, I)$. A **compatible selection** of $\mathbf{g}$ is a linear FD flow $\gamma = \{\gamma_{t,s}\}_{(t,s) \in I^{2,\geq}}$ such that $\gamma_{t,s} \in g_{t,s}$ whenever $(t,s) \in I^{2,\geq}$. $\quad\square$

We write $CSel(\mathbf{g})$ to denote the set of all compatible selections of $\mathbf{g}$. Then $CSel(\mathbf{g})$ is a subset of the product space $P_\mathbf{g} \stackrel{\text{def}}{=} \prod_{(t,s) \in I^{2,\geq}} g_{t,s}$. Since $P_\mathbf{g}$ is a compact space, by Tichonov's theorem, and $CSel(\mathbf{g})$ is a closed subset of $P_\mathbf{g}$—because $CSel(\mathbf{g})$ is the set of all $\gamma \in P_\mathbf{g}$ that satisfy a collection of equalities involving continuous functions on $P_\mathbf{g}$—we can conclude that $CSel(\mathbf{g})$ is compact.

Remark 2.4. In view of our previous definitions, the condition that γ is a linear FD flow means that $\gamma_{t,t} = \mathbb{I}_{T_{\xi(t)} X}$ for each $t \in I$, and $\gamma_{t,s} \gamma_{s,r} = \gamma_{t,r}$ whenever $(t,s,r) \in I^{3,\geq}$. $\quad\square$

2.3.2. *Fields of variational vectors and adjoint covectors*

Definition 2.7. Assume that $\mathbf{g} = \{g_{t,s}\}_{(t,s)\in I^{2,\geq}}$ is a linear FD multiflow on an FDRLS SSB $(\mathcal{Y}, I) = (\{Y_t\}_{t\in I}, I)$. A ***field of variational vectors*** of $\mathbf{g}$ is a selection $I \ni t \mapsto v(t) \in Y_t$ such that $v_t \in g_{t,s}v_s$ whenever $(t, s) \in I^{2,\geq}$.

A ***field of adjoint covectors*** (also called, simply, an ***adjoint covector***, or even an ***adjoint vector***) of $\mathbf{g}$ is a selection $I \ni t \mapsto \omega(t) \in Y_t^\dagger$ of the dual bundle $\mathcal{Y}^\dagger = \{Y_t^\dagger\}_{t\in I}$. such that $\omega_s \in g_{t,s}^\dagger \omega_t$ whenever $(t, s) \in I^{2,\geq}$, where $g_{t,s}^\dagger = \{\gamma^\dagger : \gamma \in g_{t,s}\}$. $\qquad\square$

The following result is an easy consequence of the compactness of $CSel(\mathbf{g})$.

Proposition 2.1. *Assume that* $\mathbf{g} = \{g_{t,s}\}_{(t,s)\in I^{2,\geq}}$ *is a linear FD multiflow on an FDRLS SSB* $(\mathcal{Y}, I) = (\{Y_t\}_{t\in I}, I)$. *Assume that* $I \ni t \mapsto v(t) \in Y_t$ *(resp.* $I \ni t \mapsto \omega(t) \in Y_t^\dagger$*) is a selection of* $\mathcal{Y}$ *(resp.* $\mathcal{Y}^\dagger$*). Then* v *is a field of variational vectors (resp.* ω *is a field of adjoint covectors) of* $\mathbf{g}$ *if and only if there exists a compatible selection* $\gamma = \{\gamma_{t,s}\}_{(t,s)\in I^{2,\geq}}$ *of* $\mathbf{g}$ *such that* $v_t = \gamma_{t,s}v_s$ *(resp.* $\omega_s = \gamma_{t,s}^\dagger \omega_t$*) whenever* $(t, s) \in I^{2,\geq}$. $\qquad\square$

3. Variations, impulse variations, summability

3.1. *Variations of set-valued maps*

Definition 3.1. Assume that F is a set-valued map and P is a FDRLS. A ***variation of*** F ***with ambient parameter space*** P is a family $V = \{V_p\}_{p\in C}$ such that

(1) C is a closed convex cone in P with nonempty interior;
(2) each V_p is a SVM such that $\mathrm{So}(V_p) = \mathrm{So}(F)$ and $\mathrm{Ta}(V_p) = \mathrm{Ta}(F)$;
(3) $\mathrm{Gr}(V_0) \subseteq \mathrm{Gr}(F)$.

If F' is another set-valued map such that $\mathrm{So}(F') = \mathrm{So}(F)$, $\mathrm{Ta}(F') = \mathrm{Ta}(F)$, and $\mathrm{Gr}(F) \subseteq \mathrm{Gr}(F')$, we say that V is a ***variation in*** F' if the inclusion $\mathrm{Gr}(V_p) \subseteq \mathrm{Gr}(F')$ holds for every $p \in C$, i.e., if $V_p(x) \subseteq F'(x)$ whenever $p \in C$ and $x \in \mathrm{So}(F')$. $\qquad\square$

If F, P, V are as in Definition 3.1, then the cone C is the ***parameter cone of*** V, and the dimension of C (or of P) is the ***number of parameters*** of V. We will use $\tilde{V}$ to denote the SVM with source $P \times \mathrm{So}(F)$ and target $\mathrm{Ta}(F)$ such that $\tilde{V}(p, x) = V_p(x)$ for all $p \in P$, $x \in \mathrm{So}(V_0)$. (In particular, $\tilde{V}(p, x) = \emptyset$ if $p \in P\backslash C$.)

3.2. *Infinitesimal impulse variations*

Definition 3.2. Assume that $\mathcal{X} = (\mathbf{X}, I) = (\{X_t\}_{t \in I}, I)$ is a C^1 state-space bundle, $\mathcal{F} = (\mathcal{X}, \mathbf{f})$ is a flow, and $\xi \in Traj(\mathcal{F})$. An *infinitesimal impulse variation* (abbr, IIV) *for* $(\mathcal{F}, \xi)$ is a triple (v, t, σ) such that $t \in I$, $v \in T_{\xi(t)} X_t$, and σ is one of the symbols $+$, $-$. $\qquad\square$

Remark 3.1. The purpose of including σ in the above definition is to distinguish between "left" impulse variations, which will be labelled $(v, t, -)$, and "right" impulse variations, labelled $(v, t, +)$. Left and right impulse variations will differ in the way the concept of "carrier" of an IIV (v, t, σ) is defined, which will depend strongly on σ. $\qquad\square$

3.3. *Summability*

Definition 3.3. Assume that $\mathcal{X} = (\mathbf{X}, I) = (\{X_t\}_{t \in I}, I)$ is a C^1 state-space bundle, $\mathcal{F} = (\mathcal{X}, \mathbf{f})$ is a flow, and $\xi \in Traj(\mathcal{F})$. If (v, t, σ) is an IIV for $(\mathcal{F}, \xi)$, we say that (v, t, σ) is *carried* by a subinterval J of I if $t \in J$ and one of the following two conditions holds: (i) $\sigma = +$ and there exists a $t_* \in J$ such that $t < t_*$, (ii) $\sigma = -$ and there exists a $t_* \in J$ such that $t_* < t$.

If $\mathbf{V}$ is a set of IIVs for $(\mathcal{F}, \xi)$, we say that $\mathbf{V}$ is *carried* by J if every member of $\mathbf{V}$ is carried by J. $\qquad\square$

If $\mathbf{V}$ is a finite set of IIVs for $(\mathcal{F}, \xi)$, we let $\mathbb{R}^{\mathbf{V}}$, $\mathbb{R}^{\mathbf{V}}_+$ denote, respectively, the set of all families $\vec{p} = \{p^V\}_{V \in \mathbf{V}}$ of real numbers, and the set of all $\vec{p} = \{p^V\}_{V \in \mathbf{V}} \in \mathbb{R}^{\mathbf{V}}$ such that $p^V \geq 0$ for all $V \in \mathbf{V}$. (Hence, if m is the cardinality of $\mathbf{V}$, and $\mathbf{V} = \{(v^1, t^1, \sigma^1), \ldots, (v^m, t^m, \sigma^m)\}$, the spaces $\mathbb{R}^{\mathbf{V}}$, $\mathbb{R}^{\mathbf{V}}_+$, can be identified with $\mathbb{R}^m$, $\mathbb{R}^m_+$, by identifying each family $\vec{p} = \{p^V\}_{\varepsilon \in \mathbf{V}}$ with the m-tuple $(\tilde{p}^1, \ldots, \tilde{p}^m)$, where $\tilde{p}^j = p^{(v^j, t^j, \sigma^j)}$ for $j = 1, \ldots, m$.)

If $\mathbf{g} = \{g_{t,s}\}_{(t,s) \in I^{2,\geq}}$ is an AGDQ of $\mathcal{F}$ along ξ, $\gamma = \{\gamma_{t,s}\}_{(t,s) \in I^{2,\geq}}$ is a compatible selection of $\mathbf{g}$, $a, b \in I$, $a \leq b$, and $\mathbf{V}$ is carried by $[a, b]$, we define a linear map $L^{\mathbf{V}, \gamma, a, b} : \mathbb{R}^{\mathbf{V}} \times T_{\xi(a)} X_a \mapsto T_{\xi(b)} X_b$ by letting

$$L^{\mathbf{V}, \gamma, a, b}(\vec{p}, w) = \gamma_{b,a}(w) + \sum_{(v,t,\sigma) \in \mathbf{V}} p^{(v,t,\sigma)} \gamma_{b,t}(v) \quad \text{for } \vec{p} \in \mathbb{R}^{\mathbf{V}}, \ w \in T_{\xi(a)} X_a.$$

We let $\Lambda^{\mathbf{V}, \mathbf{g}, a, b}$ be the set of all the maps $L^{\mathbf{V}, \gamma, a, b}$, for all $\gamma \in CSel(\mathbf{g})$. Then $\Lambda^{\mathbf{V}, \mathbf{g}, a, b}$ is the image of $CSel(\mathbf{g})$ under the continuous map $CSel(\mathbf{g}) \ni \gamma \mapsto L^{\mathbf{V}, \gamma, a, b} \in Lin(\mathbb{R}^{\mathbf{V}} \times T_{\xi(a)} X_a, T_{\xi(b)} X_b)$, so $\Lambda^{\mathbf{V}, \mathbf{g}, a, b}$ is a compact subset of $Lin(\mathbb{R}^{\mathbf{V}} \times T_{\xi(a)} X_a, T_{\xi(b)} X_b)$.

Definition 3.4. Assume that $\mathcal{X} = (\mathbf{X}, I) = (\{X_t\}_{t \in I}, I)$ is a C^1 state-space bundle, $\mathcal{F} = (\mathcal{X}, \mathbf{f})$ is a flow, $\xi \in Traj(\mathcal{F})$, $\mathbf{g} = \{g_{t,s}\}_{(t,s) \in I^{2,\geq}}$ is an AGDQ of $\mathcal{F}$ along ξ, and $\mathcal{F}' = (\mathcal{X}, \mathbf{f}') = (\mathcal{X}, \{f'_{t,s}\}_{(t,s) \in I^{2,\geq}})$ is a super-flow of $\mathcal{F}$. Let $\mathcal{V}$ be a set of IIVs for $(\mathcal{F}, \xi)$. We say that $\mathcal{V}$ is **g-AGDQ-summable within** $\mathcal{F}'$ if the following is true:

- *for every finite subset* $\mathbf{V}$ *of* $\mathcal{V}$, *and every pair* $(a, b) \in I \times I$ *such that* $a < b$ *and* $\mathbf{V}$ *is carried by the closed interval* $[a, b]$, *there exists a variation* $W = \{W_{\vec{p}}\}_{\vec{p} \in \mathbb{R}_+^{\mathbf{V}}}$ *of* $f_{b,a}$ *in* $f'_{b,a}$ *such that the set* $\Lambda^{\mathbf{V}, \mathbf{g}, a, b}$ *is an AGDQ of the map* $\tilde{W}$ *at* $((0, \xi(a)), \xi(b))$ *along* $\mathbb{R}_+^{\mathbf{V}} \times X_a$. $\qquad \square$

4. The AGDQ maximum principle

We now state and prove a general maximum principle in the setting of AGDQ theory. Instead of working with a control system $\dot{x} = f(x, u, t)$ and a reference trajectory-control pair (ξ_*, η_*), we consider the more general situation of a pair $(\mathcal{F}, \mathcal{F}')$ of flows such that $\mathcal{F}$ is a subflow of $\mathcal{F}'$. We assume that $\mathcal{F}$ and $\mathcal{F}'$ are defined on a common state space bundle $\mathcal{X} = (\mathbf{X}, I) = (\{X_t\}_{t \in I}, I)$, which is of class C^1, in the sense that the X_t are manifolds of class C^1.

In the control system case, (i) the time set I is a compact subinterval of $\mathbb{R}$, (ii) all the state spaces X_t coincide, so there is a manifold X of class C^1 such that $X_t = X$ for all $t \in I$, (iii) the domain of the reference control η_* is I, (iv) $\mathcal{F} = (\mathcal{X}, \mathbf{f})$ is the reference flow, i.e., the flow determined by the reference control η_*, so that, if $\mathbf{f} = \{f_{t,s}\}_{(t,s) \in I^{2,\geq}}$, then $f_{t,s}(x)$, for $x \in X$, $(t, s) \in I^{2,\geq}$, is the set given by

$$f_{t,s}(x) = \{\xi(t) : \xi \in \mathrm{Traj}(\eta_*, f, s, t), \, \xi(s) = x\},$$

where, if $\mathcal{U}$ is the class of admissible controls, then for any $\eta \in \mathcal{U}$ we use $\mathrm{Traj}(\eta, f, s, t)$ to denote the set of all $\xi \in W^{1,1}([s, t], X)$ such that $\dot{\xi}(\tau) = f(\xi(\tau), \eta(\tau), \tau)$ for a. e. $\tau \in [s, t]$, and $W^{1,1}([s, t], X)$ is the set of all absolutely continuous maps from $[s, t]$ to X, (v) $\mathcal{F}' = (\mathcal{X}, \mathbf{f}')$ is the flow of the full control system, so that, if $\mathbf{f}' = \{f'_{t,s}\}_{(t,s) \in I^{2,\geq}}$, then $f'_{t,s}(x)$, for $x \in X$ and $(t, s) \in I^{2,\geq}$, is the reachable set from x over the interval $[s, t]$, so $f'_{t,s}(x)$ is given by

$$f'_{t,s}(x) = \{\xi(t) : (\exists \eta \in \mathcal{U})(\xi \in \mathrm{Traj}(\eta, f, s, t)), \, \xi(s) = x\}.$$

(Notice that the maps $f_{t,s}$ are single-valued—that is, each set $f_{t,s}(x)$ is either empty or consists of a single member—if the ordinary differential equation $\dot{x} = f(x, \eta_*(t), t)$ has uniqueness of trajectories, but for more general reference vector fields $(x,t) \mapsto f(x, \eta_*(t), t)$ the $f_{t,s}$ can be set-valued. On the other hand, the $f'_{t,s}$ are never single-valued, except in trivial cases.)

The flow formulation, together with the use of general totally ordered sets rather than real intervals (cf. also Remark 2.1), includes situations other than that of control systems, such as, for example, "hybrid systems" in which the state is allowed to jump at some time t from a state space X_- to a state space X_+. (This is achieved by treating $t-$ and $t+$ as different times, with $t- < t+$, and having a family $\{J_\alpha\}_{\alpha \in A}$ of—possibly set-valued—jump maps from X_- to X_+, one of which is the reference jump map J_{α_*}. In that case, $f_{t+,t-}$ is the map J_{α_*}, and $f'_{t+,t-}$ is the map such that $f'_{t+,t-}(x) = \cup_{\alpha \in A} J_\alpha(x)$.)

Theorem 4.1. *Assume that $\mathcal{X} = (\mathbf{X}, I) = (\{X_t\}_{t \in I}, I)$ is a C^1 state-space bundle, $\mathcal{F} = (\mathcal{X}, \mathbf{f})$ is a flow, $\mathcal{F}' = (\mathcal{X}, \mathbf{f}')$ is a superflow of $\mathcal{F}$, $\mathbf{f} = \{f_{t,s}\}_{(t,s) \in I^{2,\geq}}$, $\mathbf{f}' = \{f'_{t,s}\}_{(t,s) \in I^{2,\geq}}$, $\xi \in Traj(\mathcal{F})$, and $\mathbf{g} = \{g_{t,s}\}_{(t,s) \in I^{2,\geq}}$ is an AGDQ of $\mathcal{F}$ along ξ. Let $\mathbf{V}$ be a set of infinitesimal impulse variations for $(\mathcal{F}, \xi)$ which is $\mathbf{g}$-AGDQ-summable within $\mathcal{F}'$. Let $a, b \in I$ be such that $a < b$, and let S be a subset of X_b such that $\xi(b) \in S$. Let $\mathcal{C}$ be an AGDQ-approximating multicone of S at $\xi(b)$. Assume that $f'_{b,a}(\xi(a)) \cap S = \{\xi(b)\}$. Then for every nonzero linear functional μ on $T_{\xi(b)} X_b$ there exist (i) a compatible selection $\gamma = \{\gamma_{t,s}\}_{a \leq s \leq t \leq b}$ of $\mathbf{g}$, (ii) covectors $\bar{\pi}, \tilde{\pi} \in T^*_{\xi(b)} X_b$, and (iii) a nonnegative real number π_0, such that $\pi_0 \mu = \bar{\pi} + \tilde{\pi}$, $(\pi_0, \bar{\pi}, \tilde{\pi}) \neq (0,0,0,)$, $\tilde{\pi} \in \mathcal{C}^\dagger$, and $\pi(t) \cdot v \leq 0$ for every $(v, t, \sigma) \in \mathcal{V}$ which is carried by $[a, b]$, where $\pi(t) = \bar{\pi} \circ \gamma_{b,t}$ for $a \leq t \leq b$.*

Proof. Fix a $\mu \in T^*_{\xi(b)} X_b \setminus \{0\}$. Let $\mathbf{V}_0$ be a finite subset of $\mathbf{V}$. Using the summability of $\mathbf{V}$, pick a variation $\{W_{\vec{p}}\}_{\vec{p} \in \mathbb{R}^{\mathbf{V}_0}_+}$ of $f_{b,a}$ in $f'_{b,a}$ such that the set $\Lambda^{\mathbf{V}_0, \mathbf{g}, a, b}$ is an AGDQ of the map $\tilde{W}$ at $((0, \xi(a)), \xi(b))$ along $\mathbb{R}^{\mathbf{V}_0}_+ \times X_a$. For each compatible selection γ of $\mathbf{g}$, let $\hat{L}^\gamma$ be the linear map $\mathbb{R}^{\mathbf{V}_0} \ni \vec{p} \mapsto L^{\mathbf{V}_0, \gamma, a, b}(\vec{p}, 0)$, so that $\hat{L}(\vec{p}) = \sum_{(v,t,\sigma) \in \mathbf{V}_0} p^{(v,t,\sigma)} \gamma_{b,t}(v)$. Let $\hat{\Lambda} = \{\hat{L}^\gamma : \gamma \in CSel(\mathbf{g})\}$. Then $\hat{\Lambda}$ is an AGDQ of the set-valued map $\mathbb{R}^{\mathbf{V}_0} \ni \vec{p} \mapsto \tilde{W}(\vec{p}, \xi(a)) \subseteq X_b$ at $(0, \xi(b))$ in the direction of $\mathbb{R}^{\mathbf{V}_0}_+$. Since $\tilde{W}(\vec{p}, \xi(a)) \subseteq f'_{b,a}(\xi(a))$, the set $\mathcal{M} = \{\hat{L}^\gamma \cdot \mathbb{R}^{\mathbf{V}_0}_+ : \gamma \in CSel(\mathbf{g})\}$ is an AGDQ-approximating multicone of the set $f'_{b,a}(\xi(a))$ at $\xi(b)$. Since $f'_{b,a}(\xi(a)) \cap S = \{\xi(b)\}$, Corollary 1.1 implies that there exists a decomposition $\pi_0 \mu = \bar{\pi} + \tilde{\pi}$, where $\bar{\pi} \in M^\dagger$ for some $M \in \mathcal{M}$, $\tilde{\pi} \in C^\dagger$ for some $C \in \mathcal{C}$, $\pi_0 \geq 0$, and $(\pi_0, \bar{\pi}, \tilde{\pi}) \neq (0,0,0)$. Since $M \in \mathcal{M}$, we can pick a $\gamma \in CSel(\mathbf{g})$ such that $M = \hat{L}^\gamma \cdot \mathbb{R}^{\mathbf{V}_0}_+$. Then the condition that $\bar{\pi} \in M^{\dagger}$ implies that $\langle \bar{\pi}, \hat{L}^{\gamma}(\vec{p}) \rangle \leq 0$ for every $\vec{p} \in \mathbb{R}_+^{\mathbf{V}_0}$. Therefore $\langle \bar{\pi}, \sum_{(v,t,\sigma) \in \mathbf{V}_0} p^{(v,t,\sigma)} \gamma_{b,t}(v) \rangle \leq 0$ for every $\vec{p} \in \mathbb{R}_+^{\mathbf{V}_0}$. This implies that $\langle \bar{\pi}, \gamma_{b,t}(v) \rangle \leq 0$—i.e., that $\langle \bar{\pi} \circ \gamma_{b,t}, v \rangle \leq 0$—for every $(v,t,\sigma) \in \mathbf{V}_0$. Furthermore, the fact that $\tilde{\pi} \in C^{\dagger}$ implies that $\tilde{\pi} \in C^{\dagger}$.

It follows that the 4-tuple $(\bar{\pi}, \tilde{\pi}, \pi_0, \gamma)$ satisfies all our conditions, except only for the fact that the inequality $\langle \bar{\pi} \circ \gamma_{b,t}, v \rangle \leq 0$ has only been shown to hold for (v,t,σ) in a finite subset $\mathbf{V}_0$ of $\mathbf{V}$. To prove the existence of a 4-tuple $(\bar{\pi}, \tilde{\pi}, \pi_0, \gamma)$ that satisfies $\langle \bar{\pi} \circ \gamma_{b,t}, v \rangle \leq 0$ for all $(v,t,\sigma) \in \mathbf{V}$, we use a familiar compactness argument. Fix a norm $\| \cdot \|$ on $T^*_{\xi_*(b)} X_b$. Let $\mathcal{Q}$ be the set of all 4-tuples $(\bar{\pi}, \tilde{\pi}, \pi_0, \gamma)$ such that $\bar{\pi} \in T^*_{\xi_*(b)} X_b$, $\tilde{\pi} \in T^*_{\xi_*(b)} X_b$, $\pi_0 \in \mathbb{R}$, $\pi_0 \geq 0$, $\pi_0 + \|\bar{\pi}\| + \|\tilde{\pi}\| = 1$, and $\gamma \in CSel(\mathbf{g})$. Then $\mathcal{Q}$ is a compact topological space, using on $CSel(\mathbf{g})$ the topology induced by the product topology of $\prod_{(t,s) \in I^{2,\geq}} g_{t,s}$. For each subset $\mathbf{U}$ of $\mathbf{V}$, let $\mathcal{Q}^{\mathbf{U}}$ be the set of those $(\bar{\pi}, \tilde{\pi}, \pi_0, \gamma) \in \mathcal{Q}$ such that $\tilde{\pi} \in C^{\dagger}$ and $\langle \bar{\pi} \circ \gamma_{b,t}, v \rangle \leq 0$ for all $(v,t,\sigma) \in \mathbf{U}$. Then every $\mathcal{Q}^{\mathbf{U}}$ is compact, and we have shown that $\mathcal{Q}^{\mathbf{U}}$ is nonempty if $\mathbf{U}$ is finite. Furthermore, it is clear that, if $\{\mathbf{U}_j\}_{j \in \{1,\dots,m\}}$ is a finite family of finite subsets of $\mathbf{V}$, then $\mathcal{Q}^{\mathbf{U}_1} \cap \cdots \cap \mathcal{Q}^{\mathbf{U}_m} = \mathcal{Q}^{\mathbf{U}_1 \cup \cdots \cup \mathbf{U}_m}$, so $\mathcal{Q}^{\mathbf{U}_1} \cap \cdots \cap \mathcal{Q}^{\mathbf{U}_m} \neq \emptyset$. If $\mathcal{U}$ is the set of all finite subsets of $\mathbf{V}$, we have shown that every finite intersection of members of the family $\{\mathcal{Q}^{\mathbf{U}}\}_{\mathbf{U} \in \mathcal{U}}$ is nonempty. Therefore the set $\bigcap \{\mathcal{Q}^{\mathbf{U}} : \mathbf{U} \in \mathcal{U}\}$ is nonempty. But $\bigcap \{\mathcal{Q}^{\mathbf{U}} : \mathbf{U} \in \mathcal{U}\} = \mathcal{Q}^{\mathbf{V}}$. So $\mathcal{Q}^{\mathbf{V}}$ is nonempty, concluding our proof. $\qquad\square$

5. Generalized Bianchini-Stefani IIVs and the summability theorem

We now present a class of IIVs that are infinitesimal generators of high-order variations in a sense that generalizes the definition proposed by Bianchini and Stefani.

We assume that we are given

(D1) a pair $(\mathcal{F}', \mathcal{F})$ of flows, where

    (D1. i) $\mathcal{F} = (\mathcal{X}, \mathbf{f})$ and $\mathcal{F}' = (\mathcal{X}, \mathbf{f}')$,

    (D1. ii) $\mathcal{X} = (\mathbf{X}, I)] = (\{X_t\}_{t \in I}, I)$ is a C^1 state-space bundle,

    (D1.iii) $\mathbf{f} = \{f_{t,s}\}_{(t,s) \in I^{2,\geq}}$ and $\mathbf{f}' = \{f'_{t,s}\}_{(t,s) \in I^{2,\geq}}$ are flows on the state=space bundle $\mathcal{X}$,

    (D1.iv) $\mathcal{F}$ is a subflow of $\mathcal{F}'$,

(D2) a "reference trajectory" $\xi_* \in Traj(\mathcal{F})$,

(D3) an AGDQ $\mathbf{g} = \{g_{t,s}\}_{(t,s) \in I^{2,\geq}}$ of $\mathcal{F}$ along ξ_*.

5.1. *Times of right and left regularity*

Definition 5.1. Given $\mathcal{F}'$, $\mathcal{F}$, ξ_*, $\mathbf{g}$ as above, a *time of right* (resp. *left*) *regularity* of $(\mathcal{F}', \mathcal{F}, \xi_*, \mathbf{g})$ is a time $\bar{t} \in I$ such that there exists a pair (t_*, X) for which

  (1) $t_* \in I$ and $\bar{t} <_I t_*$ (resp. $t_* <_I \bar{t}$),
  (2) X is a manifold of class C^1,
  (3) if we let $J \overset{\text{def}}{=} [\min(\bar{t}, t_*), \max(\bar{t}, t_*)]_I$, then

  (3.a) $X_t = X$ for all $t \in J$,
  (3.b) J is[†] a compact subinterval of $\mathbb{R}$,
  (3.c) the map $J \ni t \mapsto \xi_*(t) \in X$ is continuous,
  (3.d) the family $\{g_{t,s}\}_{s,t \in J, s \leq t}$ is a uniform AGDQ of the reference flow maps $f_{t,s}$, for $s, t \in J$, $s \leq t$, at $(\xi_*(s), \xi_*(t))$, in the direction of X,
  (3.e) $\lim_{t \downarrow \bar{t}} \sup\{\|\gamma - \mathbb{I}_{T_{\xi_*(\bar{t})} X}\| : \gamma \in g_{t,s}, \ s \in [\bar{t}, t]\} = 0$ (resp. $\lim_{s \uparrow \bar{t}} \sup\{\|\gamma - \mathbb{I}_{T_{\xi_*(\bar{t})} X}\| : \gamma \in g_{t,s}, \ t \in [s, \bar{t}]\} = 0$ (cf. Remark 5.1 below). $\qquad\square$

Remark 5.1. Condition (3.e) of the above definition is interpreted as follows: let κ be a coordinate chart of X such that, for some $\tilde{t}_* \in J \backslash \{\bar{t}\}$, the interval $\tilde{J} = [\min(\bar{t}, \tilde{t}_*), \max(\bar{t}, \tilde{t}_*)]_I$ is such that $\xi_*(t) \in \mathrm{Do}(\kappa)$ for every $t \in \tilde{J}$ (such a chart exists because of Condition (3.c)); we can then identify all the tangent spaces $T_x X$. for $x \in \mathrm{Do}(\kappa)$, with $\mathbb{R}^{\dim X}$; then, if $s, t \in \tilde{J}$ and $s \leq t$, all the members γ of $g_{t,s}$ are linear maps from $\mathbb{R}^{\dim X}$ to $\mathbb{R}^{\dim X}$, and so is $\mathbb{I}_{T_{\xi_*(\bar{t})} X}$, so the difference $\gamma - \mathbb{I}_{T_{\xi_*(\bar{t})} X}$ and its norm $\|\gamma - \mathbb{I}_{T_{\xi_*(\bar{t})} X}\|$ make sense. $\qquad\square$

5.2. *GBS IIVs*

Definition 5.2. Given $\mathcal{F}'$, $\mathcal{F}$, ξ_*, $\mathbf{g}$ as above, and a positive real number λ, a triple $(v, \bar{t}, +)$ such that $\bar{t} \in I$ and $v \in T_{\xi_*(\bar{t})} X_{\bar{t}}$ is a *generalized Bianchini-Stefani (abbr. GBS) right infinitesimal impulse variation of order $\frac{1}{\lambda}$ of $(\mathcal{F}', \mathcal{F}, \xi_*, \mathbf{g})$ at time $\bar{t}$* if

  (i) $\bar{t}$ is a time of right regularity of $(\mathcal{F}', \mathcal{F}, \xi_*, \mathbf{g})$

[†]We literally mean "is," rather than just "can be identified with." The reason is that, when we consider several impulse variations with the same time $\bar{t}$, we will not want the map identifying a right or left neighborhood of $\bar{t}$ with a real interval to depend on the variation.

(ii) if t_*, X, J are as in Definition 5.1, then there exists a 6-tuple $(\alpha, \beta, \bar{c}, \bar{\varepsilon}, \varphi, \mathcal{N})$ (called a **generator** of $(v, \bar{t}, +)$) such that

(ii.1) $0 < \alpha < \beta$, $\bar{c} > 0$, and $\bar{\varepsilon} > 0$,

(ii.2) $\varphi = \{\varphi_{c,\varepsilon}\}_{(c,\varepsilon) \in [0,\bar{c}] \times]0,\bar{\varepsilon}]}$ is a two-parameter family of set-valued maps from X to X,

(ii.3) $\mathcal{N}$ is an open neighborhood of $\xi_*(\bar{t})$ in X,

(ii.4) the set-valued map

$$\mathcal{N} \times [0, \bar{c}] \ni (x, c) \mapsto \varphi_\varepsilon(c, x) \stackrel{\text{def}}{=} \varphi_{c,\varepsilon}(x) \subseteq X$$

is Cellina continuosuly approximable for each $\varepsilon \in]0, \bar{\varepsilon}]$,

(ii.5) the maps φ_ε satisfy

$$\varphi_\varepsilon(0, x) \subseteq f_{\bar{t}+\beta\varepsilon^\lambda, \bar{t}+\alpha\varepsilon^\lambda}(x) \quad \text{for} \quad x \in \mathcal{N}, \tag{1}$$

$$\varphi_\varepsilon(c, x) \subseteq f'_{\bar{t}+\beta\varepsilon^\lambda, \bar{t}+\alpha\varepsilon^\lambda}(x) \quad \text{for} \quad (c, x) \in [0, \bar{c}] \times \mathcal{N}, \tag{2}$$

as well as the asymptotic conditions

$$\lim_{\varepsilon \downarrow 0, x \to \xi(\bar{t})} \varphi_\varepsilon(c, x) = \xi_*(\bar{t}) \quad \text{u.w.r.t.} \quad c \in [0, \bar{c}], \tag{3}$$

$$\varphi_\varepsilon(c, \xi_*(\bar{t} + \alpha\varepsilon^\lambda) + h) = \xi_*(\bar{t} + \beta\varepsilon^\lambda) + h + \varepsilon\, c\, v + o(\varepsilon + \|h\|)$$
$$\text{as} \quad \varepsilon \downarrow 0, h \to 0, \tag{4}$$

(cf. Remarks 5.2, 5.3), where "u.w.r.t." stands for "uniformly with respect to." $\qquad\square$

The definition of what it means for a triple $(v, \bar{t}, -)$ to be a *GBS left IIV of order* λ *of* $(\mathcal{F}', \mathcal{F}, \xi_*, \mathbf{g})$ *at time* $\bar{t}$ is similar, with obvious modifications.

Remark 5.2. Equation (3) is interpreted as follows: given any neighborhood U of $\xi_*(\bar{t})$ in X, there exist a positive number ε_* and a neighborhood U' of $\xi_*(\bar{t})$ in X such that $\varphi_\varepsilon(c, x) \subseteq U$ whenever $0 < \varepsilon \leq \varepsilon_*$, $x \in U$, and $c \in [0, \bar{c}]$. $\qquad\square$

Remark 5.3. In order to interpret Equation (4) precisely, we first agree, for each coordinate chart κ of X near $\xi_*(\bar{t})$ such that $\text{Do}(\kappa) \subseteq \mathcal{N}$, to write x^κ for the coordinate representation $\kappa(x)$ of a point $x \in \text{Do}(\kappa)$, and w^κ for the coordinate representation of a tangent vector $w \in T_x X$ (so that $w^\kappa = \kappa_*(w) = D\kappa(x) \cdot v \in \mathbb{R}^{\dim X}$). Then (3) implies—using Remark 5.2, with $U = \text{Do}(\kappa)$—that there exists a positive number $\varepsilon_* = \varepsilon_*(\kappa, \varphi)$ having the following properties:

P1. $0 < \varepsilon_* \leq \bar{\varepsilon}$,

P2. $\xi(t) \in \mathrm{Do}(\kappa)$ and $\bar{\mathbb{B}}(\xi_*(t)^\kappa, \varepsilon_*) \subseteq \mathrm{Im}(\kappa)$ whenever $\bar{t} \leq t \leq \bar{t} + \beta\varepsilon_*^\lambda$,

P3. $\varphi_\varepsilon(c, x) \subseteq \mathrm{Do}(\kappa)$ whenever $\bar{t} \leq t \leq \bar{t} + \beta\varepsilon_*^\lambda$, $0 < \varepsilon \leq \varepsilon_*$, $c \in [0, \bar{c}]$, and $x \in \kappa^{-1}(\bar{\mathbb{B}}(\xi_*(t)^\kappa, \varepsilon_*))$.

We then let $\varphi_\varepsilon^\kappa$ be, for $\varepsilon \in]0, \varepsilon_*]$, the set-valued map from $[0, \bar{c}] \times \bar{\mathbb{B}}(\xi_*(\bar{t})^\kappa, \varepsilon_*)$ to $\mathrm{Im}(\kappa)$ such that $\varphi_\varepsilon^\kappa(c, y) = (\varphi_\varepsilon(c, x))^\kappa$—i.e., $\varphi_\varepsilon^\kappa(c, y) = \{z^\kappa : z \in \varphi_\varepsilon(c, x)\}$—whenever $\varepsilon \in]0, \varepsilon_*]$, $c \in [0, \bar{c}]$, $x \in \mathrm{Do}(\kappa)$, $y \in \bar{\mathbb{B}}(\xi_*(\bar{t})^\kappa, \varepsilon_*)$ are such that $y = x^\kappa$ and $\varphi_\varepsilon(c, x) \subseteq \mathrm{Do}(\kappa)$. We then define the error E^κ by

$$E^\kappa(c, \varepsilon, h, y) = y - \xi_*(\bar{t} + \beta\varepsilon^\lambda)^\kappa - h - \varepsilon c v^\kappa,$$

for $y \in \mathrm{Im}(\kappa)$, $\varepsilon \in]0, \varepsilon_*]$, $c \in [0, \bar{c}]$, and $h \in \mathbb{R}^{\dim X}$, and observe that $E^\kappa(c, \varepsilon, h, y)$ belongs to $\mathbb{R}^{\dim X}$.

Then (4) is interpreted as asserting that

$$\lim_{\varepsilon \downarrow 0, h \to 0} \frac{\sup\left\{\|E^\kappa(c, \varepsilon, h, y)\| : c \in [0, \bar{c}], y \in \varphi_\varepsilon^\kappa\left(c, \xi_*(\bar{t} + \alpha\varepsilon^\lambda)^\kappa + h\right)\right\}}{\varepsilon + \|h\|} = 0.$$

It is easy to see that if this condition holds for some chart κ such that $\mathrm{Do}(\kappa) \subseteq \mathcal{N}$, then it holds for every such chart. $\square$

5.3. *The summability theorem for GBS IIVs*

The following result is then our summability theorem.

Theorem 5.1. *Let $\mathcal{F}', \mathcal{F}, \xi_*, \mathbf{g}$ be data as in (D1-2-3) above. Let $\mathcal{V}$ be the set of all generalized Bianchini-Stefani infinitesimal impulse variations of $(\mathcal{F}', \mathcal{F}, \xi_*, \mathbf{g})$. Then $\mathcal{V}$ is $\mathbf{g}$-AGDQ-summable within $\mathcal{F}'$.*

6. Proof of Theorem 5.1

We have to prove that, if $\mathbf{V}$ is a finite set of GBS IIVs of $(\mathcal{F}', \mathcal{F}, \xi_*, \mathbf{g})$, and a, b are such that $\mathbf{V}$ is carried by $[a, b]$, then there exists a variation $W = \{W_{\vec{p}}\}_{\vec{p} \in \mathbb{R}_+^{\mathbf{V}}}$ of $f_{b,a}$ in $f'_{b,a}$ such that the set $\Lambda^{\mathbf{V}, \mathbf{g}, a, b}$ is an AGDQ of $\tilde{W}$ at $((0, \xi_*(a)), \xi_*(b))$ along $\mathbb{R}_+^{\mathbf{V}} \times X_a$. It clearly suffices to consider the case when $\mathbf{V}$ is a nonempty finite set of GBS right IIVs at a point $\bar{t} \in I$, and to take $a = \bar{t}$.

Since $\mathbf{V}$ is nonempty, $\bar{t}$ is a time of right regularity for $(\mathcal{F}', \mathcal{F}, \xi_*, \mathbf{g})$, so we may pick t_*, X such that the conditions of Definition 5.1 hold. Clearly, we may restrict t_* further, and assume that $t_* \leq b$. Furthermore, we may assume that all the points $\xi_*(t)$, for $\bar{t} \leq t \leq t_*$, belong to the domain Ω of a coordinate chart κ of X.

Let the members of $\mathbf{V}$ be listed as $(v_1, \bar{t}, +), \ldots, (v_m, \bar{t}, +)$, in such a way that the inverse orders $\lambda_1, \ldots, \lambda_m$ satisfy $\lambda_1 \geq \lambda_2 \geq \ldots \geq \lambda_{m-1} \geq \lambda_m$. Then pick for each j a 6-tuple $(\alpha_j, \beta_j, \bar{\varepsilon}_j, \bar{c}_j, \varphi^j, \mathcal{N}^j)$ which is a generator of $(v_j, \bar{t}, +)$ in the sense of Definition 5.2. It is then easy to see that

(*) *without loss of generality, we may assume that*

    A1. *all the $\bar{\varepsilon}_j$ are equal to a positive number $\bar{\varepsilon}$ such that $\bar{\varepsilon} \leq 1$,*
    A2. *all the $\bar{c}_j$ are equal to a positive number $\bar{c}$,*
    A3. *the inequalities*

$$\beta_j \varepsilon^{\lambda_j} \leq \alpha_{j+1} \varepsilon^{\lambda_{j+1}} \quad \text{for all } j \in \{1, \ldots, m-1\}, \ \varepsilon \in]0, \bar{\varepsilon}] . \tag{5}$$

        are satisfied,
    A4. *the sets $\mathcal{N}^j$ all coincide;*
    A5. $\bar{t} + \beta_m \bar{\varepsilon}^{\lambda_m} \leq t_*$.

To see this, first replace each $\bar{\varepsilon}_j$ by $\min(\bar{\varepsilon}_j, 1)$, so all the $\bar{\varepsilon}^j$ are ≤ 1. Next, pick a particular j. Then if ρ is small enough,

$$\beta_j \rho^{\lambda_j} \varepsilon^{\lambda_j} \leq \alpha_{j+1} \varepsilon^{\lambda_{j+1}} \quad \text{for all } \varepsilon \in]0, \bar{\varepsilon}_j] , \tag{6}$$

because (i) $\beta_j \rho^{\lambda_j} \leq \alpha_{j+1}$ for small enough ρ (since λ_j, α_{j+1} and β_j are positive), and then (ii) the inequalities $\beta_j \rho^{\lambda_j} \varepsilon^{\lambda_j} \leq \alpha_{j+1} \varepsilon^{\lambda_j} \leq \alpha_{j+1} \varepsilon^{\lambda_{j+1}}$ hold for $0 < \varepsilon \leq \bar{\varepsilon}_j$, because $\lambda_j \geq \lambda_{j+1}$ and $\bar{\varepsilon}_j \leq 1$. Then we may pick a ρ such that (6) holds, and replace the numbers α_j, β_j, and the family $\varphi^j = \{\varphi^j_{c,\varepsilon}\}_{c \in [0, \bar{c}_j], \varepsilon \in]0, \bar{\varepsilon}_j]}$ by the numbers α_j^{new}, β_j^{new} and the family $\varphi^{j,new} = \{\varphi^{j,new}_{c,\varepsilon}\}_{c \in [0, \bar{c}_j^{new}], \varepsilon \in]0, \bar{\varepsilon}_j^{new}]}$, where $\beta_j^{new} = \beta_j \rho^{\lambda_j}$, $\alpha_j^{new} = \alpha_j \rho^{\lambda_j}$, $\bar{c}_j^{new} = \rho \bar{c}_j$, $\bar{\varepsilon}_j^{new} = \min(1, \rho^{-1} \bar{\varepsilon}_j)$, and $\varphi^{j,new}_{c,\varepsilon} = \varphi^j_{\rho^{-1}c, \rho\varepsilon}$ whenever $c \in [0, \bar{c}_j^{new}]$ and $\varepsilon \in]0, \bar{\varepsilon}_j^{new}]$. Then, if we let $\varphi^{j,new}_\varepsilon(c, x) = \varphi^{j,new}_{c,\varepsilon}(x)$, it follows that

$$\varphi^{j,new}_\varepsilon(c, \xi_*(\bar{t} + \alpha_j^{new} \varepsilon^{\lambda_j}) + h) = \xi_*(\bar{t} + \beta_j^{new} \varepsilon^{\lambda_j}) + h + (\rho\varepsilon)(\rho^{-1}c)v + o(\varepsilon + \|h\|) ,$$

so that

$$\varphi^{j,new}_\varepsilon(c, \xi_*(\bar{t} + \alpha_j^{new} \varepsilon^{\lambda_j}) + h) = \xi_*(\bar{t} + \beta_j^{new} \varepsilon^{\lambda_j}) + h + \varepsilon c v + o(\varepsilon + \|h\|) .$$

This means that the 6-tuple $(\alpha_j^{new}, \beta_j^{new}, \bar{c}_j^{new}, \bar{\varepsilon}_j^{new}, \varphi^{j,new}, \mathcal{N}^j)$ is also a generator of $(v_j, \bar{t}, +)$ and, after $(\alpha_j, \beta_j, \bar{c}_j, \bar{\varepsilon}_j, \varphi^j, \mathcal{N}^j)$ is replaced by $(\alpha_j^{new}, \beta_j^{new}, \bar{c}_j^{new}, \bar{\varepsilon}_j^{new}, \varphi^{j,new}, \mathcal{N}^j)$, the desired inequality (5) holds for our particular j. To get the inequality to hold for all j, we just carry out the replacements recursively, starting with $j = m - 1$ and moving backwards up to $j = 1$. Finally, when this is finished, we replace all the $\bar{\varepsilon}_j$ by their minimum, and do the same for the $\bar{c}_j$, thus obtaining a new family $\{(\alpha_j, \beta_j, \bar{\varepsilon}_j, \bar{c}_j, \varphi^j, \mathcal{N}^j)\}_{j=1,\ldots,m}$ of generators of the $(v_j, \bar{t}, +)$ that satisfy

(A1,2,3). To get (A4) and (A5) to hold as well, we let $\mathcal{N}^{new} = \Omega \cap (\cap_{j=1}^{m} \mathcal{N}^j)$, and replace each $\mathcal{N}^j$ by $\mathcal{N}^{new}$ and each family φ^j by the family $\hat{\varphi}^j$ of the restrictions of the φ_{ε}^j to $[0, \bar{c}] \times \mathcal{N}^{new}$. We then observe that the 6-tuples $(\alpha_j, \beta_j, \bar{\varepsilon}_j, \bar{c}_j, \hat{\varphi}^j, \mathcal{N}^{new})$ are also generators of the $(v_j, \bar{t}, +)$ that satisfy (A1,2,3,4). Finally, we make t_* smaller, if necessary, to guarantee that the set $\{\xi_*(t) : \bar{t} \leq t \leq t_*\}$ is contained in $\mathcal{N}^{new}$, and then make $\bar{\varepsilon}$ smaller, if necessary, to satisfy (A5).

We then use κ to identify Ω with an open subset of $\mathbb{R}^n$—where $n = \dim X$. Then all the tangent spaces $T_x X$, for all $x \in \Omega$, are identified with $\mathbb{R}^n$. Since

$$\limsup_{t \downarrow \bar{t}} \left\{ \|\gamma - \mathbb{I}_{\mathbb{R}^n}\| : \gamma \in g_{t,s}, \ s \in [\bar{t}, t] \right\} = 0, \tag{7}$$

we may assume, after making t_* and $\bar{\varepsilon}$ even smaller, that

$$\|\gamma\| \leq 2 \quad \text{whenever} \ \gamma \in g_{t,s} \ \text{and} \ \bar{t} \leq s \leq t \leq t_* . \tag{8}$$

We now use the fact that $\{g_{t,s}\}_{\bar{t} \leq s \leq t \leq t_*}$ is a uniform AGDQ of the maps $f_{t,s}$ at the points $(\xi_*(s), \xi_*(t))$ in the direction of X to choose a function $\theta \in \Theta$ which is an AGDQ modulus for all the 4-tuples $(f_{t,s}, \xi_*(s), \xi_*(t), X)$, for all s, t such that $\bar{t} \leq s \leq t \leq t_*$. We then fix a real number $\tilde{\varepsilon}$ such that (i) $0 < \tilde{\varepsilon} \leq \bar{\varepsilon}$, (ii) $\theta(\tilde{\varepsilon}) \leq 1$, and (iii) the closed ball $\bar{\mathbb{B}}^n(\xi_*(s), \tilde{\varepsilon})$ is contained in Ω for all $s \in [\bar{t}, t_*]$. We then choose, for each $\varepsilon \in \,]0, \tilde{\varepsilon}]$ and each pair (s, t) such that $\bar{t} \leq s \leq t \leq t_*$, a CCA map $A_{t,s}^{\varepsilon} : \bar{\mathbb{B}}^n(\xi_*(s), \varepsilon) \longmapsto Aff(\mathbb{R}^n, \mathbb{R}^n)$, taking values in $g_{t,s}^{(\theta(\varepsilon), \varepsilon)}$, such that $\xi_*(t) + A(h) \in f_{t,s}(\xi_*(s) + h)$. whenever $h \in \bar{\mathbb{B}}^n(0, \varepsilon)$ and $A \in A_{t,s}^{\varepsilon}(\xi_*(s) + h)$. We define set-valued maps $\hat{A}_{t,s}^{\varepsilon} : \bar{\mathbb{B}}^n(\xi_*(s), \varepsilon) \longmapsto \mathbb{R}^n$, for $\varepsilon \in \,]0, \tilde{\varepsilon}]$, $\bar{t} \leq s \leq t \leq t_*$, by letting

$$\hat{A}_{t,s}^{\varepsilon}(\xi_*(s) + h) = \xi(t) + A_{t,s}^{\varepsilon}(h)(h) \tag{9}$$

(that is, $\hat{A}_{t,s}^{\varepsilon}(\xi_*(s) + h) = \{\xi(t) + A(h) : A \in A_{t,s}^{\varepsilon}(\xi_*(s) + h)\}$) for $h \in \bar{\mathbb{B}}^n(0, \varepsilon)$. It is then clear that $\hat{A}_{t,s}^{\varepsilon} \in CCA(\bar{\mathbb{B}}^n(\xi_*(s), \varepsilon), \mathbb{R}^n)$, and the estimate

$$\|y - \xi_*(t)\| \leq 4\|x - \xi_*(s)\| \quad \text{whenever} \ x \in \bar{\mathbb{B}}^n(\xi_*(s), \varepsilon) , \ y \in \hat{A}_{t,s}^{\varepsilon}(x) \tag{10}$$

holds. In particular,

$$\hat{A}_{t,s}^{\varepsilon}\left(\bar{\mathbb{B}}^n(\xi_*(s), \rho)\right) \subseteq \bar{\mathbb{B}}^n(\xi_*(t), 4\rho) \subseteq \Omega \quad \text{if} \ 0 < \rho \leq \varepsilon \leq \frac{\tilde{\varepsilon}}{4} . \tag{11}$$

In addition, it is clear that

$$\hat{A}_{t,s}^{\varepsilon}(x) \subseteq f_{t,s}(x) \quad \text{whenever} \ x \in \bar{\mathbb{B}}^n(\xi_*(s), \varepsilon) \ \text{and} \ 4\varepsilon \leq \tilde{\varepsilon} . \tag{12}$$

Next, we pick positive numbers $\varepsilon_{*,j} = \varepsilon_{*,j}(\kappa, \varphi^j)$ that satisfy the properties of Remark 5.3 for the φ^j, and are such that $\varepsilon_{*,j} \leq \tilde{\varepsilon}$. We let $\varepsilon_* = \min\{\varepsilon_{*,j} : j = 1, \ldots, m\}$. It then follows that

$$\varphi^j_\varepsilon(c, x) \subseteq \Omega \text{ if } \varepsilon \leq \varepsilon_*, \; 0 \leq c \leq \bar{c}, \; \bar{t} \leq t \leq \bar{t} + \beta_j \varepsilon_*^{\lambda_j}, \; x \in \bar{\mathbb{B}}(\xi_*(t), \varepsilon_*). \quad (13)$$

We then define the errors E_j by

$$E_j(c, \varepsilon, h, y) = y - \xi_*(\bar{t} + \beta \varepsilon^\lambda) - h - \varepsilon c v_j.$$

for $y \in \mathbb{R}^n$, $\varepsilon \in \,]0, \varepsilon_{*,j}(\kappa, \varphi^j)]$, $c \in [0, \bar{c}]$, and $h \in \mathbb{R}^n$, so $E_j(c, \varepsilon, h, y) \in \mathbb{R}^n$. We then let $\zeta_*(\varepsilon)$, for $0 < \varepsilon \leq \varepsilon_*$, be the supremum of the numbers $\|E_j(c, \rho, h, y)\|$ taken over all $c \in [0, \bar{c}]$, $j \in \{1, \ldots, m\}$, $h \in \mathbb{R}^n$, $\rho \in \,]0, \varepsilon]$, such that $\|h\| \leq \varepsilon$, and $y \in \varphi^j(c, \xi_*(\bar{t} + \alpha_j \varepsilon^{\lambda_j}) + h)$. We define $\theta_*(\varepsilon) = \sup\{\rho^{-1}\zeta_*(\rho) : 0 < \rho \leq \varepsilon\}$ for $0 < \varepsilon \leq \varepsilon_*$, and $\theta_*(\varepsilon) = +\infty$ for $\varepsilon > \varepsilon_*$. We then observe that the function θ_* belongs to Θ.

Now, if $\varepsilon \in \,]0, \varepsilon_*]$, $c \in [0, \bar{c}]$, $0 < \rho \leq \varepsilon$, $x \in \bar{\mathbb{B}}(\xi_*(\bar{t} + \alpha_j \varepsilon^{\lambda_j}), \rho)$, and $y \in \varphi^j_\varepsilon(c, x)$, we have $\|y - \xi_*(\bar{t} + \beta_j \varepsilon^{\lambda_j}) - h - \varepsilon c v_j\| \leq \zeta_*(\varepsilon) \leq \varepsilon \theta_*(\varepsilon)$, where $h = x - \xi_*(\bar{t} + \alpha_j \varepsilon^{\lambda_j})$. Since $\|h\| \leq \rho$, we conclude that

$$\|y - \xi_*(\bar{t} + \beta_j \varepsilon^{\lambda_j})\| \leq \rho + \varepsilon(\bar{c}\|v_j\| + \theta_*(\varepsilon)).$$

We fix $\varepsilon_\#$ such that $0 < \varepsilon_\# \leq \varepsilon_*$ and $\theta_*(\varepsilon_\#) \leq 1$, and let $C = \bar{c}\max(\|v_1\|, \ldots, \|v_m\|) + \theta_*(\varepsilon_\#)$. Then

$$\left(\varepsilon \in \,]0, \varepsilon_\#] \wedge c \in [0, \bar{c}] \wedge x \in \bar{\mathbb{B}}(\xi_*(\bar{t} + \alpha_j \varepsilon^{\lambda_j}), \rho) \wedge y \in \varphi^j_\varepsilon(c, x)\right) \Rightarrow$$

$$\|y - \xi_*(\bar{t} + \beta_j \varepsilon^{\lambda_j})\| \leq \rho + C\varepsilon$$

so

$$\left(\varepsilon \in \,]0, \varepsilon_\#] \wedge c \in [0, \bar{c}] \wedge x \in \bar{\mathbb{B}}(\xi_*(\bar{t} + \alpha_j \varepsilon^{\lambda_j}), \rho)\right) \Rightarrow$$

$$\varphi^j_\varepsilon(c, x) \subseteq \bar{\mathbb{B}}(\xi_*(\bar{t} + \beta_j \varepsilon^{\lambda_j}), \rho + C\varepsilon). \quad (14)$$

In order to construct our variation W, we first define, for $0 < \varepsilon \leq \varepsilon_\#$,

$$\Xi^0_\varepsilon(x) = \hat{A}^\varepsilon_{\bar{t}+\alpha_1\varepsilon^{\lambda_1},\bar{t}}(x), \quad \Psi^1_\varepsilon(c_1, x) = \varphi^1_\varepsilon(c_1, \Xi^0_\varepsilon(x))$$

(that is, $\Psi^1_\varepsilon(c_1, x) = \bigcup\{\varphi^1_\varepsilon(c_1, y) : y \in \Xi^0_\varepsilon(x)\}$) and then define, recursively,

$$\Xi^j_\varepsilon = \hat{A}^\varepsilon_{\bar{t}+\alpha_{j+1}\varepsilon^{\lambda_{j+1}},\bar{t}+\beta_j\varepsilon^{\lambda_j}},$$

$$\Psi^{j+1}_\varepsilon(c_1, \ldots, c_{j+1}, x) = \varphi^{j+1}_\varepsilon\left(c_{j+1}, \Xi^j_\varepsilon\left(\Psi^j_\varepsilon(c_1, \ldots, c_j, x)\right)\right)$$

(that is, $\Psi^{j+1}_\varepsilon(c_1, \ldots, c_{j+1}, x)$ is the union of the sets $\varphi^{j+1}_\varepsilon(c_1, \ldots, c_j, y)$ for all $y \in \Xi^j_\varepsilon(\Psi^j_\varepsilon(c_1, \ldots, c_j, x)))$ for $j = 1, \ldots, m - 1$.

Next, we define

$$\Upsilon_\varepsilon(c_1,\ldots,c_m,x) = \hat{A}^\varepsilon_{t_*,\bar{t}+\beta_m\varepsilon^{\lambda_m}}\left(\Psi^m_\varepsilon(c_1,\ldots,c_m,x)\right).$$

Successive applications of (11) and (14) show that

- if $0 < \rho$ and $4\rho \le \varepsilon$, then $\Xi^0_\varepsilon\left(\bar{\mathbb{B}}(\xi_*(\bar{t}),\rho)\right) \subseteq \bar{\mathbb{B}}(\xi_*(\bar{t}+\alpha_1\varepsilon^{\lambda_1}),4\rho)$,
- if $0 < \rho$ and $4\rho + C\varepsilon \le \varepsilon$, then

$$\Psi^1_\varepsilon\left([0,\bar{c}]\times\bar{\mathbb{B}}(\xi_*(\bar{t}),\rho)\right) \subseteq \bar{\mathbb{B}}\left(\xi_*(\bar{t}+\beta_1\varepsilon^{\lambda_1}),4\rho+C\varepsilon\right),$$

- if $0 < \rho$ and $16\rho + 4C\varepsilon \le \varepsilon$, then

$$\Xi^1_\varepsilon\left(\bar{\mathbb{B}}(\xi_*(\bar{t}+\beta_1\varepsilon^{\lambda_1}),4\rho+C\varepsilon)\right) \subseteq \bar{\mathbb{B}}(\xi_*(\bar{t}+\alpha_2\varepsilon^{\lambda_2},16\rho+4C\varepsilon),$$

- if $0 < \rho$ and $16\rho + 5C\varepsilon \le \varepsilon$, then

$$\Psi^2_\varepsilon\left([0,\bar{c}]^2\times\bar{\mathbb{B}}(\xi_*(\bar{t}),\rho)\right) \subseteq \bar{\mathbb{B}}(\xi_*(\bar{t}+\beta_1\varepsilon^{\lambda_1}),16\rho+5C\varepsilon),$$

- if $0 < \rho$ and $16\rho + 4C\varepsilon \le \varepsilon$, then

$$\Xi^2_\varepsilon\left(\bar{\mathbb{B}}(\xi_*(\bar{t}+\beta_2\varepsilon^{\lambda_2}),4\rho+C\varepsilon)\right) \subseteq \bar{\mathbb{B}}(\xi_*(\bar{t}+\alpha_3\varepsilon^{\lambda_3}),64\rho+20C\varepsilon),$$

and so on, so that, for every $j \in \{1,\ldots,m\}$, if we let $G_j = 3^{-1}(4^j - 1)$, then

- if $0 < \rho$ and $4^j\rho + G_jC\varepsilon \le \varepsilon$, then

$$\Psi^j_\varepsilon\left([0,\bar{c}]^j\times\bar{\mathbb{B}}(\xi_*(\bar{t}),\rho)\right) \subseteq \bar{\mathbb{B}}(\xi_*(\bar{t}+\beta_j\varepsilon^{\lambda_j}),4^j\rho+G_jC\varepsilon).$$

In particular,

- if $0 < \rho$ and $4^m\rho + G_mC\varepsilon \le \varepsilon$, then

$$\Psi^m_\varepsilon\left([0,\bar{c}]^j\times\bar{\mathbb{B}}(\xi_*(\bar{t}),\rho)\right) \subseteq \bar{\mathbb{B}}(\xi_*(\bar{t}+\beta_m\varepsilon^{\lambda_m}),4^m\rho+G_mC\varepsilon).$$

We now choose $\varepsilon_\#$ and $\bar{c}$ so that $C \le \frac{1}{2G_m}$, and conclude that

- if $0 < 4^m\rho \le \frac{\varepsilon}{2}$ and $\varepsilon \le \varepsilon_\#$, then

$$\Psi^j_\varepsilon\left([0,\bar{c}]^j\times\bar{\mathbb{B}}(\xi_*(\bar{t}),\rho)\right) \subseteq \bar{\mathbb{B}}(\xi_*(\bar{t}+\beta_j\varepsilon^{\lambda_j}),\varepsilon).$$

¿From now on, for each $\varepsilon \in]0,\varepsilon_\#]$ we fix $\rho = \rho(\varepsilon) = 2^{-1-2m}\varepsilon$, so $0 < 4^m\rho \le \frac{\varepsilon}{2}$, define $Q^j_\varepsilon = [0,\bar{c}]^j \times \bar{\mathbb{B}}(\xi_*(\bar{t}),\rho)$, and let $\hat{\Psi}^m_\varepsilon$, $\hat{\Upsilon}^m_\varepsilon$, be the restrictions of Ψ^m_ε, Υ_ε, to Q^m_ε. Then $\hat{\Psi}^m_\varepsilon$ and $\hat{\Upsilon}_\varepsilon$ are set-valued maps from Q_ε to

352

$\bar{\mathbb{B}}(\xi_*(\bar{t} + \beta_m \varepsilon^{\lambda_m}), \varepsilon)$ and $\bar{\mathbb{B}}(\xi_*(\bar{t} + \beta_m \varepsilon^{\lambda_m}), 4\varepsilon)$, respectively. Furthermore, $\hat{\Psi}_\varepsilon^m$ and $\hat{\Upsilon}_\varepsilon$ are composites of CCA maps, so

$$\hat{\Psi}_\varepsilon^m \in CCA\Big(Q_\varepsilon, \bar{\mathbb{B}}(\xi_*(\bar{t} + \beta_m \varepsilon^{\lambda_m}), \varepsilon)\Big),$$

$$\hat{\Upsilon}_\varepsilon \in CCA\Big(Q_\varepsilon, \bar{\mathbb{B}}(\xi_*(t_*), 4\varepsilon)\Big).$$

In addition, it is easy to see that

$$\hat{\Upsilon}_\varepsilon(c_1, \ldots, c_m, x) \subseteq f'_{t_*, \bar{t}}(x) \quad \text{whenever } 0 < \varepsilon \leq \varepsilon_\#, \ (c_1, \ldots, c_m, x) \in Q_\varepsilon,$$

$$\hat{\Upsilon}_\varepsilon(0, x) \subseteq f_{t_*, \bar{t}}(x) \quad \text{whenever } 0 < \varepsilon \leq \varepsilon_\#, \ x \in \bar{\mathbb{B}}(\xi_*(\bar{t}), \rho).$$

We now define $Q_\varepsilon^0 = \bar{\mathbb{B}}(\xi_*(\bar{t}), \rho(\varepsilon))$, $\sigma_0 = \bar{t}$. For $j = 1, \ldots, m$, we write

$$\tau_j = \bar{t} + \alpha_j \varepsilon^{\lambda_j}, \quad \sigma_j = \bar{t} + \alpha_j \varepsilon^{\lambda_j}, \quad \vec{c}_j = (c_1, \ldots, c_j), \quad \vec{c}_j \cdot v = c_1 v_1 + \cdots + c_j v_j.$$

Lemma 6.1. *There exists a family* $\mathbf{\Gamma} = \{\Gamma_\varepsilon^j\}_{j=0,1,\ldots,m}$ *of CCA maps* $\Gamma_\varepsilon^j :$ $Q_\varepsilon^j \longmapsto Aff(\mathbb{R}^n, \mathbb{R}^n)$ *such that if* $(\vec{c}_j, x) \in Q_\varepsilon^j$ *then*

$$\Psi_\varepsilon^j(\vec{c}_j, x) = \xi_*(\sigma_j) + \Gamma_\varepsilon^j(\vec{c}_j, x)(x - \xi_*(\bar{t}) + \varepsilon(\vec{c}_j \cdot v)). \tag{15}$$

Furthermore, $\mathbf{\Gamma}$ *can be chosen so that*

> $(\#)$ *there exists a family* $\{\theta_j\}_{j=0,1,\ldots,m}$ *of members of* Θ *such that, for each* j, $\Gamma_\varepsilon^j(\vec{c}_j, x) \subseteq g_{\sigma_j, \bar{t}}^{(\theta_j(\varepsilon), \varepsilon)}$ *for all* $(\vec{c}_j, x) \in Q_\varepsilon^j$.

Proof of Lemma 6.1. We define the set-valued maps Γ_ε^j and the functions θ_j, for $j = 0, 1, \ldots, m$, recursively. We first let $\Gamma_\varepsilon^0 : Q_\varepsilon^0 \longmapsto Aff(\mathbb{R}^n, \mathbb{R}^n)$ be the map such that $\Gamma_\varepsilon^0(x) = \mathbb{I}_{\mathbb{R}^n}$ for each x, and take θ_0 to be any member of Θ (for example, $\theta_0(\varepsilon) \equiv \varepsilon$).

Next, we carry out the inductive step. We pick a $j \in \{1, \ldots, m\}$ and assume that Γ_ε^{j-1} has been defined. To construct Γ_ε^j, we begin by letting

$$\mathcal{P} = Aff(\mathbb{R}^n, \mathbb{R}^n) \times Aff(\mathbb{R}^n, \mathbb{R}^n) \times Aff(\mathbb{R}^n, \mathbb{R}^n) \times \mathbb{R}^n \times \mathbb{R}^n,$$

and defining $\mathcal{M}_\varepsilon^j$ be the set-valued map from Q_ε^j to $\mathcal{P}$ that sends each point $(\vec{c}_j, x) \in Q_\varepsilon^j$ to the subset $\mathcal{M}_\varepsilon^j(\vec{c}_j, x)$ of $\mathcal{P}$ that consists of all the 5-tuples (A_0, A_1, A_2, u, w) for which

$$A_0 \in \Gamma_\varepsilon^{j-1}(\vec{c}_{j-1}, x), \tag{16}$$

$$u = \xi_*(\sigma_{j-1}) + A_0(x - \xi_*(\bar{t}) + \varepsilon(\vec{c}_{j-1} \cdot v)) \tag{17}$$

$$A_1 \in A_{\tau_j, \sigma_{j-1}}^\varepsilon(u), \tag{18}$$

$$w = \xi_*(\tau_j) + A_1(u - \xi_*(\sigma_{j-1})), \tag{19}$$

$$A_2 \in A_{\sigma_j, \tau_j}^\varepsilon(w - \xi_*(\tau_j)). \tag{20}$$

We then define $\Gamma_\varepsilon^j(\vec{c}_j, x)$ to be the set of all $A \in Aff(\mathbb{R}^n, \mathbb{R}^n)$ such that

$$A = A_2 \circ A_1 \circ A_0 + affm_{0,z}$$

for some $z \in \varphi_\varepsilon^j(c_j, w) - \xi_*(\sigma_j) - (A_2 \circ A_1 \circ A_0)(x - \xi_*(\bar{t}) + \varepsilon(\vec{c}_j \cdot v))$ and some $(A_0, A_1, A_2, u, w) \in \mathcal{M}_\varepsilon^j(\vec{c}_j, x)$. (Recall that if $L \in Lin(\mathbb{R}^p, \mathbb{R}^q)$ and $z \in \mathbb{R}^q$ then $affm_{L,z}$ is the affine map $\mathbb{R}^p \ni x \mapsto L \cdot x + z \in \mathbb{R}^q$.)

If $A \in \Gamma_\varepsilon^j(\vec{c}_j, x)$, then there exist A_0, A_1, A_2, u, w, z, y, w, such that $(A_0, A_1, A_2, u, w) \in \mathcal{M}_\varepsilon^j(\vec{c}_j, x)$, $w = \xi_*(\tau_j) + A_1(u - \xi_*(\sigma_{j-1}))$, $z = y - \xi_*(\sigma_j) - (A_2 \circ A_1 \circ A_0)(x - \xi_*(\bar{t}) + \varepsilon(\vec{c}_j \cdot v))$, $y \in \varphi_\varepsilon^j(c_j, w)$, and $A = A_2 \circ A_1 \circ A_0 + affm_{0,z}$. It follows that

$$\begin{aligned}
A(x - \xi_*(\bar{t}) + \varepsilon\vec{c}_j \cdot v) &= A_2(A_1(A_0(x - \xi_*(\bar{t}) + \varepsilon(\vec{c}_j \cdot v)))) + z \\
&= (A_2 \circ A_1 \circ A_0)(x - \xi_*(\bar{t}) + \varepsilon(\vec{c}_j \cdot v)) + y \\
&\quad -\xi_*(\sigma_j) - (A_2 \circ A_1 \circ A_0)(x - \xi_*(\bar{t}) + \varepsilon(\vec{c}_j \cdot v)) \\
&= y - \xi_*(\sigma_j)\,,
\end{aligned}$$

so that $A(x - \xi_*(\bar{t}) + \varepsilon(\vec{c}_j \cdot v)) \in \varphi_\varepsilon^j(c_1, w) - \xi_*(\sigma_j)$. Furthermore, since $w = \xi_*(\tau_j) + A_1(u - \xi_*(\sigma_{j-1}))$ and $A_1 \in A_{\tau_j, \sigma_{j-1}}^\varepsilon(u)$, we see that $w \in \Xi_\varepsilon^j(u)$. So $A(x - \xi_*(\bar{t}) + \varepsilon(\vec{c}_j \cdot v)) \in \varphi_\varepsilon^j(c_j, \Xi_\varepsilon^j(u)) - \xi_*(\sigma_j)$. Since $A_0 \in \Gamma_\varepsilon^{j-1}(\vec{c}_{j-1}, x)$, and (15) holds for $j - 1$, so that

$$\xi_*(\sigma_{j-1}) + \Gamma_\varepsilon^{j-1}(\vec{c}_{j-1}, x)(x - \xi_*(\bar{t}) + \varepsilon(\vec{c}_{j-1} \cdot v)) = \Psi_\varepsilon^{j-1}(\vec{c}_{j-1}, x)\,,$$

it follows from (17) that $u \in \Psi_\varepsilon^{j-1}(\vec{c}_{j-1}, x)$. Therefore $A(x - \xi_*(\bar{t}) + \varepsilon(\vec{c}_j \cdot v)) \in \varphi_\varepsilon^j\left(c_j, \Xi_\varepsilon^j(\Psi_\varepsilon^{j-1}(\vec{c}_{j-1}, x))\right) - \xi_*(\sigma_j)$, so

$$A(x - \xi_*(\bar{t}) + \varepsilon(\vec{c}_j \cdot v)) \in \Psi_\varepsilon^j(\vec{c}_j, x) - \xi_*(\sigma_j)\,,$$

and then $\xi_*(\sigma_j) + A(x - \xi_*(\bar{t}) + \varepsilon(\vec{c}_j \cdot v)) \in \Psi_\varepsilon^j(\vec{c}_j, x)$. Since A is an arbitrary member of $\Gamma_\varepsilon^j(\vec{c}_j, x)$, we conclude that

$$\xi_*(\sigma_j) + \Gamma_\varepsilon^j(\vec{c}_j, x)(x - \xi_*(\bar{t}) + \varepsilon(\vec{c}_j \cdot v)) \subseteq \Psi_\varepsilon^j(\vec{c}_j, x)\,. \tag{21}$$

To prove the opposite inclusion, we pick $y \in \Psi_\varepsilon^j(\vec{c}_j, x)$, and find $u \in \Psi_\varepsilon^{j-1}(\vec{c}_{j-1}, x)$ and $w \in \Xi_\varepsilon^j(u)$ such that $y \in \varphi_\varepsilon^j(c_j, w)$. Since

$$\Xi_\varepsilon^j(u) = \hat{A}_{\tau_j, \sigma_{j-1}}(u) = \xi_*(\tau_j) + A_{\tau_j, \sigma_{j-1}}(u)(u - \xi_*(\sigma_{j-1}))\,,$$

we can find $A_1 \in A_{\tau_j, \sigma_{j-1}}(u)$ such that $w = \xi_*(\tau_j) + A_1(u - \xi_*(\sigma_{j-1}))$. Since $u \in \Psi_\varepsilon^{j-1}(\vec{c}_{j-1}, x)$, the inductive hypothesis (i.e., that (15) holds for $j - 1$) implies that we can pick $A_0 \in \Gamma_\varepsilon^{j-1}(\vec{c}_{j-1}, x)$ such that $u = \xi_*(\sigma_{j-1}) + A_0(x - \xi_*(\bar{t}) + \varepsilon(\vec{c}_{j-1} \cdot v))$. Pick an arbitrary member A_2 of $A_{\sigma_j, \tau_j}(w - \xi_*(\tau_j))$. Then the 5-tuple (A_0, A_1, A_2, u, w) belongs to $\mathcal{M}_\varepsilon^j(\vec{c}_j, x)$. Let $z = y - \xi_*(\sigma_j) - (A_2 \circ A_1 \circ A_0)(x - \xi_*(\bar{t}) + \varepsilon(\vec{c}_j \cdot v))$, and define an affine map A by letting $A = A_2 \circ A_1 \circ A_0 + \textit{affm}_{0,z}$. Then A belongs to $\Gamma^j_\varepsilon(\vec{c}_j, x)$, and $y = \xi_*(\sigma_j) + A(x - \xi_*(\bar{t}) + \varepsilon(\vec{c}_j \cdot v))$, so y is a member of $\xi_*(\sigma_j) + \Gamma^j_\varepsilon(\vec{c}_j, x)(x - \xi_*(\bar{t}) + \varepsilon(\vec{c}_j \cdot v))$.

Since y was an arbitrary member of $\Psi^j_\varepsilon(\vec{c}_j, x)$, we have shown that $\Psi^j_\varepsilon(c_1, x) \subseteq \xi_*(\sigma_j) + \Gamma^j_\varepsilon(\vec{c}_j, x)(x - \xi_*(\bar{t}) + \varepsilon(\vec{c}_j \cdot v))$. This fact, together with (21), implies that

$$\Psi^j_\varepsilon(c_1, x) = \xi_*(\sigma_j) + \Gamma^j_\varepsilon(\vec{c}_j, x)(x - \xi_*(\bar{t}) + \varepsilon(\vec{c}_j \cdot v)). \tag{22}$$

This completes the inductive construction of the Γ^j_ε, and the proof that (15) holds.

We now prove (#), also by induction. We assume that θ_{j-1} has been defined in such a way that $\theta_{j-1} \in \Theta$ and ($\#_{j-1}$) holds.

Let $A \in \Gamma^j_\varepsilon(\vec{c}_j, x)$. Write $A = A_2 \circ A_1 \circ A_0 + A_{0,z}$ as before, and let $A_0 = \textit{affm}_{L_0, z_0}$, $A_1 = \textit{affm}_{L_1, z_1}$, $A_2 = \textit{affm}_{L_2, z_2}$. Then $A = \textit{affm}_{L, \hat{z}}$, where $L = L_2 L_1 L_0$, $\hat{z} = L_2 L_1 z_0 + L_2 z_1 + z_2 + z$. On the other hand, we know from the inductive hypothesis that $L_0 \in g^{\theta_{j-1}(\varepsilon)}_{\sigma_{j-1}, \bar{t}}$ and $\|z_0\| \leq \theta_{j-1}(\varepsilon)\varepsilon$, and we also know that $L_1 \in g^{\theta(\varepsilon)}_{\tau_j, \sigma_{j-1}}$, $L_2 \in g^{\theta(\varepsilon)}_{\sigma_j, \tau_{j-1}}$, $\|z_1\| \leq \theta(\varepsilon)\varepsilon$, and $\|z_2\| \leq \theta(\varepsilon)\varepsilon$. Then (8) implies that $\|L_0\| \leq 2 + \theta_{j-1}(\varepsilon)$, $\|L_1\| \leq 2 + \theta(\varepsilon)$, and $\|L_2\| \leq 2 + \theta(\varepsilon)$, so

$$L_2 L_1 L_0 \in g^{\tilde{\theta}_j(\varepsilon)}_{\sigma_j, \bar{t}},$$

where $\tilde{\theta}_j \in \Theta$. (Precisely, $\tilde{\theta}_j = 8\theta + 4\theta_{j-1} + 4\theta^2 + 8\theta\theta_{j-1} + 3\theta^2\theta_{j-1}$.)

Also, $\|L_2 L_1 z_0 + L_2 z_1 + z_2\| \leq \hat{\theta}_j(\varepsilon)\varepsilon$, where $\hat{\theta}_j$ belongs to Θ. (Precisely, $\hat{\theta}_j = 3\theta + 4\theta_{j-1} + \theta^2 + 4\theta\theta_{j-1} + \theta^2\theta_{j-1}$.) As for z, we can estimate it as follows: we have

$$z = y - \xi_*(\sigma_j) - (A_2 \circ A_1 \circ A_0)(x - \xi_*(\bar{t}) + \varepsilon(\vec{c}_j \cdot v))$$

and also $y = E_j(c_j, \varepsilon, h, y) + \xi_*(\sigma_j) + h + \varepsilon c_j v_j$, where

$$h = w - \xi_*(\tau_j) = (A_1 \circ A_0)(x - \xi_*(\bar{t}) + \varepsilon(\vec{c}_{j-1}v)). \tag{23}$$

Then $y - \xi_*(\sigma_j) = E_j(c_j, \varepsilon, h, y) + h + \varepsilon c_j v_j$, so

$$\begin{aligned}
z &= E_j(c_j, \varepsilon, h, y) + h + \varepsilon_j v_j - (A_2 \circ A_1 \circ A_0)(x - \xi_*(\bar{t}) + \varepsilon(\vec{c}_j \cdot v)) \\
&= E_j(c_j, \varepsilon, h, y) + h + \varepsilon_j v_j \\
&\quad - A_2\Big((A_1 \circ A_0)(x - \xi_*(\bar{t}) + \varepsilon(\vec{c}_{j-1} \cdot v)) + (A_1 \circ A_0)(\varepsilon c_j v_j)\Big) \\
&= E_j(c_j, \varepsilon, h, y) + h + \varepsilon_j v_j - A_2\Big(h + (A_1 \circ A_0)(\varepsilon c_j v_j)\Big) \\
&= E_j(c_j, \varepsilon, h, y) + (\mathbb{I}_{\mathbb{R}^n} - A_2)h + (\mathbb{I}_{\mathbb{R}^n} - (A_2 \circ A_1 \circ A_0))(\varepsilon c_j v_j).
\end{aligned}$$

Let $\omega(\varepsilon) = \sup\{\|\mathbb{I}_{\mathbb{R}^n} - L\| : L \in g_{t,s}, \bar{t} \leq s \leq t \leq \bar{t} + \beta_m \varepsilon^{\lambda_m}\}$. Then $\lim_{\varepsilon \downarrow 0} \omega(\varepsilon) = 0$, because of (7). We then have

$$\|(\mathbb{I}_{\mathbb{R}^n} - A_2)h\| = \|(\mathbb{I}_{\mathbb{R}^n} - L_2)h - z_2\| \leq (\omega(\varepsilon) + \theta(\varepsilon))\|h\| + \theta(\varepsilon)\varepsilon.$$

Also,

$$\|(\mathbb{I}_{\mathbb{R}^n} - (A_2 \circ A_1 \circ A_0))(\varepsilon c_j v_j)\|$$
$$= \|(\mathbb{I}_{\mathbb{R}^n} - (L_2 \circ L_1 \circ L_0))(\varepsilon c_j v_j) - L_2 L_1 z_0 - L_2 z_1 - z_2\|.$$

Furthermore,

$$\|\mathbb{I}_{\mathbb{R}^n} - L_2 L_1 L_0\| = \|\mathbb{I}_{\mathbb{R}^n} - L_2 + L_2 - L_2 L_1 + L_2 L_1 - L_2 L_1 L_0\|$$
$$\leq \|\mathbb{I}_{\mathbb{R}^n} - L_2\| + \|L_2\|\,\|\mathbb{I}_{\mathbb{R}^n} - L_1\| + \|L_2\|\|L_1\|\|\mathbb{I}_{\mathbb{R}^n} - L_0\|$$
$$\leq \check{\theta}_j(\varepsilon),$$

where we may take $\check{\theta}_j(\varepsilon) = (2\omega(\varepsilon) + \theta(\varepsilon) + \theta_{j-1}(\varepsilon))(1 + (1 + \omega(\varepsilon) + \theta(\varepsilon))^2$ (because $\|\mathbb{I}_{\mathbb{R}^n} - L_2\| \leq \omega(\varepsilon) + \theta(\varepsilon)$, $\|\mathbb{I}_{\mathbb{R}^n} - L_1\| \leq \omega(\varepsilon) + \theta(\varepsilon)$, and $\|\mathbb{I}_{\mathbb{R}^n} - L_0\| \leq \omega(\varepsilon) + \theta_{j-1}(\varepsilon)$). Therefore

$$|(\mathbb{I}_{\mathbb{R}^n} - (L_2 \circ L_1 \circ L_0))(\varepsilon c_j v_j)\| \leq \check{\theta}_j(\varepsilon)\bar{c}\|v\|_j \varepsilon.$$

Since $\|L_2 L_1 z_0 + L_2 z_1 + z_2\| \leq \hat{\theta}_j(\varepsilon)\varepsilon$, we have

$$\|(\mathbb{I}_{\mathbb{R}^n} - (A_2 \circ A_1 \circ A_0))(\varepsilon c_j v_j)\| \leq (\check{\theta}_j(\varepsilon)\bar{c}\|v\|_j + \hat{\theta}_j(\varepsilon))\varepsilon.$$

Finally, $\|E_j(c_j, \varepsilon, h, y)\|$ is bounded by $\theta_*(\max(\varepsilon, \|h\|))(\varepsilon + \|h\|)$ so we get the bound

$$\|z\| \leq \theta_j^{\&}(\varepsilon, \|h\|)(\varepsilon + \|h\|) \tag{24}$$

where $\theta_j^{\&}(\varepsilon, \delta) = \omega(\varepsilon) + 2\theta(\varepsilon) + \check{\theta}_j(\varepsilon)\bar{c}\|v\|_j + \hat{\theta}_j(\varepsilon) + \theta_*(\max(\varepsilon, \delta))$. It then follows that

$$\|\hat{z}\| = \|L_2 L_1 z_0 + L_2 z_1 + z_2 + z\| \leq \hat{\theta}_j(\varepsilon)\varepsilon + \theta_j^{\&}(\varepsilon, \|h\|)(\varepsilon + \|h\|). \tag{25}$$

To conclude, we obtain an estimate for $\|h\|$. We use the identity (23), from which it follows that $h = (L_1 L_0)(x - \xi_*(\bar{t}) + \varepsilon(\bar{c}_{j-1}v)) + L_1 z_0 + z_1$. Since $\|L_0\| \leq 1 + \omega(\varepsilon) + \theta_{j-1}(\varepsilon)$, $\|L_1\| \leq 1 + \omega(\varepsilon) + \theta(\varepsilon)$, $\|x - \xi_*(\bar{t})\| \leq \varepsilon$, $\|z_0\| \leq \theta_{j-1}(\varepsilon)\varepsilon$, and $\|z_1\| \leq \theta(\varepsilon)\varepsilon$, we find that $\|h\| \leq \eta_j(\varepsilon)\varepsilon$, where

$$\eta_j(\varepsilon) = (1 + \omega(\varepsilon) + \theta(\varepsilon))\Big((1 + mC)(1 + \omega(\varepsilon) + \theta_{j-1}(\varepsilon)) + \theta(\varepsilon) + \theta_{j-1}(\varepsilon)\Big).$$

Therefore $\|\hat{z}\| \leq \theta_j^{\$}(\varepsilon)\varepsilon$, where

$$\theta_j^{\$}(\varepsilon) = \hat{\theta}_j(\varepsilon) + \theta_j^{\&}(\varepsilon, \eta_j(\varepsilon)\varepsilon)(1 + \eta_j(\varepsilon)). \tag{26}$$

Hence, if we define $\theta_j(\varepsilon) = \max(\tilde{\theta}_j^b(\varepsilon), \theta_j^\&(\varepsilon))$, it is clear that $\theta_j \in \boldsymbol{\Theta}$, and we have shown that $(\#_j)$ holds, completing the proof of Lemma 6.1.

We now define, for $\vec{p} = (p_1, \ldots, p_m) \in \mathbb{R}_+^m$, $x \in \bar{\mathbb{B}}(\xi_*(\bar{t}), \rho(\varepsilon))$

$$W_{\vec{p}}(x) = \tilde{W}(\vec{p}, x) = \{\hat{\Upsilon}_\varepsilon(\varepsilon^{-1}p_1, \ldots, \varepsilon^{-1}p_m, x) : \varepsilon \geq \bar{c}^{-1} \max(p_1, \ldots, p_m)\},$$

so each $W_{\vec{p}}$ is a set-valued map, $\mathrm{Gr}(W_0) \subseteq \mathrm{Gr}(f_{t_*, \bar{t}})$, and $\mathrm{Gr}(W_p) \subseteq \mathrm{Gr}(f'_{t_*, \bar{t}})$. Then $W = \{W_{\vec{p}}\}_{\vec{p} \in \mathbb{R}_+^m}$ is a variation of $f_{\tau, \bar{t}}$ in $f'_{\tau, \bar{t}}$. Also, given any positive ε, the map

$$[0, \bar{c}\varepsilon]^m \times \bar{\mathbb{B}}(\xi_*(\bar{t}), \rho(\varepsilon)) \ni (\vec{p}, x) \mapsto Z_\varepsilon(\vec{p}, x) \overset{\text{def}}{=} \hat{\Upsilon}_\varepsilon(\varepsilon^{-1}\vec{p}, x)$$

is a CCA map whose graph is contained in that of $\tilde{W}$. In addition, if we let $\tilde{Z}_\varepsilon$ be the set-valued map that sends each point $(\vec{p}, x) \in [0, \bar{c}\varepsilon]^m \times \bar{\mathbb{B}}(\xi_*(\bar{t}), \rho(\varepsilon))$ to the set

$$\{(A, B) : B \in \Gamma_\varepsilon(\varepsilon^{-1}\vec{p}, x), A \in A_{t_*, \sigma_m}(\xi_*(\sigma_m) + B(x - \xi_*(\bar{t}) + \vec{p}_m \cdot v))\},$$

(where $\sigma_m = \bar{t} + \beta_m \varepsilon^{\lambda_m}$, as before), and use μ to denote the map $Aff(\mathbb{R}^n, \mathbb{R}^n) \times Aff(\mathbb{R}^n, \mathbb{R}^n) \ni (A, B) \mapsto A \circ B \in Aff(\mathbb{R}^n, \mathbb{R}^n)$, then the composite map $\mathcal{Z}_\varepsilon = \mu \circ \tilde{Z}_\varepsilon$ is a CCA map such that

$$Z_\varepsilon(\vec{p}, x) = \{\xi_*(t_*) + M(x - \xi_*(\bar{t}) + \vec{p} \cdot v) : M \in \mathcal{Z}_\varepsilon(\vec{p}, x)\}.$$

Let us now recall that, if γ is a compatible selection of $\mathbf{g}$, and $\mathbf{V} = \{(v_1, \bar{t}, +), \ldots, (v_m, \bar{t}, +)\}$, then $L^{\mathbf{V}, \gamma, \bar{t}, t_*}$ is the linear map $\gamma_{t_*, \bar{t}} \circ \hat{L}$, where $\hat{L}$ is the map $(\vec{p}, h) \mapsto h + \vec{p} \cdot v$.

We also recall that $\Lambda^{\mathbf{V}, \mathbf{g}, \bar{t}, t_*}$ is the set of all maps $L^{\mathbf{V}, \gamma, \bar{t}, t_*}$, for all $\gamma \in CSel(\mathbf{g})$. The definition of Z_ε can be rewritten as

$$Z_\varepsilon(\vec{p}, x) = \{\xi_*(t_*) + M(\hat{L}(x - \xi_*(\bar{t}), \vec{p})) : M \in \mathcal{Z}_\varepsilon(\vec{p}, x)\}.$$

If $M \in \mathcal{Z}_\varepsilon(\vec{p}, x)$, then M is the composite of a member B of $\Gamma_\varepsilon^m(\varepsilon^{-1}\vec{p}, x)$ followed by a member A of $A_{t_*, \sigma_m}(y)$, where $y = \xi_*(\sigma_m) + B(x - \xi_*(\bar{t}) + \vec{p}_m \cdot v)$. If we write $B = affm_{B_1, b_1}$, we know that $B_1 \in g_{\sigma_m, \bar{t}}^{\theta_m(\varepsilon)}$ and $\|b_1\| \leq \theta_m(\varepsilon)\varepsilon$. Also, if $A = affm_{A_1, a_1}$, we know that $A_1 \in g_{t_*, \sigma_m}^{\theta(\varepsilon)}$ and $\|a_1\| \leq \theta(\varepsilon)\varepsilon$. It follows that, if $M\hat{L} = affm_{K, k}$, then $K = A_1 B_1 \hat{L}$ and $k = a_1 + A_1 b_1$. Therefore $K \in (g_{t_*, \bar{t}} \circ \hat{L})^{\theta^\$(\varepsilon)}$ and $\|k\| \leq \theta^\$(\varepsilon)$, where $\theta^\$(\varepsilon)$ is an easily computable member of $\boldsymbol{\Theta}$. (Precisely, we may take $\theta^\$ = 2(1 + \|\hat{L}\|)(\theta + \theta_m + \theta\theta_m)$.) So $M\hat{L} \in g_{t,s}^{(\theta^\$(\varepsilon), \varepsilon)}$.

Now, the set $g_{t_*, \bar{t}} \circ \hat{L}$ is precisely $\Lambda^{\mathbf{V}, \mathbf{g}, \bar{t}, t_*}$. This shows that $\Lambda^{\mathbf{V}, \mathbf{g}, \bar{t}, t_*}$ is an AGDQ of the map $\tilde{W}$ at $((0, \xi_*(\bar{t})), \xi_*(t_*))$ in the direction of $\mathbb{R}_+^m \times X$. If, for $b \geq t_*$, we define $W^b = \{W_{\vec{p}}^b\}_{\vec{p} \in \mathbb{R}_+^m}$, by letting $W_{\vec{p}}^b(x) = \tilde{W}^b(\vec{p}, x) =$

$(f_{b,t_*} \circ W_{\bar{p}})(x))$, then it is clear that W^b is a variation of $f_{b,\bar{t}}$ in $f'_{b,\bar{t}}$, and $\Lambda^{V,\mathbf{g},\bar{t},b}$ is an AGDQ of $\tilde{W}^b$ at $((0,\xi_*(\bar{t})),\xi_*(b))$ in the direction of $\mathbb{R}^m_+ \times X$. This completes our proof. $\qquad\square$

References

1. R. M. Bianchini, Variational approach to some optimization control problems, in *Geometry in nonlinear control and differential inclusions (Warsaw, 1993)*, eds. B. Jacubczyk, W. Respondek and T. Rzeżuchowski, Banach Center Publ. Vol. 32 (Polish Acad. Sci. Inst. of Math., Warsaw, 1995), pp. 83–94.

2. R. M. Bianchini and G. Stefani, *SIAM J. Control and Optim.* **31**, 900 (1993)

3. R. M. Bianchini and G. Stefani, Good needle-like variations, in *Differential geometry and control (Boulder, CO, 1997)*, Proc. Symposia Pure Math. **64** (American Mathematical Society, Providence, RI, 1999), pp. 91-101.

4. G. Stefani, Higher order variations: how can they be defined in order to have good properties?, in *Non-smooth Analysis and geometric methods in deterministic optimal control (Minneapolis, MN, 1993)*, eds. B. S. Mordukhovich and H. J. Sussmann, IMA Vol. Math. Appl. Vol. 78 (Springer, New York, 1996), pp. 227–237.

5. H.J. Sussmann, Résultats récents sur les courbes optimales, in *15e Journée Annuelle de la Société Mathémathique de France (SMF)* (Publications de la SMF, Paris, 2000), pp. 1-52.

6. H. J. Sussmann, *New theories of set-valued differentials and new version of the maximum principle of optimal control theory*, in *Nonlinear Control in the Year 2000, Vol. 2*, Lecture Notes in Control and Inform. Sci., 259, eds. A. Isidori, F. Lamnabhi-Lagarrigue and W. Respondek (Springer-Verlag, London, 2000), pp. 487–526.

7. H.J. Sussmann, Combining high-order necessary conditions for optimality with nonsmoothness, in *Proc. 43rd IEEE Conf. Decision and Control (Atlantis, Paradise Island, Bahamas, 2004)*.

8. H.J. Sussmann, Optimal control of nonsmooth systems with classically differentiable flow maps, in *Proceedings of the Sixth IFAC Symposium on Nonlinear Control Systems (NOLCOS 2004), Vol. 2*, (Stuttgart, Germany, September 1-3, 2004), pp. 609-704.

9. H.J. Sussmann, Generalized differentials, variational generators, and the maximum principle with state contraints, in *Nonlinear and Optimal Control Theory* (Lectures given at the C.I.M.E. Summer School held in Cetraro (Cosenza), June 21–29, 2004), Lecture Notes in Mathematics, eds. G. Stefani and P. Nistri (Springer-Verlag (Fondazione C.I.M.E.), to appear).

10. H.J. Sussmann, *J. Differential Equations* **243**, 448 (2007).

LIST OF PARTICIPANTS

1. ANDREI AGRACHEV [Trieste (IT)] .... agrachev@sissa.it
2. FABIO ANCONA [Bologna (IT)] .... ancona@ciram.unibo.it
3. ZVI ARTSTEIN [Rehovot (IL)] .... zvi.artstein@weizmann.ac.il
4. ANDREA BACCIOTTI [Torino (IT)] .... andrea.bacciotti@polito.it
5. MARTINO BARDI [Padova (IT)] .... bardi@math.unipd.it
4. RICHARD BARNARD [Baton Rouge, LA (US)] .. rbarnard@math.lsu.edu
6. JERROLD BEBERNES [Boulder, CO (US)] ...... bebernes@coloraodo.edu
7. UGO BOSCAIN [Trieste (IT)] .... boscain@sissa.it
8. ALBERTO BRESSAN [University Park, PA (US)] .. bressan@math.psu.edu
9. ROGER BROCKETT [Cambridge, MA (US)] brockett@deas.harvard.edu
10. FABIO CAMILLI [L'Aquila (IT)] .... camilli@ing.univaq.it
11. PIERMARCO CANNARSA [Roma (IT)] .... cannarsa@mat.uniroma2.it
12. ITALO CAPUZZO DOLCETTA [Roma (IT)] .... capuzzo@mat.uniroma1.it
13. PIERRE CARDALIAGUET [Brest (FR)] Pierre.Cardaliaguet@univ-brest.fr
18. MARCO CASTELPIETRA [Roma (IT)] ... castelpi@axp.mat.uniroma2.it
15. ARRIGO CELLINA [Milano (IT)] .... arrigo.cellina@unimib.it
16. FRANCESCA CERAGIOLI [Torino (IT)] francesca.ceragioli@polito.it
17. FRANCESCA CHITTARO [Firenze (IT)] .... chittaro@math.unifi.it
18. FRANCIS CLARKE [Lyon (FR)] .... clarke@math.univ-lyon1.fr
19. GIOVANNI COLOMBO [Padova (IT)] .... colombo@math.unipd.it
20. LUCA CONSOLINI [Parma (IT)] .... luca.consolini@polirone.mn.it
21. JEAN-MICHEL CORON [Paris (FR)] jean-michel.coron@math.u-psud.fr
22. MARCO CZARNECKI [Montpellier (FR)] .... marco@univ-montp2.fr
23. WIJESURIYA DAYAWANSA [Lubbock, TX (US)] ....
24. ASEN DONTCHEV [Ann Arbor, MI (US)] .... ald@ams.org
25. BRUNO FRANCHI [Bologna (IT)] .... franchib@dm.unibo.it
26. HELENE FRANKOWSKA [Paris (FR)] .... franko@shs.polytechnique.fr
27. DANIEL GESSUTI [Trieste (IT)] .... gessuti@sissa.it
28. LARS GRÜNE [Bayreuth (DE)] .... lars.gruene@uni-bayreuth.de
29. ALVARO GUEVARA [Baton Rouge, LA (US)] ... aguevara@math.lsu.edu
30. HENRY HERMES [Boulder, CO (US)] .... hankhermes@msn.com
31. MATTHIAS KAWSKI [Tempe, AZ (US)] .... kawski@asu.edu
32. EYUP KIZIL [Trieste (IT)] .... kizil@sissa.it
33. ARTHUR J. KRENER [Davis, CA (US)] .... ajkrener@ucdavis.edu
34. YURI LEDYAEV [Kalamazoo, MI (US)] .... ledyaev@wmich.edu
35. SOFIA LOPES [Guimarães (PT)] .... sofialopes@mct.uminho.pt

36. CLAUDIO MARCHI [Cosenza (IT)] .... marchi@mat.unical.it
37. ANTONIO MARIGONDA [Pavia (IT)] ...... antonio.marigonda@unipv.it
38. PAOLO MASON [Nancy (FR)] .... Paolo.Mason@iecn.u-nancy.fr
39. FABIO MORBIDI [Siena (IT)] .... morbidi@dii.unisi.it
40. JACQUELINE MORGAN [Napoli (IT)] .... morgan@unina.it
41. MONICA MOTTA [Padova (IT)] .... motta@math.unipd.it
42. CHADI NOUR [Byblos (LB)] .... cnour@lau.edu.lb
43. NORMA ORTIZ [Richmond, VA (US)] .... nlortiz@vcu.edu
44. BENEDETTO PICCOLI [Roma (IT)] .... b.piccoli@iac.cnr.it
45. LAURA POGGIOLINI [Firenze (IT)] .... laura.poggiolini@unifi.it
46. FABIO PRIULI [Trondheim (NO)] .... priuli@math.ntnu.no
47. FRANCO RAMPAZZO [Padova (IT)] .... rampazzo@math.unipd.it
48. LUDOVIC RIFFORD [Nice (FR)] .... rifford@math.unice.fr
49. VINICIO RIOS [Maracaibo (VE)] .... rios@math.lsu.edu
50. R. TYRRELL ROCKAFELLAR [Seattle, WA (US)] rtr@math.washington.edu
51. SERGIO RODRIGUES [Trieste (IT)] .... srodrigs@sissa.it
52. FRANCESCO ROSSI [Trieste (IT)] .... rossifr@sissa.it
53. REBECCA SALMONI [Paris (FR)] .... rebecca.salmoni@lss.supelec.fr
54. CATERINA SARTORI [Padova (IT)] .... caterina.sartori@unipd.it
55. ANDREY SARYCHEV [Firenze (IT)] .... asarychev@unifi.it
56. ULYSSE SERRES [Nancy (FR)] .... ulysse.serres@iecn.u-nancy.fr
57. CARLO SINESTRARI [Roma (IT)] .... sinestra@mat.uniroma2.it
58. MARCO SPADINI [Firenze (IT)] .... marco.spadini@unifi.it
59. GIANNA STEFANI [Firenze (IT)] .... gianna.stefani@unifi.it
60. HECTOR SUSSMANN [Piscataway, NJ (US)] sussmann@math.rutgers.edu
61. GABRIELE TERRONE [Padova (IT)] .... gabter@math.unipd.it
62. MARIO TOSQUES [Parma (IT)] .... mario.tosques@unipr.it
63. RICHARD VINTER [London (GB)] .... r.vinter@imperial.ac.uk
64. PETER WOLENSKI [Baton Rouge, LA (US)] .... wolenski@math.lsu.edu

AUTHOR INDEX

Series on Advances in Mathematics for Applied Sciences

Aims and Scope

This Series reports on new developments in mathematical research relating to methods, qualitative and numerical analysis, mathematical modeling in the applied and the technological sciences. Contributions related to constitutive theories, fluid dynamics, kinetic and transport theories, solid mechanics, system theory and mathematical methods for the applications are welcomed.

This Series includes books, lecture notes, proceedings, collections of research papers. Monograph collections on specialized topics of current interest are particularly encouraged. Both the proceedings and monograph collections will generally be edited by a Guest editor.

High quality, novelty of the content and potential for the applications to modern problems in applied science will be the guidelines for the selection of the content of this series.

Instructions for Authors

Submission of proposals should be addressed to the editors-in-charge or to any member of the editorial board. In the latter, the authors should also notify the proposal to one of the editors-in-charge. Acceptance of books and lecture notes will generally be based on the description of the general content and scope of the book or lecture notes as well as on sample of the parts judged to be more significantly by the authors.

Acceptance of proceedings will be based on relevance of the topics and of the lecturers contributing to the volume.

Acceptance of monograph collections will be based on relevance of the subject and of the authors contributing to the volume.

Authors are urged, in order to avoid re-typing, not to begin the final preparation of the text until they received the publisher's guidelines. They will receive from World Scientific the instructions for preparing camera-ready manuscript.